P. D. Panagiotopoulos

Hemivariational Inequalities

Applications in Mechanics
and Engineering

With 129 Figures

Springer-Verlag
Berlin Heidelberg New York
London Paris Tokyo
Hong Kong Barcelona Budapest

Prof. Dr. Panagiotis D. Panagiotopoulos

Department of Civil Engineering
Aristotle University
GR - 54006 Thessaloniki, Greece

Faculty of Mathematics and Physics
RWTH Aachen
D - 52062 Aachen

ISBN 3-540-54963-3 Springer-Verlag Berlin Heidelberg New York
ISBN 0-387-54963-3 Springer-Verlag New York Berlin Heidelberg

Library of Congress-in-Publication Data
Panagiotopoulos, P. D., 1950 - Hemivariational inequalities: applications in mechanics
and engineering / P. D. Panagiotopoulos. p. cm.
Includes bibliographical references and index.
ISBN 0-387-54963-3 (alk. paper: U.S.)
1. Hemivariational inequalities. 2. Engineering mathematics.
I. Title. QA316.P33 1993 620'.001'51564 - - dc20 93-29025 CIP

This work is subject to copyright. All rights are reserved, whether the whole or part of the material is concerned, specifically the rights of translation, reprinting, reuse of illustrations, recitation, broadcasting, reproduction on microfilm or in other ways, and storage in data banks. Duplication of this publication or parts thereof is permitted only under the provisions of the German Copyright Law of September 9, 1965, in its current version, and permission for use must always be obtained from Springer-Verlag. Violations are liable for prosecution act under German Copyright Law.

© Springer-Verlag Berlin Heidelberg 1993
Printed in the United States of America

The use of general descriptive names, registered names, trademarks, etc. in this publication does not imply, even in the absence of a specific statement, that such names are exempt from the relevant protective laws and regulations and therefore free for general use.

Typesetting: Camera ready by author
61/3020-5 4 3 2 1 0 - Printed on acid -free paper

To my parents

Preface

The field of Inequality Problems has seen a considerable development in Mathematics, Mechanics and Engineering Sciences in a remarkably short time. This is mainly due to the fact that new, very efficient, mathematical tools used in the area of Inequality Problems, or, more generally, the field of Nonsmooth Mechanics, proved beneficial to the promotion of scientific thought and methodology; open problems have been treated and entirely new categories of interesting problems in Mathematics, Applied Mechanics and several branches of the Engineering Sciences have been mathematically formulated, studied and/or numerically treated. In the area of Inequality Problems we can distinguish two main directions: that of Variational Inequalities, which already has a research "life" of about 30 years and is mainly connected with convex energy functions, and that of Hemivariational Inequalities which is more "young"-the idea of hemivariational inequalities was born only 10 years ago - and is connected with nonconvex energy functions.

The theory and the applications of Hemivariational Inequalities is the subject of the present book.

This book is an outgrowth of ten years of seminars and courses on the theory and applications of Hemivariational Inequalities delivered to a variety of audiences in the Technical University of Aachen, the Aristotle University of Thessaloniki, the University of Hamburg and the Pontificia Univ. Catolica in Rio de Janeiro. The book is intended for a wide range of readers. Primarily it is addressed to people working on Applied Mechanics and Engineering (Civil, Aeronautical and Mechanical) both theoretically oriented and those dealing with research, analysis and design, as well as to Applied Mathematicians by pointing out important applications which need a deeper mathematical treatment, or by introducing innovative numerical methods which need further mathematical investigation.

I would like to acknowledge the great assistance I have received from Dr. P. Zervas, Dipl. Ing. Th. Nikolaidis and Dipl. Ing. G. Nikolaidis who prepared, in a very diligent way, the final text with the LaTeX program, Dr. A. Al-Fahed, Dr. E. Koltsakis, Dr. E. Mistakidis, Dr. O. Panagouli, Dr. G. Stavroulakis and Dr. M. Tzaferopoulos for the programming of the numerical applications of Chapters 9 to 15, as well as for the proofreading and the preparation of the final figures. I also wish to acknowledge the helpful comments received from Dr. C. Bisbos, Prof. V. Demyanov, Dr. E. Koltsakis and Prof. Z. Naniewicz for critically proofreading some parts of the book. Many thanks are also due to my editors in Springer-Verlag for their friendly assistance to my efforts and for their cooperation during this project, and all those who contributed for this book to come to pass. I would like to apologize to those whose work was inadvertently neglected in compiling the literature for this book. I shall welcome all comments and corrections from readers.

P.D. Panagiotopoulos
Thessaloniki-Aachen April 1993

Table of Contents

Introduction VII

Guidelines for the Reader. Abbreviations X

I INTRODUCTORY TOPICS

1 Elements of Nonsmooth Analysis 3
   1.1 Convexity and Subdifferential 3
   1.2 Generalized Gradient and Related Calculus 9
   1.3 Minimization Problems. Duality of Convex Functionals 18
   1.4 Miscellanea: Fans, Quasidifferentials, Codifferentials 24

II MECHANICAL THEORY

2 Nonsmooth Mechanics I 33
   2.1 Convex Superpotentials 33
   2.2 Nonconvex Superpotentials 41
   2.3 Boundary Conditions Expressed via Convex Superpotentials . . 44
   2.4 Boundary Conditions Expressed via Nonconvex Superpotentials 51
   2.5 Extensions to Function Spaces 58

3 Nonsmooth Mechanics II 65
   3.1 Material Laws Expressed via Convex Superpotentials. An Overview 65
   3.2 Material Laws Expressed via Nonconvex Superpotentials I . . . 72
   3.3 Material Laws Expressed via Nonconvex Superpotentials II . . . 81
   3.4 Loading and Unloading Problems. The Advantage of the Use of Superpotentials 88
   3.5 Material Laws and Boundary Conditions Expressed via Fans, Quasidifferentials and Codifferentials 91

X   Table of Contents

4 Hemivariational Inequalities 99
   4.1 The Derivation of Hemivariational Inequalities in Mechanical
        Problems. 99
   4.2 Hemivariational and Variational-Hemivariational Inequalities . . 105
   4.3 Substationarity Problems for the Potential or the Complementary Energy 116
   4.4 Loading and Unloading Problems, Eigenvalue Problems for Hemivariational Inequalities and Dynamic Problems 121
   4.5 On the F-superpotential and the V-superpotential. Quasidifferentiability in Mechanics 126

5 Multivalued Boundary Integral Equations 135
   5.1 The Indirect and the Direct Method for Nonmonotone Boundary
        Conditions. 135
   5.2 Complement for Adhesively Bonded Cracks 143

III MATHEMATICAL THEORY

6 Static Hemivariational Inequalities 155
   6.1 Coercive Hemivariational Inequalities 155
   6.2 Semicoercive Hemivariational Inequalities 163
   6.3 On the Substationarity of the Energy 167
   6.4 Variational Hemivariational Inequalities 169
   6.5 Applications to Engineering Problems 174

7 Eigenvalue and Dynamic Problems 179
   7.1 On the Eigenvalue Problem for Hemivariational Inequalities . . 179
   7.2 Dynamic Hemivariational Inequalities 190
   7.3 Applications to Engineering Problems: Von Kármán Plates and
        Thermoelasticity 204

8 Optimal Control and Identification Problems 223
   8.1 Formulation of the Problem 223
   8.2 Mathematical Study of the Optimal Control Problem Governed
        by Hemivariational Inequalities 226
   8.3 Applications to Engineering Problems 234

IV NUMERICAL APPLICATIONS

9 On the Numerical Treatment of Hemivariational Inequalities 239
   9.1 The First Numerical Attempts and the Questions of Stability and
        Uniqueness 240
   9.2 The Microspring Approximation Method of the Decreasing Branch 245

9.3 The Method of Decreasing Branch Approximation by Monotone Laws 249
9.4 Application I: Cleavage in Laminated Composites and the Nonmonotone Unilateral Contact Problem 257
9.5 Application II: The Nonmonotone Friction Problem and the Combined Unilateral Contact Problem with Nonmonotone Friction 269

10 On the Approximation of Hemivariational Inequalities by Variational Inequalities 281
10.1 General Formulation of the Method 281
10.2 Application III: Nonmonotone Friction Interface Conditions with Debonding 284
10.3 Application IV : Adhesive Joints in Structural Mechanics 290
10.4 Application V : Comparison with the Path Following Method . 297
10.5 Application VI : Nonmonotone Stress-Strain Laws. The Sawtooth Behaviour of Composites 300
10.6 Application VII : Shear Connectors in Composite Beams 306

11 The Method of Substationary Point Search 317
11.1 General Formulation of the Method 317
11.2 On the Numerical Implementation of the Algorithm. 326
11.3 Application VIII: Delamination and Adhesive Joints in Structural Mechanics 331
11.4 Application IX: Semirigid Connections in Steel Structures ... 338

12 On a Decomposition Method into Two Convex Problems 345
12.1 General Formulation of the Method 345
12.2 Application X: The Stamp Problem and the Interfacial Debonding in Composites 351

13 Dynamic Hemivariational Inequalities and Crack Problems 361
13.1 Application XI: Numerical Treatment of Dynamic Hemivariational Inequalities 361
13.2 Application XII: The Unilateral Contact and Nonmonotone Friction Problem in Cracks 371
13.3 Application XIII: Fracture of Cracks Repaired by an Adhesive Material 376

14 Applications of the Theory of Hemivariational Inequalities in Robotics 377
14.1 Application XIV: Adhesive Grasping Problem in Robotics ... 377
14.2 Application XV: On the Optimal Control of the Adhesive Grasping Problem in Robotics 387

15 Addenda: Hemivariational Inequalities, Fractals and Neural Networks 393
15.1 Fractals in Mechanics. An Introduction 393
15.2 Application XVI: Hemivariational Inequalities for Fractal Interfaces 401

15.3 The Neural Network Approach to Hemivariational Inequalities . 406
15.4 Application XVII: D.C.B. Specimen Modelling. The Neural Network Approach 409
15.5 Application XVIII: The Inverse Delamination Problem as a Supervised Learning Problem for a Neural Network. Extensions . . 412

References 417

Subject Index 449

Introduction

The scope of the present book is the study of problems in Mechanics and Engineering Science whose variational formulations are hemivariational inequalities. These variational forms involve nonconvex energy functions and express the principle of virtual work in its inequality form. The treatment of such problems, which constitute the most interesting category of the inequality or unilateral problems, differs fundamentally from that of inequality problems whose variational forms are variational inequalities, due to the nonconvexity of the energy functions involved, and of course, from that of the classical bilateral or equality problems. In most cases the nonconvex energy functions are nonsmooth and, therefore, the methods of Nonsmooth Analysis are employed for the mathematical study and the numerical treatment of the hemivariational inequalities.

The book is divided into four parts: The "Introductory Topics" (Chapter 1) the "Mechanical Theory" (Chapters 2 to 5), the "Mathematical Theory" (Chapters 6 to 8) and the "Numerical Applications" (Chapters 9 to 15). In the Part I we give the necessary mathematical background concerning convexity and subdifferential, generalized gradient and duality, elements of the theory of fans and quasidifferentiability. Part II deals with the mechanical aspects of the theory of hemivariational inequalities. In this part we define the notions of convex and nonconvex superpotentials and, by means of these notions, we introduce boundary conditions and material laws expressed through convex and nonconvex superpotentials. Moreover, we present the general method for the derivation of hemivariational and variational-hemivariational inequalities and we give a first account of the relation between hemivariational inequalities and substationarity of the potential or complementary energy. Special attention is paid also to the unloading problems, the eigenvalue problems for hemivariational inequalities, the dynamic hemivariational inequalities and the multivalued boundary integral equations which are equivalent to (boundary) hemivariational inequalities. We also introduce the fuzzy material laws and boundary conditions and the nonconvex dissipation superpotentials. Then the corresponding class of generalized standard materials with nonconvex energy functions is studied. Moreover, material laws and boundary conditions expressed by means of fans and quasidifferentials are introduced and the corresponding variational expressions are formulated.

Part III deals with the mathematical theory of hemivariational and variational-hemivariational inequalities, as well as with their exact relation to substationarity problems. Moreover, the eigenvalue problem for hemivariational inequalities is studied along with dynamic hemivariational inequalities arising in the theory of von Kármán laminated plates and thermoelasticity. The mathematical part of the book concludes with the formulation and study of the optimal control problem of systems governed by hemivariational inequalities. In this part of the book, where existence and approximation results for the solution(s) of hemivariational inequalities are proved, the mathematical rigour is not sacrificed to the easiness of understanding.

Part IV is devoted to the numerical applications and takes the largest part of the book. We present there, numerical applications related to real engineering problems. Most of the problems treated cannot be accurately solved by other, more classical, numerical techniques due to the strong nonlinearities arising from nonmonotone and multivalued stress-strain or reaction-displacement behaviour. The Chapters of this part of the book are fairly independent from the rest of the book, since they describe numerical techniques and point to concrete engineering applications. Finally in the last Chapter we study hemivariational inequalities defined on fractal geometries and we attempt to adapt the numerical techniques for hemivariational inequalities to a neurocomputing environment.

Guidelines for the Reader. Abbreviations

The choice of the material of Chapter 1 is governed by the requirements of the subsequent Chapters. All propositions of nonsmooth analysis (convex and nonconvex) needed in this book are given in Chapter 1. We expect the mathematically oriented reader to have some knowledge of basic functional analysis especially concerning the norms and certain elementary properties of Sobolev spaces, the Lax-Milgram theorem and the trace theorem. However this functional analysis is needed only for the Chapters 6, 7 and 8. The engineer could skip these chapters. However even these chapters are equally well accessible to a reader unfamiliar with functional analysis who is interested only in mechanics and applications. In this case, proofs should be skipped and the reader should understand the variational expressions in the "usual engineering sense" assuming that spaces $[H^1(\Omega)]^3, [H^{1/2}(\Gamma)]^3$ etc. are simple three-dimensional spaces and the duality pairings $\langle \cdot, \cdot \rangle$ denote inner products.

We intentionally gather all numerical applications at the last and largest part of the book. This part is accessible to everyone with modest knowledge of the classical numerical techniques of structural analysis. Care was taken for each chapter of the numerical part of the book to remain as selfcontained as possible.

Certain notations and abbreviations used throughout the text are listed here. All notations defined in the text are not given here. Throughout the book the summation convention with respect to a repeated index is employed, unless otherwise stated. Bold face letters denote vectors and matrices of discretized problems.

B.V.P.	Boundary Value Problem
F.E.M.	Finite Element Method
B.E.M.	Boundary Element Method
B.I.E.	Boundary Integral Equation
L.C.P.	Linear Complementarity Problem
Q.P.P.	Quadratic Programming Problem
C.P.P.	Convex Programming Problem
$\fint$	Cauchy principal value
$\bar{A}$	Closure of a set A
$\emptyset$	the empty set
$\forall$	for every

$\mathbb{R}$	the set of real numbers
$\mathbb{R}_+$	the set of positive reals
$\bar{\mathbb{R}}$	the set of real numbers including $\pm\infty$
$\mathbb{R}^n$	Euclidean n-dimensional space
$\|x\| = (\sum_{i=1}^n x_i^2)^{1/2}$	length of $x \in \mathbb{R}^n$
$x \wedge y$	vector product in $\mathbb{R}^n$
$D(\Omega)$	space of infinitely differentiable functions with compact support in Ω
$D'(\Omega)$	space of distributions on Ω.
a.e.	almost every (everywhere)
μ − a.e.	almost every (everywhere) with respect to a measure μ
ker T	kernel of T
epi f	epigraph of a functional f
co K	convex hull of a set K
l.s.c., u.s.c.	lower semicontinuous, upper semicontinuous
$\partial f(x)$	subdifferential of f at x
$\bar{\partial} f(x)$	generalized gradient of f at x

Part I
INTRODUCTORY TOPICS

1. Elements of Nonsmooth Analysis

The aim of Chapter 1 is to provide some notions and propositions of Nonsmooth Analysis that will be used in the next Chapters for the study of engineering problems leading to hemivariational inequalities. The propositions are given here without proofs. In this Chapter we primarily rely on the books and monographs by Moreau [Mor69], Rockfellar [Rock60,68,70,79,80], Göpfert [Göp], Ekeland and Temam [Eke], Aubin [Aub77,79a,84], Aubin and Francowska [Aub90], Clarke [Clar83] and Panagiotopoulos [Pan85]. The reader is referred there for the proofs of the propositions.

1.1 Convexity and Subdifferential

Let X be a Hilbert space and K a subset of X. The set K is said to be convex if

$$\lambda x_1 + (1-\lambda)x_2 \in K \tag{1.1.1}$$

whenever $x_1 \in K, x_2 \in K$ and $0 < \lambda < 1$. All linear subspaces of X (including X) are convex. By convention, the empty set $\emptyset$ is convex. Given a set $K_1 \subset X$, the set of all finite linear combinations $\sum_i \lambda_i x_i, x_i \in K_1$, with $\sum_i \lambda_i = 1$, $i = 1, 2, \ldots, n$, is called the affine hull of K_1. If, additionally, $\lambda_i \geq 0$, $i = 1, 2, \ldots, n$, this set is called the convex hull of K_1 and is denoted by co K_1; it is the smallest convex subset of X which contains K_1. Of special interest are convex cones. A set $K \subset X$ is a cone if, for $x \in K, \lambda x \in K$ for every $\lambda \geq 0$. Moreover K is a convex cone if $x_1 + x_2 \in K$ for $x_1 \in K$ and $x_2 \in K$. A real-valued functional $f : K \to \mathbb{R}$ is convex (resp. strictly convex) on K if for each $x_1 \in K$, $x_2 \in K$ and $0 < \lambda < 1$

$$f(\lambda x_1 + (1-\lambda)x_2) \leq (\text{resp.} <)\lambda f(x_1) + (1-\lambda)f(x_2). \tag{1.1.2}$$

A functional f is said to be concave (resp. strictly concave) if and only if $-f$ is convex (resp. strictly convex). A linear functional is at the same time convex and concave, but not strictly. We deal here with functionals taking values in the extended real line $\bar{\mathbb{R}} = \mathbb{R} \cup \{\pm\infty\} = [-\infty, +\infty]$. A functional $f : K \to \bar{\mathbb{R}}$ is defined to be convex on K if for every $x_1 \in K$ and $x_2 \in K$ (1.1.2) holds whenever the right-hand side can be defined; this is not obviously the case if $f(x_1) = -f(x_2) = \pm\infty$. Because a convex functional may have infinite values, we can consider functionals defined on all of X. Indeed, if $f : K \to \mathbb{R}$ is convex on K, we can define the extension $\bar{f}$ of f to all of X by setting $\bar{f}(x) = f(x)$ for $x \in K$ and $\bar{f}(x) = \infty$ for $x \notin K$.

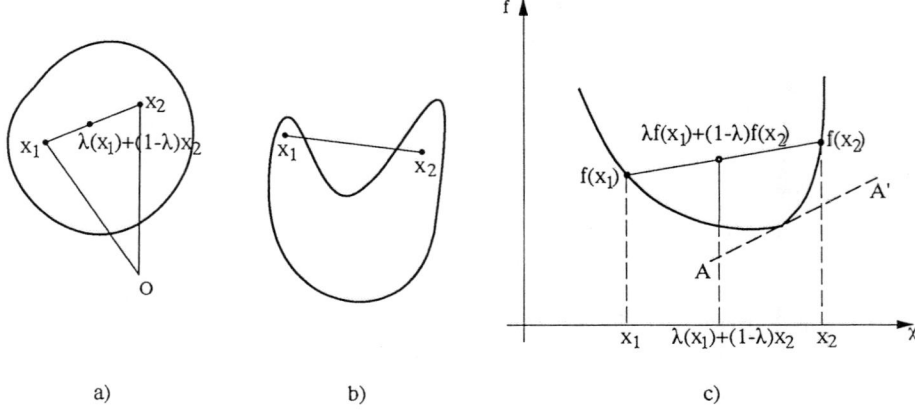

Fig. 1.1.1. Geometrical interpretation of the definition of convexity: a) a convex set b) a nonconvex set c) a convex function.

A convex functional $I_K : X \to \bar{\mathbb{R}}$, defined by

$$I_K(x) = \begin{cases} 0 & \text{for } x \in K \\ \infty & \text{for } x \notin K \end{cases} \quad (1.1.3)$$

and called the indicator of K can be associated to every convex set K. With respect to a convex functional $f : X \to \bar{\mathbb{R}}$ the set

$$\{(x, \lambda) \big| f(x) \leq \lambda, \lambda \in \mathbb{R}, x \in X\} \quad (1.1.4)$$

is introduced. It is called the epigraph of f is denoted by epi f and it is a convex set in $X \times \mathbb{R}$. An equivalent definition of convexity of a functional $f : X \to \bar{\mathbb{R}}$ is the following: We define $f : X \to \bar{\mathbb{R}}$ to be convex, whenever epi f is a convex subset of $X \times \mathbb{R}$. The effective domain $D(f)$ of a convex functional f on X is defined by

$$D(f) = \{x \big| x \in X, f(x) < \infty\}. \quad (1.1.5)$$

A functional f is called proper if $f : X \to (-\infty, +\infty]$ and $f \not\equiv \infty$. If f is convex, $\lambda f (\lambda \geq 0)$ is convex. For f_1 and f_2 convex, $f_1 + f_2$ is convex as well (here we assume that $(f_1 + f_2)(x) = +\infty$ for $f_1(x) = -f_2(x) = \pm\infty$, cf. [Eke] p.67). A functional $f : X \to \bar{\mathbb{R}}$ is called lower semicontinuous (l.s.c.) on X if for every $\lambda \in \mathbb{R}$ the set

$$\{x \big| x \in X, f(x) \leq \lambda\} \quad (1.1.6)$$

is closed in X. For f l.s.c., $-f$ is upper semicontinuous (u.s.c.), and conversely. Similarly a functional $f : X \to \bar{\mathbb{R}}$ is l.s.c. if and only if epi f is a closed subset of $X \times \mathbb{R}$.

As we associate with a functional f its epigraph epi f, so we may associate with a set K the indicator I_K. It is easily shown that K is closed if and only if its indicator is l.s.c.

The following result on the continuity of convex functionals deserves mention. In a Hilbert space X, a convex, l.s.c. functional $f : X \to \bar{\mathbb{R}}$ is continuous on int $D(f)$.

Let us denote further by X' the dual space to X and let $\langle \cdot, \cdot \rangle$ be the duality pairing between X and X'.

Convex functionals $f : X \to \bar{\mathbb{R}}$ are not necessarily everywhere differentiable. Then the supporting hyperplanes (cf. AA' in Fig. 1.1.1) to the epi f describe the differential properties of f. This leads to the notion of subdifferential. The vector $x' \in X'$, for which

$$f(x_1) - f(x) \geq \langle x', x_1 - x \rangle, \quad \forall x_1 \in X \tag{1.1.7}$$

holds, where $f(x)$ is finite at $x \in X$, is called the subgradient of f at x. The set of all $x' \in X'$ satisfying (1.1.7) is called the subdifferential of f at x and is denoted by $\partial f(x)$. We then write

$$x' \in \partial f(x). \tag{1.1.8}$$

The set $\{x \mid \partial f(x) \neq \emptyset\}$ is denoted by $D(\partial f)$, and is called the domain of ∂f.

The mapping $\partial f : X \to X'$ is multivalued and is called the subdifferential of f. If $\partial f(x) \neq \emptyset$, f is said to be subdifferentiable at x, $\partial f(x) = \emptyset$ for $x \notin D(f)$ and $f \not\equiv \infty$. From (1.1.7), it follows that a necessary and sufficient condition in order that x_0 minimize f on X is that

$$0 \in \partial f(x_0), \tag{1.1.9}$$

because then $f(x_0) \leq f(x) \; \forall x \in X$. This fact shows the close relation of the subdifferentiability to the optimization theory. The affine function $x_1 \to L(x_1) = f(x) + \langle x', x_1 - x \rangle$ is called the supporting hyperplane of epi f at $\{x, f(x)\}$. Thus (1.1.7) states that for $f(x)$ finite the supporting hyperplane to epi f at $\{x, f(x)\}$ is nonvertical. It can be shown that $\partial f(\cdot)$ is for every $x \in$ int $D(f)$ a convex closed set of the dual Hilbert space X'. Actually if X is more generally a locally convex Hausdorff topological vector space and X' its dual space then $\partial f(\cdot)$ is closed with respect to the weak topology $\Sigma(X', X)$ (see e.g. [Eke]).

In the case of convex functionals the existence of subdifferentials is ensured by means of the following result.

Proposition 1.1.1 Let $f : X \to \bar{\mathbb{R}}$ be convex, and suppose that f is finite and continuous at $x_0 \in X$. Then $\partial f(x_0) \neq \emptyset$. Moreover $\partial f(x)$ is nonempty for every $x \in$ int $D(f)$.

The case $f = I_K$, where K is a nonempty convex subset of X, is important. Then

$$\partial I_K(x) = \{x' \mid I_K(x_1) - I_K(x) \geq \langle x', x_1 - x \rangle, \forall x_1 \in X\} \qquad (1.1.10)$$

or, equivalently,

$$\partial I_K(x) = \{x' \mid \langle x', x_1 - x \rangle \leq 0, \forall x_1 \in K, \text{ for } x \in K\}. \qquad (1.1.11)$$

The geometrical meaning of the variational inequality

$$\langle x', x_1 - x \rangle \leq 0, \quad \forall x_1 \in K, \quad x \in K \qquad (1.1.12)$$

is that x' is an outward normal vector to K at x. In general, the set of all vectors x' satisfying (1.1.12) forms an outward normal cone to K at x. This cone (a) is empty for $x \notin K$ (b) has at least the zero element for $x \in K$, and (c) has only the zero element if $x \in \text{relint } K$ (here K is regarded as a subset of its affine hull. To explain better this fact we consider in $\mathbb{R}^3$ three points defining a set K. Then the int K will be understood by considering K as a subset of the plane defined by the three points in $\mathbb{R}^3$ and it is called the relative interior of K (relint K)).

Subdifferentiability is closely related to the notion of "one-sided Gâteaux-differentiability". This provides a method for the "construction" of the subdifferential for a given functional.

A functional $f : X \to \bar{\mathbb{R}}$, where X is a H-space is said to be one-sided directional Gâteaux-differentiable at x_0 if there exists $\tilde{f}'(x_0, h)$ such that

$$\lim_{\mu \to 0_+} \frac{f(x_0 + \mu h) - f(x_0)}{\mu} = \tilde{f}'(x_0, h), \quad \forall h \in X. \qquad (1.1.13)$$

It should be noted that $+\infty$ and $-\infty$ are allowed as limits in (1.1.13). Functional $h \to \tilde{f}'(x_0, h)$ is the one-sided directional Gâteaux-differential of f at x_0 with respect to the direction h. It can be shown that $\tilde{f}'(x_0, \cdot)$ is a convex, positively homogeneous function of h. If $h \to \tilde{f}'(x_0, h)$ is continuous and linear, then f is Gâteaux-differentiable at x_0. One important property of convex functionals is their one-sided directional Gâteaux-differentiability.

Proposition 1.1.2 Assume that $f : X \to \bar{\mathbb{R}}$ is convex. Then f is one-sided directional Gâteaux-differentiable at every $x \in X$ with $f(x) \neq \pm\infty$. Moreover the following properties hold

$$f(x_1) - f(x) \geq \tilde{f}'(x, x_1 - x), \quad \forall x_1 \in X \qquad (1.1.14)$$

and

1.1 Convexity and Subdifferential   7

$$\tilde{f}'(x, x_1 - x) \geq -\tilde{f}'(x, -(x_1 - x)), \quad \forall x_1 \in X. \tag{1.1.15}$$

If moreover f is bounded on a neighborhood of $x_0 \in X$, then

$$\tilde{f}'(x_0, h) = \max\{\langle x', h \rangle \big| x' \in \partial f(x_0)\}, \quad \forall h \in X. \tag{1.1.16}$$

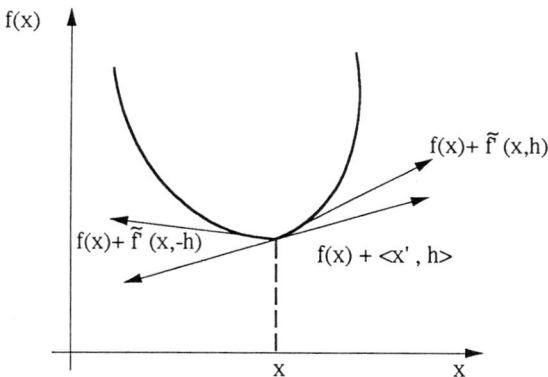

Fig. 1.1.2. On the geometrical meaning of (1.1.14) and (1.1.15).

This last proposition permits a simple construction (cf. Fig. 1.1.2) of the set $\partial f(x)$: if f maps $\mathbb{R}$ into $\bar{\mathbb{R}}$ then the subgradients x' are the slopes of the nonvertical lines through $(x, f(x))$, which have no point in common with intepi f. From $\tilde{f}'(x, 1) = f'_+(x)$ and $\tilde{f}'(x, -1) = -f'_-(x)$ (right and left derivatives), it follows that $f'_-(x) \leq x' \leq f'_+(x)$. Assume that f is a convex, l.s.c., proper functional on $\mathbb{R}$. In this case the right and left derivatives f'_+ and f'_- can be extended, when $x \notin D(f)$, by setting $f'_+ = f'_- = \infty$ (resp. $f'_+ = f'_- = -\infty$) for points lying to the right (resp. to the left) of $D(f)$. We may then write for $f : \mathbb{R} \to \bar{\mathbb{R}}$

$$\partial f(x) = \{x' \in \mathbb{R} \big| f'_-(x) \leq x' \leq f'_+(x)\}. \tag{1.1.17}$$

Proposition 1.1.3 Let $f : X \to \bar{\mathbb{R}}$ be convex and suppose that $\operatorname{grad} f(x)$ exists at x. Then $\partial f(x) = \{\operatorname{grad} f(x)\}$. Conversely, if f is finite and continuous at x and if $\partial f(x)$ has only one element, then $\operatorname{grad} f$ exists at x and $\partial f(x) = \{\operatorname{grad} f(x)\}$.

Simple examples illustrating the notion of the subdifferential are given to [Rock, Pan85]. Here we give one only example (Fig. 1.1.3).

8   1. Elements of Nonsmooth Analysis

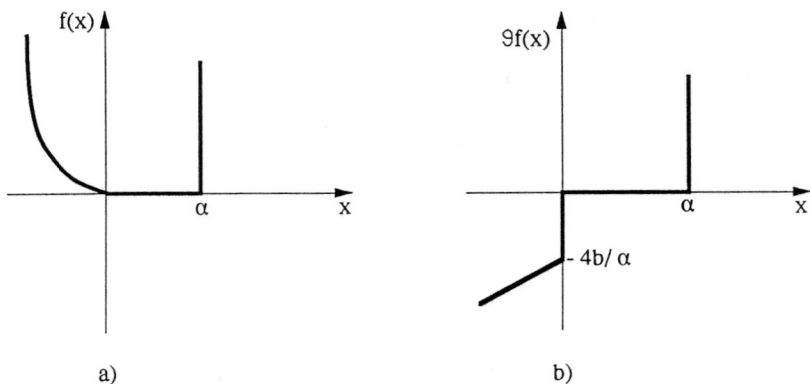

Fig. 1.1.3   The graphs of (1.1.18) and (1.1.19).

Let $f : \mathbb{R} \to \bar{\mathbb{R}}$ be given by ($b > 0$)

$$f(x) = \begin{cases} \dfrac{4bx}{a}\left(\dfrac{x}{a} - 1\right) & \text{if } x \leq 0 \\ 0 & \text{if } 0 \leq x \leq a \\ \infty & \text{if } x > a \end{cases} \quad (1.1.18)$$

Then

$$\partial f(x) = \begin{cases} \dfrac{4b}{a}\left(\dfrac{2x}{a} - 1\right) & \text{if } x < 0 \\ [-\dfrac{4b}{a}, 0] & \text{if } x = 0 \\ 0 & \text{if } 0 < x < a. \\ [0, \infty) & \text{if } x = a \\ \emptyset & \text{if } x > a \end{cases} \quad (1.1.19)$$

Now some propositions from the subdifferential calculus are given.

Proposition 1.1.4 Let $f : X \to \bar{\mathbb{R}}$ and $\lambda > 0$. Then for every $x \in D(\partial f)$

$$\partial(\lambda f)(x) = \lambda \partial f(x) \quad (1.1.20)$$

Proposition 1.1.5 Let $f_1 : X \to \bar{\mathbb{R}}$ and $f_2 : X \to \bar{\mathbb{R}}$. Then for every $x \in D(\partial f_1) \cap D(\partial f_2)$

$$\partial f_1(x) + \partial f_2(x) \subset \partial(f_1 + f_2)(x) \quad (1.1.21)$$

This inclusion holds as an equality of sets, if certain additional conditions given in the following propositions hold.

Proposition 1.1.6 Assume that $f_1 : X \to \bar{\mathbb{R}}$ and $f_2 : X \to \bar{\mathbb{R}}$ are convex and that grad f_2 exists at the point x. Then if $x \in D(\partial f_1)$,

$$\partial(f_1 + f_2)(x) = \partial f_1(x) + \partial f_2(x). \tag{1.1.22}$$

Proposition 1.1.7 Suppose that $f_1 : X \to (-\infty, +\infty]$ and $f_2 : X \to (-\infty, +\infty]$ are convex and l.s.c., and that a point $x_0 \in D(f_1) \cap D(f_2)$ exists at which f_1 is continuous. Then

$$\partial(f_1 + f_2)(x) = \partial f_1(x) + \partial f_2(x) \quad \forall x \in X. \tag{1.1.23}$$

The following example shows the applicability of the above addition propositions. Let us consider the minimization problem of a convex functional on a convex set $K \subset X$, where X is a H-space. We want to find a point x_0 which is a solution to the problem

$$f(x_0) = \inf\{f(x) \big| x \in K\} \text{ or } f(x_0) = \inf_K f(x). \tag{1.1.24}$$

If f achieves the infinum on K for $x = x_0 \in K$, we write

$$f(x_0) = \min\{f(x) \big| x \in K\} \text{ or } f(x_0) = \min_K f(x). \tag{1.1.25}$$

Every x_0 satisfying (1.1.25) is a solution of the optimization problem. Let K be a nonempty convex subset of X and f a convex proper functional $f : K \to \bar{\mathbb{R}}$. It is obvious that f can be extended to all of X, and hence the solution of (1.1.25) is sought in X. Thus the minimization problem of f over K is equivalent to the minimization of $f + I_K$ over all of X. For x_0 to be a solution of this problem, it is necessary and sufficient (cf. (1.1.9)) that

$$0 \in \partial(f(x_0) + I_K(x_0)). \tag{1.1.26}$$

Let us assume now that grad f exists everywhere. Then (1.1.26) is equivalent to the relation

$$-\text{grad } f(x_0) \in \partial I_K(x_0) \tag{1.1.27}$$

i.e $-\text{grad } f(x_0)$ is an element of the outward normal cone to K at x_0.

1.2 Generalized Gradient and Related Calculus

Suppose that X and X' are dual Hilbert spaces and let $\langle \cdot, \cdot \rangle$ be the duality pairing. Now let $A : X \to X'$ be a multivalued mapping characterized by its graph

$$G(A) = \{(x, y) \in X \times X' \big| y \in A(x)\} \tag{1.2.1}$$

Then the inverse mapping A^{-1} is again a multivalued mapping $A^{-1} : X' \to X$ defined by the relation

$$x \in A^{-1}(y) \iff (x,y) \in G(A). \quad (1.2.2)$$

Generally we say that a set-valued map satisfies a property (e.g closedness, measurability etc), if and only if this property is satisfied by its graph [Aub90].

The definition of the derivative f' of a function $f : \mathbb{R} \to \mathbb{R}$ at $x \in \mathbb{R}$ is given by considering the limit of the quotients $\{f(x+h)-f(x)\}/h$ when $h \in \mathbb{R}$ tends to zero. Then if this limit exists we say that $f'(x)$ exists. We recall here that the limit of a sequence may not exist, whereas the upper limit "limsup" and the lower limit "liminf" of a sequence always exist. Thus in the case of nonsmooth functions as it is obvious, the upper and lower limits must play the role which plays the limit in the case of differentiable functions. There are indeed several attempts in mathematics to generalize the notion of derivative for nonsmooth functions using the upper and lower limits. In this Section we deal with the generalized gradient of F.H.Clarke, which leads to the theory of hemivariational inequalities. Closing we would like to remark that until now there is not any optimal generalization of the notion of differentiability for nonsmooth functions. Let us recall now the notion of upper and lower limit which will be used in the sequel.

Let $f : X_1 \to \bar{\mathbb{R}}$ and $0 < \delta < 1$, where X_1 is a subset of X. We denote by $B(x,\delta)$ the ball $\{y \in X_1 \big| \|x-y\| \leq \delta\}$ for $x \in X_1$ and following Aubin [Aub77] we associate to δ the function

$$\alpha(\delta) = \inf\{f(y)\big| y \in B(x,\delta)\}, \quad (1.2.3)$$

which is a decreasing function of δ. Accordingly, the $\lim \alpha(\delta)$ as $\delta \to 0_+$ exists in $\bar{\mathbb{R}}$ and is equal to $\sup\{\alpha(\delta)\big| \delta > 0\}$ (recall the well-known theorem of classical analysis concerning the convergence of bounded monotone sequences). We give now the following definitions and relations

$$\liminf_{y \to x} f(y) = \lim_{\delta \to 0}\left(\inf_y \{f(y)\big| y \in B(x,\delta)\}\right) = \sup_{\delta > 0}\inf\{f(y)\big| y \in B(x,\delta)\} \quad (1.2.4)$$

$$\limsup_{y \to x} f(y) = \lim_{\delta \to 0}\left(\sup_y\{f(y)\big| y \in B(x,\delta)\}\right) = -\liminf_{y \to x}(-f(y)). \quad (1.2.5)$$

Analogously we may write that

$$\liminf_{n \to \infty} f(x_n) = \lim_{n \to \infty}\left(\inf_{p \geq 0} f(x_{n+p})\right) = \sup_n \inf_{p \geq 0} f(x_{n+p}) \quad (1.2.6)$$

$$\limsup_{n \to \infty} f(x_n) = \lim_{n \to \infty}\left(\sup_{p \geq 0} f(x_{n+p})\right) = -\liminf_{n \to \infty}(-f(x_n)). \quad (1.2.7)$$

From the above definition we can easily write the following inequalities for a function $f : X_1 \to \bar{\mathbb{R}}$ and for every $x \in X_1$.

$$\inf\{f(y)\big| y \in X_1\} \leq \liminf_{y \to x} f(y) \quad (1.2.8)$$
$$\leq f(x) \leq \limsup_{y \to x} f(y) \leq \sup\{f(y)\big| y \in X_1\}.$$

Since we have defined the "liminf" and the "limsup" we can give now an equivalent definition of the lower semicontinuity.

Proposition 1.2.1 A functional $f : X \to \bar{\mathbb{R}}$ is l.s.c if and only if.

$$\forall x_0 \in X \quad \liminf_{x \to x_0} f(x) \geq f(x_0). \tag{1.2.9}$$

Note that if (1.2.9) holds only at a point $x_0 \in X$ then f is called l.s.c. at x_0. Note at this point that if $f : X \to \bar{\mathbb{R}}$ is a convex functional such that $f(x) = \{\tilde{f}(x)$ for $x \in D(\tilde{f})$, ∞ otherwise$\}$, then the lower semicontinuity of $\tilde{f}$ on $D(\tilde{f})$ does not necessarily imply the lower semicontinuity of f on X (cf. e.g. the function $f(x) = \{x$ for $x > 0$, ∞ for $x \leq 0\}$). Let now C be a nonempty subset of the Hilbert space X and let B denote the unit ball with center at zero. Then $x_0 + \delta B$ denotes the ball $B(x_0, \delta)$ defined previously. By writing $x_n \to_C x$ we mean that x_n converges to x in the set C. The subset K_C defined by

$$K_C(x) = \bigcap_{\delta > 0} \bigcap_{\alpha > 0} \bigcup_{0 < \mu \leq \alpha} (\frac{1}{\mu}(C - x) + \delta B) \tag{1.2.10}$$

is called the contingent cone to C at the point x. The contingent cone is closed. From this definition it results [Aub84] that $y \in K_C(x)$ if and only if a sequence of positive numbers μ_n and a sequence $\{u_n\} \in X$ exist such that as $n \to \infty$

$$u_n \to y, \mu_n \to 0 \quad \text{and} \quad \forall n \geq 0 \quad x + \mu_n u_n \in C. \tag{1.2.11}$$

Another equivalent definition of the contingent cone uses the distance function of x from the set C

$$d_C(x) = \inf\{\|x - y\| \big| y \in C\}. \tag{1.2.13}$$

It can be shown that

$$y \in K_C(x) \iff \liminf_{\mu \to 0+} \frac{d_C(x + \mu y)}{\mu} = 0. \tag{1.2.14}$$

One can easily verify that for $x \in \text{int } C$, $K_C(x) = X$. Related to the contingent cone is the notion of the tangent cone. In order to give a definition of it, it is necessary to introduce the lower limit "liminf" of a multifunction. Let X_1 be a subset of X and let F be a set-valued function from X_1 to Y, where Y is a Hilbert space. We define that for $x \in D(F)$

$$\liminf_{x' \to x} F(x') = \{v \in Y \big| \lim d_{F(x')}(v) = 0 \text{ as } x' \to x \quad x' \in D(F)\}. \tag{1.2.15}$$

The multifunction F is called by definition l.s.c. at x, if

$$\liminf_{x' \to x} F(x') = F(x). \tag{1.2.16}$$

A useful result is that F is l.s.c. at $x \in D(F)$, if and only if for any $y \in F(x)$ and for any sequence $\{x_n\} \in D(F)$ with $x_n \to x$, there exists a sequence $y_n \in F(x_n)$

which converges to y. For a set $C \in X$ and for $x \in X$, the tangent cone $T_C(x)$ to C at x is by definition

$$T_C(x) = \liminf_{\substack{\tilde{x} \to_C x \\ \mu \to 0_+}} \frac{1}{\mu}(C - \{\tilde{x}\}); \qquad (1.2.17)$$

Equivalently,

$$T_C(x) = \{y | y \in X, \text{ for } \mu_n \to 0_+, \text{ and } x_n \to_C x, \qquad (1.2.18)$$
$$\text{there exists } y_n \to y \text{ with } y_n + \mu_n x_n \in C\}.$$

and

$$y \in T_C(x) \iff \lim_{\substack{\tilde{x} \to_C x \\ \mu \to 0_+}} \frac{d_C(\tilde{x} + \mu y)}{\mu}. \qquad (1.2.19)$$

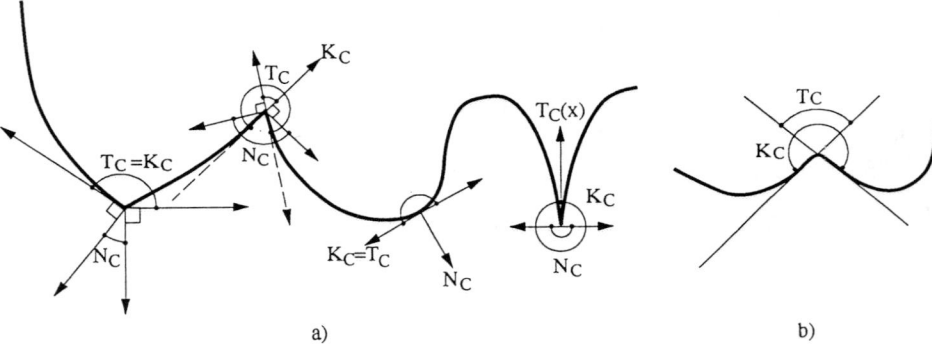

Fig. 1.2.1. Contingent, tangent and normal cones.

It may be proved that $T_C(x)$ is a convex closed cone which always contains 0. Note that always $T_C(x)$ is contained in $K_C(x)$, which is always contained in the closure of the set $\cup_{\mu>0} \frac{1}{\mu}(C - \{x\})$. For $x \in$ int C, $T_C(x) = X$, and for every $x \in X$, $T_X(x) = X$. By definition $T_\emptyset(x) = \emptyset$. The normal cone $N_C(x)$ to C at x is defined as

$$N_C(x) = \{x' | x' \in X', (z, x') \le 0, \ \forall z \in T_C(x)\}. \qquad (1.2.20)$$

Obviously

$$T_C(x) = \{y | y \in X, (y, z') \le 0, \ \forall z' \in N_C(x)\}. \qquad (1.2.21)$$

If C is convex then $K_C(x) = T_C(x)$ and if the boundary of C is continuously differentiable around a boundary point x, then $K_C(x) = T_C(x)$ and they are identified with the usual tangent vector space to C at this point. In Fig. 1.2.1 we give the geometical forms of the aforementioned cones with respect to certain types of set boundaries. We say that a set C is regular at x_0 if $T_C(x_0) = K_C(x_0)$.

In the present book we shall define first the generalized gradient only for Lipschitz functions according to the initially developed theory of Clarke [Clar73,83], and then we shall consider the more general case of $f : X \to \bar{\mathbb{R}}$. We recall that $f : X \to \mathbb{R}$ is Lipschitz at x, or locally Lipschitz at x, if a neighborhood U of x exists such that f is finite on U and

$$|f(x_1) - f(x_2)| \leq c||x_1 - x_2|| \quad \forall x_1, x_2 \in U, \qquad (1.2.22)$$

where c is a positive constant depending on U. If f is locally Lipschitz at every $x \in X_1 \subset X$, then f is called Lipschitz on X_1. Note that f is Lipschitzian at x if it is continuously differentiable at x, or if it is convex (or concave) and finite at x, or if it is the linear combination of Lipschitzian functions at x.

Let f be locally Lipschitz at $x \in X$ and let y be any other vector in X. The directional differential in the sense of Clarke of f at x in the direction y, denoted $f^0(x, y)$ is defined by the relation

$$f^0(x, y) = \limsup_{\substack{\mu \to 0_+ \\ h \to 0}} \frac{f(x + h + \mu y) - f(x + h)}{\mu}. \qquad (1.2.23)$$

$f^0(x,y)$ is also called generalized directional differential and has the following properties.

Proposition 1.2.2 Let f be locally Lipschitz at x. Then i) $g : y \to f^0(x,y)$ is finite, convex, l.s.c., positively homogeneous and satisfies the inequality

$$|f^0(x, y)| \leq c||y||, \qquad (1.2.24)$$

where c is the constant of (1.2.22) depending on the neighborhood U.
ii) $(x, y) \to f^0(x, y)$ is u.s.c., and $g : y \to f^0(x, y)$ is locally Lipschitz at y.
iii) The following relations hold

$$\operatorname{epi} g = T_{\operatorname{epi} f}(x, f(x)) \qquad (1.2.25)$$

$$f^0(x, -y) = (-f)^0(x, y). \qquad (1.2.26)$$

By means of the generalized directional differential $f^0(x,y)$ we can now define the new notion of the generalized gradient $\bar{\partial} f(x)$.

Let $f : X \to \mathbb{R}$ be a locally Lipschitz functional at $x \in X$. The following two equivalent definitions of the generalized gradient $\bar{\partial} f(x) : X \to X'$ (multivalued mapping) are given now

i) $\quad \bar{\partial} f(x) = \{x' | x' \in X', f^0(x, x_1 - x) \geq (x', x_1 - x) \quad \forall x_1 \in X\} \qquad (1.2.27)$

and

ii) $\quad \bar{\partial} f(x) = \{x' | x' \in X', (x', -1) \in N_{\operatorname{epi} f}(x, f(x))\}. \qquad (1.2.28)$

The following propositions hold.

Proposition 1.2.3 Let $f : X \to \mathbb{R}$ be a locally Lipschitz functional at $x \in X$. Then $\bar{\partial} f(x)$ is a nonempty convex, closed and bounded subset of X'.

Proposition 1.2.4 Let $f : X \to \mathbb{R}$ be a locally Lipschitz functional at $x \in X$. Then
$$f^0(x,y) = \max\{\langle y, x'\rangle \big| x' \in \bar{\partial} f(x)\}. \tag{1.2.29}$$

Until now we have defined the generalized gradient only for Lipschitz function. If the definition (1.2.27) holds, then (1.2.28) is a property of the generalized gradient. This property can be used to define the generalized gradient $\bar{\partial} f(x)$ for any type of function $f : X \to \bar{\mathbb{R}}$ which is finite at the point x: The set $\bar{\partial} f(x)$ is the set of all $x' \in X'$ such that (1.2.28) holds. Note that $\bar{\partial} f(x)$ may be empty. The above definition of $\bar{\partial} f(x)$ for any function $f : X \to \bar{\mathbb{R}}$ makes sense, because the normal cone $N_C(x)$ can be defined with respect to any set epi f. Let us define now the generalized directional differential $f^\uparrow(x;y)$ at x in the direction y by the relation

$$f^\uparrow(x,y) = \sup\{\langle y, x'\rangle \big| x' \in \bar{\partial} f(x)\}. \tag{1.2.30}$$

Obviously we can write that

$$\bar{\partial} f(x) = \{x' \big| x' \in X', f^\uparrow(x, x_1 - x) \geq \langle x', x_1 - x\rangle \forall \ x_1 \in X\}. \tag{1.2.31}$$

The directional differential $f^\uparrow(x;y)$ is called in this book also directional differential in the sense of Rockafellar, who has given another equivalent definition of it [Rock79,80]. Note that $\bar{\partial} f(x) = \emptyset$ if $f^\uparrow(x,0) = -\infty$, and if $f^\uparrow(x,y)$ is finite for every y then $\bar{\partial} f(x) \neq \emptyset$. The following propositions may be proved [Rock80].

Proposition 1.2.5 Let $f : X \to \bar{\mathbb{R}}$ and let $f(x)$ be finite. Then

i) $\bar{\partial} f(x)$ is a convex, closed subset of X'.

ii) Function $g : y \to f^\uparrow(x;y)$ is convex, l.s.c. and positively homogeneous when $f^\uparrow(x;y) > -\infty$ for all $y \in X$, and (1.2.25) is valid.

If f is convex, then

$$f^\uparrow(x,y) = \liminf_{\tilde{y} \to y} \tilde{f}'(x, \tilde{y}) \quad \forall y \in X, \tag{1.2.32}$$

where $\tilde{f}'(\cdot, \cdot)$ denotes the one-sided directional Gâteaux differential (see relation (1.1.13)). If f is locally Lipschitz at x then

$$f^\uparrow(x,y) = f^0(x,y) \quad \forall y \in X \tag{1.2.33}$$

and if f is continuously differentiable at x, then

$$\bar{\partial} f(x) = \{\operatorname{grad} f(x)\}. \tag{1.2.34}$$

The indicator function I_C of a set C is defined as in the convex case, i.e. $I_C(x) = \{0 \text{ if } x \in C, \infty \text{ otherwise}\}$. It is shown [Rock79,80] that

$$\bar{\partial} I_C(x) = N_C(x) \tag{1.2.35}$$

and

$$I_C^\uparrow(x, y) = I_{T_{C(x)}}(y). \tag{1.2.36}$$

For f is convex (resp. concave and bounded below on a neighborhood of x)

$$\bar{\partial} f(x) = \partial f(x) \tag{1.2.37}$$

resp.

$$\bar{\partial} f(x) = -\partial(-f)(x) \tag{1.2.38}$$

at every x where f is finite. The following proposition is important.

Proposition 1.2.6 If f has at x_0 a finite local minimum, then

$$0 \in \bar{\partial} f(x_0). \tag{1.2.39}$$

Moreover due to (1.2.37) and the convexity of $f^\uparrow(x,\cdot)$ we may write that

$$x' \in \bar{\partial} f(x) \iff x' \in \bar{\partial} f^\uparrow(x,0) = \partial f^\uparrow(x,0). \tag{1.2.40}$$

Let us suppose that $f, g : X \to \mathbb{R}$ are Lipschitz functions; then

$$\bar{\partial}(f+g)(x) \subset \bar{\partial} f(x) + \bar{\partial} g(x) \tag{1.2.41}$$

and

$$\bar{\partial}(\lambda f)(x) = \lambda \bar{\partial} f(x) \text{ for } \lambda \in \mathbb{R}. \tag{1.2.42}$$

Of importance is also the finite dimensional case $X \equiv \mathbb{R}^n$. Then for f locally Lipschitz at x a definition equivalent to the definition (1.2.27) is the following: $\bar{\partial} f(x)$ is the convex hull of all points $x' \in \mathbb{R}^n$ of the form

$$x' = \lim_{i \to \infty} \text{grad } f(x_i), \tag{1.2.43}$$

where x_i converges as $i \to \infty$ to x avoiding the nondifferentiability points and any other points of a set of measure zero (in the sense of Lebesgue). At this point we would like also to recall Rademacher's theorem stating that a Lipschitz function f on an open subset of $\mathbb{R}^n$ is almost everywhere (a.e.) in the sense of Lebesgue differentiable.

An important notion is the notion of the substationarity [Rock79] of a functional $f : -X \to \bar{\mathbb{R}}$ at a point x_0. We call x_0 a substationarity point of f if

$$0 \in \bar{\partial} f(x_0). \tag{1.2.44}$$

Equivalent to this definition if the statement that

$$f^\uparrow(x_0, y) \geq 0 \quad \forall y \in X. \tag{1.2.45}$$

Substationarity points are all the classical stationarity points, all the local minima, a large class of local maxima (e.g. if at x_0 there is y such that. $\limsup\{[f(x' + \mu y') - f(x')]/\mu\} < \infty$, where $x' \to x_0, f(x') \to f(x_0), y' \to y, \mu \to 0_+$; then f is called locally Lipschitz at x_0 in the direction y), as well as all the saddle points. Obviously if at x_0, f is locally Lipschitz and has a local maximum, then (1.2.44) holds and x_0 is a substationarity point. We say that x is a substationarity point of f with respect to a set C if $f + I_C$ is substationary at x. The notion of substationarity plays an important role in the theory of hemivariational inequalities because it permits the formulation of the propositions of substationary potential and complementary energy which generalize the corresponding classical minimum propositions.

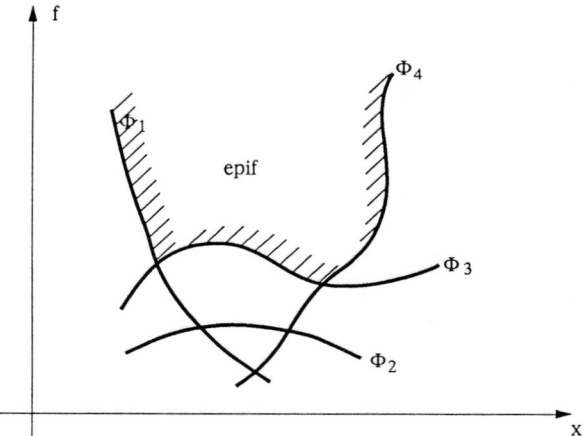

Fig. 1.2.2. A maximum type function.

Suppose now that f is a maximum-type function i.e., $f = \max\{\varphi_i, \ldots, \varphi_m\}$ where $\varphi_i = \varphi_i(x)$, $i = 1, \ldots, m$, $x \in \mathbb{R}^n$ are continuously differentiable functions. We denote the sets $\{x|\varphi_i = f\}$ by A_i. It is easy to verify that f is a locally Lipschitz function and that

$$\bar{\partial} f(x) = \{\text{grad } \varphi_i(x)\} \quad \text{if } x \in A_i, \ x \notin A_i \cap A_j \text{ etc.} \quad (1.2.46a)$$

$$\bar{\partial} f(x) = \text{co}\{\text{grad } \varphi_i(x), \text{grad } \varphi_j(x)\} \quad \text{if} \quad (1.2.46b)$$

$$x \in A_i \cap A_j, x \notin (A_i \cap A_j) \cap A_k \text{ etc.}$$

and

$$\bar{\partial} f(x) = \text{co}\{\text{grad } \varphi_i(x), \text{grad } \varphi_j(x), \text{grad } \varphi_k(x)\} \quad \text{if} \quad (1.2.46c)$$

$$x \in (A_i \cap A_j) \cap A_k, x \notin ((A_i \cap A_j) \cap A_k) \cap A_l \text{ etc.}$$

Suppose finally that $C = \{x \in \mathbb{R}^n | f(x) \leq 0\}$. Then at a point x_0 with $f(x_0) = 0$

$$N_C(x_0) \subset \{\lambda x' | x' \in \mathbb{R}^n, \lambda \geq 0, x' \in \bar{\partial} f(x_0\}, \quad (1.2.47)$$

whenever f is Lipschitzian on a neighborhood of x_0 and $0 \notin \bar{\partial} f(x_0)$. If for a locally Lipschitz function

$$f^0(x,y) = \tilde{f}'(x,y) \qquad (1.2.48)$$

at a point x for every $y \in X$, f is called $\bar{\partial}$-regular at the point x. This definition is equivalent to the statement that epi f is regular at $(x, f(x))$. For instance, a convex function and a maximum type function are $\bar{\partial}$-regular at a point x where they take finite values. If f and g are $\bar{\partial}$-regular at x then (1.2.41) holds as an equality. Similarly (1.2.47) holds as an equality if f is $\bar{\partial}$-regular at x_0. The combination of (1.2.46) with (1.2.47) yields for $f = \max\{\varphi_1, \ldots, \varphi_m\}$ the relation

$$N_C(x_0) = \bar{\partial} I_C(x_0) = \qquad (1.2.49)$$
$$\{z | z = \sum_{i=1}^{m} \lambda_i \operatorname{grad} \varphi_i(x_0), \lambda_i \geq 0, \varphi_i(x_0) \leq 0 \; \lambda_i \varphi_i(x_0) = 0\},$$

if $0 \notin \bar{\partial} f(x_0)$, which permits the extension of the Lagrange multiplier rule for optimization problems subjected to the nonconvex inequality constraints $\varphi_i(x) \leq 0$, $i = 1, \ldots, m$. This becomes obvious, e.g. if one considers the search for a local minimum problem of a continuously differentiable function $g : \mathbb{R}^n \to \mathbb{R}$ over $C = \{x \in \mathbb{R}^n | \varphi_i(x) \leq 0 \; i = 1, \ldots, m\}$. A necessary condition is $0 \in \bar{\partial}(g + I_C)(x)$ which implies that

$$-\operatorname{grad} g(x) \in \bar{\partial} I_C(x) \qquad (1.2.50)$$

which together with (1.2.49) leads to the Lagrange multiplier rule.

Further we give an application of the generalized gradient which is useful in the theory of hemivariational inequalities. Suppose that $\beta : \mathbb{R} \to \mathbb{R}$ is a function such that $\beta \in L^{\infty}_{loc}(\mathbb{R})$, i.e. a function essentially bounded on any bounded interval of $\mathbb{R}$. For any $\rho > 0$ and $\xi \in \mathbb{R}$ let us define

$$\bar{\beta}_\rho(\xi) = \operatorname*{ess\,inf}_{|\xi_1 - \xi| \leq \rho} \beta(\xi_1) \text{ and } \bar{\bar{\beta}}_\rho(\xi) = \operatorname*{ess\,sup}_{|\xi_1 - \xi| \leq \rho} \beta(\xi_1). \qquad (1.2.51)$$

Obviously the monotonicity properties of $\rho \to \bar{\beta}_\rho(\xi)$ and $\rho \to \bar{\bar{\beta}}_\rho(\xi)$ imply that the limits as $\rho \to 0_+$ exist and therefore we may write

$$\bar{\beta}(\xi) = \lim_{\rho \to 0_+} \bar{\beta}_\rho(\xi) \text{ and } \bar{\bar{\beta}}(\xi) = \lim_{\rho \to 0_+} \bar{\bar{\beta}}_\rho(\xi). \qquad (1.2.52)$$

Let us define the multivalued function

$$\tilde{\beta}(\xi) = [\bar{\beta}(\xi), \bar{\bar{\beta}}(\xi)] \qquad (1.2.53)$$

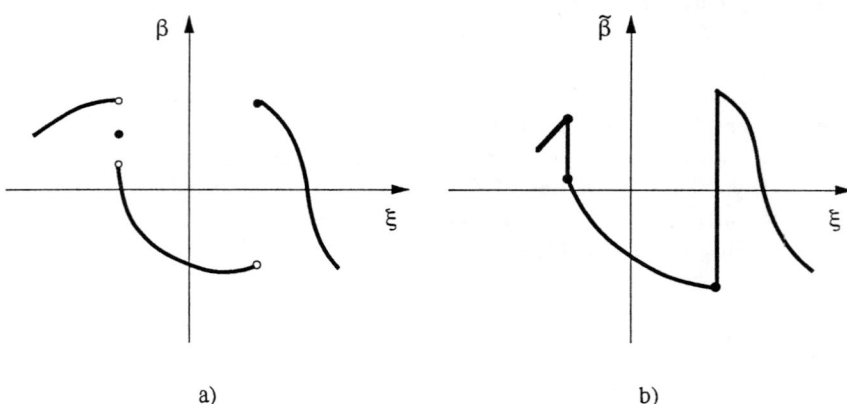

Fig. 1.2.3. On the definition of β and $\tilde{\beta}$.

where $[\cdot,\cdot]$ denotes the interval. Roughly speaking (Fig. 1.2.3) $\tilde{\beta}$ results from the generally discontinuous function β by "filling in the gaps". For instance if at ξ, $\beta(\xi_+) > \beta(\xi_-)$ (resp. $\beta(\xi_+) < \beta(\xi_-)$ then $\tilde{\beta}(\xi) = [\beta(\xi_-), \beta(\xi_+)]$ (resp. $\tilde{\beta}(\xi) = [\beta(\xi_+), \beta(\xi_-)]$). It was proved by Chang [Ch] that a locally Lipschitz function $\tilde{j}$ can be determined up to an additive constant by the relation

$$\tilde{j}(\xi) = \int_0^\xi \beta(\xi_1) d\xi_1 \qquad (1.2.54)$$

such that $\bar{\partial}\tilde{j}(\xi) \subset \tilde{\beta}(\xi)$. If moreover $\beta(\xi_\pm)$ exist for each $\xi \in \mathbb{R}$ then

$$\bar{\partial}\tilde{j}(\xi) = \tilde{\beta}(\xi). \qquad (1.2.55)$$

1.3 Minimization Problems. Duality of Convex Functionals

In the previous two Sections we have given some results connecting the search for a local or global minimum with the solution of a multivalued equation or inclusion of the type $0 \in \bar{\partial} f(x)$. In this Section we give some additional results concerning the minimization problems, as well as certain results on the notion of duality of convex functionals. Let us consider the minimum problem (1.1.25): Find $x_0 \in K$ such that

$$f(x_0) = \min\{f(x) | x \in K\} \qquad (1.3.1)$$

where K is a nonempty convex closed subset of a Hilbert space X and $f : X \to (-\infty, +\infty]$, $f(x) \not\equiv \infty$, is a convex l.s.c. functional. The following proposition concerns the existence of a minimum over K.

1.3 Minimization Problems. Duality of Convex Functionals

Proposition 1.3.1 Let $||.||$ be the norm of the space X and let

$$\lim f(x) = \infty \text{ when } ||x|| \to \infty, x \in K \subset X, \tag{1.3.2a}$$

or let

$$K \text{ be bounded.} \tag{1.3.2b}$$

The problem (1.3.1) admits at least one solution. If f is strictly convex, the solution is unique.

The solutions of problem (1.3.1) constitute a convex closed subset of X. Some variational inequalities equivalent to problem (1.3.1) will now be obtained. X is a Hilbert space and $\langle \cdot, \cdot \rangle$ denotes the dualtiy pairing with the dual space X'.

Proposition 1.3.2 Let $f = f_1 + f_2$ be a proper functional, where f_1 and f_2 are convex. l.s.c. functionals on K and suppose, that grad f_1 exists on X. For $x_0 \in K$, the following conditions are equivalent to each other:

$$f(x_0) = \inf_K f(x); \tag{1.3.3}$$

$$\langle \text{grad } f_1(x_0), x - x_0 \rangle + f_2(x) - f_2(x_0) \geq 0, \quad \forall x \in K; \tag{1.3.4}$$

and

$$\langle \text{grad } f_1(x), x - x_0 \rangle + f_2(x) - f_2(x_0) \geq 0, \quad \forall x \in K. \tag{1.3.5}$$

Obviously (1.3.4) is by definition equivalent to the inclusion

$$-\text{grad } f_1(x_0) \in \partial(f_2 + I_K)(x_0) \tag{1.3.6}$$

as it can be easily verified.

For any functional $f : X \to \bar{\mathbb{R}}$ there arises the question as to whether an affine continuous function $x \to \langle x', x \rangle - \mu, \mu \in \mathbb{R}$, can be determined which is a minorant of f, i.e

$$f(x) \geq \langle x', x \rangle - \mu \quad \forall x \in X. \tag{1.3.7}$$

Necessary and sufficient condition for it is that

$$\mu \geq \sup\{\langle x', x \rangle - f(x) | x \in X\}. \tag{1.3.8}$$

This relation introduces the conjugate functional f^c, which is defined on X' by the relation

$$f^c(x') = \sup_{x \in X} (\langle x', x \rangle - f(x)). \tag{1.3.9}$$

Obviously the supremum in (1.3.9) may be taken only over $D(f)$. f^c can be regarded as the pointwise supremum of the family of affine continuous functionals $g(\cdot) = \langle \cdot, x \rangle - \mu$ with $(x, \mu) \in \text{epi } f$. Let us denote by $\Gamma(X)$ the set of functions $f : X \to \bar{\mathbb{R}}$, which are the pointwise suprema of a family of affine continuous functionals $\langle x', \cdot \rangle + \alpha, \alpha \in \mathbb{R}$ on X. The following propositions hold.

Proposition 1.3.3 The class $\Gamma(X)$ consists exactly of the convex, l.s.c., proper functionals $f: X \to \bar{\mathbb{R}}$ and of the constants $+\infty$ and $-\infty$.

Further we denote by $\Gamma_0(X)$ the set of functionals $f \in \Gamma(X)$ such that $f \not\equiv \pm\infty$. It is obvious that $\Gamma_0(X)$ consists precisely of the convex, l.s.c. and proper functionals on X.

Proposition 1.3.4 Suppose that f is a convex functional on X. Then f^c is a convex, l.s.c. functional on X'. If, in addition f is proper, then f^c is proper as well, and conversely.

The conjugacy operation $f \to f^c$ can be considered as a one-to-one correspondence between $\Gamma_0(X)$ and $\Gamma_0(X')$ and is called Fenchel transformation (also Fenchel-Young or Legendre-Fenchel or polarity transformation). The following proposition concerns the relation between ∂f and ∂f^c.

Proposition 1.3.5 Assume that f is a convex, proper functional on X. The following conditions are equivalent to one another

(i) $\quad x' \in \partial f(x);$ \hfill (1.3.10)

(ii) $\quad \sup_{z \in X} (\langle x', z \rangle - f(z))$ is achieved at $z = x;$ \hfill (1.3.11)

(iii) $\quad f(x) + f^c(x') \leq \langle x', x \rangle;$ and \hfill (1.3.12)

(iv) $\quad f(x) + f^c(x') = \langle x', x \rangle.$ \hfill (1.3.13)

If additionally, f is l.s.c, the above conditions are equivalent to:

(v) $\quad x \in \partial f^c(x');$ and \hfill (1.3.14)

(vi) $\quad \sup_{z' \in X'} (\langle z', x \rangle - f^c(z'))$ is achieved at $z' = x'$ \hfill (1.3.15)

Suppose that $f = I_K$, where K is a nonempty, convex subset of X. Then I_K^c is given on X' by

$$I_K^c(x') = \sup_{x \in K} \langle x', x \rangle. \tag{1.3.16}$$

I_K^c is called the support function of K. Now let K be a linear subspace M of $\mathbb{R}^n$. Then the supremum is ∞, unless $\langle x', x \rangle = 0$, $\forall x \in M$. Accordingly, $I_K^c = I_{M^\perp}$, where $M^\perp$ is the orthogonal complement of M.

Let f be a nonconvex function and let epi f be its epigraph. We construct first the convex hull of epi f and then its closure, i.e. the closed convex hull of the epi f. This convex closed hull is the epigraph of a functional f_1 which is called the Γ-regularization of f. f_1 is the largest minorant of f in $\Gamma(X)$ and is the pointwise supremum of the affine continuous functions which are for every x less than f. Obviously if $f \in \Gamma(X)$ then $f_1 \equiv f$. Let us further define the conjugate functional f^{cc} of f^c defined on X by the relation $f^{cc}(x) = (f^c)^c(x)$. Clearly f^{cc} is the Γ-regularization of f and for $f \in \Gamma(X)$ $f^{cc} = f$. Note that for f nonconvex, f^c and f^{cc} are convex functionals. If $f_1 \leq f_2$ on X, then $f_1^c \geq f_2^c$

on X' and since f^{cc} is the Γ-regularization of f we may write that $f \geq f^{cc}$ on X, which implies that $f^c \leq f^{ccc}$. On the other hand

$$f^{ccc}(x') = \sup_{x \in X} \left\{ \langle x', x \rangle - f^{cc}(x) \right\} \leq f^c(x') \tag{1.3.17}$$

and therefore $f^{ccc} = f^c$ for every functional $f : X \to \bar{\mathbb{R}}$.

Further we shall give some propositions from the duality theory of minimization problems according to Ekeland and Temam [Eke]. Let again f be a convex l.s.c. and proper functional on a Hilbert space X and let X' be the dual space of X with the duality pairing $\langle \cdot, \cdot \rangle$. We consider the problem

$$\min \left\{ f(x) \big| x \in X \right\} \tag{1.3.18}$$

which includes the problem (1.3.1) as a special case. Indeed we may set in the problem (1.3.1) $f(x) = \infty$ for $x \notin K$ and we reduce it in the problem (1.3.18) which is called the primal problem or problem I.

Let us further introduce a Hilbert space Y, and let Y' be its dual space. For $p \in Y$ and $p' \in Y'$ we denote by $\langle\langle p, p' \rangle\rangle$ the duality pairing between Y and Y'. Then a functional $F : X \times Y \to \bar{\mathbb{R}}$ is defined such that

$$F(x^*, 0) = f(x^*). \tag{1.3.19}$$

Then the problem

$$\inf_{x^* \in X} F(x^*, p), \tag{1.3.20}$$

also called problem I_p, is considered. Obviously, for $p = 0$ problem I_p coincides with problem I. Further, let F^c be the conjugate functional of F defined on the space $X' \times Y'$. We consider the problem

$$\sup_{p'^* \in Y'} -F^c(0, p'^*), \tag{1.3.21}$$

also called problem I^c. Problem I_p is the "perturbed" form of problem I, and I^c is called the dual problem of I. Let us further denote by $\inf I$ and $\sup I^c$ the infimum and the supremum of problems I and I^c respectively. Inf I and sup I^c are the real numbers $f(x)$ and $\sup \left\{ -F^c(0, p'^*) \big| p'^* \in Y' \right\}$. It may easily be verified by means of (1.3.9) that

$$-F^c(0, p'^*) \leq F(x^*, 0), \quad \forall p'^* \in Y', \quad \forall x^* \in X \tag{1.3.22}$$

and, therefore, that

$$-\infty \leq \sup I^c \leq \inf I < \infty. \tag{1.3.23}$$

For $p \in Y$, we denote by g the function

$$g(p) = \inf I_p = \inf_{x^* \in X} F(x^*, p) \tag{1.3.24}$$

and we assume that

$$F(x^*, p) \in \Gamma_0(X \times Y). \tag{1.3.25}$$

It can be shown that if (1.3.25) holds then $g: Y \to \bar{\mathbb{R}}$ is convex. Let us now define the conjugate functional g^c of g. Then

$$g^c(p') = F^c(0, p'), \quad \forall p' \in Y' \tag{1.3.26}$$

and

$$\sup I^c = \sup_{p'^* \in Y'} (-g^c(p'^*)) = g^{cc}(0). \tag{1.3.27}$$

Indeed both relations result from (1.3.21) by noting that for every p'

$$g^c(p') = \sup_{p \in Y} \left[\langle\langle p', p \rangle\rangle - g(p) \right] = \sup_{p \in Y} \sup_{x^* \in X} \left[\langle\langle p', p \rangle\rangle - F(x^*, p) \right] = F^c(0, p') \tag{1.3.28}$$

and by recalling the definition of g^{cc}. Now we associate with I^c the "perturbed" problem $\sup\{-F^c(x', p'^*) \mid p'^* \in Y'\}$. Hence the dual problem of I^c with respect to the perturbation x' reads

$$\inf_{x^* \in X} \{F^{cc}(x^*, 0)\} \tag{1.3.29}$$

and is denoted by I^{cc}. On the assumption (1.3.25), $F^{cc}(x^*, 0) = F(x^*, 0)$ for every $x^* \in X$, and thus problem I^{cc} coincides with the primal problem I. Moreover, since $P^{ccc} = P^c$, problem I^{ccc}, which would result by continuing the dualization procedure, is identical to I^c. The following propositions give a condition under which $\inf I = \sup I^c$.

Proposition 1.3.6 On the assumption (1.3.25), the following three conditions are equivalent to each other:

(i) $-\infty < \inf I = \sup I^c < \infty$;

(ii) $g(0)$ is finite and g is l.s.c. at $p = 0$ (normality property of problem I).

(iii) Problem I^c is normal.

Proposition 1.3.7 Suppose that solutions to problems I and I^c exist and that

$$-\infty < \inf I = \sup I^c < \infty \tag{1.3.30}$$

Then any solution x of I and any solution p' of I^c satisfy the relation

$$(0, p') \in \partial F(x, 0). \tag{1.3.31}$$

Conversely, if x and p' satisfy (1.3.31), then x is a solution of I, p' is a solution of I^c and (1.3.30) holds.

Condition (1.3.31) is called the extremality condition of the problem and may equivalently be written as (cf. (1.3.13))

$$F(x, 0) + F^c(0, p') = 0. \tag{1.3.32}$$

1.3 Minimization Problems. Duality of Convex Functionals

Let X and X' be two vector spaces and A a mapping from X into the power set (set of all subsets) $\mathcal{P}(X')$ of X'. The mapping A is called a multivalued operator or multivalued mapping or multifunction. In this case, from $A(x_1) = x_1'$ and $A(x_1) = y_1'$ it does not follow that $x_1' = y_1'$, as happens with single-valued functions. Considering A as a subset of $X \times X'$, we can write $A(x) = \{y \in X' | (x, y) \in A\}$. The set $D(A) = \{x | x \in X, A(x) \neq \emptyset\}$ is called the domain of A and the set $R(A) = \bigcup_x A(x), x \in X$, the range of A. Because A is multivalued, we will write $y \in A(x)$, where $x \in D(A)$ and $y \in X'$. If A and B are two multivalued operators on X, then $\lambda A + \mu B$, $\lambda, \mu \in \mathbb{R}$, is a multivalued operator mapping x into $\lambda A(x) + \mu B(x) = \{\lambda y + \mu z | y \in A(x), z \in B(x)\}$. Moreover, $D(\lambda A + \mu B) = D(A) \cap D(B)$. Suppose further that X and X' are dual H-spaces with duality pairing $\langle x', x \rangle$ for $x \in X$, $x' \in X'$. The multivalued mapping $A: X \to \mathcal{P}(X')$ is said to be monotone if

$$\langle y_1 - y_2, x_1 - x_2 \rangle \geq 0, \tag{1.3.33}$$

$$\forall x_1, x_2 \in D(A), \quad \forall y_1 \in A(x_1), \quad \forall y_2 \in A(x_2).$$

If $\geq$ is replaced by $>$, then A is said to be strictly monotone.

Let f be a convex proper functional on X. Then it can be shown that ∂f is a monotone multivalued function from X into $\mathcal{P}(X')$. The graph of the multivalued operator $A: X \to \mathcal{P}(X')$ is a set $\mathcal{Q}(A) = \{(x, y) | (x, y) \in D(A) \times X', y \in A(x)\}$. Then $\mathcal{Q}(A_1) \subset \mathcal{Q}(A_2)$, if and only if $A_1(x) \subset A_2(x) \forall x \in X$. The set $\mathcal{A}$ of the monotone operators from X into $\mathcal{P}(X')$ can be partially ordered by graph inclusion. It can be shown, furthermore, that every totally ordered subset of $\mathcal{A}$ has an upper bound. Then by means of the Zorn Lemma $\mathcal{A}$ contains at least one maximal element, which is called a maximal monotone operator. Accordingly, a monotone operator $A: X \to \mathcal{P}(X')$ is called maximal monotone if and only if $\mathcal{Q}(A) \subset \mathcal{Q}(B)$ implies that $A = B$, where $B: X \to \mathcal{P}(X')$ is an arbitrary monotone operator, i.e., if and only if $\mathcal{Q}(A)$ is not properly contained in any other monotone subset of $X \times X'$. From the above we obtain equivalently that an operator $A: X \to \mathcal{P}(X')$ is called maximal monotone if and only if i) A is monotone and ii) for every $x \in X$ and $y \in X'$ such that

$$\langle y - y_1, x - x_1 \rangle \geq 0, \quad \forall x_1 \in D(A), \quad \forall y_1 \in A(x_1) \tag{1.3.34}$$

the relation

$$y \in A(x) \tag{1.3.35}$$

holds.

The following proposition relates the theory of maximal monotone operators to subdifferentiation.

Proposition 1.3.8 *The subdifferential ∂f of a convex, l.s.c., proper functional f on X, where X is a H-space, is a maximal monotone operator.*

The class of the monotone operators $\beta: \mathbb{R} \to \mathcal{P}(\mathbb{R})$ is subsequently considered. A complete nondecreasing curve in $\mathbb{R}^2$ is the graph $\mathcal{Q}(\beta)$ of a maximal monotone mapping $\beta : \mathbb{R} \to \mathcal{P}(\mathbb{R})$. In a Cartesian coordinate system such a graph is similar to the graph of a continuous nondecreasing function, with the difference that it may contain vertical segments as well. The maximal monotone graphs in $\mathbb{R}^2$ are used for the formulation of unilateral boundary conditions. A proposition now follows relating the complete nondecreasing curves in $\mathbb{R}^2$ and the subdifferentials ∂f of convex, l.s.c and proper functionals on $\mathbb{R}$.

Proposition 1.3.9 Let $\beta : \mathbb{R} \to \mathcal{P}(\mathbb{R})$ be a maximal monotone mapping. A convex, l.s.c., proper functional $f : \mathbb{R} \to \bar{\mathbb{R}}$ can be determined up to an additive constant such that

$$\beta = \partial f. \tag{1.3.36}$$

Accordingly the graphs of the subdifferentials ∂f, where $f \in \Gamma_0(\mathbb{R})$, are precisely the complete nondecreasing curves of $\mathbb{R}^2$.

1.4 Miscellanea: Fans, Quasidifferentials, Codifferentials

The definition (1.2.27) of the generalized gradient does not permit the description of the nonsmoothness or smoothness properties of a function in a detailed manner. This fact becomes obvious, e.g., because we need for a differentiable function the property of continuous differentiability to conclude that $\bar{\partial} f(x) = \{\text{grad} f(x)\}$, unless f is convex around the point x. In this last case we have near to x according to Prop. 1.1.3 that $\bar{\partial} f(x) = \partial f(x) = \{\text{grad} f(x)\}$.

Therefore, there exist several attempts to develop a different approach towards nonsmoothness. Some of these approaches are important in mechanical problems, either from the theoretical or from the numerical point of view. Let us begin with the notion of "fan" introduced by Ioffe [Iof81] which permits, as we shall see, to generalize even further the variational inequality expressions of mechanics. One can easily recognize in the definition of the fan certain of the properties of the generalized gradient. Let X and Y be two Hilbert spaces and X' and Y' their dual spaces. The duality pairing between X and X' is denoted by $\langle x', x \rangle$ and between Y and Y' by $\langle y', y \rangle$.

Let $F : X \to Y$ be a multivalued function such that

$$0 \in F(0) \tag{1.4.1}$$

$$F(\lambda x) = \lambda F(x) \quad \forall x \in X \quad \text{and for} \quad \lambda > 0 \tag{1.4.2}$$

$$F(x) \text{ is convex} \quad \forall x \in X \tag{1.4.3}$$

$$F(x_1 + x_2) \subset \overline{F(x_1) + F(x_2)} \quad \forall x_1, x_2 \in X. \tag{1.4.4}$$

We call F a fan and the set $D(F) = \{x \big| F(x) \neq \emptyset\}$ its domain. Then the function $g : Y' \times X \to \bar{\mathbb{R}}$ defined by the relation

$$g(y', x) = \sup\left\{\langle y', y\rangle \big| y \in F(x)\right\} \qquad (1.4.5)$$

is called the support function of the fan F. The following proposition holds:

Proposition 1.4.1 A function $g : Y' \times X \to \bar{\mathbb{R}}$ is the support function of a fan $F : X \to Y$ if and only if g is convex and positively homogeneous of degree one in each one of the variables (bisublinear mapping) and $g(., x)$ is a weakly l.s.c. function such that $g(., x) > -\infty$ if $x \in D(F)$. Then we may write that

$$F(x) = \left\{y \in Y \big| g(y', x) \geq \langle y', y\rangle \quad \forall y' \in Y'\right\}. \qquad (1.4.6)$$

In [Pan 87b] we have used (1.4.6) as the definition of the fan in order to point out its relation to the definitions of $\bar{\partial} f$ and ∂f and to the virtual work inequality. The fan F is called odd if

$$F(-x) = -F(x) \quad \forall x \in X \qquad (1.4.7)$$

The fans constitute a generalization of the linear operators. A comparison of their properties with the properties of the linear operators can be found in [Iof82]. Here we give some interesting cases of fans:

i) If $A : X \to Y$ is a linear operator then $F(x) = \{Ax\}$ is an odd fan

ii) If B is a convex set of linear operators $A : X \to Y$ then $F(x) = \{y | y = Ax$ for certain $A \in B\}$ is an odd fan generated by the set of operators B.

iii) Let K be a closed convex cone in Y and let $R = \{(x_1, x_2) | x_1 \in X, x_2 \in X$ $x_1 - x_2 \in K\}$ be a "partial ordering" on X through K. We say that $x_1 \prec x_2$ if $\{x_1, x_2\} \in R$ i.e. if $x_1 - x_2 \in K$. A mapping $A : X \to Y$ is K-sublinear if it is positively homogeneous and $A(x_1 + x_2) \prec A(x_1) + A(x_2)$. Let $K^c = \{y' | \langle y', y\rangle \geq 0 \quad \forall y \in K\}$ and let $A : X \to Y$ be a K-sublinear mapping. Then the multivalued function

$$x \to F(x) = \left\{y \in Y \big| y \prec A(x)\right\} \qquad (1.4.8)$$

is a K^c-fan from X into Y.

iv) Let $P : X \to \mathbb{R}$ and $Q : X \to \mathbb{R}$ be two functionals. P and $-Q$ are convex and positively homogeneous of degree 1 (i.e. sublinear) such that $P(x) \geq Q(x) \quad \forall x \in X$. Then

$$x \to F(x) = \left\{y \in \mathbb{R} \big| Q(x) \leq y \leq P(x)\right\} \qquad (1.4.9)$$

is a fan and

$$x \to F(x) = \left\{y \in \mathbb{R} \big| - P(-x) \leq y \leq P(x)\right\} \qquad (1.4.10)$$

is an odd fan. (The same result holds for P and Q taking values in a vector lattice Y. with $\leq$ replaced by $\prec$).

26   1. Elements of Nonsmooth Analysis

The notion of quasidifferential has been introduced by V.F.Demyanov and studied by Demyanov, Polyakova and Rubinov (cf. e.g. [Dem83,85,86a,b,89a,b] in order to describe more accurately than with the generalized gradient the directional properties of a function at a point. Moreover the quasidifferential gives rise to a calculus involving equalities analogous to the differential calculus; thus we avoid the inclusions of the calculus of the generalized gradient. Let X be a Hilbert space, X' its dual space and let $\langle x', x \rangle$ be the duality pairing. Let A be an open subset of X and let $f : A \to \mathbb{R}$. The functional f is said to be quasidifferentiable at $x \in A$ if it is directionally differentiable at x (i.e. $\tilde{f}'(x, h)$ exists for every $h \in X$, cf. (1.1.13)) and if two convex weakly compact sets $\underline{\partial}' f(x) \subset X'$ and $\bar{\partial}' f(x) \subset X'$ can be determined such that

$$\tilde{f}'(x, h) = \max_{x_1} \left\{ \langle x_1, h \rangle \big| x_1 \in \underline{\partial}' f(x) \right\} \qquad (1.4.11)$$
$$+ \min_{x_2} \left\{ \langle x_2, h \rangle \big| x_2 \in \bar{\partial}' f(x) \right\} \quad \forall h \in X$$

Eq. (1.4.11) can be written equivalently in the form

$$\tilde{f}'(x, h) = \min_{x_2} \max_{x_1} \left\{ \langle x_1 + x_2, h \rangle \big| x_1 \in \underline{\partial}' f(x), x_2 \in \bar{\partial}' f(x) \right\}. \qquad (1.4.12)$$

The pair of sets

$$Df(x) = \left\{ \underline{\partial}' f(x), \bar{\partial}' f(x) \right\} \qquad (1.4.13)$$

is called a quasidifferential of f at x. We call the sets $\underline{\partial}' f(x)$ and $\bar{\partial}' f(x)$ ∂'-subdifferential and ∂'-superdifferential of f at x respectively. The prime is used to avoid a possible confusion with the subdifferential $\partial f(x)$. From (1.4.11) it is obvious that $Df(x)$ is not uniquely determined at x since any pair of sets $\left\{ \underline{\partial}' f(x) + A, \bar{\partial}' f(x) - A \right\}$ where A is any convex weakly compact subset of X' is also a quasidifferential of f at x. If f is continuously differentiable on A then f is quasidifferentiable at every $x \in A$ and

$$Df(x) = \left\{ \operatorname{grad} f(x), 0 \right\} \quad \text{or} \quad Df(x) = \left\{ 0, \operatorname{grad} f(x) \right\}. \qquad (1.4.14)$$

Thus f is both ∂'-subdifferentiable and ∂'-superdifferentiable at x.

If f is convex defined on A convex, then (1.4.11) is written in the form (1.1.16). Accordingly

$$Df(x) = \left\{ \partial f(x), 0 \right\}, \qquad (1.4.15)$$

where $\partial f(x)$ is the subdifferential of convex analysis. In a similar way, if f is concave and A is convex then $f_1 = -f$ is convex and we obtain that

$$Df(x) = \left\{ 0, \tilde{\partial} f(x) \right\}, \qquad (1.4.16)$$

where

$$\tilde{\partial} f(x) = \left\{ x' \in X' \big| f(x_1) - f(x) \leq \langle x', x_1 - x \rangle \quad \forall x_1 \in X \right\} \qquad (1.4.17)$$

is the "superdifferential" of the concave function f at x.

1.4 Miscellanea: Fans, Quasidifferentials, Codifferentials

Analogously we may find for a $\bar{\partial}$-regular function (cf. (1.2.48)) that (cf. (1.2.29))

$$\tilde{f}'(x,h) = f^\circ(x,h) = \max\{\langle x_1, h\rangle | x_1 \in \bar{\partial}f(x)\} \tag{1.4.18}$$

and thus

$$Df(x) = \{\bar{\partial}f(x), 0\}, \tag{1.4.19}$$

where $\bar{\partial}f(x)$ is the generalized gradient. For instance, if f is a maximum type function, e.g. $f = \max\{\Phi_1, \ldots, \Phi_m\}$, where $\Phi_i = \Phi_i(x)$, $i = 1, \ldots, m$, $x \in \mathbb{R}^n$, are smooth functions, then we know that f is $\bar{\partial}$-regular and thus

$$Df(x) = \{\text{co}\{\text{grad}\Phi_i(x), \ldots, \text{grad}\Phi_k(x)\}, 0\} \tag{1.4.20}$$

if $x \in A_1 \cap \ldots \cap A_k$ (cf. 1.2.46b). We consider further pairs of sets $\Gamma_i = \{A_i, B_i\}$, $i = 1, 2$, where $A_i \subset X'$ and $B_i \subset X'$ and we define their addition by

$$\Gamma_1 + \Gamma_2 = \{A_1 + A_2, B_1 + B_2\} \tag{1.4.21}$$

and the multiplication by $\lambda \in \mathbb{R}$ as

$$\lambda \Gamma_i = \begin{cases} \{\lambda A_i, \lambda B_i\} & \text{if } \lambda \geq 0 \\ \{\lambda B_i, \lambda A_i\} & \text{if } \lambda < 0. \end{cases} \tag{1.4.22}$$

By means of these definitions we can show that if $f_i, i = 1, \ldots, n$, are quasidifferentiable functions at x, then $g = \sum_i a_i f_i$ is also quasidifferentiable at x ($a_i \in \mathbb{R}$) and $Dg = \sum_i a_i Df_i$. Moreover it can be proved that if f_1 and f_2 are quasidifferentiable at x then $g = f_1 f_2$ has the same property and

$$Dg(x) = f_1(x)Df_2(x) + f_2(x)Df_1(x). \tag{1.4.23}$$

Moreover for $f_2(x) \neq 0$, $g = f_1/f_2$ is quasidifferentiable at x and

$$Dg(x) = [f_2(x)Df_1(x) - f_1(x)Df_2(x)]/f_2^2(x). \tag{1.4.24}$$

The following properties are of importance: With Φ_i, $i = 1, \ldots, n$, $f = \max_i\{\Phi_i\}$ is quasidifferentiable at x and for $I(x) = \{i | \Phi_i(x) = f(x)\}$

$$\underline{\partial}'f(x) = \text{co}\left\{\underline{\partial}'\Phi_k(x) - \sum_{\substack{i \in I(x) \\ i \neq k}} \bar{\partial}'\Phi_i(x) \bigg| k \in I(x)\right\} \tag{1.4.25}$$

$$\bar{\partial}'f(x) = \sum_{k \in I(x)} \bar{\partial}'\Phi_k(x). \tag{1.4.26}$$

Analogous formula holds for a minimum type function f. In this case $\bar{\partial}'f(x)$ (resp $\underline{\partial}'f(x)$) is given by (1.4.25) (resp. (1.4.26)) by replacing in the right hand side $\underline{\partial}'$ by $\bar{\partial}'$ and $\bar{\partial}'$ by $\underline{\partial}'$. If $f = f_1 + f_2$ where f_1 (resp. f_2) is convex (resp. concave) and $f_1(x), f_2(x)$ are finite then f is quasidifferentiable and we have that [Poly 86]

$$Df(x) = \{\partial f_1(x), \bar{\partial} f_2(x)\} = \{\partial f_1(x), -\partial(-f_2(x))\}. \tag{1.4.27}$$

The above formula leads for $f = f_1 - f_2$, where f_1 and f_2 are convex and $f_1(x), f_2(x)$ are finite to the quasidifferential (see also [Ell])

$$Df(x) = \{\partial f_1(x), -\partial f_2(x)\}. \tag{1.4.28}$$

If f is quasidifferentiable on X then

$$f(x_0) = \min\{f(x) | x \in X\} \tag{1.4.29}$$

(resp.

$$f(\tilde{x}_0) = \max\{f(x) | x \in X\}) \tag{1.4.30}$$

implies that

$$-\bar{\partial}' f(x_0) \subset \underline{\partial}' f(x_0) \tag{1.4.31}$$

(resp.

$$-\underline{\partial}' f(\tilde{x}_0) \subset \bar{\partial}' f(\tilde{x}_0)). \tag{1.4.32}$$

A point x_0 (resp. $\tilde{x}_0$) satisfying (1.4.31) (resp. (1.4.32) is called an inf-stationary (resp. a sup-stationary) point of f on X. It is also verified that if x_0 does not satisfy (1.4.31) and $X = \mathbb{R}^n$ then the direction

$$\begin{aligned}\xi &= -\frac{v_0 + w_0}{||v_0 + w_0||}, \max_w \min_v \{||v + w|| \, | \, w \in \bar{\partial}' f(x_0), v \in \underline{\partial}' f(x_0)\} \\ &= ||v_0 + w_0||\end{aligned} \tag{1.4.33}$$

is a direction of steepest descent (not unique) of f at x_0, and if x_0 does not satisfy (1.4.32) then the direction

$$\begin{aligned}\xi &= \frac{\tilde{v}_0 + \tilde{w}_0}{||\tilde{v}_0 + \tilde{w}_0||}, \max_v \min_w \{||v + w|| \, | \, w \in \bar{\partial}' f(x_0), v \in \underline{\partial}' f(x_0)\} \\ &= ||\tilde{v}_0 + \tilde{w}_0||\end{aligned} \tag{1.4.34}$$

is a direction of steepest ascent (not unique) of f at x_0. Some results on the relation of the quasidifferential with the generalized gradient have been proved recently by Liqun Qi [Liq91]. Note that the quasidifferential as well as the subdifferential at a point x are not continuous in the general case. The notion of codifferential introduced by V. Demyanov [Dem89a] has ameliorated continuity properties for a large class of nonsmooth functions. In the same functional framework as in the case of quasidifferentiability we consider a function $f : X \to \mathbb{R}$. We say that f is codifferentiable at x if there exist weakly compact convex sets $\underline{d}f(x) \subset X' \times \mathbb{R}$ and $\bar{d}f(x) \subset X' \times \mathbb{R}$ such that

$$\begin{aligned}f(x+h) &= f(x) + \max_{(v,a) \in \underline{d}f(x)} \{a + \langle v, h \rangle\} \\ &+ \min_{(w,b) \in \bar{d}f(x)} \{b + \langle w, h \rangle\} + O(h) \quad \forall h \in X\end{aligned} \tag{1.4.35}$$

where $a, b \in \mathbb{R}$, $v, w \in X$ and $\frac{O(ah)}{a} \to 0$ as $a \to 0_+$. The pair of sets

$$\bar{D}(f(x)) = \{\underline{d}f(x), \bar{d}f(x)\} \quad (1.4.36)$$

is called the codifferential of f at x. The set $\underline{d}f(x)$ (resp. $\bar{d}f(x)$) is called the hypo-(resp. hyper-) differential of f at x. A function is called continuously codifferentiable at x, if it is codifferentiable on some neighborhood of x and $x \to \bar{D}f(x)$ is continuous (in the Hausdorff metric [Sen90]) at x. The class of codifferentiable functions coincides with the class of quasidifferentiable functions. We should note that continuously codifferentiable functions lead to a "better" approximation of a given function in comparison with the approximation obtained by the use of a quasidifferential (see also [Pan92a] for examples of quasidifferentials and codifferentials). The continuously differentiable functions, the convex and concave functions at a point in which they take finite values, the maximum and minimum type functions e.g. $\max\{\Phi_1, \ldots, \Phi_n\}$ or $\min\{\Phi_1, \ldots, \Phi_n\}$, where $\Phi_i \in C^1$, $i = 1, \ldots, n$, are continuously codifferentiable functions. The calculus of codifferentiable functions is the same as the calculus of quasidifferentiable functions.

Part II
MECHANICAL THEORY

2. Nonsmooth Mechanics I

In this Chapter we explain the origins of Nonsmooth Mechanics and of the Inequality Problems. To do this we use the two notions of convex and of nonconvex superpotentials. We consider boundary conditions and material laws resulting from convex or nonconvex, nonsmooth energy functions using the concept of subdifferential or of generalized gradient. For additional information on these subjects the reader is referred to the monographs and books of Duvaut and Lions [Duv72], Panagiotopoulos [Pan85], Hlavaček et al. [Hl88], Moreau, Panagiotopoulos, Strang [Mor88a,b], Antes, Panagiotopoulos [Ant92], as well as to [Mor68,86,88c] and [Ger74].

2.1 Convex Superpotentials

We first recall some notions from continuum mechanics, [Beck, Ger73a,b,c, Maug, Tru66], which will be used in the sequel. Let Ω be an open subset of $\mathbb{R}^3$ with boundary Γ. Ω is occupied by a deformable body and is referred to a fixed orthogonal Cartesian coordinate system $OX_1X_2X_3$. Thus a one-to-one correspondence between the material particles X of the body and the point $\{X_1, X_2, X_3\}$ (material coordinates of X) is established. Henceforth we will refer to the body simply as body Ω. With respect to another orthogonal Cartesian coordinate system $\bar{0}x_1x_2x_3$ we may consider the coordinate transformation.

$$x_k = x_k(X_1, X_2, X_3), \quad k = 1, 2, 3. \tag{2.1.1}$$

Any deformation process may generally be described by means of the trajectory of each material particle X, i.e. by

$$x_i = \chi_i(X_k, t) \quad i, k = 1, 2, 3, \tag{2.1.2}$$

where $t \in [0, T]$ is the time variable. The point $x = \chi(X, t)$ is the place occupied by X at time t and let $x \equiv X$ for $t = 0$. The coordinates $x_k, k = 1, 2, 3$, are called the spatial coordinates of X. A mapping $\chi = \{\chi_i\} : \Omega \times [0, T] \to \Omega_t \subset \mathbb{R}^3$ is called "motion" of Ω. Ω_t denotes the subset of $\mathbb{R}^3$ occupied by Ω at time t. We shall assume that χ, χ^{-1} exist and are appropriately regular functions. Let $A = A(X, t)$ be a function describing a quantity A. We call it the material (or Lagrangian) description of A, whereas $A = A(x, t)$ is the spatial (or Eulerian) description. We define the local or spatial derivative $\partial A(x,t)/\partial t$ of A, and its material derivative $\partial A(X,t)/\partial t$, which we will denote simply by dA/dt. Between material and local derivatives there holds the relation

$$\frac{dA}{dt} = \frac{\partial A(x,t)}{\partial t} + \frac{\partial A}{\partial x_i}\frac{\partial x_i(X,t)}{\partial t}, \quad i=1,2,3. \qquad (2.1.3)$$

The velocity $v = v(X,t)$ is obtained by differentiating (2.1.2) with respect to t keeping X unchanged, i.e.

$$v_i(X,t) = \frac{\partial \chi_i(X,t)}{\partial t}, \qquad (2.1.4)$$

whereas the acceleration $\gamma = \gamma(X,t)$ is given by

$$\gamma_i = \frac{\partial v_i(X,t)}{\partial t} = \frac{dv_i}{dt} = \frac{\partial v_i(x,t)}{\partial t} + \frac{\partial v_i}{\partial x_j}v_j. \qquad (2.1.5)$$

Inverting (2.1.2) implies $v = v(x,t)$. In the spatial description we consider the velocity gradient $L = \{L_{ij}\} = \{v_{i,j}\}$. Its symmetric and antisymmetric parts are called the rate of deformation and spin; thus

$$D = \frac{1}{2}(L + L^T) = \text{sym}(\text{grad } v), \qquad (2.1.6)$$
$$W = \frac{1}{2}(L - L^T) = \text{asym}(\text{grad } v).$$

The instantaneous position x of a material particle is related to the initial position X by means of the displacement vector u, i.e. $x = X + u$. We call $F = \{F_{i,j}\} = \{\partial x_i/\partial X_j\}$ the deformation gradient. Then $F^T F$ is the right Cauchy-Green tensor (T denotes the transpose of a matrix) and

$$E = \frac{F^T F - I}{2} \qquad (2.1.7)$$

is the Green strain tensor. Hereafter, for a vector a_i the derivative $\partial a_i/\partial X_j$ will be denoted by $a_{i,j}$. The deformation gradient rate takes the form

$$\frac{\partial F_{ik}(X,t)}{\partial t} = v_{i,j}x_{j,k} \quad \text{or} \quad \frac{dF}{dt} = LF. \qquad (2.1.8)$$

Thus the rate $\dot{E}$ of the Green strain tensor reads

$$\dot{E} = \frac{dE}{dt} = \frac{1}{2}\left(\frac{dF^T}{dt}F + F^T\frac{dF}{dt}\right) = \frac{1}{2}(F^T L^T F + F^T LF) = F^T DF. \qquad (2.1.9)$$

After some manipulations we obtain that

$$E_{i,j} = \frac{1}{2}(u_{i,j} + u_{j,i} + u_{k,i}u_{k,j}) \qquad (2.1.10)$$

and

$$\frac{dE_{ij}}{dt} = \frac{1}{2}(v_{i,j} + v_{j,i} + v_{k,i}u_{k,j} + u_{k,i}v_{k,j}). \tag{2.1.11}$$

The Cauchy stress tensor $\sigma = \sigma(x,t)$ is used in the case of spatial description. From the tensor σ the (second) Piola-Kirchhoff stress tensor $\Sigma = \Sigma(X,t)$ is defined by

$$\Sigma = (\det F)F^{-1}\sigma(F^T)^{-1}. \tag{2.1.12}$$

If the displacement gradients are small enough we may write that

$$E_{ij} \simeq \varepsilon_{ij} = \frac{1}{2}(u_{i,j} + u_{j,i}). \tag{2.1.13}$$

The tensor $\varepsilon = \{\varepsilon_{ij}\}$ is called the (small) strain tensor. The replacing of E by ε is the geometric or kinematic linearization.

Let $\Omega_t \subset \mathbb{R}^3$ be an open, subset of $\mathbb{R}^3$ occupied by a body Ω at time t. Ω_t is referred to an orthogonal Cartesian coordinate system attached to the body. Let P be a point of Ω_t having coordinates $\{x_{p_1}, x_{p_2}, x_{p_3}\}$. A velocity field such that

$$v(x_p, t) = v_0(t) + \omega(t) \wedge x_p, \quad \forall x_p = \{x_{p_i}\} \in \Omega_t, \quad i = 1, 2, 3, \tag{2.1.14}$$

is called a rigidifying velocity field, because it corresponds to a rigid motion of the body. By v_0 we denote the velocity of the origin of the coordinate system. It is readily seen that such a velocity field results from a motion described by the equation

$$x = Q(t)X + x_0(t) \tag{2.1.15}$$

where $Q^TQ = QQ^T = I$ and $\det Q = 1$. In (2.1.15) we have set $X \equiv x_p$ in order to consider the velocity field in (2.1.14) as resulting from a motion of point X to the position x. We have also the relation

$$\omega_k = \frac{1}{2}\varepsilon_{kij}\Omega_{ji}$$

where Ω is a skew-symmetric tensor given by $\Omega = (dQ/dt)Q^T$, and where $\varepsilon_{ijk} = 0$ if any two indices are alike, $\varepsilon_{ijk} = \varepsilon_{123} = 1$ if (i,j,k) is an even permutation of $(1,2,3)$ and $\varepsilon_{ijk} = -1$ if (i,j,k) is an odd permutation.

Let x_i and $\bar{x}_i$, i=1,2,3, be the coordinates of the same material point with respect to two orthogonal Cartesian coordinate systems which rotate and translate each with respect to the other arbitrarily. Then a relation of the form

$$\bar{x} = \tilde{Q}(t)x + \tilde{x}_0(t) \tag{2.1.16}$$

holds, where $\tilde{Q}$ has the same properties as Q and $\tilde{x}_0$ is an arbitrary vector.

A tensor field is said to be frame-indifferent, or objective, if it transforms in the well-known tensorial manner, whenever (2.1.16) is considered as a classical coordinate transformation. A mapping between tensor fields is objective or frame-indifferent if all dependent and independent variables transform in the previous manner.

We now consider a mechanical system Σ corresponding to the body Ω and we assume that all the admissible velocity fields, which may occur in a time interval at which the observation takes place, are known. The admissibility is understood with respect to the kinematical or geometrical constraints imposed to the body. Let us denote by U_0, the space of velocity fields assumed to be a Hilbert space, and by U_1 the subset of kinematically admissible velocity fields. The forces f acting on Σ constitute a Hilbert space F and let $\langle v, f \rangle$, $v \in U$, $f \in F$ be a bilinear form with the properties:

i) for each $v \neq 0$ in U there exists $f \in F$ such that $\langle v, f \rangle \neq 0$ and (2.1.17)

ii) for each $f \neq 0$ in F there exists $u \in U$ such that $\langle v, f \rangle \neq 0$. (2.1.18)

Accordingly (cf. [Aub79a]) $\langle v, f \rangle$ places the Hilbert spaces U and F in duality. Thus the force f is a linear continuous functional on the space U. In the terminology of Mechanics U is the space of virtual velocities and $\langle v, f \rangle$ $v \in U$, $f \in F$ is the virtual power produced by f. One can say that a force field f acting on Σ is given if the virtual power $\langle v, f \rangle$ is defined on U such as to satisfy (i) and (ii).

In [Pan85] the above situation is generalized to cover the case of topological vector spaces and the very important relationship of the notion of virtual power with the notion of weak topology (and the Mackey-Ahrens theorem) is discussed. In order to create a framework for the study of continuous systems we place this system in an inertial frame of reference without any kinematical constraints and we define the Hilbert space U of virtual velocities. Let us denote by U_0 the subspace of rigidifying velocity fields and let us apply the following postulate P_1 [Ger73b].

P_1: The virtual power Π_i of the internal forces of the body is zero for any rigidifying velocity field at any time.

As we have defined before the virtual power Π_i is a linear function of v, i.e. the value $\Pi_i(v)$ at v of the virtual power remains unchanged if v is replaced by $v + v_0, v_0 \in U_0$. Further, we write $\Pi_i(v)$ in the form $\langle v, f \rangle$, and P_1 is equivalent to the statement that $\Pi_i(v) = \langle v, f \rangle = 0 \quad \forall v \in U_0$. We are thus led to consider the quotient space $\bar{U} = U/U_0$, which is called the space of objective virtual velocities. The internal forces are by definition the continuous linear functionals on $\bar{U}$. We denote their space by $\bar{F}$. Obviously, the elements of both $\bar{U}$ and $\bar{F}$ are objective quantities, as can easily be seen by considering the invariance of the duality mapping. For the complete formulation of a continuum theory, we have to choose in addition to the space U, the precise form of the linear mapping $v \to \langle v, f \rangle$ expressing the virtual power of the internal forces of the system. Hence in the framework of a local theory we may consider that the successive gradients of the vector field v, i.e. $v_{i,j}, v_{i,jk}$, etc. are involved in the mapping

$v \to \langle v, f \rangle$, and thus we are led to first-, second-, etc. order gradient theories respectively. Thus we may consider as a space U the cartesian product space $U^{(0)} \times U^{(1)} \times \ldots \times U^{(m)}$ where $U^{(m)} = \{v_{i,jk} \ldots (m-\text{ spatial derivations })\}$. Let be Ω the body considered. Then in the first-order gradient theory, which is the most common, the power of the continuous system depends on the velocities v_i and the first gradients $v_{i,j}$. On the assumption that the power Π_i of the internal forces acting in Ω can be expressed in the integral form,

$$\Pi_i(v) = -\int_\Omega p_i(v) d\Omega \qquad (2.1.19)$$

we may show [Ger73b] that at any point of Ω

$$p_i(v) = t_{ij} D_{ij}, \quad D_{ij} = \frac{1}{2}(v_{i,j} + v_{j,i}). \qquad (2.1.20)$$

Indeed, $p_i(v)$ may be written at any point of Ω in the general form

$$p_i(v) = t_{ij} D_{ij} + r_{ij} \Omega_{ij} + q_i v_i, \quad \Omega_{ij} = \frac{1}{2}(v_{i,j} - v_{j,i}). \qquad (2.1.21)$$

Obviously, t_{ij} is a component of a symmetric tensor. If $q_i \neq 0$ in a neighborhood of a point $M \in \Omega_1 \subset \Omega$, then one can easily determine a subsystem containing M and a translational virtual velocity such that $p_i(v)$ is not zero on this neighborhood. But this contradicts P_1. Analogously, by means of a rotational virtual velocity we find that $r_{ij} = 0$. Here $t = \{t_{ij}\}$ is called the intrinsic stress tensor. The symmetric tensor $D = \{D_{ij}\}$ is the rate of deformation (or stretching) tensor. For the theory of the second gradient and the respective theories of materials with microstructure, the reader is referred to [Ger73b,c].

Let us now give the main postulate of mechanics, which governs the motion of any body. It is the "principle of virtual power" and reads.

P_2: At any time and for any field of kinematically admissible virtual velocities (i.e. elements of U_1), the virtual power of all the internal and external forces impressed on the system is equal to the virtual power of the inertial (or d'Alembert) forces.

Further, we pay some attention to the virtual power of the external and inertial forces in Ω. The external forces are volume forces (e.g., gravity, electromagnetic forces, etc.), or boundary forces (e.g., contact forces) acting on the boundary Γ of Ω. In the first case they are defined as continuous linear functionals on the space U, and in the second as continuous linear functionals on a vector space U_Γ defined on the boundary Γ of Ω. The latter space is assumed to exist and is such that it includes the traces $v|_\Gamma$ of the elements v of U. The elements of both U and U_Γ are not necessarily objective. Finally, the virtual power of the inertial forces is a continuous linear functional on $U^{(0)}$ whose elements are not objective. If we denote by Π_v, Π_c and Π_α the three virtual powers of volume, contact and inertial forces, then P_2 implies that at any time t_0.

$$\Pi_i + \Pi_v + \Pi_c = \Pi_\alpha \quad \forall v \in U_1. \qquad (2.1.22)$$

Eq. (2.1.22) may also be formulated for any subsystem of Ω occupying the domain $\Omega_1 \subset \Omega$ at time t. Assuming that $U_1 \equiv U$, i.e. that the body is not constrained, we can obtain the possible forms of the terms Π_v, Π_c and Π_α. In the context of a first-order gradient theory and with regard to Ω_1, we assume that $\Pi_i(v), \Pi_v(v)$ and $\Pi_\alpha(v)$ (resp. $\Pi_c(v|_\Gamma)$) can be expressed as integrals over Ω_1 (resp. over Γ_1) of $p_i(v), p_v(v)$ and $p_\alpha(v)$ (resp. $p_c(v|_\Gamma)$). Then, as is readily verified, the most general forms of $p_v(v)$ and $p_c(v|_\Gamma)$ are respectively

$$p_v(v) = f_i v_i + b_{ij} D_{ij} + c_{ij} \Omega_{ij}, \quad b_{ij} = b_{ji}, \quad c_{ij} = -c_{ji} \quad (2.1.23)$$

and

$$p_c(v|_\Gamma) = S_i(v|_\Gamma)_i, \quad (2.1.24)$$

where $\{f_i\}, \{b_{ij}\}$ and $\{c_{ij}\}$ are respectively the volume force, the double symmetric force, and the couple tensor in Ω_1, and $\{S_i\}$ is the stress vector on the boundary Γ_1 of Ω_1. In a mechanical theory (i.e. without Maxwelliam fields [Maug80]) the density of the inertial force power is

$$p_\alpha(v) = \rho \frac{d\hat{v}_i}{dt} v_i, \quad (2.1.25)$$

where ρ is the density of the body and $d\hat{v}/dt$ denotes the material derivative of the real velocity $\hat{v}$. If all the quantities are sufficiently smooth we may apply the Green-Gauss theorem, and since Ω_1 is arbitrary we obtain the equations (cf. [Ger73a,b,c])

$$\sigma_{ij,j} + f_i = \rho \frac{d\hat{v}_i}{dt}, \quad \sigma_{ij} - \sigma_{ji} + c_{ij} = 0 \quad \text{in} \quad \Omega_1 \quad (2.1.26)$$

$$S_i = \sigma_{ij} n_j \quad \text{on} \quad \Gamma_1, \quad (2.1.27)$$

where $n = \{n_i\}$ is the outward unit normal vector to Γ_1,

$$\sigma_{ij} = t_{ij} - b_{ij} - c_{ij} \quad \text{in} \quad \Omega_1 \quad (2.1.28)$$

and $\sigma = \{\sigma_{ij}\}$ is the Cauchy stress tensor. In the framework of a classical continuum theory we have $b_{ij} = c_{ij}$. This is not the case, e.g. in polar continua and in electromaqnetic continua.

Until now we have seen how the application of the postulates P_1 and P_2 permit us to derive the basic equations of the mechanics of continua. For the study of the dynamic behaviour of a given continuous structure we have to take into account the kinematical and the statical constraints imposed on the structure in the application of the principle of virtual power. The foregoing investigation has shown that we may consider the space $\bar{U}$ (resp. $\bar{F}$) as consisting of the tensors D (resp. σ). Obviously higher order gradient theories would give rise to higher order tensors [Ger73c]. Accordingly we may extend the classical notion of force and call the elements of F generalized forces. Analogously the elements of U are called generalized velocities. For static problems the principle of virtual power takes a form known as principle of virtual work. Then U is the space of generalized displacements u and F is the space of the corresponding generalized forces f. For a complete information on the principles of virtual work and power and on the other postulates of mechanics we refer to Hamel [Ham67] and to Lanczos [Lanc66]. Note that the principle of virtual power does not imply the first and second principle of thermodynamic, which must be postulated separately. Only if in the velocity variation the variation of time is included, the principle of virtual power implies the energy-balance equation [Hein70] [Pana80].

Let us consider now a mechanical system Σ on which the triplet $\{U, \langle \cdot, \cdot \rangle, F\}$, is defined and suppose that only certain subsets of U and F are admissible for the mechanical system. Then a multivalued mapping $A : U \to F$ such that

$$f \in A(v) \quad \forall v \in X \subset U \tag{2.1.29}$$

introduces a law or a constraint on Σ. In equilibrium problems v will be replaced by the generalized diplacement u.

Let us consider further the equilibrium of a system Σ subjected to certain forces f_i, $i = 1, \ldots, n$, and laws or constraints which are defined on X_j by the operators A_j, $j = 1, 2, \ldots, m$. From the principle of virtual work we find that at the position of equilibrium

$$\sum_{i=1}^{n} f_i + \sum_{j=1}^{m} f_j = 0 \tag{2.1.30}$$

$$f_j \in A_j(u) \quad \forall u \in X_j \quad j = 1, 2, \ldots, m.$$

Accordingly, at the position of equilibrium $u \in \cap_j X_j$ and

$$-\sum_{i=1}^{n} f_i \in \sum_{j=1}^{m} A_j(u). \tag{2.1.31}$$

Of special interest is the case of subdifferential laws or constraints. Then $A = -\partial \Phi$ where Φ is a convex, l.s.c. and proper functional on U. Φ is called convex superpotential, after Moreau [Mor68]. Then (2.1.29) takes the form

$$-f \in \partial \Phi(u). \tag{2.1.32}$$

Obviously (cf. Sect. 1.1) Φ needs to be defined only on a convex closed subset of U. By means of the conjugate functional Φ^c, which is also convex, l.s.c. and proper on F, we can write (2.1.32) equivalently as

$$u \in \partial \Phi^c(-f), \tag{2.1.33}$$

$$\Phi(u) + \Phi^c(-f) + \langle u, f \rangle = 0 \tag{2.1.34}$$

$$\Phi(u) + \Phi^c(-f) + \langle u, f \rangle \leq 0, \quad \forall f \in F, \quad \forall u \in U. \tag{2.1.35}$$

Functionals Φ and Φ^c can be considered respectively as the potential and the complementary energy corresponding to the mechanical law or the constraint.

By definition, for $u \in U$ and for $f \in F$, (2.1.32) and (2.1.33) are equivalent to the variational inequalities.

$$\Phi(u^*) - \Phi(u) \geq -\langle f, u^* - u \rangle, \quad \forall u^* \in U \tag{2.1.36}$$

and

$$\Phi^c(-f^0) - \Phi^c(-f) \geq -\langle u, f^0 - f \rangle, \quad \forall f^0 \in F. \tag{2.1.37}$$

Suppose further that Φ is the indicator I_K of a convex closed subset K of U. Then

$$-f \in \partial I_K(u) \tag{2.1.38}$$

and the constraint corresponding to (2.1.38) is called an "ideal unilateral constraint". According to Sect. 1.1 $-f$ is an element of the outward normal cone to K at u. The term unilateral results if one considers for $u \in K$ the variational inequality

$$\langle f, u^* - u \rangle \geq 0, \quad \forall u^* \in K, \tag{2.1.39}$$

which results by definition from (2.1.38). Indeed, if $u^* - u$ is an admissible variation of u (in the sense that it satisfies (2.1.39)), then the same does not hold for the variation $u - u^*$. Only if K is a linear subspace of U, (2.1.39) holds as an equality and thus $f \in K^\perp$. In order to illustrate (2.1.39) let us assume that a material point with mass m is subjected to a force $f \in \mathbb{R}^3$ and is constrained to belong in a convex closed subset K of $\mathbb{R}^3$. If $K = \{x | x \in \mathbb{R}^3, F(x) \leq 0\}$, where F is a continuously differentiable function on $\mathbb{R}^3$ referred to a Cartesian coordinate system $0x_1x_2x_3$ and the contact of the material point with the boundary of K is frictionless, then the reaction force R is given by the relation

$$R = -\lambda \operatorname{grad} F(x), \quad \lambda > 0, \tag{2.1.40}$$

where λ is an unknown proportionality factor. If the material point is in $\operatorname{int} K$, then the reaction force is zero, i.e., $\lambda = 0$; otherwise $\lambda \geq 0$. This type of constraint is described by the relation

$$-R \in \partial I_K(x). \tag{2.1.41}$$

The material point is in equilibrium, if and only if

$$f + R = 0. \tag{2.1.42}$$

Thus

$$f \in \partial I_K(x), \tag{2.1.43}$$

and conversely. For $x \in K$, (2.1.43) is equivalent to the variational inequality

$$f_i(x_i^* - x_i) \leq 0, \quad \forall x^* \in K, \tag{2.1.44}$$

which is the expression of the principle of virtual work.

We first consider a system Σ acted upon by forces f_i where $i = 1, \ldots, n$, and reactions $f_j, j = 1, 2, \ldots, m$, which are derived (see (2.1.32)) from the superpotentials Φ_j defined on the space U of generalized displasements. Then, the condition of equilibrium (2.1.31) reads

$$\sum_{i=1}^{n} f_i \in \sum_{j=1}^{m} \partial \Phi_j(u). \tag{2.1.45}$$

A solution $u \in D(\partial \Phi_1) \cap \ldots \cap D(\partial \Phi_m)$ of (2.1.45) satisfies

$$\sum_{i=1}^{n} f_i \in \partial \Phi_0(u), \quad \Phi_0(u) = \sum_{i=1}^{m} \Phi_j(u) \tag{2.1.46}$$

the converse being generally not true. A combination of Props. 1.1.6 and 1.1.7 supplies the following sufficient condition for the equivalence of (2.1.46) and (2.1.45): if (i) the gradients of $l(0 \leq l \leq m)$ of the superpotentials Φ_j, exist for every $u \in U$ and if (ii) a $u_0 \in U$ exists such that from the remaining $m - l$ functionals $m - l - 1$ are finite and continuous at u_0 and (iii) the $(m - l)$-th functional is finite at u_0, then every solution of (2.1.45) is a solution of (2.1.46), and conversely. Then (2.1.46) is equivalent to the problem

$$\Pi(u) = \min\{\Pi(u^*)|u^* \in U\}, \tag{2.1.47}$$

where

$$\Pi(u^*) = \Phi_0(u^*) - \sum_{i=1}^{n} (f_i, u^*) \tag{2.1.48}$$

is the potential energy of the system considered. Note that if U is the space of generalized velocities and u in (2.1.45) is replaced by v, then (2.1.46) and (2.1.47) describe the motion of the mechanical system Σ where the corresponding inertial forces have been neglected.

2.2 Nonconvex Superpotentials

With respect to a mechanical system Σ characterized by the triplet $\{U, \langle u, f \rangle, F\}$, a mechanical law or constraint is considered between the generalized forces f and the generalized displacements u of the form

$$-f \in \bar{\partial}\Phi(u), \tag{2.2.1}$$

where Φ is an extended real-valued functional defined on U. We shall call Φ a nonconvex superpotential. This mechanical law is by definition equivalent to the inequality

$$\Phi^{\uparrow}(u, u^* - u) \geq \langle -f, u^* - u \rangle, \quad \forall u^* \in U \tag{2.2.2}$$

for $u \in U$, which will henceforth be called a hemivariational inequality, and to the inclusion

$$(-f, -1) \in N_{\text{epi}\,\Phi}(u, \Phi(u)). \tag{2.2.3}$$

Obviously, if Φ is convex, (2.2.2) coincides with (2.1.32). For Φ Lipschitzian, $\Phi^{\uparrow}$ in (2.2.2) is replaced by Φ^0. The conjugacy theory of convex analysis cannot be extended to nonconvex energy functions. However, we shall consider mechanical laws of the form

$$u \in \bar{\partial}\Phi_1(-f). \tag{2.2.4}$$

Note that (2.2.3) and (2.2.4) are not generally related. (2.2.4) leads to a hemivariational inequality with respect to the generalized forces:

$$\Phi_1^\uparrow(-f, -f^0 + f) \geq \langle -u, f^0 - f \rangle, \quad \forall f^0 \in F, \quad f \in F. \tag{2.2.5}$$

If Φ is the indicator I_C of a set C, then (2.2.1) implies that

$$-f \in \bar{\partial} I_C(u) = N_C(u), \tag{2.2.6}$$

i.e., $-f$ is an element of the normal cone to C at u. In order to explain (2.2.6) let us place ourselves in the framework of (2.1.40). We assume now that $K = \{x | f_i(x) \leq 0 \ i = 1, \ldots, m\}$ is a closed but not convex subset of $\mathbb{R}^3$; the f_i's are assumed to be continuously differentiable. K can be written as $\{x | F(x) \leq 0\}$, where $F = \max\{f_1, \ldots, f_m\}$. F is Lipschitzian and $\bar{\partial}$-regular. As is obvious, the reaction force R satisfies (2.1.40) on any smooth part of the boundary of K. At any corner of K, $-R$ is a nonnegative linear combination of the gradients of the f_i's that are zero on the corner. Accordingly, (1.2.47) imply that if $0 \notin \bar{\partial} F(x)$, the relation

$$-R \in N_K(x) \tag{2.2.7}$$

holds. As a result, a necessary and sufficient condition for equilibrioum is that the inclusion

$$f \in \bar{\partial} I_K(x) \tag{2.2.8}$$

holds. Relation (2.2.8) is equivalent to the inequality (cf. (1.2.36))

$$I_{T_K(x)}(x^* - x) \geq f_i(x_i^* - x_i), \quad \forall x^* \in K, \quad \text{and for} \quad x \in K. \tag{2.2.9}$$

Inequality (2.2.9) reduces to (2.1.44) if $(x^* - x) \in T_K(x)$. At a re-entrant corner of K, $I_{T_K(x)}(x^* - x)$ becomes ∞ if $(x^* - x) \notin T_K(x)$.

Suppose now that forces f_i, $i = 1, \ldots, n$, act on the system Σ which is subjected to the nonconvex superpotential laws or constraints $-\bar{f}_j \in \bar{\partial}\Phi_j(u)$ $j = 1, \ldots, m$. Then the condition of equilibrium (2.1.30) implies that

$$\sum_{i=1}^n f_i \in \sum_{j=1}^m \bar{\partial}\Phi_j(u). \tag{2.2.10}$$

Obviously, if

$$0 \in \bar{\partial}\Pi(u), \quad \Pi(u) = -\sum_{i=1}^n (f_i, u) + \sum_{j=1}^m \Phi_j(u), \tag{2.2.11}$$

then (2.2.10) holds, but the converse is not always true; u is a substationarity point of Π, where Π is the potential energy of the system Σ. Accordingly, for the present mechanical system, u is an equilibrium configuration, if u is a substationarity point of Π. Equivalent to (2.2.11) is the inequality (cf. (1.2.45))

$$\Pi^{\uparrow}(u, u^*) \geq 0 \quad \forall u^* \in U. \tag{2.2.12}$$

We recall here that any local minimum of the potential energy Π is a substationarity point and thus corresponds to an equilibrium configuration, as well as any classical stationary point, any saddle point and finally any local maximum, if Π is locally Lipschitz at this local maximum (cf. Sect. 1.2 and [Rock79]).

As we have mentioned, there does not exist a duality theory for nonconvex functional because we cannot extend the conjugacy theory of convex functionals to nonconvex functionals. However it should be mentioned here that such an extension is possible for quasi-convex functionals [Cro77] and for functionals which can be expressed as the difference of two convex superpotentials [Tol79] [Stav91,93a]. In Sect. 1.2 we have given the generalized gradients of several types of functions, as e.g. the maximum type functions, the concave functions etc. Such type of functions permit the formulation of several classes of mechanical problems in terms of nonconvex superpotentials and thus in terms of hemivariational inequalities. Also, the superpotentials resulting by integrating discontinuous functions $\beta \in L^{\infty}_{\text{loc}}(\mathbb{R})$ (cf. (1.2.53)÷(1.2.55)) play an important role in the formulation of hemivariational inequalities for several types of mechanical problems. The superpotential law which will be given next further illustrates the possibilities offered to Mechanics by the introduction of the notion of the generalized gradient. Following Rockafellar [Rock79] we define the following nonconvex superpotential: Let l be an open subset of the real line $\mathbb{R}$ and let M be a measurable subset of l such that for every open and nonempty subset I of l, $\text{mes}\,(I \cap M)$ and $\text{mes}\,(I \cap (l - M))$ are positive. Let

$$g(u) = \begin{cases} +b_1 & \text{if } u \in M \\ -b_2 & \text{if } u \notin M \end{cases} \tag{2.2.13}$$

and

$$f(u) = \int_0^u g(u^*) du^*. \tag{2.2.14}$$

Then f is Lipschitzian and it can be verified that

$$\bar{\partial} f(u) = [-b_2, b_1] \tag{2.2.15}$$

for every $u \in l$, i.e., we obtain an infinite number of jumps in l (Fig. 2.2.1). Modifications of this law are also possible. For instance, b_1 and b_2 may depend on u. In this context see also the Sections that follow. We would like to point out that the aforementioned nonconvex superpotential laws and constraints also hold in the framework of generalized velocities. Closing this Section we would like to remark a deficiency of the notion of generalized gradient which becomes obvious especially in the framework of mechanics. The superpotential law (2.2.1) should reduce to a classical potential law of the type.

$$-f = \text{grad}\,\Phi(u) \tag{2.2.16}$$

at any point u, where Φ is differentiable. This is however not possible for Φ nonconvex; one needs the C^1-continuity of Φ at u according to the theory of the generalized gradient.

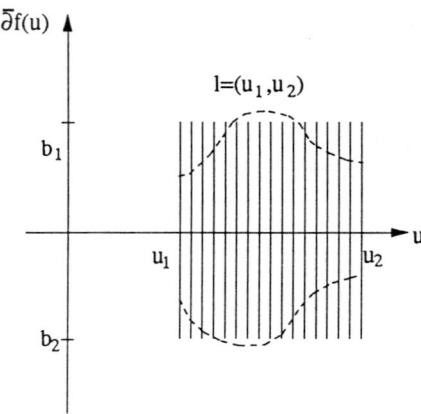

Fig. 2.2.1. The graph of (2.2.15).

2.3 Boundary Conditions Expressed via Convex Superpotentials

Using the definitions of Sect. 2.1 we may introduce subdifferential boundary conditions for a deformable body. These boundary conditions include the boundary conditions of classical elasticity as special cases. We denote by Ω an open, bounded subset of $\mathbb{R}^3$ which is occupied by a deformable body. The forthcoming definitions hold both for small and large deformation theory. The boundary of Ω is denoted by Γ. The points $x \in \Omega, x = \{x_i\}, i = 1, 2, 3$, are referred to a Cartesian coordinate system. We denote by $S = \{S_i\}$ the stress vector on Γ. It is $S_i = \sigma_{ij} n_j$, where $\sigma = \{\sigma_{ij}\}$ is an appropriately defined stress tensor depending on the framework chosen and $n = \{n_i\}$ is the outward unit normal vector on Γ. The vector S is decomposed into a normal component S_N and a tangential component S_T with respect to Γ

$$S_N = \sigma_{ij} n_j n_i \quad \text{and} \quad S_{T_i} = \sigma_{ij} n_j - (\sigma_{ij} n_i n_j) n_i. \tag{2.3.1}$$

Analogously to S_N and S_T, u_N and u_T denote the normal and the tangential components of the displacement vector u with respect to Γ. S_N and u_N are considered as positive if they are parallel to n.

A maximal monotone operator $\beta_i : \mathbb{R} \to \mathcal{P}(\mathbb{R})$ is introduced and a boundary condition of the form

$$-S_i \in \beta_i(u_i) \tag{2.3.2}$$

is considered in the i-th direction. Here $U = F = \mathbb{R}$. Then (Prop. 1.3.9) a convex, l.s.c. and proper functional j_i on $\mathbb{R}$ may be determined up to an additive constant such that

$$\beta_i = \partial j_i. \tag{2.3.3}$$

Then (2.3.2) is written as

$$-S_i \in \partial j_i(u_i). \tag{2.3.4}$$

This relation is a subdifferential boundary condition and is understood pointwise, i.e., as a relation between $-S_i(x) \in \mathbb{R}$ and $u_i(x) \in \mathbb{R}$ at every point $x \in \Gamma$. Obviously, (2.3.4) may also be written in the inverse form

$$u_i \in \partial j_i^c(-S_i) \tag{2.3.5}$$

and

$$u_i \in \beta_i^c(-S_i), \tag{2.3.6}$$

where $\beta_i^c = \partial j_i^c$ is again a maximal monotone operator on $\mathbb{R}$ and is the inverse operator of β_i. The graph of β_i, referred to a Cartesian system Oxy, is a complete nondecreasing curve in $\mathbb{R}^2$ which is generally multivalued; thus the graph may include segments parallel to both coordinate axes. "Superpotential" j_i (resp. j_i^c) is a local superpotential (resp. conjugate superpotential) and expresses the potential (resp. complementary energy) of the constraint [Mor68]. The boundary condition (2.3.2) may be considered as the material law of a fictive spring of zero length at x in the ith-direction.

Analogously to (2.3.2), a boundary condition of the form

$$-S_N \in \beta_N(u_N) = \partial j_N(u_N) \tag{2.3.7}$$

may be defined. Assume now that $U = F = \mathbb{R}^3$ and that j is a convex, l.s.c., proper functional on $\mathbb{R}^3$. Then a boundary condition of the form

$$-S \in \partial j(u) \tag{2.3.8a}$$

is defined pointwise on Γ, i.e., as a monotone relation between $S(x)$ and $u(x)$. Equivalently to (2.3.8), we may write

$$u \in \partial j^c(-S) \tag{2.3.8b}$$

and

$$j(u) + j^c(-S) = -u_i S_i. \tag{2.3.8c}$$

Similarly to (2.3.8), a subdifferential law

$$-S_T \in \partial j_T(u_T) \tag{2.3.9}$$

may be considered.

In dynamic mechanical problems, similar boundary conditions may be defined between S and the partial time derivative of the displacement $\partial u/\partial t$, or the velocity v.

We give some examples to illustrate these boundary conditions.

i) The classical boundary conditions $u_i = 0$ can be put in the form (2.3.2) through the operator

$$\beta_i(u_i) = \begin{cases} \mathbb{R} & \text{if } u_i = 0 \\ \emptyset & \text{otherwise} \end{cases}, \qquad (2.3.10)$$

or through the functional $j_i(u_i) = \{0 \text{ if } u_i = 0 \text{ and } \infty \text{ otherwise}\}$. The boundary conditions $S_i = C_i$ is written in the form (2.3.2) or (2.3.4) with $\beta_i(u_i) = -C_i$ (C_i given) or $j_i(u_i) = -C_i u_i$ (no summation) for every $u_i \in \mathbb{R}$.

ii) The Winkler contact boundary condition

$$-S_N = ku_N, \quad k \text{ const} > 0 \qquad (2.3.11)$$

may be expressed in the form (2.3.7) by setting

$$\beta_N(u_N) = ku_N, \quad j_N(u_N) = \frac{1}{2}ku_N^2.$$

This law, describes in a simplified manner the interaction between a deformable body and the soil and is used in practical civil engineering.

iii) The foregoing boundary condition does not describe the case in which the body loses contact with the support [Pan75,80]. To do so we should consider the following law:

$$\text{if } u_N < 0, \quad \text{then} \quad S_N = 0; \qquad (2.3.12a)$$

$$\text{if } u_N \geq 0, \quad \text{then} \quad S_N + ku_N = 0; \quad k \text{ const} > 0. \qquad (2.3.12b)$$

Relation (2.3.12a) corresponds to the case of lack of contact and (2.3.12b) to the case of contact. The regions of contact and noncontact are not known a priori; thus (2.3.12a,b) lead to a free B.V.P. The respective operator β_N (resp. j_N) is given by

$$\beta_N(u_N) = \begin{cases} ku_N & \text{if } u_N \geq 0 \\ 0 & \text{if } u_N < 0 \end{cases} \qquad (2.3.12c)$$

and

$$j_N(u_N) = \begin{cases} \dfrac{1}{2}ku_N^2 & \text{if } u_N \geq 0 \\ 0 & \text{if } u_N < 0 \end{cases}. \qquad (2.3.12d)$$

Note that $j_N(u_N)$ can be written also as $\frac{1}{2}ku_{N+}^2$, where u_{N+} denotes the positive part of u_N, i.e., $u_{N+} = \sup\{0, u_N\}$. Relations (2.3.12) are called conditions of unilateral contact (Fig.2.3.1) for a linear Winkler law, whereas (2.3.11) is the condition of bilateral contact. In Fig.2.3.1 the graph AOE corresponds to (2.3.12a,b)

2.3 Boundary Conditions Expressed via Convex Superpotentials

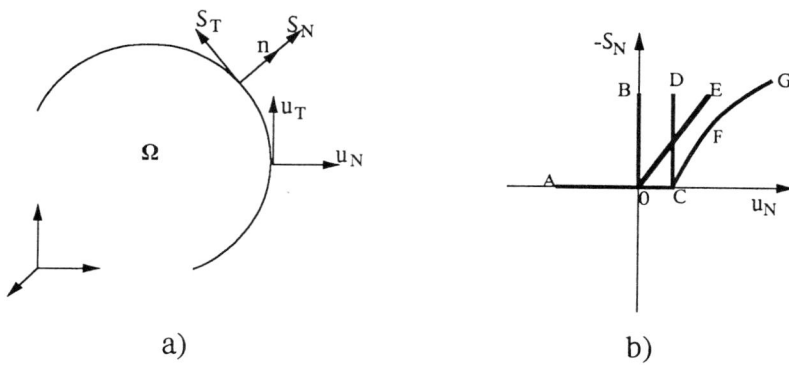

Fig. 2.3.1. Unilateral contact boundary conditions.

We can consider generally the operators

$$\beta_N(u_N) = \begin{cases} \beta_1(u_N - h) & \text{if } u_N \geq h \\ 0 & \text{if } u_N < h \end{cases}. \quad (2.3.13)$$

Here β_1 is assumed to be a maximal monotone operator on $\mathbb{R}$ such that $0 \in \beta_1(0)$. Eq. (2.3.13) leads to unilateral contact boundary conditions, but with a nonlinear Winkler law and a support at a given distance $h = h(x)$ from the body under consideration. In Fig. 2.3.1 the diagramm AOCFG corresponds to (2.3.13), where the segment OC has a length h. The relations are not sufficient to formulate a B.V.P., but they must be combined with a boundary condition concerning S_T or u_T or both, e.g. $S_T = C_T$, where $C_T = C_T(x)$ is given, or $u_T = 0$, or, more generally, (2.3.9). It is also possible for β_N to change from point to point, in which case $\beta_N = \beta_N(u_N(x), x)$. Note that the uncoupling of the contact conditions in the tangential and in the normal directions is a considerable simplification of the mechanical problem. The case of coupled contact conditions will be examined at the end of this Section.

iv) If the support is rigid, then the boundary conditions of Signorini hold [Fich63,64,72;Duv72]. They read

$$\begin{aligned} &\text{if } u_N < 0, \quad \text{then } S_N = 0; \\ &\text{if } u_N = 0, \quad \text{then } S_N \leq 0, \end{aligned} \quad (2.3.14)$$

or equivalently

$$S_N \leq 0, \quad u_N \leq 0, \quad \text{and} \quad S_N u_N = 0. \quad (2.3.15)$$

This last form is called linear complementarity form. In Fig.2.3.1 the graph AOB corresponds to the boundary conditions (2.3.14). The respective operator β_N is

48   2. Nonsmooth Mechanics I

$$\beta_N(u_N) = \begin{cases} 0 & \text{if } u_N < 0 \\ [0, +\infty) & \text{if } u_N = 0 \\ \emptyset & \text{if } u_N > 0 \end{cases}, \qquad (2.3.16)$$

and the corresponding superpotential

$$j_N(u_N) = \begin{cases} 0 & \text{if } u_N \le 0 \\ \infty & \text{if } u_N > 0 \end{cases} \qquad (2.3.17)$$

If the support is at a distance h from the boundary of the body, then u_N has to be replaced by $u_N - h$ (cf. in Fig. 2.3.1 the graph AOCD) to describe the contact with the possibility of debonding (or detachment) between two deformable bodies we consider an interface condition analogous to (2.3.14) on the assumption that the boundary displacements are sufficiently small. As the two bodies cannot penetrate one another, we assume that the sum of the displacements $u_N^{(1)}$ and $u_N^{(2)}$ of the two bodies and of the existing normal distance between them $h = h(x)$ must be greater than, or equal to the approach u^0 of the two bodies in the normal direction due to a rigid body displacement. We denote by $\bar{u}_N$ the quantity

$$u_N^{(1)} + u_N^{(2)} + h - u^0,$$

and let R_N be the respective contact force. The contact conditions read:

$$\begin{aligned} &\text{if } \bar{u}_N > 0, \text{ then } R_N = 0; \\ &\text{if } \bar{u}_N = 0, \text{ then } R_N \ge 0. \end{aligned} \qquad (2.3.18)$$

Concerning the noninterpenetration condition etc. for discretized bodies and/or for finite boundary displacements we refer to [Bis86,90;Kik,Sim,Wri,Hug,Has82].

v) The next example concerns the static version of Coulomb's friction boundary condition [Duv71,Mor86]. We consider the following boundary conditions for $U = F = \mathbb{R}^2$ (if $\Omega \subset \mathbb{R}^3$):

if $|S_T| < \mu|S_N|$, then $u_{T_i} = 0$, $i = 1, 2, 3$ \qquad (2.3.19a)

if $|S_T| = \mu|S_N|$, then there exists

$\lambda \ge 0$ such that $u_{T_i} = -\lambda S_{T_i}$, $i = 1, 2, 3$. \qquad (2.3.19b)

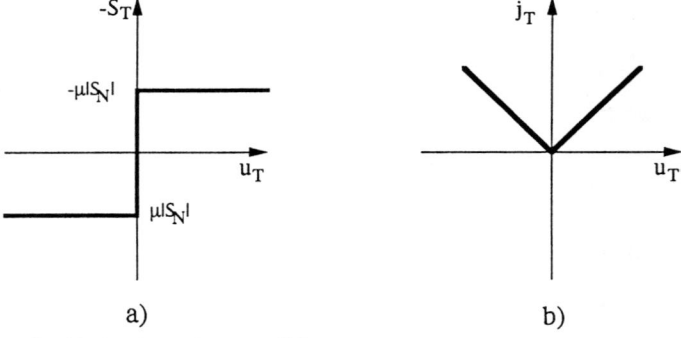

Fig. 2.3.2. The friction boundary condition.

2.3 Boundary Conditions Expressed via Convex Superpotentials 49

Here $\mu = \mu(x) > 0$ denotes the coefficient of friction and $|\cdot|$ the usual $\mathbb{R}^3$-norm. If Ω is a two-dimensional body, then Γ is a curve, and thus S_T, u_T may be referred to a local right-handed coordinate system (n, τ) on Γ where τ denotes the unit vector tangential to Γ. Then (2.3.19a,b) can be put in the form

$$-S_T \in \beta_T(u_T), \qquad (2.3.20)$$

where (Fig.2.3.2)

$$\beta_T(u_T) = \begin{cases} [-\mu|S_N|, +\mu|S_N|] & \text{if } u_T = 0 \\ \mu|S_N| & \text{if } u_T > 0 \\ -\mu|S_N| & \text{if } u_T < 0 \end{cases}. \qquad (2.3.21)$$

Assume further that $S_N = C_N$, where C_N is given, and denote $\mu|C_N|$ by S_{T_0}. Then

$$\beta_T(u_T) = \partial(S_{T_0}|u_T|). \qquad (2.3.22)$$

If Ω is a three-dimensional body, then (2.3.19a,b) can be written only in the form (2.3.9) with

$$j_T(u_T) = S_{T_0}|u_T|. \qquad (2.3.23)$$

We can verify that

$$j_T^c(-S_T) = \begin{cases} 0 & \text{if } |S_T| \leq S_{T_0} \\ \infty & \text{otherwise} \end{cases}, \qquad (2.3.24)$$

and thus (2.3.19a,b) take equivalently the inverse form

$$u_T \in \partial j_T^c(-S_T). \qquad (2.3.25)$$

Note that the two conjugate subdifferential formulations give rise, for $u_T, S_T \in \mathbb{R}^3$ to the variational inequalities

$$j_T(u_T^\star) - j_T(u_T) \geq -S_{T_i}(u_{T_i}^\star - u_{T_i}), \quad \forall u_T^\star \in \mathbb{R}^3 \qquad (2.3.26)$$

and

$$-u_{T_i}(S_{T_i}^0 - S_{T_i}) \leq 0, \quad \forall S_T^0 \in \mathbb{R}^3 \text{ such that } |S_T^0| \leq S_{T_0} \qquad (2.3.27)$$

for $|S_T| \leq S_{T_0}$.

In dynamic or quasistatic problems, a friction law of the form

$$-S_T \in \partial j_T(v_T) = \partial(S_{T_0}|v_T|) \qquad (2.3.28)$$

can be considered (Coulomb's law of friction). Here v_T denotes the tangential velocity which is equal to $\partial u_T/\partial t$ if the displacements are sufficiently small.

It is possible to combine the friction boundary condition with the unilateral contact boundary condition. Then we obtain the following relations:

$$\begin{aligned} &\text{if } u_N < 0, \text{ then } S_N = 0, \quad S_{T_i} = 0, \quad i = 1, 2, 3 \\ &\text{if } u_N \geq 0, \text{ then } S_N + k u_N = 0, \end{aligned} \qquad (2.3.29)$$

where k is a constant > 0 and (2.3.19a,b) hold. In this case, it is not possible to write the boundary conditions in the subdifferential form. A generalization of (2.3.25) is obtained if the superpotential j_T takes the form

$$j_T(u_T) = |C_N|(\mu_a^2 u_{Ta}^2 + \mu_b^2 u_{Tb}^2)^{1/2}. \tag{2.3.30}$$

Here a and b are two orthogonal directions termed orthotropy directions, which are defined at every point on the surface of the body, u_{Ta} and u_{Tb} are the components of the displacement u_T in purely static problems (cf. Duv72) with respect to a local coordinate system (a, b), and μ_a and μ_b are the two corresponding friction coefficients. In quasistatic or dynamic genuine friction B.V.Ps u_T must be replaced by the corresponding velocity component v_T. The resulting friction law is called orthotropic friction law (cf. in this context also [Pan85]). Note that if we have two deformable bodies in contact, the interface friction condition can be described by the same laws given above with the only difference that the tangential displacement u_T must be replaced by the relative tangential displacement $[u_T]$, or relative velocity $[v_T]$. In the context of unilateral contact with friction we refer also to [Has82,84], [Gas88a,b], [Gol84,88], [Jea85,87], [Jin], [Ja83,84], [Duv80], [Fei], [An], [Bat], [Böh], [Cam], [Co], [Cur86,88], [De], [Ju], [Ka88a,b,90], [Kla84,87,88,90a,b,c], [Kik], [Kwa88;91], [Li86,87,88,89,91], [Mar86,87], [Mit83,87,91,93], [Neč80], [Ode81,83,85], [Pdp92c], [Spe82a,b,85, 87a,b], [Ta], [Tel], [Bis86,90,92], [Zh88;89] and to the references given there.

vi) Further, we give some subdifferential boundary conditions arising in the theory of plates. Ω is here an open bounded subset $\mathbb{R}^2$ defined by the middle surface of the plate. Γ denotes the boundary of Ω. The points of Ω are referred to a fixed Cartesian coordinate system $Ox_1x_2x_3$. The x_1- and x_2-axes coincide with the middle surface of the plate, and the x_3-axis with the direction of the normal to the middle surface. The positive direction of the x_3-axis is upwards. The displacements of the plate in its plane are denoted by u_1, u_2 and vertical to its plane by w. By M_n and K_n we denote respectively the bending moment and the total or Kirchhoff shearing force [Gir] on the boundary of the plate, and we introduce boundary conditions of the form

$$M_n \in \beta_1\left(\frac{\partial w}{\partial n}\right) = \partial j_1\left(\frac{\partial w}{\partial n}\right) \tag{2.3.31}$$

$$-K_n \in \beta_2(w) = \partial j_2(w) \tag{2.3.32}$$

for applications of (2.3.31),(2.3.32) see [Pan85].

vii) Another type of subdifferential relation can be formulated in the theory of plates. Assume that the load vector f at every point $x \in \Omega_0 \subset \Omega$ consists of a part $\bar{f}$, which is given, and of another part $\bar{\bar{f}}$ related to the displacement of that point by a relation of the form

$$-\bar{\bar{f}} \in \beta_3(w) = \partial j_3(w). \tag{2.3.33}$$

Here β_3 and j_3 have the same properties as β_i and j_i in (2.3.2), (2.3.3). As an application let us consider a plate which at points $x_0 \in \Omega_0 \subset \Omega$, $\bar{\Omega}_0 \cap \Gamma = \emptyset$, is at a distance $h = h(x)$ from a deformable support. It is assumed that the support causes a reaction force which is proportional to its deformation (Winkler support). We may then write the relation

$$-\bar{\bar{f}} \in \beta(w) \quad \text{in} \quad \Omega_0 \subset \Omega,$$

and

$$\bar{\bar{f}} = 0 \quad \text{in} \quad \Omega - \Omega_0. \qquad (2.3.34a)$$

β is a maximal monotone operator defined by

$$\beta(w) = \begin{cases} k(w-h), \; k \text{ const} > 0 & , \text{ if } w \geq h \\ 0 & , \text{ if } w < h. \end{cases} \qquad (2.3.34b)$$

(viii) Note here that besides the boundary conditions (2.3.7) and (2.3.9), where the actions normally and tangentially to the boundary of a deformable body $\Omega \subset \mathbb{R}^3$ are considered separately, laws of the form

$$-S_N \in \partial j_N(u_N; S_T) \qquad (2.3.35)$$

$$-S_T \in \partial j_T(u_T; S_N) \qquad (2.3.36)$$

can be also considered. For instance, the unilateral contact with friction is a boundary condition of this type. The numerical treatment of such boundary conditions is made possible by means of a decomposition technique introduced in [Pan75] for the unilateral contact problem with friction. The same technique holds also for nonconvex superpotentials. We assume in the first step that $S_T = S_T^{(1)}$, where $S_T^{(1)}$ is given and we solve the problem with the boundary conditions $-S_N \in \partial j_N(u_N; S_T^{(1)})$, $S_T = S_T^{(1)}$. The solution of this problem yields the values for S_N, say $S_N^{(2)}$. Then the second step problem is considered with the boundary conditions $-S_T \in \partial j_T(u_T; S_N^{(2)})$, $S_N = S_N^{(2)}$. The numerical solution of this problem yields the value of S_T, say $S_T^{(3)}$. This procedure continues until the differences $|S_N^{(i+1)} - S_N^{(i)}|$, $|S_T^{(i+1)} - S_T^{(i)}|$, are made sufficiently small.

2.4 Boundary Conditions Expressed via Nonconvex Superpotentials

In this Section we deal with boundary conditions expressed in the forms

$$-S_N \in \bar{\partial} j_N(u_N), \quad -S_T \in \bar{\partial} j_T(u_T) \qquad (2.4.1)$$

or

$$-S \in \bar{\partial} j(u). \qquad (2.4.2)$$

Here $\bar{\partial}$ is the generalized gradient and j_N, j_T and j is a locally Lipschitz functional defined on $\mathbb{R}$, $\mathbb{R}^2$ and $\mathbb{R}^3$ respectively. Relations (2.4.1), (2.4.2) may be considered both in the framework of a small or a large deformation theory. Of course one can also consider the general definition of the generalized gradient for any type of function, i.e. not necessarily locally Lipschitz. The nonconvex superpotentials j_N in $\Omega \subset \mathbb{R}^3$ and both the j_N and j_T in $\Omega \subset \mathbb{R}^2$ can be formulated by integrating an appropriate function $\beta \in L_{loc}^{\infty}(\mathbb{R})$, as it is shown in eqs. (1.2.51)÷(1.2.55). The more general nonconvex superpotentials j and j_T for $\Omega \subset \mathbb{R}^3$ are formulated, by "extending" to $\mathbb{R}^3$ or $\mathbb{R}^2$ certain onedimensional nonmonotone multivalued laws, e.g. by considering maximum type functions etc. Let us first give some simple examples.

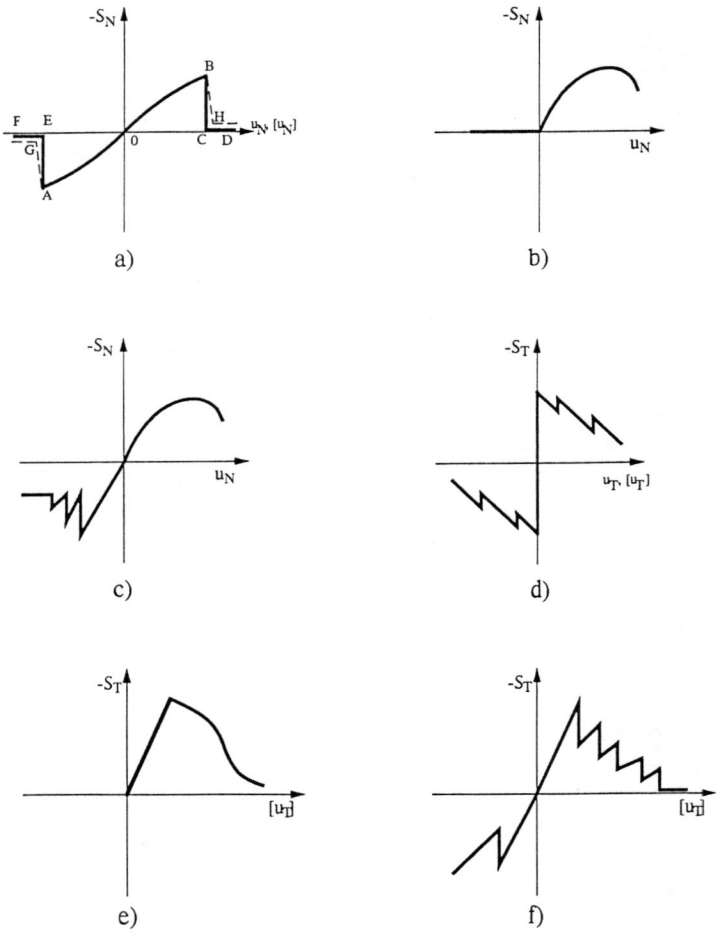

Fig. 2.4.1. Nonmonotone multivalued boundary conditions.

2.4 Boundary Conditions Expressed via Nonconvex Superpotentials 53

i) The diagram depicted in Fig. 2.4.1a concerns the behaviour of the normal forces at any interface point in an adhesive contact problem. The adhesive material between the body and the support may sustain a small tensile force before rupture. Then debonding takes place, which may obey the brittle type diagram FEABCD or the semibrittle diagram FGAOBHD. Note that the vertical branches (i.e. the multivaluedness) are complete, i.e. for an appropriate loading the reaction and the normal boundary displacement u_N (resp. interface relative displacement $[u_N]$) can assume a value on the vertical branch.

ii) In Fig. 2.4.1b a normal contact law between a deformable body and a support of a granular material or concrete is depicted. If debonding takes place the stresses are zero.

iii) In Fig. 2.4.1c the same contact law this time without debonding is presented. Here the support is assumed to be a reinforced concrete support which obeys in tension Scanlon's zig-zag law [Flo]. An analogous multivalued zig-zag diagram in tension holds for a composite material. The fact that we may have an equilibrium state for which the stress and strain of an interface point may assume a value on the complete vertical branch of the boundary stress-strain law has been experimentally verified (cf. e.g. [Paip],[Bam85] [Ond] [Moys] and the literature given there in).

iv) In Fig. 2.4.1d,e certain nonmonotone friction laws are depicted. The first comes from geomechanics and rock interface analysis, whereas the second arises between reinforcement and concrete in a concrete structure. Finally the law of Fig. 2.4.1f appears in the tangential direction of an adhesive interface and describes the partial cracking and crushing of the adhesive bonding material.

All these laws can be put in the form (2.4.1) where j_N and j_T are nonconvex superpotentials resulting by integrating the diagrams describing the boundary conditions (cf. eq. (1.2.54)). In the aforementioned boundary conditions j_N and j_T are locally Lispchitz and thus the generalized gradients $\bar{\partial} j_N$ and $\bar{\partial} j_T$ are described by means of the directional differentials $j_N^0(\cdot,\cdot)$ and $j_T^0(\cdot,\cdot)$. Nonmonotone laws may hold either between a deformable body and a rigid support or between two deformable bodies. In the latter case u, u_N and u_T must be replaced by the corresponding relative displacements $[u]$, $[u_N]$ and $[u_T]$, e.g. in the nonmonotone friction laws of Fig. 2.4.1d,e,f.

v) We also may have nonmonotone boundary conditions including infinite branches. For instance the law of Fig. 2.4.2a describes the adhesive contact with a rubber support, which presents an ideally locking effect in compression (the infinite branch AB). The law of Fig. 2.4.2b describes the adhesive contact of two deformable bodies. The noninterpenetration effect gives rise to the infinite branch AB on the assumption that the interface may sustain infinite compressive normal forces.

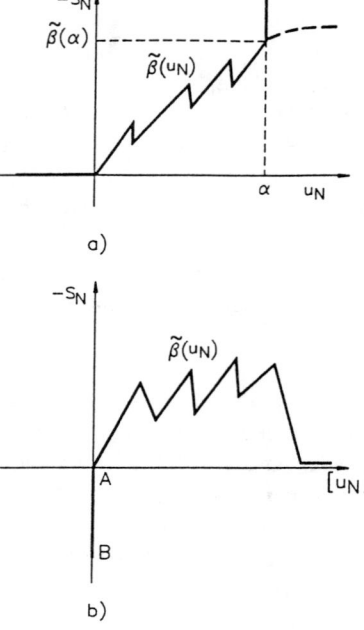

Fig. 2.4.2. Nonmonotone multivalued boundary conditions with infinite branches.

Obviously the law of Fig. 2.4.2a takes the form

$$\begin{aligned} &\text{if} \quad u_N < \alpha \quad \text{then} \quad -S_N \in \tilde{\beta}(u_N) \\ &\text{if} \quad u_N = \alpha \quad \text{then} \quad -\infty < -S_N \leq \tilde{\beta}(\alpha) \\ &\text{if} \quad u_N > \alpha \quad \text{then} \quad S_N = \emptyset \end{aligned} \qquad (2.4.3)$$

or equivalently the form

$$-S_N \in \tilde{\beta}(u_N) + N_K(u_N) = \bar{\partial} j_N(u_N) + \partial I_K(u_N) \qquad (2.4.4)$$

where $K = \{u_N | u_N \leq \alpha\}$, $\bar{\partial} j_N$ results from $\tilde{\beta}$ as in (1.2.54) (here we assume that $\tilde{\beta}$ results from $\beta \in L^\infty_{\text{loc}}(\mathbb{R})$ as in Fig. 1.2.3 and in eqs. (1.2.51)÷(1.2.55)), N_K is the normal cone to K at u_N (cf. also eq. (1.1.12)) and I_K is the indicator of the set K. Recall that (cf. Sect 1.1) $N_K(u_N) = \{0 \text{ if } u_N < \alpha, \ [0, \infty) \text{ for } u_N = \alpha, \emptyset \text{ for } u_N > \alpha\}$. Relation (2.4.4) implies that

$$\begin{aligned} u_N \in \mathbb{R}, \quad & j_N^o(u_N, u_N^\star - u_N) + I_K(u_N^\star) - I_K(u) \\ & \geq -\langle S_N, u_N^\star - u_N \rangle \, \forall u_N \in \mathbb{R} \end{aligned} \qquad (2.4.5)$$

thus giving rise to a variational hemivariational inequality. In this special case (2.4.5) becomes also

$$u_N \in K, \quad j_N^o(u_N, u_N^\star - u_N) \geq -\langle S_N, u_N^\star - u_N \rangle \quad \forall u_N^\star \in K. \qquad (2.4.6)$$

Analogously we treat the problem of Fig. 2.4.2b.

2.4 Boundary Conditions Expressed via Nonconvex Superpotentials 55

vi) The aforementioned case v) can be extended to cover threedimensional contact problems with an ideally locking support having a nonconvex locking criterion. We assume that $K = \{u \in \mathbb{R}^3 | f_1(u) \leq 0, \ldots, f_n(u) \leq 0\}$, where $f_i, i = 1, \ldots, n$, are continuously differentiable functions such that K is a closed, but not necessarily convex subset of $\mathbb{R}^3$. Thus, we may consider a generalization of (2.4.4) of the type

$$-S \in \bar{\partial} j(u) + \bar{\partial} I_K(u), \qquad (2.4.7)$$

where $\bar{\partial} I_K(u)$ is given by (1.2.49) for $u \in K$ and $\bar{\partial} I_K(u) = \emptyset$ if $u \notin K$. We leave as an exercise for the reader the derivation of the interesting formula obtained from (2.4.7) if $j(u)$ is a quadratic function of u (cf. also the nonconvex yield functions in Ch.3).

vii) Suppose that in the contact boundary condition depicted in Fig. 2.4.2 the interface may sustain only bounded compressive normal forces. More specifically let us assume that in the small interval $-l < [u_N] < 0$ the traction $-S_N$ may take any value between 0 and α (Fig. 2.4.3a).

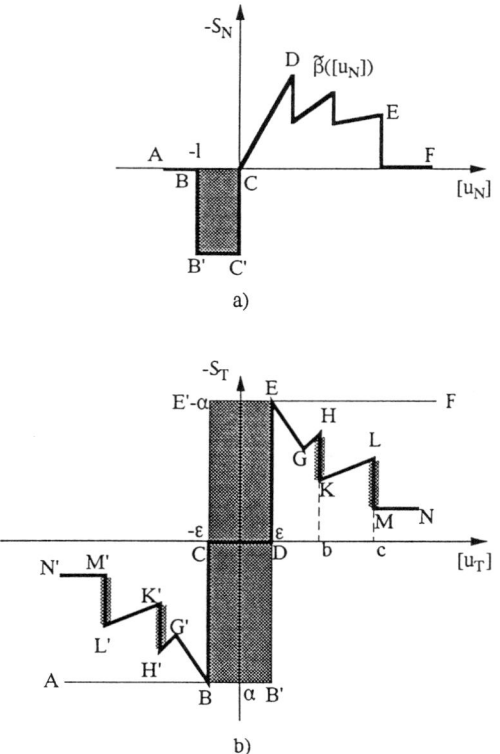

Fig. 2.4.3. Nonmonotone laws with nonfully determined values (fuzzy law).

We may write this law in the form.

56   2. Nonsmooth Mechanics I

$$\begin{array}{ll} \text{if } u_N \geq 0 & -S_N \in \bar{\beta}([u_N]) \\ \text{if } -l < u_N < 0 & \alpha \geq S_N > 0 \\ \text{if } u_N \leq -l & S_N = 0 \end{array} \qquad (2.4.8)$$

or equivalently in the form

$$-S_N \in \bar{\partial}j([u_N]) + \bar{\partial}f([u_N]) \qquad (2.4.9)$$

where j results from the multivalued graph ABCDEF through integration (cf. eqs. (1.2.54), (1.2.55)) and f is defined $(-l, 0)$, by the relations (2.2.13)÷(2.2.15) where $b_2 = \alpha$ and $b_1 = 0$. Thus (2.4.9) describes the law of Fig. 2.4.3a, which in the interval BC has nonfully determined values. Analogously we may express the laws depicted in Fig. 2.4.3b. Indeed the law ABB'EE'F may be written in the form (fuzzy law)

$$-S_T \in \partial j([u_T]) + \bar{\partial}f([u_T]), \qquad (2.4.10)$$

where j is convex and results from the maximal monotone graph $ABCDEF$ according to Prop. 1.3.9, and f is given by the relations (2.2.13)÷(2.2.15) with $b_1 = b_2 = \alpha$ and $l = (-\varepsilon, \varepsilon)$. Analogously is written the law N'M'L'K'H'G'BB'E'EGH KLMN. Now $\partial j([u_T])$ is replaced in (2.4.10) by $\bar{\partial}j([u_T])$, where j is nonconvex and results from the graph N'M'L'K'H'G'BCDEGHKLMN by means of the relation (1.2.54), (1.2.55). Physically, the above diagrams correspond to monotone and nonmonotone interface friction laws with a nonfully determined the behaviour around the adhesive friction region (fuzzy adhesive behaviour). Such nonfully determined regions may be considered around the complete vertical segments of the nonmonotone law (see the shaded areas around HK and LM in Fig. 2.4.3b). The expression of such nonfully determined laws in terms of nonconvex superpotentials permits the formulation of a variational theory for this class of problems.

viii) Let us put ourselves in the framework of (2.3.30). It leads to the friction law [Michal], [Pan85]

$$\text{if } \left[\left(\frac{S_{T_a}}{\mu_a}\right)^2 + \left(\frac{S_{T_b}}{\mu_b}\right)^2\right]^{\frac{1}{2}} < |C_N| \text{ then } u_{T_a} = u_{T_b} = 0 \qquad (2.4.11a)$$

$$\text{if } \left[\left(\frac{S_{T_a}}{\mu_a}\right)^2 + \left(\frac{S_{T_b}}{\mu_b}\right)^2\right]^{\frac{1}{2}} = |C_N| \text{ then there exists } \lambda \geq 0 \qquad (2.4.11b)$$

$$\text{such that } u_{T_a} = -\lambda \frac{S_{T_a}}{\mu_a^2}, u_{T_b} = -\lambda \frac{S_{T_b}}{\mu_b^2}.$$

In (2.4.11a,b) we have an elliptic limit friction condition. Then the orthotropic friction law can be written in the form

$$u_T = -\{u_{T_a}, u_{T_b}\} \in \partial I_K(S_{T_a}, S_{T_b}) \qquad (2.4.12)$$

where

$$K = \left\{\{S_{T_a}, S_{T_b}\} \middle| \left[\left(\frac{S_{T_a}}{\mu_a}\right)^2 + \left(\frac{S_{T_b}}{\mu_b}\right)^2\right]^{\frac{1}{2}} \leq |C_N|\right\}. \qquad (2.4.13)$$

2.4 Boundary Conditions Expressed via Nonconvex Superpotentials

Now we can consider any type of closed set K in the $\{S_{T_a}, S_{T_b}\}$-space, e.g.

$$K = \{S_{T_a}, S_{T_b}\} | f(S_{T_a}, S_{T_b}) \leq 0 \tag{2.4.14}$$

where f is a C^1-function of S_{T_a}, S_{T_b}. Then instead of (2.4.12) we may write the law

$$-u_T \in N_K(S_{T_a}, S_{T_b}) = \bar{\partial} I_K(S_{T_a}, S_{T_b}). \tag{2.4.15}$$

Then (1.2.49) implies that for $f(S_{T_a}, S_{T_b}) = 0$ a $\lambda \geq 0$ exists such that

$$u_{T_a} = -\lambda \frac{\partial f}{\partial S_{T_a}}, \quad u_{T_b} = -\lambda \frac{\partial f}{\partial S_{T_b}}, \tag{2.4.16}$$

provided that at $\{S_{T_a}, S_{T_b}\}$, grad $f(S_{T_a}, S_{T_b}) \neq 0$. This new friction law generalizes the orthotropic friction law for nonconvex smooth limit friction conditions. We may have also a set K of the form.

$$K = \{\{S_{T_a}, S_{T_b}\} | f_i(S_{T_a}, S_{T_b}) \leq 0 \quad i = 1, \ldots, m\} \tag{2.4.17}$$

where the f_i's are C^1-type functions. Then we proceed exactly as in (1.2.49) introducing the function $\max_i \{f_i(S_{T_a}, S_{T_b})\}$. The reader could try to generalize in this spirit all the anisotropic friction and wear models of elastic frictional contact presented in [Mroz92].

ix) In the framework of heat-conduction problems, nonmonotone boundary conditions between the heat flux vector $q = \{q_i\}$ and the absolute temperature θ on the boundary of a body can be formulated. They read

$$q_i n_i \in \tilde{\beta}(\theta) \tag{2.4.18}$$

where $\tilde{\beta}$ results from $\beta \in L^\infty_{loc}(\mathbb{R})$ as in (1.2.51)÷(1.2.53). More generally, we may assume that $\beta = \beta(\theta(x,t), x, t)$. From Fourier's law,

$$q_i = -k_{ij}\theta_{,j}, \tag{2.4.19}$$

where $k = \{k_{ij}\}$, with $k_{ij} = k_{ji}$, is the thermal conductivity tensor. For an isotropic material, $k_{ij} = k\delta_{ij}$. Thus (2.4.8) can be written in the form

$$-\frac{\partial \theta}{\partial n_k} \in \tilde{\beta}(\theta) = \bar{\partial}\tilde{j}(\theta) \tag{2.4.20}$$

where $\tilde{j}$ is a locally Lipschitz functional corresponding to $\tilde{\beta}$ according to (1.2.55) and $\partial \theta / \partial n_k = k_{ij} n_i \theta_{,j}$.

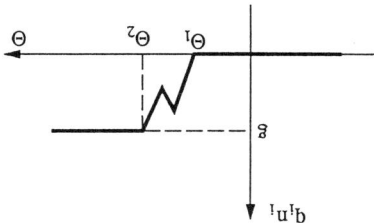

Fig. 2.4.4. Nonmonotone boundary condition in heat conduction problem.

The following example illustrates the boundary condition (2.4.20). The boundary condition corresponding to the graph of Fig. 2.4.4 appears in temperature control problems. The temperature θ on Γ must always remain smaller than θ_1. For $\theta > \theta_1$, heat flux $q_i n_i$ takes place outwards from the body in order to maintain $\theta \leq \theta_1$. This heat flux is a nonmonotone function of θ for $\theta_1 \leq \theta \leq \theta_2$, whereas for $\theta \geq \theta_2$ the heat flux remains constant and equal to g.

2.5 Extensions to Function Spaces

Let us consider first the monotone multivalued boundary conditions. Until now we have considered subdifferential boundary conditions of a pointwise nature i.e. the superpotentials define a relation, at every point of the boundary Γ, between the value of S_N and u_N etc. It is, however, necessary for the formulation and the study of Boundary Value Problems (B.V.Ps) connected with subdifferential boundary conditions to consider extensions of these boundary conditions to function spaces. Roughly speaking, we are concerned with the following important question: if $j(u(x)) \in \Gamma_0(\mathbb{R}^3)$, (cf. Sect.1.1) what will the properties of $\int_\Gamma j(u(x)) d\Gamma$ be?

Assume generally that $(\Gamma, \mathcal{B}, \mu)$ is a positive measure space with $\mu(\Gamma) < \infty$. If $A : H \to \mathcal{P}(H)$ is a multivalued operator and H is a Hilbert space identified with its dual, then we can define the operator $\bar{A}$ on $L^2(\Gamma, H)$ (extension of A to $L^2(\Gamma, H)$) by setting

$$f \in \bar{A}(u) \iff f(x) \in A(u(x)) \quad \mu - \text{a.e. on } \Gamma \tag{2.5.1}$$

It is shown in [Bréz73] that if A is maximal monotone, $\bar{A}$ is too. The following proposition holds [Bréz72]

Proposition 2.5.1 Let $A = \partial \varphi$, where φ is a convex, l.s.c., proper functional on H. For $u \in L^2(\Gamma, H)$, we define the functional

$$\Phi(u) = \begin{cases} \int_\Gamma \varphi(u(x)) d\mu & \text{if } \varphi(u) \in L^1(\Gamma) \\ \infty & \text{otherwise} \end{cases}, \tag{2.5.2}$$

Then Φ is convex, l.s.c., and proper on $L^2(\Gamma, H)$, and $\bar{A} = \partial \Phi$, i.e., $\partial \Phi$ is the extension of $\partial \varphi$ to $L^2(\Gamma, H)$.

For the proof of this proposition we refer to [Bréz72, Pan85]. The proof is based on the equivalence of the following three conditions: for $u, f \in L^2(\Gamma, H)$,

i) $\Phi(v) - \Phi(u) \geq \int_\Gamma (f, v - u)_H d\mu \quad \forall v \in L^2(\Gamma, H);$ \hfill (2.5.3)

ii) $\varphi(v(x)) - \varphi(u(x)) \geq (f(x), v(x) - u(x))_H$ \hfill (2.5.4)

$$\mu - \text{a.e. on } \Gamma, \, \forall v \in L^2(\Gamma, H).$$

iii) there exists $\Gamma_1 \subset \Gamma$ with $\mu(\Gamma_1) = 0$ such that

$$\varphi(v) - \varphi(u(x)) \geq (f(x), v - u(x))_H, \quad \forall x \in \Gamma - \Gamma_1, \forall v \in H. \quad (2.5.5)$$

In order to extend the subdifferential contact boundary conditions we apply Prop. 2.5.1 for the usual Lebesgue measure on Γ and $H = \mathbb{R}^3$, if (2.3.8a) is considered, or $H = \mathbb{R}$ in the case of (2.3.7) etc. For instance, in the case of (2.3.8a) we ask for the extension of $-S(x) \in \partial j(u(x))$ to $[L^2(\Gamma)]^3$. Thus we introduce the functional

$$\Phi(u) = \begin{cases} \int_\Gamma j(u(x)) d\Gamma & \text{if } j(u) \in L^1(\Gamma) \\ \infty & \text{otherwise} \end{cases}, \quad (2.5.6)$$

which is convex, l.s.c. and proper on $[L^2(\Gamma)]^3$ and we apply Prop. 2.5.1. For $u, S \in [L^2(\Gamma)]^3$ the relation

$$-S \in \partial \Phi(u) \quad (2.5.7)$$

holds, if and only if

$$-S(x) \in \partial j(u(x)) \quad \text{a.e. on } \Gamma. \quad (2.5.8)$$

These two relations are obviously equivalent to the inverse relations

$$u \in \partial \Phi^c(-S) \quad (2.5.9)$$

and

$$u(x) \in \partial j^c(-S(x)) \quad \text{a.e. on } \Gamma, \quad (2.5.10)$$

as they are to the relations

$$\Phi(u) + \Phi^c(-S) = -\int_\Gamma u_i(x) S_i(x) d\Gamma \quad (2.5.11)$$

and

$$j(u(x)) + j^c(-S(x)) = -u_i(x) S_i(x) \quad \text{a.e. on } \Gamma, \quad (2.5.12)$$

where

$$\Phi^c(-S) = \begin{cases} \int_\Gamma j^c(-S(x)) d\Gamma & \text{if } j^c(-S) \in L^1(\Gamma) \\ \infty & \text{otherwise} \end{cases}. \quad (2.5.13)$$

In deformable bodies usually the displacement field $u \in [H^1(\Omega)]^3$. Then $u \to \Phi(u|_\Gamma)$ is a convex, l.s.c. and proper functional on $[H^1(\Omega)]^3$ due to the continuity of the trace application $u \to u|_\Gamma$ (the trace is denoted also as γu). Let us denote by $\bar{\Phi}$ the restriction $\Phi|_{[H^{1/2}(\Gamma)]^3}$ of Φ to $[H^{1/2}(\Gamma)]^3$, assuming that $\bar{\Phi}$ on $H^{1/2}(\Gamma)$ is not identically equal to $+\infty$. Then $\partial \Phi$ and $\partial \bar{\Phi}$ define maximal monotone graphs on

$$[L^2(\Gamma)]^3 \times [L^2(\Gamma)]^3 \quad \text{and} \quad [H^{1/2}(\Gamma)]^3 \times [H^{-1/2}(\Gamma)]^3$$

respectively. In many B.V.Ps we shall encounter the boundary condition written for $u \in [H^1(\Omega)]^3$ and $S \in [H^{-1/2}(\Gamma)]^3$ in the form

$$\Phi(\gamma v) - \Phi(\gamma u) \geq \langle -S, \gamma v - \gamma u \rangle, \quad \forall v \in [H^1(\Omega)]^3, \tag{2.5.14}$$

where $\langle .,. \rangle$ denotes the duality pairing between $[H^{1/2}(\Gamma)]^3$ and $[H^{-1/2}(\Gamma)]^3$. Since the trace application $v \to \gamma v$ is surjective from $[H^1(\Omega)]^3$ onto $[H^{1/2}(\Gamma)]^3$, (2.5.14) yields that

$$\bar{\Phi}(v) - \bar{\Phi}(\gamma u) \geq \langle -S, v - \gamma u \rangle, \quad \forall v \in [H^{1/2}(\Gamma)]^3, \tag{2.5.15}$$

which is equivalent to

$$-S \in \partial \bar{\Phi}(\gamma u) \quad \text{or} \quad \gamma u \in \partial \bar{\Phi}^c(-S). \tag{2.5.16}$$

Recall that the duality pairing between $[H^{1/2}(\Gamma)]^3$ and $[H^{-1/2}(\Gamma)]^3$ coincides with the integral $\int_\Gamma S_i(v_i - u_i) d\Gamma$ if $S \in [L^2(\Gamma)]^3$. Henceforth, the bar on Φ and the γ will be omitted if no ambiguity occurs. The expression (2.5.16) is a weak formulation of the boundary condition (2.3.8a); indeed only if $S \in [L^2(\Gamma)]^3$ the complete equivalence is ensured.

Further the weak formulations of some contact boundary conditions will be studied. We denote by H_T the space

$$H_T = \{v | v \in [H^{1/2}(\Gamma)]^3, \; v_i n_i = 0 \text{ a.e. on } \Gamma\} \tag{2.5.17}$$

and we recall ([Pan85] p.32) that if $a = \{a_i\} \in [H^{1/2}(\Gamma)]^3$, and $a_N = a_i n_i$, $a_T = \{a_{T_i}\}$ where $a_{T_i} = a_i - a_N n_i$, then the mapping $a \to \{a_N, a_T\}$ is an isomorphism from $[H^{1/2}(\Gamma)]^3$ onto $H^{1/2}(\Gamma) \times H_T$. In the dual spaces, a'_N and a'_T are uniquely determined by the relation

$$\langle a', a \rangle = \langle a'_N, a_N \rangle_{1/2} + \langle a'_T, a_T \rangle_{H_T} \quad \forall a \in [H^{1/2}(\Gamma)]^3, \tag{2.5.18}$$

where $\langle \cdot, \cdot \rangle_{1/2}$ and $\langle \cdot, \cdot \rangle_{H_T}$ denote the duality pairings on $H^{1/2}(\Gamma) \times H^{-1/2}(\Gamma)$ and $H'_T \times H_T$. Obviously $a' \to \{a'_N, a'_T\}$ is again an isomorphism from $([H^{1/2}(\Gamma)]^3)'$ onto $H^{-1/2}(\Gamma) \times H'_T$. For all the previous results it is sufficient that Γ be $C^{1,1}$-regular ($C^{0,1}$-regularity, i.e. a Lipschitz boundary is also possible [Has82]).

Suppose now that Φ is a proper functional on $[H^{1/2}(\Gamma)]^3$. Φ is called a decomposable energy functional if two proper functionals Φ_N on $H^{1/2}(\Gamma)$ and Φ_T on $H_T(\Gamma)$ can be determined such that

$$\Phi(v) = \Phi_N(v_N) + \Phi_T(v_T), \quad \forall v \in [H^{1/2}(\Gamma)]^3, \tag{2.5.19}$$

where

$$v = \{v_i\}, \quad v_i = v_N n_i + v_{T_i}.$$

Proposition 2.5.2 Suppose that Γ is $C^{1,1}$-regular and that the functional Φ is proper and decomposable on $[H^{1/2}(\Gamma)]^3$. If Φ is convex (resp. l.s.c.), then Φ_N and Φ_T are convex (resp. l.s.c.) as well, and conversely. Moreover,

$$\Phi^c(T) = \Phi_N^c(T_N) + \Phi_T^c(T_T), \quad \forall T \in [H^{-1/2}(\Gamma)]^3, \tag{2.5.20}$$

where
$$\langle T, v \rangle = \langle T_N, v_N \rangle_{1/2} + \langle T_T, v_T \rangle_{H_T}, \quad \forall v \in [H^{1/2}(\Gamma)]^3. \tag{2.5.21}$$

For the proof cf. [Pan85] p.109 and [Hün]. Further some boundary conditions are examined.

i) Suppose now that $C_T \in H'_T$ and $c_N \in H^{1/2}(\Gamma)$ are given and let us consider the boundary conditions
$$u_N = c_N \quad \text{and} \quad S_T = C_T. \tag{2.5.22}$$

We define a functional Φ on $[H^{1/2}(\Gamma)]^3$ by setting
$$\Phi_T(v_T) = -\langle C_T, v_T \rangle_{H_T} \quad \text{for } v_T \in H_T \tag{2.5.23}$$
and
$$\Phi_N(v_N) = \begin{cases} 0 & \text{if } v_N = c_N \\ \infty & \text{if } v_N \neq c_N \end{cases}, \tag{2.5.24}$$

for $v_N \in H^{1/2}(\Gamma)$. Then Φ belongs to the class $\Gamma_0([H^{1/2}(\Gamma)]^3)$ and is decomposable. By Prop. 2.5.2, (2.5.20) holds, and since
$$\Phi_T^c(T_T) = \begin{cases} 0 & \text{if } T_T = -C_T \\ \infty & \text{otherwise} \end{cases}, \tag{2.5.25}$$

for $T_T \in H'_T$ and
$$\Phi_N^c(T_N) = \langle T_N, c_N \rangle_{1/2} \quad \text{for} \quad T_N \in H^{-1/2}(\Gamma), \tag{2.5.26}$$

it results that
$$\Phi^c(T) = \begin{cases} \langle T_N, c_N \rangle_{1/2} & \text{if } T_T = -C_T \\ \infty & \text{otherwise} \end{cases}. \tag{2.5.27}$$

If $C_{T_i} \in L^2(\Gamma)$, then the conditions (2.5.22) are equivalent to the pointwise boundary conditions
$$u_N(x) = c_N(x) \quad \text{and} \quad S_{T_i}(x) = C_{T_i}(x) \quad \text{a.e. on } \Gamma. \tag{2.5.28}$$

ii) Let $C_T \in H'_T$ be given. We study the Signorini boundary condition
$$u_N \leq 0 \quad \text{a.e. on} \quad \Gamma, \quad \langle S_N, u_N \rangle_{1/2} = 0 \quad S_N \leq 0 \quad \text{in } H^{-1/2}(\Gamma) \tag{2.5.29}$$
and
$$S_T = C_T. \tag{2.5.30}$$

Recall that the last inequality in (2.5.29) is by the definition of a non-negative distribution (which is a measure) equivalent to
$$\langle S_N, v_N \rangle_{1/2} \geq 0 \quad \forall v_N \in H^{1/2}(\Gamma) \quad v_N \leq 0 \quad \text{a.e. on } \Gamma. \tag{2.5.31}$$

Further, the functional Φ is defined on $[H^{1/2}(\Gamma)]^3$ by (2.5.19), where for $v_N \in H^{1/2}(\Gamma)$

$$\Phi_N(v_N) = \begin{cases} 0 & \text{if } v_N \leq 0 \quad \text{a.e. on } \Gamma \\ \infty & \text{if } v_N > 0 \quad \text{on } \Gamma_1 \subset \Gamma \text{ with mes } \Gamma_1 > 0 \end{cases} \qquad (2.5.32)$$

and for $v_T \in H_T$

$$\Phi_T(v_T) = -\langle C_T, v_T \rangle_{H_T}. \qquad (2.5.33)$$

Thus we have that

$$\Phi_N^c(-T_N) = \begin{cases} 0 & \text{if } T_N \leq 0 \\ \infty & \text{otherwise} \end{cases} \qquad (2.5.34)$$

for $T_N \in H^{-1/2}(\Gamma)$ and

$$\Phi_T^c(T_T) = \begin{cases} 0 & \text{if } T_T = -C_T \\ \infty & \text{otherwise} \end{cases} \qquad (2.5.35)$$

for T_T in H_T'. $\Phi^c(T)$ results from (2.5.20). If $S_N \in [L^2(\Gamma)]^3$, (2.5.29) is equivalent to the pointwise condition (2.3.14) or (2.3.15).

iii) Let $S_{T_0} \in L^\infty(\Gamma)$ with essinf $S_{T_0} > 0$. We denote by A, A', A_T and A_T' the injections

$$A : H^{1/2}(\Gamma) \to L^2(\Gamma), \quad A' : L^2(\Gamma) \to H^{-1/2}(\Gamma),$$

$$A_T : H_T \to L_T = \{v \in [L^2(\Gamma)]^3, \quad v_i n_i = 0 \quad \text{a.e. on } \Gamma\}$$

$$\text{and} \quad A_T' : L_T \to H_T'.$$

Obviously, $v \to \{v_N, v_T\}$ is an isometry from $[L^2(\Gamma)]^3$ onto $L^2(\Gamma) \times L_T$. Let us now consider the following boundary conditions:

$$\text{if} \quad S_T \in H_T', \quad S_T = A_T' \tilde{S}_T, \quad \tilde{S}_T \in L_T \text{ and } |\tilde{S}_T| < S_{T_0} \qquad (2.5.36)$$

a.e. on Γ, then $u_T = 0$, with $u_T \in H_T$;

$$\text{if} \quad S_T \in H_T', \quad S_T = A_T' \tilde{S}_T, \quad \tilde{S}_T \in L_T \text{ and } |\tilde{S}_T| = S_{T_0}$$

a.e. on Γ, then there exists $\lambda \geq 0$ such that

$$u_T = -\lambda \tilde{S}_T, \quad u_T \in H_T; \qquad (2.5.37)$$

$$S_N = C_N \text{ in } H^{-1/2}(\Gamma). \qquad (2.5.38)$$

We can easily verify that (2.5.37),(2.5.38) are equivalent to

$$-\tilde{S}_T \in \partial F_T(A_T u_T), \quad S_T = A_T' \tilde{S}_T, \qquad (2.5.39)$$

where

$$F_T(v) = \int_\Gamma S_{T_0} |v| d\Gamma \quad \text{for} \quad v \in L_T. \qquad (2.5.40)$$

An equivalent formulation to (2.5.39) reads:

2.5 Extensions to Function Spaces 63

$$S_T = A'_T \tilde{S}_T, \quad |\tilde{S}_T| \leq S_{T_0} \quad \text{a.e. on } \Gamma,$$
$$S_{T_0}|u_T| + \tilde{S}_{T_i} u_{T_i} = 0 \quad \text{a.e. on } \Gamma. \tag{2.5.41}$$

The functional F_T^c is defined on L_T by

$$F_T^c(T) = \begin{cases} 0 & \text{if } \tilde{T} \in L_T, \quad |\tilde{T}| \leq S_{T_0} \text{ a.e. on } \Gamma \\ \infty & \text{otherwise} \end{cases} \tag{2.5.42}$$

The boundary conditions (2.5.36),(2.5.37) can be put in the form (2.5.16) by setting $\Phi_T = F_T \circ A_T$; then

$$\Phi_T^c(T) = \begin{cases} 0 & \text{if } T \in H'_T, \text{ with } T = A'_T\tilde{T}, \tilde{T} \in L_T \\ & \text{and } |\tilde{T}_T| \leq S_{T_0} \quad \text{a.e on } \Gamma, \\ \infty & \text{otherwise} \end{cases} \tag{2.5.43}$$

and thus (2.5.36),(2.5.37) take the form

$$-S_T \in \partial \Phi_T(u_T) \quad \text{or} \quad u_T \in \partial \Phi_T^c(-S_T) \tag{2.5.44}$$

on $H_T \times H'_T$ for $S_T = A'_T \tilde{S}_T$.

Using the special structure of the superpotential we can derive certain regularity results. Thus we can obtain from (2.5.44) that

$$\frac{S_T}{S_{T_0}} \in [L^\infty(\Gamma)]^3 \quad \text{and} \quad \left|\frac{S_T}{S_{T_0}}\right|_{L^\infty} \leq 1. \tag{2.5.45}$$

For the proof see [Duv72]p.140. The same proof is given with slight changes in [Pan85]p.112.

Analogous to the extension of the pointwise monotone multivalued boundary conditions to function spaces in the extension of the pointwise nonmonotone multivalued boundary conditions. Then the following proposition holds.

Proposition 2.5.3 Let H be a separable Hilbert space and $j : \Gamma \times H \to \mathbb{R}$ be a functional such that, (i) $x \to j(x,u)$ is measurable for each u, (ii) $u \to j(x,u)$ is a locally Lipschitz function for each x and (iii) $x \to j(x,0)$ is finitely integrable. Let $p \geq 1$ and $c \geq 0$ such that for every (x,u) and for every $f \in \bar{\partial}_u j(x,u)$ the estimate

$$|f| \leq c(1 + |u|^{p-1}) \tag{2.5.46}$$

holds. Then for $u \in L^p(\Gamma, H)$ the functional $\Phi : L^p(\Gamma, H) \to \mathbb{R}$ defined by

$$\Phi(u) = \int_\Gamma j(x, u(x)) d\Gamma \tag{2.5.47}$$

is Lipschitzian on every bounded subset of $L^{p'}(\Gamma, H)$. Every element $f \in L^{p'}(\Gamma, H)$ of $\bar{\partial} \Phi(u)$ satisfies the relation $(\frac{1}{p} + \frac{1}{p'} = 1)$

$$f(x) \in \bar{\partial}_u j(x, u(x)) \quad \text{a.e on } \Gamma, \tag{2.5.48}$$

and if $j_u^0(x,u,v) = \tilde{j}_u'(x,u,v)$ a.e. on Γ for every $v \in H$ ($\bar{\partial}$-regularity), then the converse is also true.

For the proof of this proposition, the reader is referred to [Aub79] and [Clar83]. The proposition states simply that

$$\bar{\partial}\Phi(u) \subset \int_\Gamma \bar{\partial}_u j(x,u(x)) d\Gamma, \qquad (2.5.49)$$

where the equality holds when j is $\bar{\partial}$-regular. If for instance $H = \mathbb{R}$, then (2.5.49) permits the definition of $\bar{\partial}\Phi(u_N)$ on $L^2(\Gamma)$ and by considering the restriction $\tilde{\Phi}$ of Φ to $H^{1/2}(\Gamma)$ the definition of $\bar{\partial}\tilde{\Phi}(u_N)$. Note that due to a proposition proved by Chang [Ch] the inclusion $\bar{\partial}\tilde{\Phi}(u_N) \subset \bar{\partial}\Phi(u_N)$ holds for every $u_N \in H^{1/2}(\Gamma)$.

To give an example let us place ourselves in the functional framework of the previously examined case iii) of classical friction. A nonmonotone friction law (e.g. the zig-zag law of Fig. 2.4.1d which satisfies the assumption (2.5.46)), can be put in the form

$$-\tilde{S}_T \in \bar{\partial}\tilde{F}(A_T u_T) \quad S_T = A_T' \tilde{S}_T, \qquad (2.5.50)$$

where

$$\tilde{F}(v) = \int_\Gamma j_T(v) d\Gamma \quad \text{for} \quad v \in L_T. \qquad (2.5.51)$$

Here j_T results from a nonmonotone, function $\beta_T \in L_{\text{loc}}^\infty(\mathbb{R})$ by applying the method defined by the relation (1.2.51)÷(1.2.55). Analogously we may proceed in all cases of nonmonotone pointwise boundary conditions. Note that in friction problems of quasistatic or dynamic nature, in the aforementioned relations u_T repreents the tangential boundary velocity.

3. Nonsmooth Mechanics II

This chapter constitutes a continuation of the previous chapter. First we deal with material laws derived by convex superpotentials and then with material laws derived by nonconvex superpotentials. Special attention is paid to the loading and unloading problem. After a short illustration of the advantages of the use of superpotentials we introduce material laws and boundary conditions expressed by means of the new notions of fans, quasidifferentials and codifferentials. For further reading on these subjects the reader may consult the following books and papers: [Pan85,87b,88,92], [Ant92,] [Mor88a,b], [Hal74,75], [Léné73,74], [Stav91,93a,b].

3.1 Material Laws Expressed via Convex Superpotentials. An Overview

In the previous chapter we considered monotone and nonmonotone possibly multivalued laws holding on the boundary of a deformable body between the traction vector $-S$, and the displacement (resp. velocity) vector u (resp. v) or between the normal (resp. tangential) traction $-S_N$ (resp. $-S_T$) and the corresponding displacements or velocities u_N or v_N (resp. u_T or v_T). Analogous relations can be considered between the stress tensor $\sigma = \{\sigma_{ij}\}$ and strain tensor $\varepsilon = \{\varepsilon_{ij}\}$ in a small deformation theory (after geometric linearization cf. Sect. 2.1) or between the Cauchy stress tensor $\sigma = \{\sigma_{ij}\}$ and the rate of deformation tensor $D = \{D_{ij}\}$ or between any other type of stress tensor (e.g. the Piola-Kirchhoff stress tensor) and the corresponding strain tensor. For instance, for $w \in \Gamma_0(\mathbb{R}^6)$, i.e. for w convex, l.s.c and proper on $\mathbb{R}^6$, we introduce, in the framework of small deformation theory, the superpotential law

$$\sigma \in \partial w(\varepsilon) \qquad (3.1.1)$$

whereas for general flow problems the law (Eulerian description)

$$\sigma \in \partial w(D) \qquad (3.1.2)$$

where $\sigma = \{\sigma_{ij}\}$ is the Cauchy stress tensor. Note that in the case of small displacements and small displacement gradients we may write that (cf. e.g. [Mroz73])

$$D_{ij} = \frac{\partial \varepsilon_{ij}}{\partial t} = \frac{\partial E_{ij}}{\partial t}. \tag{3.1.3}$$

Let us use the notation $\frac{\partial \varepsilon_{ij}}{\partial t} = \dot{\varepsilon}_{ij}$. Then we may consider the law

$$\sigma \in \partial w(\dot{\varepsilon}) \tag{3.1.4}$$

instead of (3.1.2). The law (3.1.1) can also be written in the inverse form

$$\varepsilon \in \partial w^c(\sigma). \tag{3.1.5}$$

Functional w (resp. w^c) is the potential (resp. the complementary) energy per unit volume, and satisfies for $\varepsilon \in \mathbb{R}^6$ (resp. for $\sigma \in \mathbb{R}^6$), the variational inequality

$$w(\varepsilon^\star) - w(\varepsilon) \geq \sigma_{ij}(\varepsilon^\star_{ij} - \varepsilon_{ij}), \quad \forall \varepsilon^\star \in \mathbb{R}^6, \tag{3.1.6}$$

resp.

$$w^c(\sigma^0) - w^c(\sigma) \geq \varepsilon_{ij}(\sigma^0_{ij} - \sigma_{ij}), \quad \forall \sigma^0 \in \mathbb{R}^6. \tag{3.1.7}$$

Along the lines of the previous chapter w and w^c can be seen as local superpotentials which define at every point of the body the law (3.1.1) (or (3.1.4)). Obviously we may define the corresponding global forms or $L^2(\Omega)$-extensions of the above convex superpotential laws according to Prop. 2.5.1 (Γ will be replaced by Ω).

In this chapter we will not insist too much on the convex superpotentials for which we refer mainly to [Pan85]. We shall only give certain important applications of the convex superpotentials.

i) Elastic ideally locking materials: let us consider a functional w defined on $\mathbb{R}^6$ by [Léné74]

$$w(\varepsilon) = w_0(\varepsilon) + I_K(\varepsilon), \quad \varepsilon = \{\varepsilon_{ij}\}, \tag{3.1.8}$$

where w_0 is a continuously differentiable convex functional and I_K is the indicator of the convex closed subset of $\mathbb{R}^6$

$$K = \{\varepsilon | Q(\varepsilon) \leq 0\}. \tag{3.1.9}$$

Q is a convex, continuously differentiable functional on $\mathbb{R}^6$ such that $0 \in K$. From (3.1.1) and (3.1.8) we obtain that (cf. Prop. 1.1.6)

$$\sigma_{ij} \in [\partial w(\varepsilon)]_{ij} = \frac{\partial w_0(\varepsilon)}{\partial \varepsilon_{ij}} + \partial I_K(\varepsilon), \tag{3.1.10}$$

or

$$\sigma_{ij} = \frac{\partial w_0(\varepsilon)}{\partial \varepsilon_{ij}} + \bar{\sigma}_{ij}, \tag{3.1.11}$$

$$\bar{\sigma}_{ij}(\varepsilon^\star_{ij} - \varepsilon_{ij}) \leq 0 \quad \text{for} \quad \varepsilon \in K \quad \text{and} \quad \forall \varepsilon^\star \in K.$$

Then $\bar{\sigma} = \{\bar{\sigma}_{ij}\}$ is an element of the outward normal cone to K at the point ε, and thus (3.1.11) may be written in the equivalent form

3.1 Material Laws Expressed via Convex Superpotentials. An Overview

$$\sigma_{ij} = \frac{\partial w_0(\varepsilon)}{\partial \varepsilon_{ij}} + \mu \frac{\partial Q(\varepsilon)}{\partial \varepsilon_{ij}}, \quad \mu \geq 0, \quad \mu Q(\varepsilon) = 0, \quad Q(\varepsilon) \leq 0. \quad (3.1.12)$$

Accordingly, if $Q(\varepsilon) < 0$, then $\mu = 0$, and the material behaves like a nonlinear elastic material. If $Q(\varepsilon) = 0$ then no finite or infinite increment of the stresses can cause an increase in the value of the function $Q(\varepsilon)$. This is called an ideal-locking effect [Prag57,58]. Two possible forms for $Q(\varepsilon)$ are [Prag58]

$$Q(\varepsilon) = \frac{1}{2}\varepsilon_{ij}^D \varepsilon_{ij}^D - k^2, \quad (3.1.13)$$

where $\varepsilon^D = \{\varepsilon_{ij}^D\}$ is the strain deviator, and

$$Q(\varepsilon) = (\varepsilon_{ii} - k^2) \quad (3.1.14)$$

In both cases, k is a positive material constant. Materials subjected to the locking criterion defined by (3.1.14) are called materials of limited compressibility. The behaviour of rubber as well as of some other types of plastic materials can be described fairly well by the stress-strain law of locking materials (cf. Fig. 3.1.1).

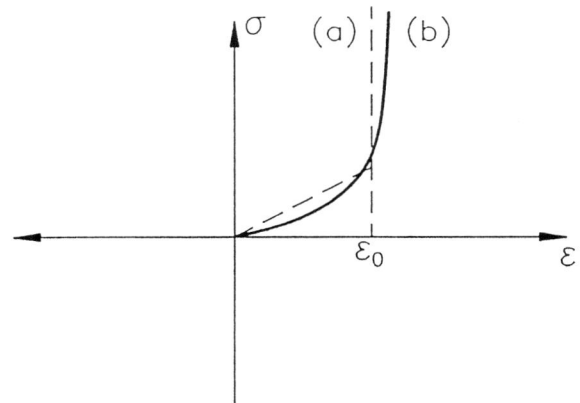

Fig. 3.1.1. The onedimensional idealized (a) and real (b) stress-strain law for rubber. Here $Q(\varepsilon) = \varepsilon - \varepsilon_0 \leq 0$.

ii) Perfectly plastic and elastic perfectly plastic materials: we define a convex, closed subset of $\mathbb{R}^6$, $K = \{\sigma | F(\sigma) \leq 0\}$ such that $0 \in K$, where F is a continuously differentiable function called the yield function of the material. Let us consider the law

$$\dot{\varepsilon}^P \in \partial I_K(\sigma), \quad (3.1.15)$$

which can also be written in the form (cf. the derivation of (3.1.12))

$$\dot{\varepsilon}_{ij}^P = \lambda \frac{\partial F}{\partial \sigma_{ij}}, \quad \lambda \geq 0, \quad F(\sigma) \leq 0, \quad \lambda F(\sigma) = 0 \quad (3.1.16)$$

68   3. Nonsmooth Mechanics II

In Fig. 3.1.2 we give the geometric characterization of $\dot{\varepsilon}^P$ which, due to (3.1.15), is an element of the outward normal cone at the point σ of the yield surface. If the boundary of the yield surface has a corner, e.g. if $K = \{\sigma | F_k(\sigma) \leq 0, \ k = 1, 2, \ldots, m\}$ then at the corner

$$\dot{\varepsilon}_{ij}^P = \sum_{k=1}^{m} \lambda_k \frac{\partial F_k}{\partial \sigma_{ij}}(\sigma), \quad \lambda_k \geq 0, \quad F_k(\sigma) \leq 0, \quad \lambda_k F_k(\sigma) = 0, \quad (3.1.17)$$

$$k = 1, \ldots, m.$$

By definition (3.1.15) is equivalent to the inequality

$$\sigma \in K, \quad \dot{\varepsilon}_{ij}^P(\sigma_{ij}^0 - \sigma_{ij}) \leq 0 \quad \forall \sigma^0 \in K \tag{3.1.18}$$

which is the well-known Drucker's stability postulate [Dru] (cf. also Hill's principle of maximum plastic work [Reck] [Hill48,50]). The Prandtl-Reuss relations for an elastic perfectly plastic material can be written in the form

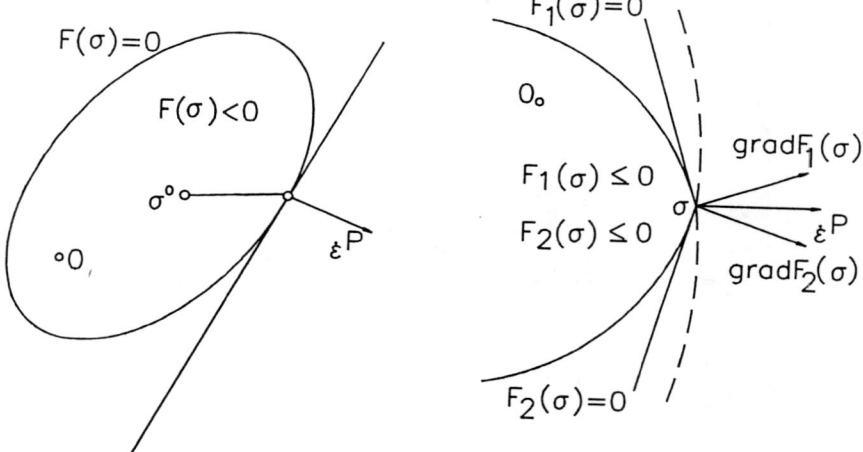

Fig. 3.1.2. On the geometric interpretation of $\dot{\varepsilon}^P$.

$$\dot{\varepsilon} = \dot{\varepsilon}^E + \dot{\varepsilon}^P \in c\dot{\sigma} + \partial I_K(\sigma) \tag{3.1.19}$$

where

$$c\dot{\sigma} = \{c_{ijhk} \frac{\partial \sigma_{hk}}{\partial t}\} \quad i, j, h, k = 1, 2, 3 \tag{3.1.20}$$

and $c = \{c_{ijhk}\}$ is the inverse Hooke's elasticity tensor. Note at this point that the Hooke's elasticity tensor is denoted by $C = \{C_{ijkh}\} \ i, j, h, k = 1, 2, 3$ and relates the stress tensor to the strain tensor by the relation

$$\sigma_{ij} = C_{ijhk} \varepsilon_{hk}. \tag{3.1.21}$$

3.1 Material Laws Expressed via Convex Superpotentials. An Overview

It is assumed that the components of the tensor $C = \{C_{ijhk}\}$, $i,j,h,k = 1,2,3$, have the symmetry property

$$C_{ijhk} = C_{jihk} = C_{khij} \tag{3.1.22a}$$

and the ellipticity property

$$C_{ijhk}\varepsilon_{ij}\varepsilon_{hk} \geq c\varepsilon_{ij}\varepsilon_{hk}, \tag{3.1.22b}$$

$$\forall \varepsilon = \{\varepsilon_{ij}\} \in \mathbb{R}^6, \quad c \text{ const} > 0.$$

Coming back to (3.1.19) we may write it in the form

$$\dot{\varepsilon}_{ij} = \dot{\varepsilon}_{ij}^E + \dot{\varepsilon}_{ij}^P = c_{ijhk}\dot{\sigma}_{hk} + \lambda\frac{\partial F(\sigma)}{\partial \sigma_{ij}} \tag{3.1.23}$$

$$\lambda \geq 0, \quad F(\sigma) \leq 0, \quad \lambda F(\sigma) = 0.$$

In order to include the case of unloading the laws (3.1.16) or (3.1.23) must be completed as it follows

$$\lambda = 0, \quad \text{if} \quad F(\sigma) < 0 \quad \text{or if} \quad F(\sigma) = 0 \quad \text{and} \quad \dot{F}(\sigma) < 0 \tag{3.1.24}$$

$$\lambda \geq 0, \quad \text{if} \quad F(\sigma) = 0 \quad \text{and} \quad \dot{F}(\sigma) = 0. \tag{3.1.25}$$

The yield surface may change during the loading process due to hardening effects. In this case we assume that the yield surface has the form $F(\sigma, h(\dot{\varepsilon}^P), A) \leq 0$ where $h(\cdot)$ is generally a tensor function and A is a scalar. Note that in the case of rigid ideally plastic flow problems (Eulerian description) the law $D \in \partial I_K(\sigma)$ holds instead of (3.1.15).

iii) Polygonal stress-strain laws. These laws appear in many engineering problems either as nonlinear elastic analoga of plasticity laws, or as simplifications of nonlinear elastic laws. For instance the laws of Fig. 3.1.3a,b are simplifications of the laws depicted with dotted lines in the same figures. Moreover the law of Fig. 3.1.3b is the elastic analogon of rigid plastic hardening material.

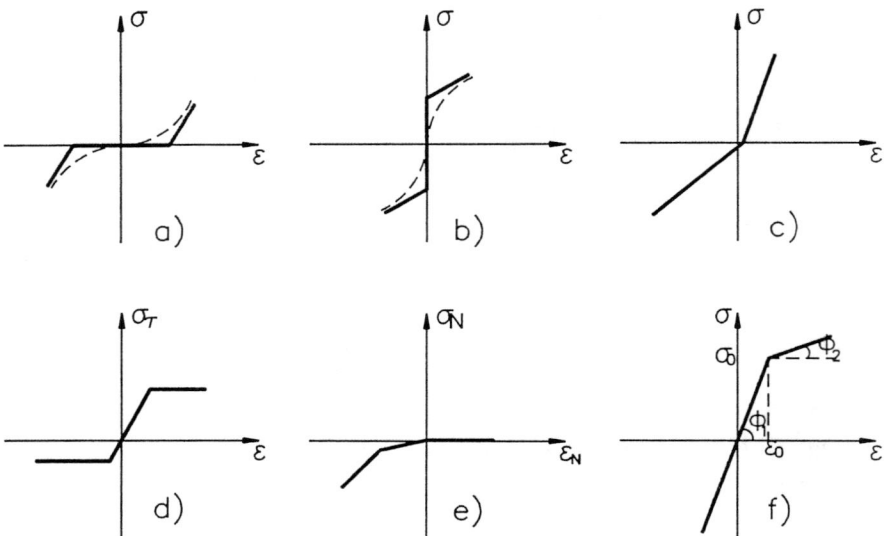

Fig. 3.1.3. Polygonal stress-strain laws. The onedimensional case.

The bimodulus elasticity law of Fig. 3.1.3c appears in fiber-reinforced materials which have different behaviour in compression and in tension due to the buckling of the fibers. The stress-strain laws of 3.1.3d,e represent the tangential and normal behaviour of a nondilatant-joint element in rock mechanics [Gha].

In order to present a generalization to the three dimensions of the above laws let us consider Fig. 3.1.3f. If $K = \{\varepsilon | \varepsilon \leq \varepsilon_0\}$ and $P_K(\varepsilon) = \{\varepsilon$ if $\varepsilon \in K$, ε_0 if $\varepsilon \notin K\}$ denotes the projection operator of $\mathbb{R}$ onto the convex set K, then the diagram of Fig. 3.1.3a can be written in the form

$$\sigma = \varepsilon \, \mathrm{tg}\phi_1 + (I - P_K)(\varepsilon)(\mathrm{tg}\phi_2 - \mathrm{tg}\phi_1), \quad (3.1.26)$$

where I denotes the identity operator. In the threedimensional case, K is a closed convex subset of the space $\mathbb{R}^6$ of the strain tensors $\varepsilon = \{\varepsilon_{ij}\}$, zero is an element of K and P_K is the projection operator[1] of $\mathbb{R}^6$ onto K. Then the superpotential

$$w(\varepsilon) = \frac{1}{2}C_{ijhk}\varepsilon_{ij}\varepsilon_{hk} + \frac{1}{2\mu}(\varepsilon_{ij} - [P_K(\varepsilon)]_{ij})(\varepsilon_{ij} - [P_K(\varepsilon)]_{ij}), \quad \mu > 0 \quad (3.1.27)$$

is considered and from this the law

---
[1]Let K be a convex closed subset of a Hilbert space X. Then the projection mapping P_K is defined by the unique solution of the problem (cf. Prop. 1.3.1) $||y - P_K y|| = \inf_z \{||y - z|| \, \big| \, z \in K\}$, $y \in X$.

3.1 Material Laws Expressed via Convex Superpotentials. An Overview

$$\sigma_{ij} = C_{ijhk}\varepsilon_{hk} + \frac{1}{\mu}[(I - P_K)\varepsilon]_{ij} \tag{3.1.28}$$

is obtained. Similarly, we may write the bimodulus elasticity law for a three-dimensional body. Let us write now (3.1.26) in the form

$$\varepsilon = \frac{1}{\text{tg}\phi_1}\sigma + (I - P_{\bar{K}})\sigma\Big(\frac{1}{\text{tg}\phi_2} - \frac{1}{\text{tg}\phi_1}\Big), \tag{3.1.29}$$

where $\bar{K} = \{\sigma \big| \sigma \leq \sigma_0\}$. This leads to a complementary energy density of the general form

$$w^c(\sigma) = \frac{1}{2}c_{ijhk}\sigma_{ij}\sigma_{hk} + \frac{1}{2\bar{\mu}}(\sigma_{ij} - [P_{\bar{R}}(\sigma)]_{ij})(\sigma_{ij} - [P_{\bar{R}}(\sigma)]_{ij}), \quad \bar{\mu} > 0, \tag{3.1.30}$$

where $c = \{c_{ijhk}\}$ is the inverse Hooke's tensor. $\bar{K}$ is a closed convex set of the space $\mathbb{R}^6$ of the stress tensor σ, and $\bar{\mu} > 0$ is a constant. $\bar{K}$ and $\bar{\mu}$ must be compatible with K and μ. If $\bar{\mu} \to 0$, the material law

$$\varepsilon \in \partial w^c(\sigma) = \partial\Big(\frac{1}{2}c_{ijhk}\sigma_{ij}\sigma_{hk} + I_{\bar{K}}(\sigma)\Big) \tag{3.1.31}$$

is obtained which generalizes for a threedimensional continuum the material law depicted in Fig. 3.1.3d. Indeed, if $\bar{K} = \{\sigma \big| F(\sigma) \leq 0\}$, where $F : \mathbb{R}^6 \to \mathbb{R}$ is continuously differentiable, then (3.1.31) becomes

$$\varepsilon_{ij} = c_{ijhk}\sigma_{hk} + \lambda\frac{\partial F}{\partial \sigma_{ij}}, \quad \lambda \geq 0, \quad F(\sigma) \leq 0, \quad \lambda F(\sigma) = 0. \tag{3.1.32}$$

This material is the elastic analogon of an elastic, perfectly plastic material. It is also called a "holonomic" elastic perfectly plastic material, since (3.1.32) does not relate the stress with the strain rates. If $c = \{c_{ijhk}\} = 0$, we obtain the stress-strain law of a Hencky-material [Henck]. The threedimensional expressions of the polygonal laws of Figs. 3.1.3b,a are obtained from (3.1.30) and (3.1.27) respectively by disregarding the elastic parts, i.e.,

$$\varepsilon_{ij} = \frac{1}{\bar{\mu}}(\sigma_{ij} - [P_{\bar{R}}(\sigma)]_{ij}) \quad \text{and} \quad \sigma_{ij} = \frac{1}{\mu}(\varepsilon_{ij} - [P_K(\varepsilon)]_{ij}). \tag{3.1.33}$$

Note that threedimensional generalizations of polygonal laws with more than one vertex are obtained as before by considering more than one convex sets and the corresponding projections.

iv) Rigid viscoplastic materials: keeping in mind the method followed in iii), and with the notation used there, we may consider the superpotential

$$J_\mu(\sigma) = \frac{1}{4\mu}|\sigma - P_K(\sigma)|^2, \quad \mu > 0, \tag{3.1.34}$$

where $K = \{\sigma \big| F(\sigma) \leq 0\}$ is a convex closed subset of $\mathbb{R}^6$ such that $0 \in K$ and such that F is continuously differentiable. The constant μ is called the viscosity coefficient. The material law

$$D \in \partial J_\mu(\sigma) \tag{3.1.35}$$

is the threedimensional generalization of the law of Fig. 3.1.4b; it describes a rigid-viscoplastic material, called also a Bingham fluid, and it reads [Duv72]

$$D = \frac{1}{2\mu}(\sigma - P_K(\sigma)). \tag{3.1.36}$$

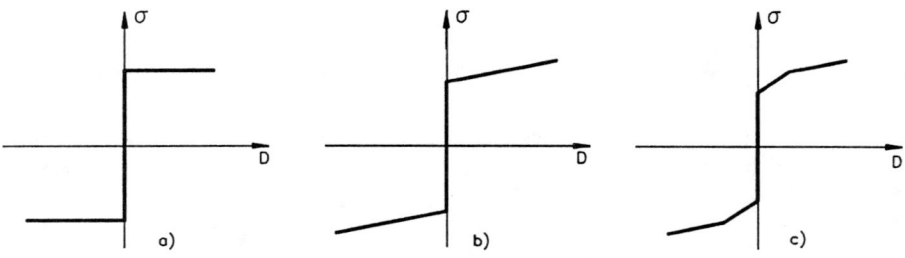

Fig. 3.1.4. Viscoplastic materials. Onedimensional behaviour
a) rigid perfectly plastic b) rigid viscoplastic (Bingham's fluid)
c) biviscous substance

For other types of material laws expressed by means of convex superpotentials (deformation theory of plasticity, certain types of viscous and viscoplastic materials etc.) we refer to [Pan85], where incremental superpotential laws of the type $\dot{\sigma} \in \partial w(\dot{\varepsilon})$ are also examined. Note finally that in general the superpotentials w and w^c may depend on both the space variables x_i and on time t.

3.2 Material Laws Expressed via Nonconvex Superpotentials I

In this Section we consider laws of the type

$$\text{i)} \quad \sigma \in \bar{\partial} w(\varepsilon), \quad \text{ii)} \quad \sigma \in \bar{\partial} w(D) \quad \text{or} \quad \text{iii)} \quad \sigma \in \bar{\partial} w(\dot{\varepsilon}) \tag{3.2.1}$$

or

$$\text{i)} \quad \varepsilon \in \bar{\partial} \tilde{w}(\sigma), \quad \text{ii)} \quad D \in \bar{\partial} \tilde{w}(\sigma) \quad \text{or} \quad \text{iii)} \quad \dot{\varepsilon} \in \bar{\partial} \tilde{w}(\sigma) \tag{3.2.2}$$

where $\bar{\partial}$ is the generalized gradient and $w, \tilde{w}$ are locally Lipschitz functionals on $\mathbb{R}^6$. Then the generalized gradient is defined by (1.2.27) (generalized gradient in the sense of Clarke). More generally we can assume that w and $\tilde{w}$ are not locally Lipschitz but are defined on $\mathbb{R}^6$ and take values in $[-\infty, +\infty]$. In this case $\bar{\partial}$ is defined by the relation (1.2.31) (generalized gradient in the sense of Rockafellar).

i) As a first example let us assume that w in the first relation of (3.2.1) is continuously differentiable. Then we obtain the law (cf. Fig. 3.2.1)

$$\sigma = \operatorname{grad} w(\varepsilon) \qquad (3.2.3)$$

corresponding to a nonlinear elastic material

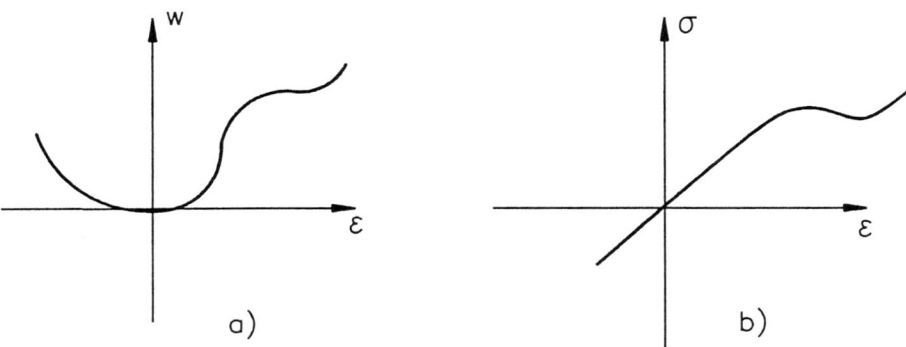

Fig. 3.2.1. A nonlinear elastic material. The onedimensional case.

ii) There is a large number of stress strain laws in engineering problems which may be expressed by means of nonconvex superpotentials. All these laws in their onedimensional form are nonmonotone and generally multivalued. Let us give some examples.

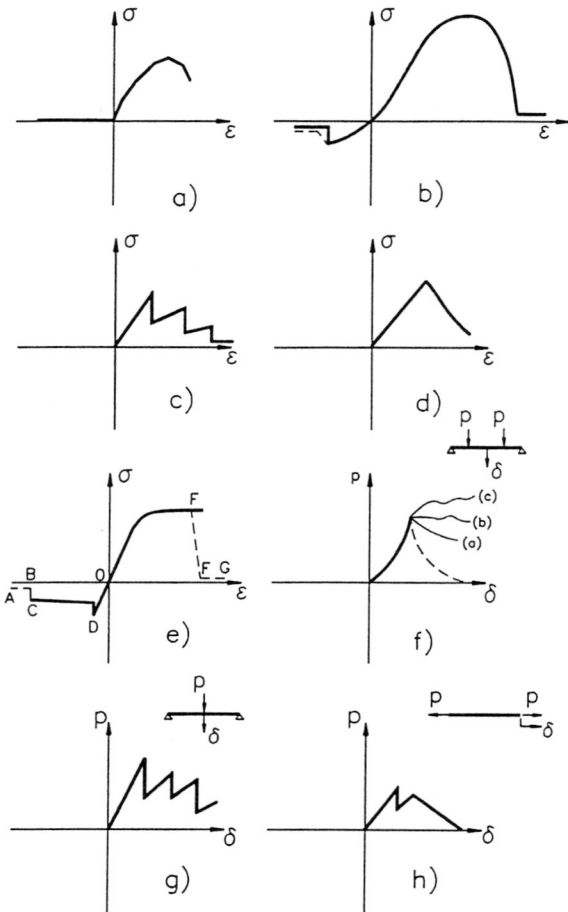

Fig. 3.2.2. Onedimensional stress-strain laws resulting from nonconvex superpotentials.

The law of Fig. 3.2.2a results by performing an experiment of uniaxial monotonic loading of a granular medium, for instance of soils or rocks. The nonmonotone and multivalued behaviour is a result of localized crack formations. The same holds for the diagram of Fig. 3.2.2b which results in uniaxial tests in concrete speciments [Groot]. The diagram of Fig. 3.2.2c describes the tension-stiffening effect of the reinforced concrete in tension [Flo] whereas the diagram of Fig. 3.2.2d describes the behaviour of a bond element between reinforcement and concrete [Groot]. The diagram of Fig. 3.2.2e may be applied for the calculation of beton which is reinforced by steel fibers according to the corresponding regulations [Dbv]. The graph CDOE (CDO in tension, OE in compression), which is proposed in [Dbv] p.23, can be completed by the parts CBA and EFG (dotted lines). Then we are able to calculate the structure for any given loading even if some elements of the structure will be completely damaged due to compression or tension and they will have strains on FG or on BA respectively. Moreover some elements may have stresses and strains on the branches EF or BC which correspond to the partial damage of the structural element under considerations. These branches are vertical (resp. nonvertical) in the case of brittle (resp. semibrittle) behaviour. The diagram of Fig. 3.2.2e results from the experimental diagram of Fig. 3.2.2f ([Dbv] p.11). Here the dotted line corresponds to a beton without fibers and the curves (a) (b) and (c) to different types of steel fiber reinforced concrete. The diagram of Fig. 3.2.2g is observed in experiments concerning the loading of glass-fiber epoxid-crossply-laminates [Moys] and the diagram of Fig. 3.2.2h in experiments concerning the pull-out test of a metallic glass ribbon from a matrix [Fels, Ond]. Note that in all the above stress-strain laws we may have stress-strain states on the vertical parts of the diagrams. These states correspond to localized partial cracking and crushing effects.

All the above onedimensional laws can be put according to (1.2.54), (1.2.55) in the general form (3.2.1i) and, as we shall see further, can be extended for threedimensional continua.

iii) Nonconvex locking criterion. Let L be a closed subset of the strain space $\mathbb{R}^6$ and let $I_L(\varepsilon)$ be the indicator of L. Then the relation

$$\sigma - C\varepsilon \in \bar{\partial} I_L(\varepsilon) = N_L(\varepsilon) \tag{3.2.4}$$

generalizes (3.1.10) (where we put $w_0(\varepsilon) = \frac{1}{2}(C\varepsilon, \varepsilon) = \frac{1}{2}C_{ijhk}\varepsilon_{ij}\varepsilon_{hk}$) for a locking criterion defined by a closed but generally nonconvex surface in the strain space. We recall that if L is given by a set of inequalities, i.e. $L = \{\varepsilon \big| \phi_i(\varepsilon) \leq 0 \ i = 1, \ldots, m\}$, where ϕ_i are continuously differentiable, then $N_L(\varepsilon)$ is given by (1.2.49) if $\varepsilon \in L$, provided that $0 \notin \bar{\partial}\{\max_i \phi_i(\varepsilon)\}$, and $N_L(\varepsilon) = \emptyset$ if $\varepsilon \notin L$. In this context cf. also the cases v) and vi) in Sect. 2.4.

As we can easily verify from the onedimensional case, the stress can take at the locking criterion boundary any arbitrarily large value without any change of the strain state. This phenomenon can be considered with respect to any law by means of the addition of the term $\bar{\partial} I_L(\varepsilon)$. Therefore the law

$$\sigma \in \bar{\partial} w(\varepsilon) + \bar{\partial} I_L(\varepsilon) \tag{3.2.5}$$

describes the behaviour of a material obeying to both the nonmonotone multivalued law $\sigma \in \bar{\partial} w(\varepsilon)$ and to the locking criterion defined by the closed set L made arbitrarily large.

iv) Nonconvex yield surfaces. There are cases in the theory of plasticity where the yield surface is nonconvex. This is the case if there exists an interaction between elastic and plastic deformation, especially if the elastic properties vary with the plastic deformation. It has been experimentally verified [Zhuk] (see also [Ilyu], [Olsz], [Palm]) that in such a case the yield surface may be locally concave. Note at this point that thermodynamic considerations do permit a star-shaped and therefore nonconvex yield surface [Gre]. Such star-shaped admissible surfaces for the stresses have been considered in soil mechanics [Sal]. A material law of the form

$$\varepsilon - c\sigma \in \bar{\partial} I_K(\sigma) \tag{3.2.6}$$

where K is a closed subset of the stress space, describes within the framework of a holonomic theory of plasticity, an elastoplastic material but with a nonconvex yield function. Relation (3.2.6) generalizes the holonomic elastic perfectly plastic law (3.1.32) when the yield criterion is nonconvex. If $K = \{\sigma | F(\sigma) \leq 0\}$ and F is continuously differentiable, then (3.2.6) implies that (cf. (1.2.49))

$$\varepsilon_{ij} - c_{ijhk}\sigma_{hk} = \lambda \left(\frac{\partial F}{\partial \sigma_{ij}} \right), \tag{3.2.7}$$

$$\lambda \geq 0, \ F(\sigma) \leq 0, \quad \lambda F(\sigma) = 0.$$

Relations (3.2.7) hold if at σ, $\text{grad} F(\sigma) \neq 0$. In the case where the yield criterion is defined by $K = \{\sigma | F_i(\sigma) \leq 0 \ i = 1, \ldots, m\}$, where the F_i's are continuously differentiable, we may directly apply the formula (1.2.49).

In the framework of flow theory of plasticity (3.2.6) becomes

$$\dot{\varepsilon}^P = \dot{\varepsilon} - c\dot{\sigma} \in \bar{\partial} I_K(\sigma) \tag{3.2.8}$$

The law (3.2.8) generalizes (3.1.15) for nonconvex yield surfaces. We note again that $\dot{\varepsilon}^P$ is normal to the boundary of the yield surface and that (3.1.17) still holds even if K is nonconvex.

v) Let w be a nonconvex locally Lipschitz functional defined on the stress space $\mathbb{R}^6$. Then we can consider material laws of the form

$$\dot{\varepsilon}^P \in \bar{\partial} w(\sigma) \tag{3.2.9}$$

For their derivation we refer to [Pan85] p.154.

vi) Fuzzy material laws: Here we apply the formulas (2.2.14), (2.2.15) in order to derscribe componentwise fuzziness in material laws. For instance suppose that a material obeying the general law

$$\sigma \in \bar{\partial} w(\varepsilon) \tag{3.2.10}$$

exhibits the following type of fuzzy behaviour for $-\delta < \varepsilon_{ij} < +\delta$, $i, j = 1, 2, 3$. Each stress tensor component, say the σ_{ij}, may take any value in the interval $[\sigma_{ij} - \bar{\bar{b}}_{ij}, \sigma_{ij} + \bar{b}_{ij}]$ where σ_{ij} satisfies (3.2.10) (cf. Fig. 3.2.3).

3.2 Material Laws Expressed via Nonconvex Superpotentials I 77

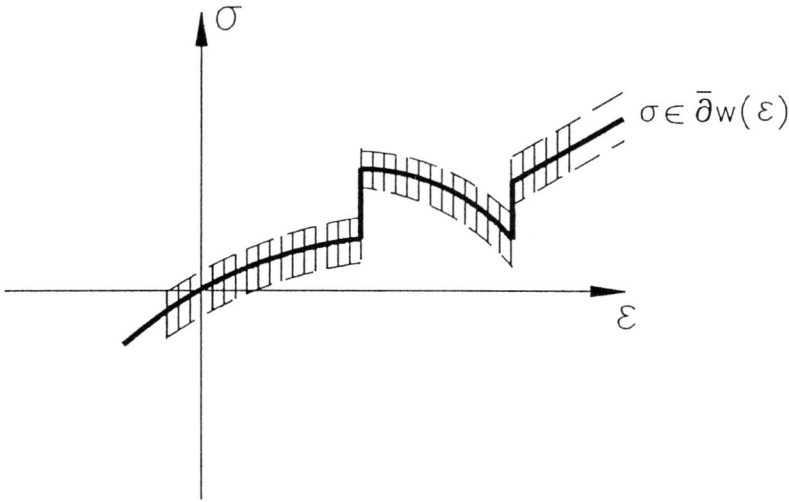

Fig. 3.2.3. A onedimensional fuzzy stress-strain law.

Thus we may write this fuzzy law in the form

$$\sigma_{ij} \in [\bar{\partial}w(\varepsilon)]_{ij} + \bar{\partial}f_{ij}(\varepsilon_{ij}) \qquad (3.2.11)$$

Here $[\,\cdot\,]_{ij}$ denotes the ij-component of $\bar{\partial}w(\varepsilon)$ and $\bar{\partial}f_{ij}(\varepsilon_{ij}) = [-\bar{\bar{b}}_{ij}, \bar{b}_{ij}]$ is defined by relations similar to (2.2.14), (2.2.15). Note that the upper and lower bounds $\bar{b}_{ij}$ and $\bar{\bar{b}}_{ij}$ may depend on ε_{ij} and δ may be different for each strain component.

vii) Complete Laws, Composite Materials and Threedimensional Generalizations: In order to study the behaviour of a structure under a given loading we may consider stress-strain diagrams for each element defined along the whole length of the strain axis (Fig. 3.2.4a) [Ban85,87][Pan88]. Even, if only the final values ε_0 and ε'_0 are known, we may render the respective law complete by assuming a brittle (vertical lines) or a semibrittle behaviour beyond ε_0 and ε'_0. Accordingly, a complete law has as main advantage the consideration of several nonlinearities like cracking, crushing, delamination, slip etc. using a static or a quasi-static method. This approach may be applied to composite material structures, because it leads to an estimation of the behaviour of the structure through purely static methods, as it happens in several engineering theories, e.g. in the calculation of reinforced concrete structures etc. This is only possible through the use of superpotentials, since the sawtooth form is predominant in the stress-strain curves obtained experimentally in composite materials. The diagrams of Fig. 3.2.4b,c describe the progressive, almost ideally brittle failure of the fibres in a fibre reinforced material.

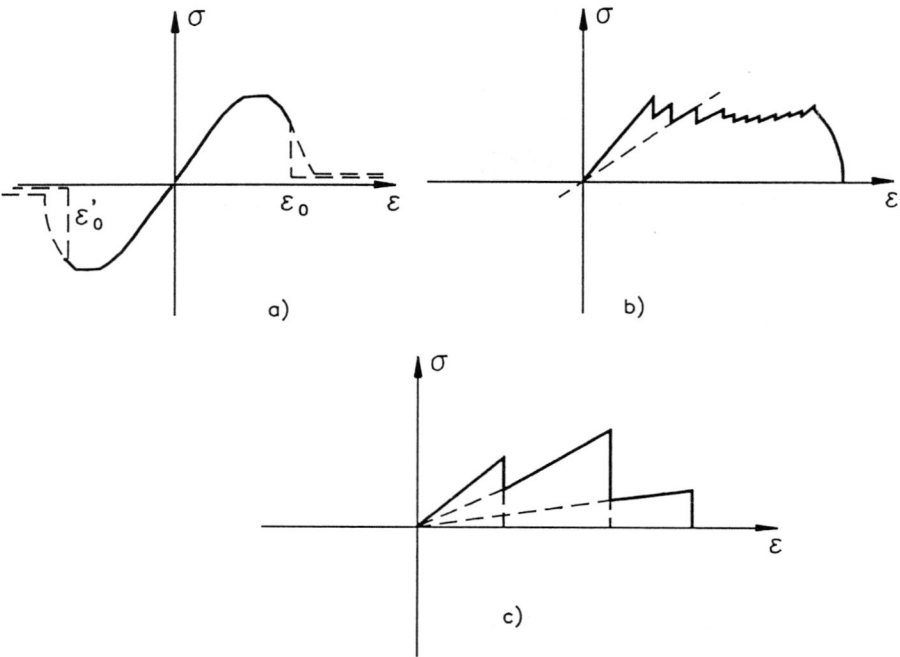

Fig. 3.2.4. Complete diagrams. Sawtooth laws for composite materials.

In the peaks of the diagram fibre failure happens; the jumps (which are complete) correspond to the load transfer to the rest of the fibres. The upward parts are due to the strength recovery after the redistribution of the stresses between fibres and matrix. For other diagrams of this kind and their interpretation the reader is referred to [Ond] [Hult] [Schw] [Ban87] [Willi82,89].

It is obvious, that all the aforementioned onedimensional phenomenological laws can be written through a nonconvex superpotential in the form $\sigma \in \bar{\partial} w(\varepsilon)$, where w denotes the area between the graph and the horizontal axis on which the strains are measured.

Generalizations of the above onedimensional examples for threedimensional continua can be constructed in analogy to the ideally plastic and locking materials. Let us e.g. construct a threedimensional analogon to the onedimensional stress-strain diagram of Fig. 3.2.4c. One method would be to consider four elastic energy surfaces which intersect each other. The re-entrant corners cause the filled-in stress jumps. Moreover the last energy surface should guarantee that $\sigma = 0$ after the last jump (cf. Fig. 3.2.5a).

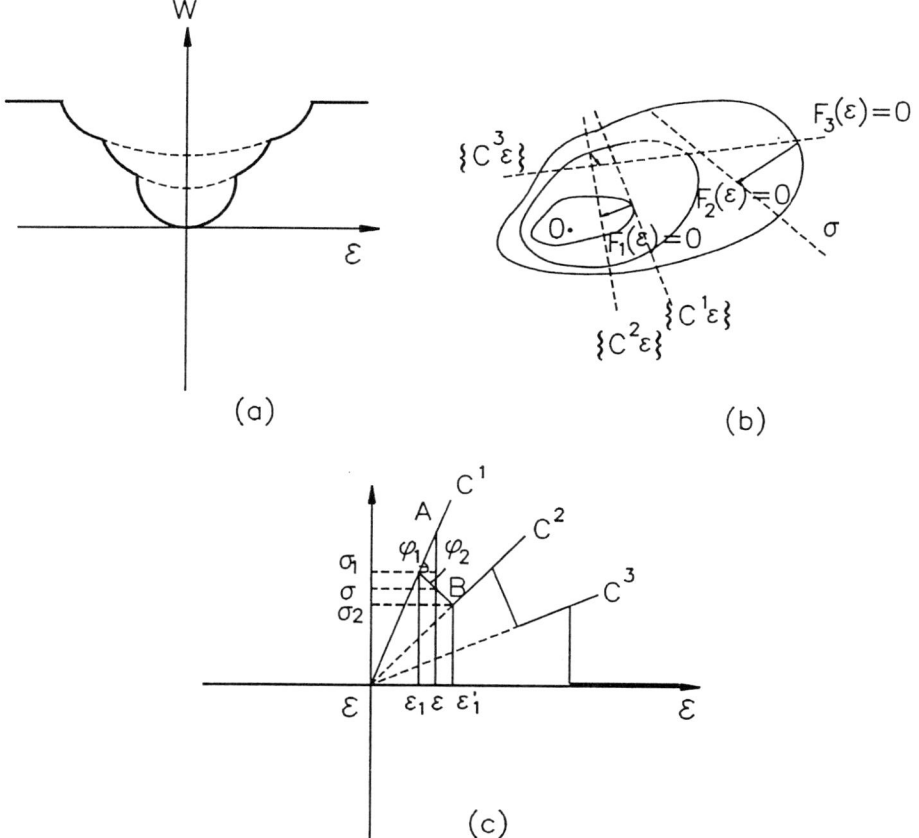

Fig. 3.2.5. Nonconvex superpotentials for threedimensional bodies.

However, a more general definition of the superpotential of a composite material is the following: Let us define in the space $\mathbb{R}^6$ of the symmetric strain tensors $\varepsilon = \{\varepsilon_{ij}\}$ certain surfaces $F_i(\varepsilon) = 0$, $i = 1, 2, \ldots, m$. We assume that the F_i's are continuously differentiable. Moreover $K_1 = \{\varepsilon \mid F_1(\varepsilon) \leq 0\}$ is a closed set. If $\varepsilon \in \{\varepsilon \mid F_1(\varepsilon) < 0\}$ then we have a linear elastic stress-strain behaviour governed by the elasticity tensor $C^1 = \{C^1_{ijhk}\}$. If ε satisfies $F_1(\varepsilon) = 0$, we have a brittle fracture (vertical segment in Fig. 3.2.4c) and we assume that the stress

$$\sigma = \{\sigma_{ij}\} = \{C^1_{ijhk}\varepsilon_{hk}\} + \lambda_1 \operatorname{grad} F_1(\varepsilon) \text{ with } \lambda_1 \in [-a_1, 0] \quad (3.2.12)$$

and the number $a_1 \geq 0$ satisfies the relation

$$\{C^1_{ijhk}\varepsilon_{hk}\} - a_1 \operatorname{grad} F_1(\varepsilon) = \{C^2_{ijhk}\varepsilon_{hk}\}, \quad (3.2.13)$$

where $C^2 = \{C^2_{ijhk}\}$ is the elasticity tensor of the subsequent linear elastic region. This region lies within the admissible strain set $K_2 = \{\varepsilon \mid F_2(\varepsilon) \leq 0\}$.

For $\varepsilon \in \{\varepsilon | F_1(\varepsilon) > 0,\ F_2(\varepsilon) < 0\}$ we have that $\sigma_{ij} = C^2_{ijhk}\varepsilon_{hk}$ and for ε lying on the boundary of K_2 we can again consider a law of the form (3.2.12). We continue in this way until the last brittle fracture surface, after which the stress becomes zero, i.e. the right-hand-side of (3.2.13) becomes zero (Fig. 3.2.5b). The aforementioned procedure can be modified to include the case of semibrittle fracture (Fig. 3.2.5c). We note that the segment AB in Fig. 3.2.5c is equal to $(\varepsilon - P_{K_1}\varepsilon)(\operatorname{tg}\varphi_1 + \operatorname{tg}\varphi_2)$, where $P_{K_1}(\varepsilon)$ denotes the projection of ε onto $K_1 = \{\varepsilon | \varepsilon \leq \varepsilon_1\}$; in our case $P_{K_1}(\varepsilon) = \varepsilon_1$. Therefore we may generalize the law of Fig. 3.2.5c in three dimensions writing that for $F_1(\varepsilon) \geq 0$ we have

$$\sigma = \{\sigma_{ij}\} = \{C^1_{ijhk}\varepsilon_{hk}\} - \mu_1\{\varepsilon_{ij} - (P_{K_1}\varepsilon)_{ij}\} \quad \mu_1 > 0. \tag{3.2.14}$$

Here we assume that all the sets K_i defined by the inequalities $F_i(\sigma) \leq 0$ are convex in order to have uniqueness of the projections. Moreover for $F_1(\varepsilon) < 0$ we have a linear elastic behaviour governed by the Hooke's tensor C^1. Relation (3.2.14) holds for all strain states $\{\varepsilon_{ij}\}$ which satisfy the inequality

$$\{C^1_{ijhk}\varepsilon_{hk}\} - \mu\{\varepsilon_{ij} - (P_{K_1}\varepsilon)_{ij}\} \geq \{C^2_{ijhk}\varepsilon_{hk}\}. \tag{3.2.15}$$

Let us denote this set by S^1. Let $\{\varepsilon'_{1ij}\}$ be the value for which (3.2.15) holds as an equality. This procedure continues until the final surface $\sigma = 0$ is reached. Note that in case of semibrittle fracture we may characterize the fracture by means of closed (and, for simplicity, convex) sets defined in the stress space by the inequalities $\bar{F}_1(\sigma) \leq 0$, $\bar{F}_2(\sigma) \leq 0$ etc. Indeed Fig. 3.2.5c implies that $\varepsilon = \sigma \cot \operatorname{g} \varphi_1 + (P_K \sigma - \sigma)(\cot \operatorname{g} \varphi_2 + \cot \operatorname{g} \varphi_1)$ where $P_K \sigma$ is now the projection of σ onto the set $\sigma \leq \sigma_1$. The generalization is obvious and reads (cf. also (3.1.33))

$$\text{if } \bar{F}_1(\sigma) < 0 \quad \text{then} \quad \varepsilon_{ij} = c^1_{ijhk}\sigma_{hk} \tag{3.2.16}$$

$$\text{if } \bar{F}_1(\sigma) \geq 0 \quad \text{then} \quad \varepsilon_{ij} = c^1_{ijhk}\sigma_{hk} - \bar{\mu}_1[\sigma_{ij} - (P_{R_1}\sigma)_{ij}] \quad \bar{\mu} > 0 \tag{3.2.17}$$

Here c^1 is the inverse tensor of C^1, $\bar{K}_1$ is the set $\{\sigma | \bar{F}_1(\sigma) \leq 0\}$. Relation (3.2.17) holds until the value $\{\sigma_{1ij}\}$ is attained satisfying the relation

$$c^2_{ijhk}\sigma'_{1hk} = c^1_{ijhk}\sigma'_{1hk} - \bar{\mu}_1[\sigma'_{1ij} - (P_{R_1}\sigma'_1)_{ij}]. \tag{3.2.18}$$

This scheme continues until the final surface $\sigma = 0$ is attained. All the aforementioned threedimensional generalizations of the sawtooth onedimensional laws for composite materials, due to their formulation, can be put in the general forms (3.2.1i) or (3.2.2i), where w or $\tilde{w}$ are nonconvex superpotentials. The form of w is given in the case of (3.2.14) by the expression

$$w(\varepsilon) = \begin{cases} \frac{1}{2}(\varepsilon_{ij}C^1_{ijhk}\varepsilon_{hk} - \mu_1||\varepsilon - P_{K_1}\varepsilon||^2)(= A_1(\varepsilon_{ij})) & \text{if } \varepsilon \in S^1 \\ \frac{1}{2}(A_1(\varepsilon'_{1ij}) + \varepsilon_{ij}C^2_{ijhk}\varepsilon_{hk} - \mu_2||\varepsilon - P_{K_2}\varepsilon||^2) & \text{if } \varepsilon \notin S^1,\ \varepsilon \in S^2 \\ \text{etc.} \end{cases}$$
$$\tag{3.2.19}$$

In the expressions, that we have derived, F_i and $\bar{F}_i$ are very general. We can, e.g., assume for them forms analogous to those of ideally locking or ideally plastic materials. In the case of brittle fracture we may assume that the mutual positions of the hyperplanes, defined by the elasticity tensors C_i, and the gradients to the surfaces $F_i(\varepsilon) = 0$ always give bounded stresses for every strain; then the nonconvex superpotential w is locally Lipschitz continuous. A further discussion would compel us to impose several restrictions on $\bar{F}_i, \bar{F}_1, C^1$ etc.; e.g. if one wants to assure the feasibility of the previous geometric constructions (i.e. nonvoid intersections of hypersurfaces, finite a_i in (3.2.13) etc.). All the above considerations hold for the case of loading. For the unloading we refer to the next section.

3.3 Material Laws Expressed via Nonconvex Superpotentials II

i) Objectivity in multivalued laws. In Sect. 2.1 we have recalled the definition of objectivity for classical mechanical laws. For multivalued laws the objectivity concept is defined as following.

Definition 3.3.1 A multivalued mapping between tensor fields is objective or frame-indifferent if all dependent and independent variables transform in the tensorial manner, whenever (2.1.16) is considered as a coordinate transformation.

Suppose, for instance, that we have a law of the form

$$\sigma \in f(D) \tag{3.3.1}$$

where σ is the Cauchy-stress tensor and D is the rate of deformation tensor. The objectivity implies that f must be such that

$$\tilde{Q}\sigma\tilde{Q}^T \in f(\tilde{Q}D\tilde{Q}^T), \tag{3.3.2}$$

for all $\tilde{Q}$'s satisfying (2.1.16).

Note that if f is such that the graph of f in (3.3.1) can be approximated in a prescribed sense by the graph of classical (i.e. nonmultivalued) functions f_ε as $\varepsilon \to 0$, then f_ε has the form

$$f_\varepsilon(D) = \varphi_{0\varepsilon}I + \varphi_{1\varepsilon}D + \varphi_{2\varepsilon}D^2, \tag{3.3.3}$$

where $\varphi_{i\varepsilon}$, $i = 0, 1, 2$, are functions of the principal invariants of D. Indeed we consider the equalities

$$\sigma = f_\varepsilon(D) \tag{3.3.4}$$

which constitute the regularized form of (3.3.1) and by applying the requirement of objectivity (cf. [Beck] p.148) we obtain (3.3.3). The objectivity invariance of the regularized forms of multivalued functions is still an open problem. Another open problem is the study of multivalued laws by the methods of rational mechanics.

ii) The fuzzy viscoplastic materials can be described by means of the law

$$D_{ij} \in [\partial J_\mu(\sigma)]_{ij} + \bar{\partial} f_{ij}(D_{ij}) \tag{3.3.5}$$

where J_μ is given by (3.1.34) and $f_{ij}(\cdot)$ is defined as in (3.2.11).

iii) Nonconvex yield surfaces may be considered in elastic viscous plastic and in elastic plastic viscoplastic materials. The first (resp. second) type of material results if the total strain rate tensor $\dot{\varepsilon}$ is the sum of an elastic part $\dot{\varepsilon}^E$ a plastic part $\dot{\varepsilon}^P$ and a viscous part $\dot{\varepsilon}^V$ (resp. a viscoplastic part $\dot{\varepsilon}^{VP}$). The plastic part $\dot{\varepsilon}^P$ results from relation (3.2.8), where K is closed but nonconvex. Thus we can write for the first (resp. second) material the constitutive relation in the form

$$\dot{\varepsilon} = \dot{\varepsilon}^E + \dot{\varepsilon}^P + \dot{\varepsilon}^V = c\dot{\sigma} + \bar{\partial} I_K(\sigma) + \frac{\sigma^D}{2\mu} \tag{3.3.6}$$

(resp.)

$$\dot{\varepsilon} = \dot{\varepsilon}^E + \dot{\varepsilon}^P + \dot{\varepsilon}^{VP} = c\dot{\sigma} + \bar{\partial} I_K(\sigma) + \partial J_\mu(\sigma) \tag{3.3.7}$$

where μ is the viscosity coefficient, σ^P is the stress deviator and J_μ is given by (3.1.34).

iv) Superpotential of dissipation. Let Ω be a body having density ρ in the configuration under consideration. We denote by $\varepsilon, \eta, \psi = \varepsilon - \theta\eta$ the specific (i.e. per unit mass) internal energy, specific entropy, and specific free energy, θ the absolute temperature, $q = \{q_i\}$ the heat flux vector and Q the heat supply per unit volume (through radiation or chemical reactions) inside Ω. The First and Second Principles of Thermodynamics, in local form, read [Ger73a]

$$\rho \frac{d\varepsilon}{dt} = \sigma_{ij} D_{ij} + Q - \text{div} q \tag{3.3.8}$$

and

$$\frac{d\eta}{dt} + \frac{1}{\rho} \text{div}\left(\frac{q}{\theta}\right) - \frac{Q}{\rho\theta} \geq 0. \tag{3.3.9}$$

Assuming that $Q = 0$, (3.3.9) yields the inequality

$$\sigma_{ij} D_{ij} - \rho \left(\frac{d\psi}{dt} + \eta \frac{d\theta}{dt}\right) - \frac{q_i \theta_{,i}}{\theta} \geq 0, \tag{3.3.10}$$

which may be written in the form

$$\bar{A} = \rho A - \frac{q_i \theta_{,i}}{\theta} \geq 0, \tag{3.3.11}$$

where

$$A = \frac{\sigma_{ij} D_{ij}}{\rho} - \left(\frac{d\psi}{dt} + \eta \frac{d\theta}{dt}\right). \quad (3.3.12)$$

A is called "intrinsic specific dissipation" and $q_i \theta_{,i}/\theta$ "thermal dissipation". On the assumption that the thermal and intrinsic dissipations are "uncoupled", we may write the inequalities

$$A \geq 0 \quad \text{and} \quad -\frac{q_i \theta_{,i}}{\theta} \geq 0. \quad (3.3.13)$$

The thermodynamic state of a material element at time $t = t_0$ exhibiting plastic and/or viscoplastic behavior cannot be completely described by assuming as state variables the Piola-Kirchhoff stress tensor $\Sigma = \Sigma(X,t)$ (cf. (2.1.12)) and the absolute temperature $\theta = \theta(X,t)$ as in the case of an elastic material. Thus either the history of these variables for $-\infty < t \leq t_0$ should be given, or certain supplementary state variables (called hidden, i.e., not observable, variables) should be introduced. The first method is not adequate for classical elastoplastic bodies because of the different material behaviour during loading and unloading. Hidden variables (e.g., the dislocation distribution in a crystal, the residual stresses between the microelements of a polycrystal or of a granular medium, certain physical and chemical properties, etc.) are not necessarily internal variables (which always have a tensorial character) as the case is for hidden variables depending on the orientation. Following Mandel [Mand] we denote by (I) the initial configuration of the body, in which we assume that the stresses are zero and the absolute temperature has a fixed value θ_0. The configuration at time t with stress σ and absolute temperature θ is denoted by (III). Let (II) denote another configuration, called a released configuration, which results from (III) by rapid, purely elastic unloading while the temperature is brought back to the initial value θ_0. If F denotes the deformation gradient of the "transformation" $(I) \to (III)$ and F_e and F_p those of the "transformations" $(II) \to (III)$ and $(I) \to (II)$ respectively, then the relation

$$F = F_e F_p \quad (3.3.14)$$

is true. With respect to the configuration (III), the rate of deformation tensor has the form ("sym" denotes the symmetric part of a tensor)

$$D = \{\frac{1}{2}(v_{i,j} + v_{j,i})\} = \text{sym}\,(\text{grad}\,v) = \text{sym}\left(\frac{dF}{dt}F^{-1}\right) \quad (3.3.15)$$
$$= \text{sym}\left(\frac{dF_e}{dt}F_e^{-1}\right) + \text{sym}\,(F_e V_p F_e^{-1}).$$

Here $V_p = (\tilde{d}F_p/dt)F_p^{-1}$ is the plastic transformation rate with respect to the configuration (II) ($\tilde{d}/dt$ is defined further). Each term of the last equality depends on changes, with time, of the orientation of the released configuration. Therefore Mandel has employed released configurations such that the director triad chosen, always has the same orientation with respect to a fixed coordinate system. The resulting configurations (II) are then called isoclinic released configurations.

We take as state variables the Green strain tensor E_e of the elastic transformation $(II) \to (III)$, the absolute temperature θ and some hidden variables $a_j, j = 1, \ldots, l$ of a tensorial nature. The specific power of the stresses is given by the expression

$$\frac{\sigma_{ij} D_{ij}}{\rho} = \text{tr}\left(\frac{\Sigma}{\rho_{II}} F_e^T D F_e\right) = \text{tr}\left(\frac{\Sigma}{\rho_{II}} \frac{\tilde{d} E_e}{dt}\right) + \text{tr}\left(\frac{\Sigma}{\rho_{II}} F_e^T F_e V_p\right), \quad (3.3.16)$$

where $\text{tr}(A)$ denotes the trace of the tensor A, $\tilde{d}/dt$ is the time derivative of a tensor in motion referred to the director triad of (II) and ρ_{II} is the density in the reference configuration (II). If $\psi = \psi(E_e, a_j, \theta)$, the first of (3.3.13) acquires the form

$$A = \text{tr}\left(\left(\frac{\Sigma}{\rho_{II}} - \frac{\partial \psi}{\partial E_e}\right) \frac{\tilde{d} E_e}{dt}\right) - \left(\frac{\partial \psi}{\partial \theta} + \eta\right) \frac{\tilde{d}\theta}{dt}$$

$$+ \text{tr}\left(\frac{\Sigma}{\rho_{II}} F_e^T F_e V_p\right) - \frac{\partial \psi}{\partial a_j} \frac{\tilde{d} a_j}{dt} \geq 0. \quad (3.1.17)$$

Next, we assume that $\tilde{d} E_e/dt$ and $\tilde{d}\theta/dt$ may assume arbitrary values, while both permanent deformations and $\tilde{d} a_j/dt$ remain bounded (e.g., in the case of a viscoplastic material). It results from (3.3.17) that

$$\frac{\Sigma}{\rho_{II}} = \frac{\partial \psi}{\partial E_e} \quad (3.3.18)$$

$$\eta = -\frac{\partial \psi}{\partial \theta_e} \quad (3.3.19)$$

and thus

$$A = \text{tr}\left(\frac{\Sigma}{\rho_{II}} F_e^T F_e V_p\right) - \frac{\partial \psi}{\partial a_j} \frac{\tilde{d} a_j}{dt} \geq 0. \quad (3.3.20)$$

From (3.3.20) we associate with the "flux" $-\tilde{d} a_j/dt$ the force $A_j = \partial \psi/\partial a_j, j = 1, 2, \ldots, l$, and with the flux V_p the "force"

$$R = F_e^T F_e \frac{\Sigma}{\rho_{II}}. \quad (3.3.21)$$

Thus we may write the intrinsic specific dissipation as

$$A = \text{tr}(R^T V_p) - A_j \frac{\tilde{d} a_j}{dt}. \quad (3.3.22)$$

Here we have used the words "flux" instead of "generalized velocity" and "force" instead of "generalized force" following the terminology of Onsager's theory in irreversible thermodynamics [Gr]. For instance a flux (resp. a corresponding "force") is the heat flux vector (resp. the temperature gradient), the chemical reaction (resp. the chemical affinity) etc. Let us denote the fluxes by v_i and the forces by f_i. Then the first inequality of (3.3.13) may be written in the form

$$A = f_i v_i \geq 0 \qquad (3.3.23)$$

where i runs over all fluxes and all corresponding forces. Inequality (3.2.23) expresses the Second Principle of Thermodynamics for irreversible processes. The case of ideal plasticity is considered as the "limit" of viscoplasticity and thus we may assume that (3.3.18) and (3.3.19) remain valid, although for ideally plastic bodies $d\tilde{E}_e/dt$, $d\tilde{\theta}/dt$, and $d\tilde{a}_j/dt$ are actually interrelated.

The simplest law between fluxes and forces has the linear form

$$f_i = A_{ij} v_j, \qquad (3.3.24)$$

where the symmetry relation $A_{ij} = A_{ji}$ of Onsager-Casimir hold. From (3.3.24) and (3.3.23) we have that

$$A_{ij} v_i v_j \geq 0. \qquad (3.3.25)$$

Relation (3.3.24) and the symmetry conditions are verified in many physical problems. There is also a serious criticism because of the difficulty to make a proper distinction between fluxes and forces in many physical problems [Tru65,66]. Ziegler has proposed a generalization of linear Onsager's relations through an orthogonality principle [Zieg62,63,81]. Moreau and Germain have extended this idea [Mor70],[Ger73a,74,83] by proposing a subdifferential relation between fluxes and forces arising from a convex superpotential.

The following hypothesis called "hypothesis of normal dissipation" generalizes (3.3.24). It is a slight modification of the analogous hypothesis in [Ger83].

(H_1) For every real thermodynamic process there exists a convex superpotential G (i.e. a convex, l.s.c and proper functional) such that

$$f \in \partial G(v). \qquad (3.3.26)$$

The functional G is called the convex superpotential of dissipation.

The following proposition holds. It results by considering the definitions of G and of the conjugate functional G^c. We denote here by V the space of fluxes and by F the space of forces assuming that they are Hilbert spaces such that $F = V'$. Then by means of the duality pairing $\langle \cdot, \cdot \rangle$ we can write the irreversibility inequality (3.3.23) in the form $\langle f, v \rangle \geq 0$,

Proposition 3.3.1 Hypothesis (H_1) is equivalent to each of the following two assertions.

(i) The flux v corresponding to a given force f is solution of the problem

$$G(v) - \langle v, f \rangle = \min_V \{G(v^*) - \langle f, v^* \rangle\}. \qquad (3.3.27)$$

ii) The force f corresponding to a given flux v is solution of the problem

$$G^c(f) - \langle v, f \rangle = \min_F \{G^c(f^\star) - \langle f^\star, v \rangle\}. \tag{3.3.28}$$

Until now we have considered convex superpotentials of dissipation. Let us now introduce nonconvex superpotentials of dissipation according to the papers [Pan81,82] of the author. The same idea has also been exploited by Kim and Oden [Kim84,85]. Let us formulate now the following more general hypothesis of normal dissipation.

(H_2) For every real thermodynamic process, an extended, real-valued functional G on V exists such that the flux $v \in V$ associated with the force $f \in F$ satisfies the relation

$$f \in \bar{\partial} G(v). \tag{3.3.29}$$

By the second principle of thermodynamics, the inequality (3.3.23) written as

$$\langle f, v \rangle \geq 0 \tag{3.3.30}$$

must always be fulfilled. Relation (3.3.29) is by definition equivalent to both the hemivariational inequality

$$G^\uparrow(v, v^\star - v) \geq \langle f, v^\star - v \rangle, \quad \forall v^\star \in V \tag{3.3.31}$$

and the relation (cf. (1.2.28))

$$(f, -1) \in N_{\mathrm{epi}\,G}(v, G(v)), \tag{3.3.32}$$

i.e., $(f, -1)$ is normal in the sense of Clarke to $\mathrm{epi}\,G$ at the point $(v, G(v))$. This fact explains the name "normal dissipation hypothesis". If, moreover $G^\uparrow(v, -v) \leq 0$, then (3.3.30) is also satisfied. The following proposition holds.

Proposition 3.3.2 Hypothesis (H_2) is equivalent to the following two statements:

i) if the force vector $f \in F$ is given, the actual flux vector $v \in V$ renders the functional

$$F(v^\star) = G(v^\star) - \langle v^\star, f \rangle \quad \text{substationary}; \tag{3.3.33}$$

i.e.

$$0 \in \bar{\partial} F(v); \quad \text{and} \tag{3.3.34}$$

ii)

$$F^\uparrow(v, v^\star) \geq 0, \quad \forall v^\star \in V. \tag{3.3.35}$$

The proof is obvious by using the definition of substationarity and (1.2.31). If the dissipation mechanism is given by a relation of the form

$$v \in \bar{\partial} G_1(f), \tag{3.3.36}$$

3.3 Material Laws Expressed via Nonconvex Superpotentials II

then for a given flux $v \in V$ the actual force $f \in F$ is a substationarity point of the functional

$$\tilde{F}(f^0) = G_1(f^0) - \langle v, f^0 \rangle. \tag{3.3.37}$$

The normal dissipation hypothesis (H_1) leads to the definition of the generalized standard materials [Hal74,75]. To this class belong plastic and viscoplastic materials related to convex superpotentials. Hypothesis (H_2) extends this class of materials for nonconvex energy functions. We thus obtain a class of materials which we call generalized standard materials related to nonconvex energy functions. Let (3.3.37) be fulfilled with $v = ((\tilde{d}F_p/dt)F_p^{-1}, -\tilde{d}\{a_j\}/dt)$ and $f = \{R, \{A_j\}\}$, i.e.,

$$\left(\frac{\tilde{d}F_p}{dt} F_p^{-1}, \frac{-\tilde{d}\{a_j\}}{dt} \right) \in \bar{\partial} G_1(R, \{A_j\}). \tag{3.3.38}$$

If G_1 is convex, l.s.c. and proper, then (3.3.38) describes a generalized standard material as defined in [Hal74,75]. Suppose that $G_1 = I_K$, where I_K is the indicator of a nonempty, convex, closed subset K of $\mathbb{R}^n$; $0 \in K$ and n is the dimension of $(R, \{A_j\})$. Then (3.3.38) is equivalent to the variational inequality

$$\operatorname{tr}\left(\left(\frac{\tilde{d}F_p}{dt} F_p^{-1} \right)^T (R - R^\star) \right) - \frac{\tilde{d}a_j}{dt}(A_j - A_j^\star) \geq 0, \quad \forall (R^\star, \{A_j^\star\}) \in K \tag{3.3.39}$$

for $(R, \{A_j\}) \in K$, which generalizes Hill's principle of maximum plastic work [Hill48] along the lines of Mandel [Mand]. This case corresponds to a plastic material. If $G_1 = I_K$, where K is a nonempty, closed subset of $\mathbb{R}^n$ with $0 \in K$, then (3.3.38) implies that

$$\left(\frac{\tilde{d}F_p}{dt} F_p^{-1}, -\frac{\tilde{d}\{a_j\}}{dt} \right) \in N_K(R, \{A_j\}), \tag{3.3.40}$$

i.e., normality in the sense of Clarke holds. In this case the inequality

$$I_{T_K(R,\{A_j\})}((R^\star, \{A_j^\star\}) - (R, \{A_j\})) \geq \operatorname{tr}\left\{ \left(\frac{\tilde{d}F_p}{dt} F_p^{-1} \right)^T (R^\star - R) \right\} \tag{3.3.41}$$

$$- \frac{\tilde{d}a_j}{dt}(A_j^\star - A_j), \quad \forall (R^\star, \{A_j^\star\}) \in \mathbb{R}^n$$

results, because of (1.2.36). Accordingly, (3.3.41) coincides with (3.3.39) if

$$(R^\star, \{A_j^\star\}) - (R, \{A_j\}) \in T_K(R, \{A_j\}). \tag{3.3.42}$$

If we exclude re-entrant corners, we may confirm the validity of Hill's principle for a nonconvex yield surface.

We further assume that $G_1(R, A) = \Phi(R, A)$, where Φ is a continuously differentiable functional on $\mathbb{R}^n$. Then (3.3.38) yields

$$\frac{\tilde{d}F_p}{dt}F_p^{-1} = \frac{\partial \Phi}{\partial R} \tag{3.3.43}$$

and

$$-\frac{\tilde{d}a_j}{dt} = \frac{\partial \Phi}{\partial A_j} \quad j = 1,\ldots,l. \tag{3.3.44}$$

Relation (3.3.43) extends the viscoplastic-potential theory of Rice, in the sense of Mandel, while the second relation concerns the behavior of the internal parameters. The function $G_1 = \Phi + I_K$ corresponds to an elastic viscoplastic material with instantaneous plastic deformations.

Let us assume for the material law (3.3.40) that $K = \{(R,\{A_j\}) \big| F_i(R,\{A_j\}) \leq 0, i = 1,\ldots,m\}$, where the F_i's are continuously differentiable functionals on $\mathbb{R}^n$. Due to (1.2.49), relation (3.3.40) is equivalent to the expressions

$$\frac{\tilde{d}F_p}{dt}F_p^{-1} = \sum_{i=1}^{m}\lambda_i \frac{\partial F_i}{\partial R}, \tag{3.3.45}$$

$$-\frac{\tilde{d}a_j}{dt} = \sum_{i=1}^{m}\lambda_i \frac{\partial F_i}{\partial A_j} \quad j = 1,\ldots,l \tag{3.3.46}$$

and

$$\lambda_i F_i = 0, \quad \lambda_i \geq 0 \quad F_i \leq 0, \quad i = 1,\ldots,m, \tag{3.3.47}$$

if $0 \notin \bar{\partial}\tilde{F}(R,A)$ ($\tilde{F} = \max\{F_1,\ldots,F_m\}$). Suppose that the deformations are sufficiently small and isothermal; then (3.3.38) reduces to the relation

$$(\dot{\varepsilon}^p, -\{\dot{a}_j\}) \in \bar{\partial}G_1(\sigma,\{A_j\}), \tag{3.3.48}$$

which generalizes the plasticity law (3.2.8) and (3.2.9). For certain interesting applications of the hypothesis (H_2) we refer to [Kim85].

3.4 Loading and Unloading Problems. The Advantage of the Use of Superpotentials

Until this point we have considered material laws and boundary conditions where the strain increases, i.e. we have considered only the case of loading. Now we will try to study the case of unloading in the stress-strain relations that we have introduced. We recall that all these relations are generally multivalued and nonmonotone. In [Pana81] a method was developed by the author for the calculation of a structure for a given path of unloading (method of "macroincrements"). Here we shall first present a quite general method which takes into account all the possibilities of loading and/or unloading (not only along a prescribed unloading path), i.e. we will completely describe the phenomenon. This is possible using appropriately defined multifunctions involving $\varepsilon, \sigma, \dot{\varepsilon}$ and $\dot{\sigma}$. The dot means as usual the time partial derivatives. Let us assume that the

3.4 Loading and Unloading Problems. The Advantage of the Use of Superpotentials

unloading is linear and that the modulus of elasticity changes with the strain. In Fig. 3.4.1 the onedimensional case is depicted, or a uniaxial law between equivalent stress and strains (e.g. for two- and threedimensional problems if the consideration of equivalent stresses and strains is possible). Note that the following thoughts also hold for multidimensional laws. We consider that the loading of the structure is applied on it in the form of a sequence of increments as the time increases. At each point along the softening branch AB we pose the question whether we have loading or unloading, i.e. whether we remain on AB, or the elastic unloading paths AC, A'C' etc. should be realized. If the stress and strain of an element are on OAB at the end of a load increment, then for the next load increment or the next time step we can write that

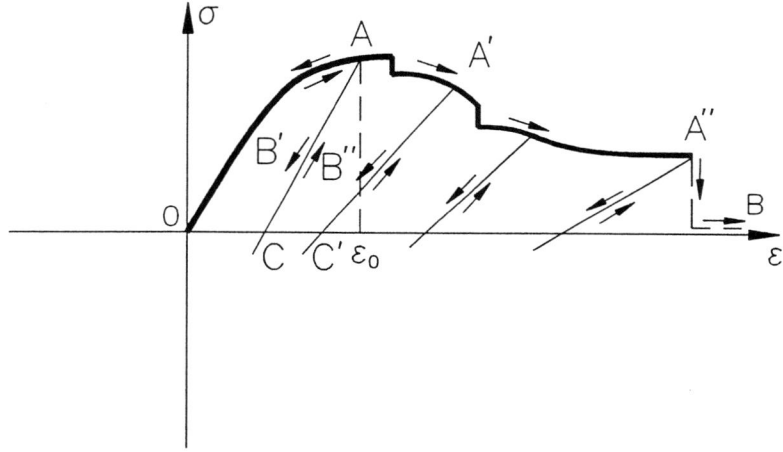

Fig. 3.4.1. The loading-unloading problem.

$$\sigma \in \bar{\partial}\varphi(\varepsilon) \quad \text{if} \quad \varepsilon < \varepsilon_0, \quad \text{or if} \quad \varepsilon \geq \varepsilon_0 \quad \text{and} \quad \dot{\varepsilon} \geq 0 \qquad (3.4.1)$$

$$\sigma = C(\varepsilon)\varepsilon \quad \text{if} \quad \varepsilon \geq \varepsilon_0 \quad \text{and} \quad \dot{\varepsilon} < 0. \qquad (3.4.2)$$

If the stress and strain of an element are, e.g., on A'C' then

$$\sigma = C(\varepsilon')\varepsilon \qquad (3.4.3)$$

where ε' denotes the strain at A'. We may easily conclude that (3.4.1)÷(3.4.3) can be written in the form

$$\sigma \in \bar{\partial}\varphi(\varepsilon)\psi(\varepsilon) + [\bar{\partial}\varphi(\varepsilon)\chi_+ + C(\varepsilon)(\varepsilon)\chi_-]q(\varepsilon,\sigma)(1-\psi(\varepsilon)) \qquad (3.4.4)$$
$$+ C(\varepsilon')(\varepsilon)(1-q(\varepsilon,\sigma))(1-\psi(\varepsilon)),$$

where $\psi(\varepsilon) = \{1 \text{ for } \varepsilon < \varepsilon_0, 0 \text{ for } \varepsilon \geq \varepsilon_0\}$ and ε_0 is the strain for which irreversible strains appear, $\chi = \dot{\varepsilon}/|\dot{\varepsilon}|$ and χ_+ and χ_- are the positive and the negative parts of χ respectively, $q(\varepsilon, \sigma) = \{1 \text{ if } (\varepsilon, \sigma) \in E, 0 \text{ otherwise}\}$ with $E = \{(\varepsilon, \sigma)\big| \sigma \in \bar{\partial}\varphi(\varepsilon)\}$. Note that in (3.4.4) $C(\varepsilon')$ is assumed as known. The general form of the law (3.4.4) is a differential inclusion of the form

$$0 \in F(\varepsilon, \sigma, \dot{\varepsilon}), \tag{3.4.5}$$

where F is an appropriately defined multifunction.

Note that the classical elastoplastic law is a multivalued law between $\dot{\varepsilon}$, $\dot{\sigma}$ and σ. Let us assume now that we know σ and ε for a given load p and let a new load $\dot{p}\delta t$ be imposed. Now for the total load $p + \dot{p}\delta t$ the stresses and strains become $\sigma + \dot{\sigma}\delta t$ and $\varepsilon + \dot{\varepsilon}\delta t$. Then an incremental relation of the form

$$0 \in F_1(\varepsilon, \sigma, \dot{\varepsilon}, \dot{\sigma}) \tag{3.4.6}$$

describes the incremental loading and unloading within each increment.

Suppose now that we have a structure which is divided into a finite number of elements and that each one of the elements obeys a law anologous to the one depicted in Fig. 3.4.1. The elements of the structure can be divided into three classes. To the first class belong those elements that have stress and strain states on the elastic part OA of the law. Thus for the next load increment these elements obey the law OAA'A''. To the second class belong those elements which have stress and strain states on the part OAA'A'' of the stress strain curve but after the point A where the elastic behaviour ceases. Then for an element having at the end of the previous load increment stress and strain states at A or at A' respectively the law for the next increment is CB'AA'A''B or C'B''A'A''B. Again we have a nonmonotone possibly multivalued law with the zero at A or at A'. Finally to the third class belong those elements which have stress and strain states on the unloading branches CB'A or C'B''A'. For an element belonging to this class, e.g. having stress and strain states at B' we can adopt as a law for the next increment the graph CB'AA'A''B. Again we have a nonmonotone possibly multivalued stress strain law having B' as the origin of the coordinates.

Since in a discretized structure each element must belong to one of the aforementioned three classes we may state the following result which is of importance for the applications.

Proposition 3.4.1 For a discretized problem the differential inclusion (3.4.6) reduces, within each load increment, to a finite number of nonmonotone multivalued relations or, equivalently, to a finite number of relations expressed in terms of nonconvex superpotentials.

As we have already indicated in Sect. 2.2 nonconvex superpotentials lead to hemivariational inequalities. Thus the following proposition obviously holds.

Proposition 3.4.2 The loading-unloading problem for a discretized structure reduces to the solution of a finite number of hemivariational inequalities within each load increment.

The use of convex or nonconvex superpotentials in Mechanics and Engineering Sciences has the following advantages:

a) The unified study of large classes of problems which in the case of convex superpotentials lead to variational inequalities and in the case of nonconvex superpotentials to hemivariational inequalities.
b) The correct (for the first time) theoretical and numerical treatment of inequalities and, more generally, of multivaluedness in material laws and boundary conditions.
c) The derivation of variational formultions for nonsmooth, convex or nonconvex energy functions.
d) The possibility to study fuzzy laws by means of a variational theory.
e) The unified derivation of minimum propositions for the potential and the complementary energy (see e.g. [Pan85]) in the case of convex superpotentials. For nonconvex superpotentials, as we shall see in the next chapter, we get, instead of minimum propositions, substationarity propositions.
f) The easy passing from micromechanical laws to macromechanical ones (cf. [Pan88]).
g) Rational numerical treatment of the combinatorial character of the inequality constraints and/or the multivaluedness for large scale problems by formulating the convex (resp. nonconvex) superpotential problems as an optimization problem (resp. as a sequence of optimization problems) with respect to the potential or the complementary energy (see Ch. 10).

3.5 Material Laws and Boundary Conditions Expressed via Fans, Quasidifferentials and Codifferentials

i) In Sect. 1.4 we have introduced the concept of fan. The inequality in (1.4.6) permits to use the fans in Mechanics by considering material laws and respectively boundary conditions of the type (small deformation theory)

$$\sigma \in F_1(\varepsilon) \tag{3.5.1}$$

and

$$-S \in F_2(u), \tag{3.5.2}$$

where F_1 and F_2 are appropriately defined fans. Then, by Prop. 1.4.1 one can determine energy functionals g_1 and g_2 such that

$$F_1(\varepsilon) = \{\sigma \big| g_1(\varepsilon^\star, \varepsilon) \geq \langle \sigma, \varepsilon^\star \rangle \quad \forall \varepsilon^\star \in \mathbb{R}^6 \} \tag{3.5.3}$$

and

$$F_2(u) = \{-S \big| g_2(u^\star, u) \geq \langle -S, u^\star \rangle \quad \forall u^\star \in \mathbb{R}^3 \}. \tag{3.5.4}$$

Here $F_1 : \mathbb{R}^6 \to \mathbb{R}^6$, $g_1 : \mathbb{R}^6 \times \mathbb{R}^6 \to \bar{\mathbb{R}}$, $F_2 : \mathbb{R}^3 \to \mathbb{R}^3$, $g_2 : \mathbb{R}^3 \times \mathbb{R}^3 \to \bar{\mathbb{R}}$, the g_1 and g_2 are convex and positively homogeneous of degree one in each one of the variables and $g_1(\cdot, \varepsilon)$, $g_2(\cdot, u)$ are l.s.c. functions with $g_1(\cdot, \varepsilon) > -\infty$,

$g_2(\cdot, u) > -\infty$, for $\varepsilon \in D(F_1)$ and $u \in D(F_2)$. The energy functionals g_1 (resp. g_2) express the virtual work at ε (resp. at u) for a variation ε^* (resp. u^*).

The fans may serve for the derivation of a variational theory for the following type of boundary conditions. Suppose that on the boundary of a deformable body S_T is given and that between $-S_N$ and u_N a boundary condition of the form

$$-S_N \in F(u_N), \qquad (3.5.5)$$

where F is given by (1.4.9) with $X \equiv \mathbb{R}$, i.e.

$$u_N \to F(u_N) = \{T \in \mathbb{R} \,|\, Q(u_N) \leq T \leq P(u_N)\}. \qquad (3.5.6)$$

In other words we have a boundary condition of the form

$$Q(u_N) \leq -S_N \leq P(u_N) \qquad (3.5.7)$$

at each point of the boundary. This type of boundary condition is analogous to the fuzzy boundary conditions we have introduced in Sect. 3.2.

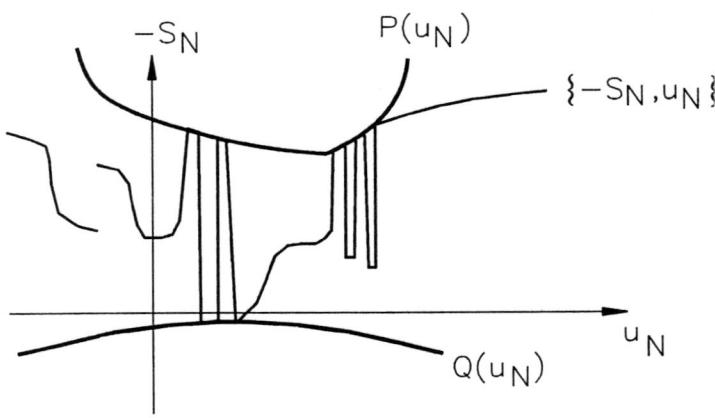

Fig. 3.5.1. The boundary condition (3.5.7).

Note that the $\{-S_N, u_N\}$ graph in Fig. 3.5.1 does not need to be a continuous line. Analogously we may define material laws. It is worth noting here that in damage problems the decreasing branch AM may be replaced by a discontinuous law ABCDEFG... (Fig. 3.5.2) which always yields a positive stiffness (cf. e.g. Sect. 9.2)

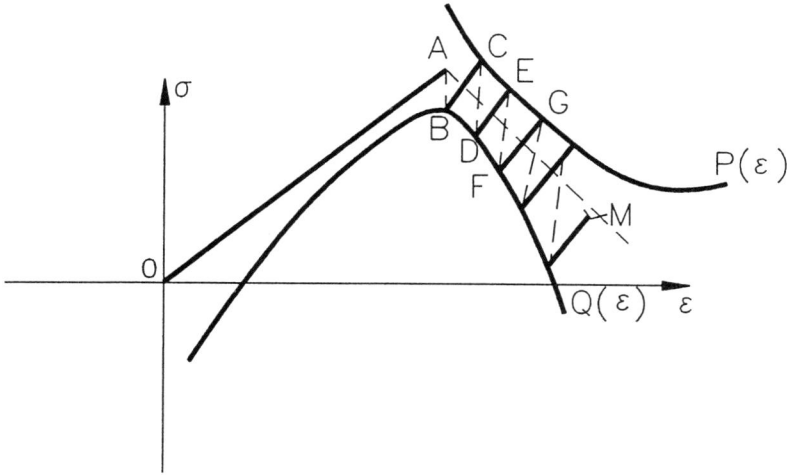

Fig. 3.5.2. Approximation of a decreasing branch.

It is obvious from Fig. 3.5.2 that this law may be expressed in terms of a fan, because it may be put in the form (3.5.7).

Besides the above lines and the results of [Pan87b] there are not other results yet on the applications of fans to mechanical problems. However, as the last two figures show, the fans may lead to important results (e.g. the fomulation of a variational theory for damage problems). Boundary conditions and material laws expressed in terms of fans are called F-superpotential laws in order to make a comparison with the nonconvex and the convex superpotential laws.

It is worth noting that we have not until now completely exploited all the statements of Prop. 1.4.1. Indeed it also states that if a mechanical law satisfies relations (1.4.1)÷(1.4.4), then there exists $g(\cdot,\cdot)$, i.e. there exists an expression for the virtual work such as to fulfill the inequality in (1.4.6). Since the proof of Prop. 1.4.1 is based on the Hahn-Banach separation theorem one can see the relation between the theorem of Hahn-Banach and the existence of a virtual work expression.

ii) We shall further give certain boundary conditions, and material laws expressed in terms of quasidifferentials.

One interesting case occurs if the energy functional F (called simply the superpotential) can be expressed as the difference of two convex functions, namely

$$F(x) = \varphi_1(x) - \varphi_2(x). \tag{3.5.8}$$

Here $\varphi_1(\cdot)$, $\varphi_2(\cdot)$, are convex, not necessarily differentiable functions. In this case the quasidifferential of function F is expressed as (cf. (1.4.27))

$$Df(x) = \{\partial\varphi_1(x), -\partial\varphi_2(x)\}, \tag{3.5.9}$$

where ∂ is the classical subdifferential of convex analysis. Relations (3.5.8),(3.5.9) lead to the following interesting property: The convex and the concave parts of F can be "treated" separately (cf. in this context also Sect. 4.5).

Let us give now an example: We consider the nonmonotone interface law which is depicted in Fig. 3.5.3. Here $[u_N]$ denotes the relative displacement in the normal direction to the interface.

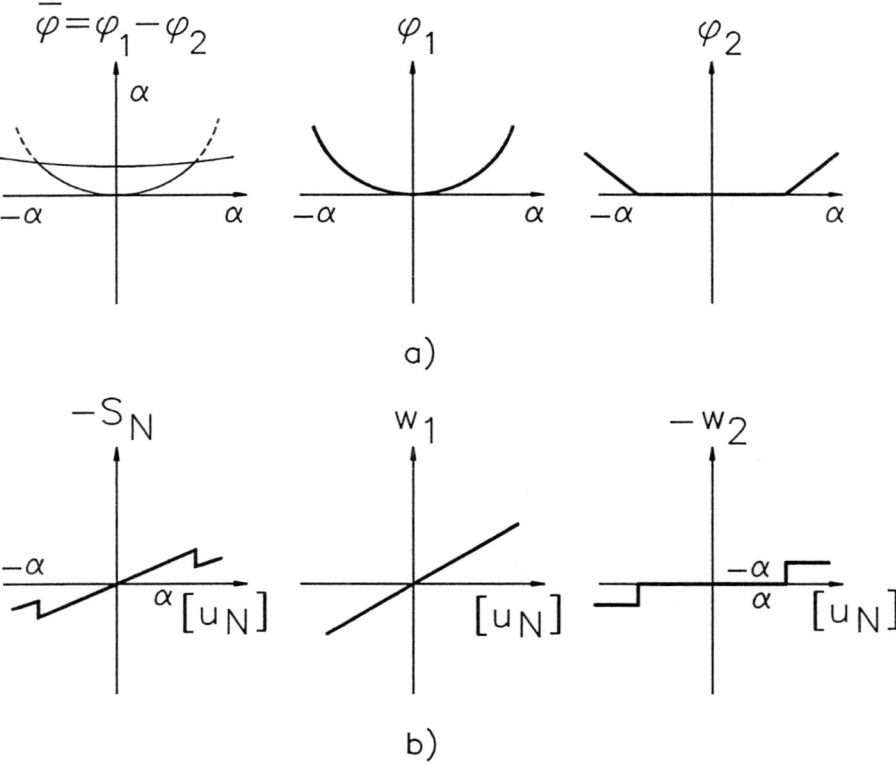

Fig. 3.5.3. The difference of two convex superpotentials.

3.5 Material Laws via Fans and Quasidifferentials

The law has the following form (here $k_1 > 0$, $k_2 > 0$, $a > 0$)

$$-S_N = \begin{cases} k_1[u_N] + k_2 & \text{if } [u] < -a \\ [-k_1 a, -k_1 a + k_2] & \text{if } [u] = -a \\ k_1[u_N] & \text{if } -a < [u] < a \\ [k_1 a, k_1 a - k_2] & \text{if } [u] = a \\ k_1[u_N] - k_2 & \text{if } [u] > a. \end{cases} \quad (3.5.10)$$

and can be written in the form (cf. (1.4.27))

$$-S_N = w_1 + w_2, w_1 \in \partial \varphi_1([u_N]), w_2 \in -\bar{\partial} \varphi_2([u_N]), \quad (3.5.11)$$

where

$$\varphi_1([u_N]) = \frac{1}{2} k_1 [u_N]^2 \quad (3.5.12)$$

and

$$\varphi_2([u_N]) = \begin{cases} -k_2([u_N] + a) & \text{if } [u_N] \le -a \\ 0 & \text{if } -a \le [u_N] \le a \\ k_2([u_N] - a) & \text{if } [u_N] \ge a. \end{cases} \quad (3.5.13)$$

Let us generalize this example. Indeed using the notion of quasidifferentiability, general nonmonotone and possibly multivalued interface laws can be expressed in the form (here S stays for S_N or S_T and $[u]$ for $[u_N]$ and $[u_T]$ respectively):

$$-S = w_1 + w_2 \quad (3.5.14)$$

with

$$\{w_1, w_2\} \in DF([u]) = \{\underline{\partial}' F([u]), \bar{\partial}' F([u])\}, \quad (3.5.15)$$

where F is an appropriately defined, quasidifferentiable superpotential. By the definition of the quasidifferential we can write the following equation, which, in the terminology of mechanics expresses the virtual work of an interface joint

$$\begin{aligned} \tilde{F}'([u], [v]) &= \langle -S, [v] \rangle = \langle w_1, [v] \rangle + \langle w_2, [v] \rangle \quad (3.5.16) \\ &= \max_{w_1^\star} \{\langle w_1^\star, [v] \rangle \big| w_1^\star \in \underline{\partial}' F([u])\} + \min_{w_2^\star} \{\langle w_2^\star, [v] \rangle \big| w_2^\star \in \bar{\partial}' F([u])\} \\ &= \max_{w_1^\star} \min_{w_2^\star} \{\langle w_1^\star + w_2^\star, [v] \rangle \big| w_1^\star \in \underline{\partial}' F([u]), w_2^\star \in \bar{\partial}' F([u])\} \\ &\forall [v] \in V. \end{aligned}$$

Here S corresponds to the displacement field u and V denotes the vector space of $[v]$. Moreover $\langle \cdot, \cdot \rangle$ denotes the appropriate duality pairing. There are also two other equivalent forms to (3.5.16). First we have that $\forall [v] \in V$

$$\tilde{F}'([u], [v]) \ge \langle w_1^\star, [v] \rangle + \min\{\langle w_2^\star, [v] \rangle \big| \forall w_2^\star \in \bar{\partial}' F([u])\}, \forall w_1^\star \in \underline{\partial}' F([u]) \quad (3.5.17)$$

and then that $\forall [v] \in V$ there exists

$$\begin{aligned} w_2 &\in \bar{\partial}' F([u]) \text{ such that } \tilde{F}'([u], [v]) \\ &\ge \langle w_1^\star, [v] \rangle + \langle w_2, [v] \rangle, \quad \forall w_1^\star \in \underline{\partial}' F([u]). \end{aligned} \quad (3.5.18)$$

96   3. Nonsmooth Mechanics II

From the above relations it results that, if e.g. F is the difference of two convex functionals, then $\bar{\partial}'$ and $\underline{\partial}'$ are classical subdifferentials and using the definition of the subdifferential yields a system of coupled variational inequalities describing the nonconvex superpotential problem (see also Sect. 4.5).

Quasidifferentiable functions include besides the difference of convex functions, the minimum and maximum-type functions as we have shown in Sect. 1.4.

For example the delamination law depicted in Fig. 3.5.4b is derived from the nonconvex and nonsmooth superpotential of Fig. 3.5.4a which is quasidifferentiable because it can be written in the min-form:

$$F([u_N]) = \min_i \{f_i([u_N])\}, \quad i = 1, 2, 3 \tag{3.5.19}$$

with $f_1([u_N]) = \frac{1}{2}k_1[u_N]^2, f_2([u_N]) = \lambda_2 + \frac{1}{2}k_2[u_N]^2, f_3([u_N]) = \lambda_1$.

Here $0 < k_2 < k_1$. Note at this point that even the classical convex superpotential of friction for twodimensional bodies $F(u_T) = |u_T|$ can be written as a maximum type function. Indeed $|u_T| = \max\{u_T, -u_T\}$.

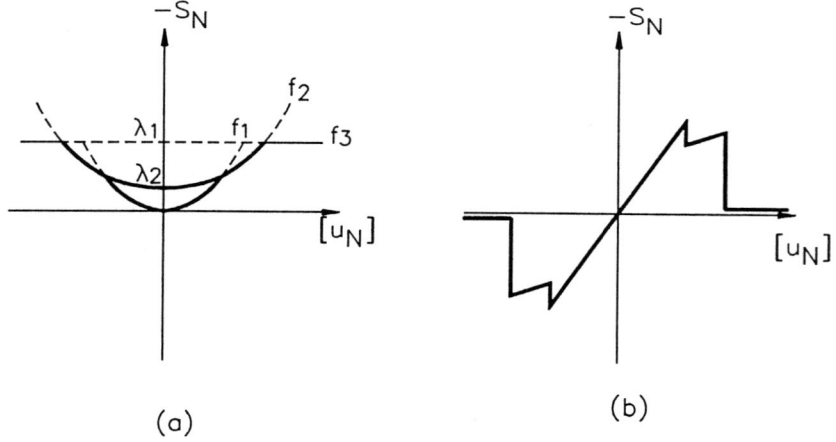

Fig. 3.5.4. A delamination law expressed via a minimum type superpotential.

Now we will study some material laws expressed through quasidifferentials. We assume that w is quasidifferentiable, i.e. that the directional differential of w exists everywhere and satisfies (1.4.11). Then we consider the laws:

$$\sigma_{ij}\varepsilon_{ij}^* = \tilde{w}'(\varepsilon, \varepsilon^*), \quad \forall \varepsilon^* = \{\varepsilon_{ij}\} \in \mathbb{R}^6 \tag{3.5.20}$$

resp.

3.5 Material Laws via Fans and Quasidifferentials

$$\sigma_{ij}\varepsilon_{ij}^{\star} \leq \tilde{w}'(\varepsilon,\varepsilon^{\star}), \quad \forall \varepsilon^{\star} = \{\varepsilon_{ij}\} \in R^6 \quad (3.5.21)$$

The material law given in (3.5.20) describes a general hyperelastic material including e.g. the possibility of smooth softening behaviour. In the material law (3.5.21) certain dissipation mechanisms are allowed in addition to occur. Since w is quasidifferentiable two convex and compact sets $A = \underline{\partial}'w(\varepsilon)$ and $B = \bar{\partial}'w(\varepsilon)$ can be determined, such that

$$\begin{aligned}\tilde{w}'(\varepsilon,\varepsilon^{\star}) &= \max\{\bar{\tau}_{ij}\varepsilon_{ij}^{\star} \big| \bar{\tau} = \{\bar{\tau}_{ij}\} \in A\} \\ &+ \min\{\bar{\bar{\tau}}_{ij}\,\varepsilon_{ij}^{\star} \big| \bar{\bar{\tau}} = \{\bar{\bar{\tau}}_{ij}\} \in B\}, \quad \forall \varepsilon^{\star} \in \mathbb{R}^6\end{aligned} \quad (3.5.22)$$

In the sequel the notation $\bar{\tau}_{ij}\varepsilon_{ij}^{\star} = (\bar{\tau},\varepsilon^{\star})$ etc. will be used. From the definition of $\tilde{w}'(\varepsilon,\varepsilon^{\star})$ we get the relation

$$\tilde{w}'(\varepsilon,\varepsilon^{\star}) \geq (\bar{\tau},\varepsilon^{\star}) + \min\{(\bar{\bar{\tau}},\varepsilon^{\star}) \big| \bar{\bar{\tau}} \in \bar{\partial}'w(\varepsilon)\}, \quad \forall \bar{\tau} \in \underline{\partial}'w(\varepsilon),\ \forall \varepsilon^{\star} \in \mathbb{R}^6 \quad (3.5.23)$$

and the equivalent formulation: there exists $\bar{\bar{\tau}} \in \bar{\partial}'w(\varepsilon)$ such that

$$\tilde{w}'(\varepsilon,\varepsilon^{\star}) \geq (\bar{\tau}+\bar{\bar{\tau}},\varepsilon^{\star}) \quad \forall \varepsilon^{\star} \in \mathbb{R}^6 \text{ and } \forall \bar{\tau} \in \underline{\partial}'w(\varepsilon). \quad (3.5.24)$$

One may consider material laws of the type (3.5.20) (resp. (3.5.21)) in which the determination of the sets A, B is straightforward. For instance, if the strain energy density w of a material is a finite function which is a sum of a convex and a concave finite part, i.e. $w = w_1 + w_2$, then (1.4.27) holds.

A second interesting case arises if the strain energy is convex with respect to certain components ε_1 of the strain tensor and concave with respect to the remaining components ε_2. In detail, let $\varepsilon \to w(\varepsilon) = w(\varepsilon_1,\varepsilon_2)$ be finite on $E = E_1 \times E_2 \subset \mathbb{R}^6$, where $E_1 \subset \mathbb{R}^\alpha$, $E_2 \subset \mathbb{R}^\beta$, $\alpha + \beta = 6$, are open and convex sets in $\mathbb{R}^\alpha$ and in $\mathbb{R}^\beta$ respectively. With respect to the latter partition let the function $w(\cdot,\varepsilon_2)$ (resp. $w(\varepsilon_1,\cdot)$) be convex (resp. concave) and let the following partial subdifferential sets be nonempty, compact and convex sets (cf. (1.4.17))

$$\partial_{\varepsilon_1}w(\cdot,\varepsilon_2) \text{ and } \tilde{\partial}_{\varepsilon_2}w(\varepsilon_1,\cdot) = -\partial_{\varepsilon_2}w(\varepsilon_1,\cdot). \quad (3.5.25)$$

It can be proved in this case that the function $\varepsilon \to w(\varepsilon_1,\varepsilon_2)$ is quasidifferentiable on $E_1 \times E_2$, (see [Dem85], where the term convex-concave function is used). Moreover the quasidifferential is easily calculated and we have at $\varepsilon = \{\varepsilon_1,\varepsilon_2\}$:

$$\underline{\partial}'w(\varepsilon_1,\varepsilon_2) = \{\partial_{\varepsilon_1}w(\varepsilon_1,\varepsilon_2),0\} \in \mathbb{R}^\alpha \times \mathbb{R}^\beta, \quad (3.5.26)$$

and

$$\bar{\partial}'w(\varepsilon_1,\varepsilon_2) = \{0,-\partial_{\varepsilon_2}w(\varepsilon_1,\varepsilon_2)\} \in \mathbb{R}^\alpha \times \mathbb{R}^\beta. \quad (3.5.27)$$

By means of the quasidifferentiability, generalizations of the interface and boundary conditions for three dimensional bodies can be formulated. Moreover the possibility of coupling between the normal and the tangential mechanical behaviour along a boundary or an interface may also be considered. Let us consider such a boundary law in the form

$$-S_i v_i = (\text{resp.} \leq) \tilde{j}'(u, v), \quad \forall v \in \mathbb{R}^3. \tag{3.5.28}$$

In (3.5.28) j is assumed to be a quasidifferentiable function. Along the lines of (3.5.28) we can write a coupled contact law of the form

$$-S_{T_i} v_{T_i} - S_N v_N = \tilde{j}'(u_{T_i}, u_N, v_{T_i}, v_N). \tag{3.5.29}$$

If the case in which e.g. the function $u_N \to j(u_T, u_N)$ is concave and finite and the function $u_T \to j(u_T, u_N)$ is convex and finite we may work along the lines of (3.5.26), (3.5.27). We thus obtain that

$$\underline{\partial}' j(u_T, u_N) = \{\partial_{u_T} j(u_T, u_N), 0\}, \bar{\partial}' j(u_T, u_N) = \{0, -\partial_{u_N} j(u_T, u_N),\}. \tag{3.5.30}$$

We refer the reader to [Pan88,92a] for certain plasticity laws using the notion of quasidifferentiability. Here we will give only one interesting result. According to a theorem by Rubinov and Yagubov [Rub86], to a pair of star-shaped sets $\{U, V\}$ a quasidifferentiable function F can be associated, having a directional derivative $\tilde{F}'(x, y)$, which is continuous in y, such that

$$\tilde{F}'(x, y) = \max\{\lambda < 0 \big| y \in -\lambda V\} + \min\{\lambda > 0 \big| y \in \lambda U\}. \tag{3.5.31}$$

We recall here that a closed set $U \in X$ is called star-shaped if $0 \in \text{int} U$ and every line of the form $\{\lambda x \big| \lambda \geq 0, x \in U\}$ intersects the boundary of U in one point at most.

Accordingly we can consider a star-shaped yield function in the stress space K_Σ and a star-shaped admissible function for the plastic strains in the strain space K_E. Both of them then define through (3.5.31) an expression of the "virtual work" produced at the stress state σ for a strain state variation ε, i.e. K_Σ and K_E imply the existence of a quasidifferentiable function F such that $\varepsilon^p \to \tilde{F}'(\sigma, \varepsilon^p)$ is continuous and

$$\tilde{F}'(\sigma, \varepsilon^p) = \max\{\lambda < 0 \big| \varepsilon^p \in -\lambda K_E\} + \min\{\lambda > 0 \big| \sigma \in \lambda K_\Sigma\}. \tag{3.5.32}$$

iii) Codifferentiability may lead also to certain formulations of material laws and boundary conditions in full analogy to the quasidifferentiability. We leave these formulations as an exercise for the reader. Note also that a and b in (1.4.35) may be considered as internal parameters which guarantee the Hausdorff continuity of the codifferential.

Closing this chapter we would like to notice that a notion of Nonsmooth Analysis which may play an important role in Nonsmooth Mechanics is the "derivative container" of Warga [Warg75,76a,76b]. In this context see also [Pan85] p. 159, where we have dealt very shortly with the applications of derivative containers to Mechanics. On this area much work remains to be done. We should remark, however, here that the derivate containers – due to their definition – do not lead to inequalities but to sequences of classical (regularized) differential equations.

4. Hemivariational Inequalities

In the present chapter we present the method for the formulation of hemivariational inequalities and related variational expressions. For static problems the classical minimum propositions for the potential and the complementary energy are extended to analogous substationarity propositions.

Subsequently dynamic problems and eigenvalue problems for hemivariational inequalities are studied, as well as certain variational formulations for the loading and unloading problem. The chapter closes with some variational inequality formulations resulting from fans, quasidifferentials and certain generalizations of the nonconvex superpotentials which we have called V-superpotentials. The present chapter is based mainly on [Pan81,83a,85,87a,87b, 88,88a,89a,90,92b] and on [Nan88,89a,89b], [Mor88b], [Ant92], [Stav91,93a,b], [Mot93], [Kar91,92].

4.1 The Derivation of Hemivariational Inequalities in Mechanical Problems

A variational formulation is a statement that a solution of an operator equation subjected to certain boundary and/or initial conditions makes an expression involving variations of the quantities of the problems equal to zero or nonnegative. Thus we may distinguish between the bilateral or equality problems and the unilateral or inequality problems. Moreover in Mechanics we call the variational formulations, variational "principles". Further we shall derive certain variational principles for a deformable body. Let $\Omega \subset \mathbb{R}^3$ be an open bounded subset occupied by a deformable body in its undeformed state. On the assumption of small deformations we can write the relation.

$$\int_\Omega \sigma_{ij}(u)\varepsilon_{ij}(v-u)d\Omega = \int_\Omega f_i(v_i - u_i)d\Omega + \int_\Gamma \sigma_{ij}n_j(v_i - u_i)d\Gamma \quad \forall v \in V \quad (4.1.1)$$

for $u \in V$. Here V denotes the function space of the displacements which will be defined further. Relation (4.1.1) is obtained from the operator equations of the problem by applying the Green-Gauss theorem, and is the expression of the principle of virtual work for the body when considered free, i.e., with no constraints on its boundary Γ. Note that for the derivation of (4.1.1) we have multiplied the equilibrium equation

100   4. Hemivariational Inequalities

$$\sigma_{ij,j} + f_i = 0 \tag{4.1.2}$$

where the f_i is the volume force vector, by $v_i - u_i$ and then we have integrated over Ω. On the assumption of "appropriately smooth" functions, we have applied the Green-Gauss theorem by taking into account the strain-displacement relation

$$\varepsilon_{ij} = \frac{1}{2}(u_{i,j} + u_{j,i}). \tag{4.1.3}$$

Let us assume further that the boby is linear elastic, i.e. that

$$\sigma_{ij} = C_{ijhk}\varepsilon_{hk} \tag{4.1.4}$$

where $C = \{C_{ijhk}\}$, $i, j, h, k = 1, 2, 3$, is the elasticity tensor which satisfies the well-known symmetry and ellipticity properties

$$C_{ijhk} = C_{jihk} = C_{khij} \tag{4.1.5}$$

$$C_{ijhk}\varepsilon_{ij}\varepsilon_{hk} \geq c\varepsilon_{ij}\varepsilon_{hk} \quad \forall \varepsilon = \{\varepsilon_{ij}\} \in \mathbb{R}^6 \quad c \text{ const} > 0. \tag{4.1.6}$$

We denote the bilinear form of linear elasticity by $\alpha(\cdot,\cdot)$, i.e.

$$\alpha(u,v) = \int_\Omega C_{ijhk}\varepsilon_{ij}(u)\varepsilon_{hk}(v)d\Omega. \tag{4.1.7}$$

Note also that instead of (4.1.1) we can write the relation

$$\int_\Omega \sigma_{ij}\varepsilon_{ij}(v-u)d\Omega = \int_\Omega f_i(v_i - u_i)d\Omega + \int_\Gamma S_N(v_N - u_N)d\Gamma \tag{4.1.8}$$

$$+ \int_\Gamma S_{T_i}(v_{T_i} - u_{T_i})d\Gamma \quad \forall v \in V.$$

Here we have splitted the last term in (4.1.1) into the work of the normal and the work of the tangential tractions to the boundary. From (4.1.1) or (4.1.8) we shall derive certain variational formulations.

Let us assume first that on Γ the classical boundary conditions $S_N = 0$ and $u_{T_i} = 0$, $i = 1, 2, 3$, hold. Then (4.1.8) with (4.1.7) leads to the following variational equality:
Find $u \in V_0 = \{v | v \in V, v_{T_i} = 0 \text{ on } \Gamma\}$ such that

$$\alpha(u,v) = \int_\Omega f_i v_i d\Omega \quad \forall v \in V_0. \tag{4.1.9}$$

Let us assume now that on Γ the general monotone multivalued boundary condition (2.3.8a) holds. Introducing into (4.1.1) the inequality

$$j(v) - j(u) \geq S_i(v_i - u_i) \quad \forall v = \{v_i\} \in \mathbb{R}^3, \tag{4.1.10}$$

holding by definition at every point of Γ due to (2.3.8a), we obtain together with (4.1.7) the following variational formulation:
Find $u \in V$ with $j(u) < \infty$, such that

4.1 The Derivation of Hemivariational Inequalities in Mechanical Problems

$$a(u, v-u) + \int_\Gamma (j(v) - j(u))d\Gamma \qquad (4.1.11)$$
$$\geq \int_\Omega f_i(v_i - u_i)d\Omega \quad \forall v \in V \text{ with } j(v) < \infty.$$

Let us assume further that on Γ the nonmonotone, possibly multivalued boundary condition (2.4.2) holds where j is a locally Lipschitz functional. Combining (4.1.1) with the inequality

$$j^0(u, v-u) \geq S_i(v_i - u_i) \quad \forall v = \{v_i\} \in \mathbb{R}^3, \qquad (4.1.12)$$

which defines on Γ the condition (2.4.2) we obtain with (4.1.7) the following variational formulation:
Find $u \in V$ such as to satisfy the inequality

$$a(u, v-u) + \int_\Gamma j^0(u, v-u)d\Gamma \geq \int_\Omega f_i(v_i - u_i)d\Omega \quad \forall v \in V. \qquad (4.1.13)$$

If instead of (2.4.2), (2.4.1) hold on Γ then (4.1.8) gives rise to the following variational expression:
Find $u \in V$ such that

$$a(u, v-u) + \int_\Gamma j_N^0(u_N, v_N - u_N)d\Gamma + \int_\Gamma j_T^0(u_T, v_T - u_T)d\Gamma \qquad (4.1.14)$$
$$\geq \int_\Omega f_i(v_i - u_i)d\Omega \quad \forall v \in V.$$

The last type of variational expressions involving $j^0(\cdot, \cdot)$ or $j_N^0(\cdot, \cdot)$ and $j_T^0(\cdot, \cdot)$ have been called by the author, who introduced them in Mechanics in [Pan81;82; 83] and studied them in [Pan85;88;Mor88a] "hemivariational inequalities" (see also [Pan91, Nan88;89a,b]). Note that in the more general case in which j or j_N and j_T are not locally Lipschitz $j^0(\cdot, \cdot)$ in (4.1.13) and $j_N^0(\cdot, \cdot), j_T^0(\cdot, \cdot)$ in (4.1.14) are replaced by $j^\uparrow(\cdot, \cdot)$ and $j_N^\uparrow(\cdot, \cdot), j_T^\uparrow(\cdot, \cdot)$.

In order to investigate in which sense the solution of the hemivariational inequalities satisfies the operator equations of the problem, e.g. the equation of equilibrium, and the boundary conditions of the problem we need a more rigorous formulation of the above variational expressions. Let us place ourselves in the mathematical framework of Sect.2.5. Note, that the reader who is not mathematically skilled, may understand the expressions $\langle \cdot, \cdot \rangle$ simply as expressions of work. For

$$u \in V = [H^1(\Omega)]^3, \quad f_i \in L^2(\Omega) \qquad (4.1.15)$$

we have instead of (4.1.1) the relation

$$\int_\Omega \sigma_{ij}\varepsilon_{ij}(v-u)d\Omega = \int_\Omega f_i(v_i - u_i)d\Omega + \langle \sigma_{ij}n_j, v_i - u_i \rangle \quad \forall v \in [H^1(\Omega)]^3 \quad (4.1.16)$$

where
$$\sigma_{ij} \in L^2(\Omega), \quad \varepsilon_{ij} \in L^2(\Omega), \quad \sigma_{ij}n_j \in [H^{-1/2}(\Omega)]^3$$
which results from (4.1.1) by extending the arising functionals by density. The last term in (4.1.16) is the work produced by the traction $S_i = \sigma_{ij}n_j$ for the displacement variation $v_i - u_i$. If $\sigma_{ij}n_j \in L^2(\Gamma)$ then this term reduces to the correponding integral in (4.1.1). Instead of (4.1.8) we have the relation (cf. (2.5.21))

$$\int_\Omega \sigma_{ij}\varepsilon_{ij}(v-u)d\Omega = \int_\Omega f_i(v_i - u_i)d\Omega + \langle S_N, v_N - u_N \rangle_{\frac{1}{2}} \quad (4.1.17)$$
$$+ \langle S_T, v_T - u_T \rangle_{H_T}, \quad \forall v \in [H^1(\Omega)]^3.$$

In the case of (4.1.9) we have $V_0 = \{v | v \in [H^1(\Omega)]^3, v_T = 0 \text{ on } \Gamma\}$. Let us now consider the boundary condition (2.3.8a). Then using the notation (2.5.6) and the inequality (4.1.10) we are led to the following variational inequality: Find $u \in [H^1(\Omega)]^3$ such that

$$a(u, v-u) + \Phi(v) - \Phi(u) \geq \int_\Omega f_i(v_i - u_i)d\Omega \quad \forall v \in [H^1(\Omega)]^3. \quad (4.1.18)$$

Obviously we can replace in (4.1.18) $\Phi(v) - \Phi(u)$ by $\bar{\Phi}(\gamma v) - \bar{\Phi}(\gamma u)$ (see (2.5.15)). Let us now discuss in which sense the solution of (4.1.18) satisfies the equations and boundary conditions of the problem. Setting in (4.1.18) $v_i - u_i = \pm\varphi_i$, $i = 1, 2, 3$, where φ is an infinitely differentiable function from the space $C_0^\infty(\Omega)$ (see List of Notations) we obtain from (4.1.18) by means of (4.1.3), (4.1.4) and the Green-Gauss theorem the relation

$$\int_\Omega (\sigma_{ij,j} + f_i)\varphi_i d\Omega = 0 \quad \forall \varphi_i \in C_0^\infty(\Omega) \quad (4.1.19)$$

which implies that (4.1.2) holds in the sense of the distributions, i.e. as an equality in the space of distributions $D'(\Omega)$ (see List of Notations). Then the assumption $f_i \in L^2(\Omega)$ implies that (4.1.2) holds in $L^2(\Omega)$. Accordingly we can multiply it by $v - u \in [H^1(\Omega)]^3$ and apply the Green-Gauss theorem. Thus we obtain (4.1.16), which along with (4.1.3) and (4.1.4), and with (4.1.18) imply the inequality

$$\Phi(v) - \Phi(u) \geq -\langle \sigma_{ij}n_j, (v_i - u_i) \rangle \quad \forall v \in [H^1(\Omega)]^3. \quad (4.1.20)$$

This inequality can be put in the form (2.5.15), i.e. it implies that the solution of the variational inequality (4.1.18) satisfies the boundary condition in the weak form (2.5.16) on $[H^{1/2}(\Gamma)]^3 \times [H^{-1/2}(\Gamma)]^3$. Only if the structure of the problem implies that $S_i = \sigma_{ij}n_j \in L^2(\Omega)$ can we say that the boundary condition on $[L^2(\Gamma)]^3 \times [L^2(\Gamma)]^3$ (2.5.7) holds, which implies the pointwise boundary condition (2.5.8).

Let us further assume that the general nonmonotone possibly multivalued boundary condition (2.4.2) holds. Using (4.1.12) we are led to the following

4.1 The Derivation of Hemivariational Inequalities in Mechanical Problems

hemivariational inequality:
Find $u \in [H^1(\Omega)]^3$ such as to satisfy

$$\alpha(u, v-u) + \int_\Gamma j^0(u, v-u)d\Gamma \geq \int_\Omega f_i(v_i - u_i)d\Omega \quad \forall v \in [H^1(\Omega)]^3 \quad (4.1.21)$$

Conversely we can show as before that a solution of (4.1.21), if any exists, satisfies the equations of equilibrium (4.1.2) in the sense of distributions and thus in the sense of $L^2(\Omega)$, because $f_i \in L^2(\Omega)$. Indeed setting in (4.1.21) $v_i - u_i = \pm\varphi_i \in C_0^\infty(\Omega)$ implies (4.1.19) etc. Then we proceed as before by applying the Green-Gauss theorem, and we obtain that the solution of (4.1.21) satisfies the inequality

$$\int_\Gamma j^0(u, v-u)d\Gamma \geq -\langle \sigma_{ij}n_j, v_i - u_i \rangle \quad \forall v \in [H^1(\Omega)]^3 \quad (4.1.22)$$

which is a weak formulation of the boundary condition (2.4.2) on $[H^{1/2}(\Gamma)]^3 \times [H^{-1/2}(\Gamma)]^3$. Suppose now that instead of (2.4.2) the boundary condition

$$u \in \bar\partial \tilde{j}(-S) \quad (4.1.23)$$

holds. Then by definition (4.1.23) is equivalent to the inequality

$$\tilde{j}^0(S, -T+S) \geq -u_i(T_i - S_i) \quad \forall T = \{T_i\} \in [H^{-1/2}(\Gamma)]^3. \quad (4.1.24)$$

Thus we may derive hemivariational inequalities with respect to the stress variation. To this end let us multiply (4.1.3) by $\tau_{ij} - \sigma_{ij}$, add according to the Einstein convention, as it is common in this book, and then integrate over Ω. On the assumption that all the functions are appropriately smooth we obtain the equality

$$\int_\Omega \varepsilon_{ij}(\tau_{ij} - \sigma_{ij})d\Omega = -\int_\Omega u_i(\tau_{ij,j} - \sigma_{ij,j})d\Omega \quad (4.1.25)$$
$$+ \int_\Gamma u_i(\tau_{ij}n_j - \sigma_{ij}n_j)d\Gamma \quad \forall \tau = \{\tau_{ij}\} \in \Sigma$$

where Σ is the admissible space of the stress tensor. Now we define the statically admissible set

$$\Sigma_0 = \{\tau \mid \tau\{\tau_{ij}\}, \tau_{ij} \in \Sigma, \tau_{ij,j} + f_i = 0 \quad \text{a.e. on} \quad \Omega\} \quad (4.1.26)$$

and (4.1.25) becomes

$$\int_\Omega \varepsilon_{ij}(\tau_{ij} - \sigma_{ij})d\Omega = \int_\Gamma u_i(\tau_{ij}n_j - \sigma_{ij}n_j)d\Gamma \quad \forall \tau \in \Sigma_0. \quad (4.1.27)$$

From (4.1.27) and (4.1.24) and the Hooke's law we obtain the hemivariational inequality:
Find $\sigma \in \Sigma_0$ such as to satisfy the inequality

$$A(\sigma, \tau - \sigma) + \int_\Gamma \bar{j}^0(-S, -T + S) d\Gamma \geq 0 \quad \forall \tau \in \Sigma_0 \qquad (4.1.28)$$

where

$$A(\tau, \sigma) = \int_\Omega c_{ijhk} \tau_{ij} \sigma_{hk} d\Omega. \qquad (4.1.29)$$

Here $c = \{c_{ijhk}\}$ is the inverse Hooke's tensor and $T = \{T_i\} = \{\tau_{ij} n_j\}$. Note that if $u_i \in H^1(\Omega)$, and $f_i \in L^2(\Omega)$ then $\sigma_{ij} \in L^2(\Omega), \sigma_{ij} n_j \in H^{-1/2}(\Gamma)$, and (4.1.27) takes the form

$$\int_\Omega \varepsilon_{ij}(\tau_{ij} - \sigma_{ij}) d\Omega = \langle u_i, \tau_{ij} n_j - \sigma_{ij} n_j \rangle \quad \forall \tau \in \Sigma_0 \qquad (4.1.30)$$

where $\Sigma = L^2(\Omega)$. Conversely arguing as in the case of (4.1.21) we may find that the solution at (4.1.28) satisfies (4.1.3) in the sense of distributions over Ω. But $\varepsilon_{ij} \in L^2(\Omega)$ because it is related to $\sigma_{ij} \in L^2(\Omega)$ by Hooke's law. Thus (4.1.3) holds as an equality in $L^2(\Omega)$. We apply again Green-Gauss theorem to (4.1.3) (after its multiplication by $\tau_{ij} - \sigma_{ij}$, addition with respect to i, j, and integration over Ω) and we obtain from (4.1.28) and (4.1.30) the inequality

$$\int_\Gamma \bar{j}^0(-S, -T + S) \geq -\langle u_i, \tau_{ij} n_j - \sigma_{ij} n_j \rangle \quad \forall \tau = \{\tau_{ij}\} \in \Sigma_0 \qquad (4.1.31)$$

which is a weak formulation over $[H^{-1/2}(\Gamma)]^3 \times [H^{1/2}(\Gamma)]^3$ of the boundary condition (4.1.23).

If j in (4.1.10) is convex, l.s.c. and proper then (2.5.8) is equivalently written in the inverse form (2.5.10). This is not possible in the nonconvex case. The hemivariational inequality (4.1.21) (resp. (4.1.28)) expresses the "principle" of virtual work (resp. of complementary virtual work) in inequality form. Note that for boundary conditions expressed via convex superpotentials the hemivariational inequalities (4.1.21) and (4.1.28) become variational inequalities (e.g. the terms involving the directional derivative $\int_\Gamma j^0(u, v) d\Gamma$ is replaced by the difference $\Phi(v) - \Phi(u)$).

A B.V.P. is called bilateral (resp. unilateral) if it leads to variational equality (resp. variational, or hemivariational, or variational-hemivariational inequality) formulations. We call the unilateral problems "inequality problems" too. Note at this point that the term "unilateral boundary conditions" has been initially used and is until now in use, in order to characterize boundary conditions involving inequalities. But as we have seen, through the introduction of the indicator, the inequality boundary conditions may be put also in the general multivalued form (2.5.16) (cf. (2.5.32) and (2.5.34)). As Fourier has noticed [Lan] the inequality form of the principle of virtual or complementary virtual work is due to the fact that the variations of certain variables involved into the problem are "irreversible". For instance, if (2.1.39) held for $u, v \in V$, where V is a vector space then the substitution $v - u = \pm w$ would lead to a variational equality. But since $u, v \in K$, where K is a closed convex set, we cannot set $v - u = \pm w$, i.e., the variation $v - u$ is irreversible. Irreversible variations are called "unilateral" variations. Unilateral are the variations also in (4.1.18), unless grad $\bar{\Phi}$ exists everywhere. Indeed in this case (4.1.18) is equivalent to the variational equality

$$\alpha(u, w) + \langle \mathrm{grad}\bar{\Phi}(\gamma u), \gamma w \rangle = \int_\Omega f_i w_i d\Omega \quad \forall w \in [H^1(\Omega)]^3 \tag{4.1.32}$$

as it results easily by setting in (4.1.18) $v = u \pm \lambda w, \lambda \to 0_+$. The converse easily results by setting in (4.1.32) $w = v - u$ and by applying the inequality ($\bar{\Phi}$ is convex)

$$\bar{\Phi}(\gamma v) - \bar{\Phi}(\gamma u) \geq \langle \mathrm{grad}\bar{\Phi}(\gamma u), \gamma(v - u) \rangle \quad \forall v \in [H^1(\Omega)]^3. \tag{4.1.33}$$

Analogously we may agrue in the case of hemivariational inequalities.

We refer also to [Pan85] concerning the relation of variational "principles" with the chosen duality pairing between the "generalized forces" and "generalized displacements" of the problem under consideration. These last ideas constitute generalizations and amelioration of analogous ideas of Tonti concerning the bilateral problems [Tont].

Closing this section let us note that the variational inequalities, the hemivariational inequalities etc., belong, due to the arising nonsmooth energy functionals, to the Nonsmooth Mechanics [Mor88a,b] as it has been called this category of problems by the author in [Pan85] p.374.

4.2 Hemivariational and Variational-Hemivariational Inequalities

In this section we shall derive certain hemivariational inequalities and variational-hemivaritional inequalities with respect to the problem of adhesive contact of linear elastic and nonlinear elastic bodies and also with respect to the delamination problem of laminated von Kármán plates subjected to unilateral boundary conditions. These two pilot problems will permit the reader to understand the method of derivation of hemivariational and of variational-hemivariational inequalities.

i) Adhesive Contact Problem

Let Ω_m, $m = 1, 2, \ldots, l$, be a set of deformable bodies, possibly with different elasticity properties, with the boundaries Γ_m, $m = 1, 2, \ldots, l$, assumed to be appropriately regular. Let $x = \{x_i\}$, $i = 1, 2, 3$, be a point of $\mathbb{R}^3$ and let $\sigma^{(m)} = \sigma_{ij}^{(m)}$ and $\varepsilon^{(m)} = \varepsilon_{ij}^{(m)}$, $i, j = 1, 2, 3$, be the stress and strain tensors of the m-body. We denote by $f^{(m)} = \{f_i^{(m)}\}$ and $u^{(m)} = \{u_i^{(m)}\}$ the volume force and the displacement vector in each body. If $n^{(m)} = \{n_i^{(m)}\}$ is the outward unit normal vector to $\Gamma^{(m)}$, the boundary force on $\Gamma^{(m)}$ is $S_i^{(m)} = \sigma_{ij}^{(m)} n_{ij}^{(m)}$ (summation convention). Let $S_N^{(m)}$ and $S_T^{(m)}$ be the normal and tangential components of it respectively. The corresponding displacement components are $u_N^{(m)}$ and $u_T^{(m)}$. The boundary $\Gamma^{(m)}$ is divided into three non-overlapping parts $\Gamma_U^{(m)}, \Gamma_F^{(m)}$ and $\Gamma_S^{(m)}$. On $\Gamma_U^{(m)}$ the displacements are given; let us take for simplicity that

$$u_i^{(m)} = 0 \quad \text{on} \quad \Gamma_U^{(m)}. \tag{4.2.1}$$

On $\Gamma_F^{(m)}$ the forces are prescribed, i.e.,

$$S_i^{(m)} = F_i^{(m)} \quad \text{on} \quad \Gamma_F^{(m)} \tag{4.2.2}$$

and on $\Gamma_S^{(m)}$-which corresponds to the interface of structure m with other substructures - nonmonotone interface conditions hold describing slip and delamination effects. We write in the general case $\Omega^{(m)} \subset \mathbb{R}^3$ the interface conditions in the form

$$-S_N^{(m)} \in \bar{\partial} j_{N(m)}(S^{(m)}; [u_N^{(m)}]) \tag{4.2.3}$$

$$-S_T^{(m)} \in \bar{\partial} j_{T(m)}(S^{(m)}; [u_T^{(m)}]) \tag{4.2.4}$$

in the normal and in the tangential direction to the interface. The superpotentials j_N and j_T are assumed to be functions of the interlayer gap $[u_N]$ and slip $[u_T]$ (locally Lipschitz continuous) respectively and of the interface traction S, which is also a function of u. Here, however, we assume that (4.2.3) and (4.2.4) are uncoupled, i.e. that $S^{(m)}$ is considered as having a given value, or that $j_{N(m)}$ and $j_{T(m)}$ do not depend on $S^{(m)}$. Then (4.2.3) (4.2.4) are equivalent to the inequalities

$$j_{N(m)}^0([u_N^{(m)}], v - [u_N^{(m)}]) \geq -S_N^{(m)}(v - [u_N^{(m)}]) \quad \forall v \in \mathbb{R} \tag{4.2.5}$$

$$j_{T(m)}^0([u_T^{(m)}], v - [u_T^{(m)}]) \geq -S_{T_i}^{(m)}(v_i - [u_T^{(m)}]) \quad \forall v_i \in \mathbb{R} \quad i = 1, 2, 3. \tag{4.2.6}$$

In the framework of small deformations and linear elastic behaviour for $\Omega^{(m)}$ $m = 1, 2, \ldots, l$, we can write the relations

$$\sigma_{ij,j}^{(m)} + f_i^{(m)} = 0, \tag{4.2.7}$$

$$\varepsilon_{ij}^{(m)} = \frac{1}{2}(u_{i,j}^{(m)} + u_{j,i}^{(m)}) = \varepsilon_{ij}(u^{(m)}), \tag{4.2.8}$$

$$\sigma_{ij}^{(m)} = C_{ijhk}^{(m)} \varepsilon_{hk}^{(m)}. \tag{4.2.9}$$

Hooke's tensor $C^{(m)} = \{C_{ijhk}^{(m)}\}$ satisfies the well-known symmetry and ellipticity conditions. We write the principle of virtual work for every body $\Omega^{(m)}$ in the form

$$\int_{\Omega^{(m)}} \sigma_{ij}^{(m)} \varepsilon_{ij}^{(m)}(v^{(m)} - u^{(m)}) d\Omega = \int_{\Omega^{(m)}} f_i^{(m)}(v_i^{(m)} - u_i^{(m)}) d\Omega \tag{4.2.10}$$

$$+ \int_{\Gamma_F^{(m)}} F_i^{(m)}(v_i^{(m)} - u_i^{(m)}) d\Gamma$$

$$+ \int_{\Gamma_S^{(m)}} \left[S_N^{(m)}(v_N^{(m)} - u_N^{(m)}) + S_{T_i}^{(m)}(v_{T_i}^{(m)} - u_{T_i}^{(m)}) \right] d\Gamma \quad \forall v \in V_{ad}^{(m)}$$

where $V_{\text{ad}}^{(m)}$ is the kinematically admissible set of $\Omega^{(m)}$, i.e.

$$V_{\text{ad}}^{(m)} = \{v^{(m)} | v^{(m)} = v_i^{(m)}, \ v_i^{(m)} \in V(\Omega^{(m)}), \ v_i^{(m)} = 0 \ \text{ on } \Gamma_U^{(m)}\}. \quad (4.2.11)$$

Here $V(\Omega^{(m)})$ denotes a space of functions defined on $\Omega^{(m)}$. Adding with respect to m all the expressions (4.2.10) and taking into account the interconnection of the bodies yields a relation of the form

$$\sum_{m=1}^{l} \int_{\Omega^{(m)}} \sigma_{ij}^{(m)} \varepsilon_{ij}^{(m)} (v^{(m)} - u^{(m)}) d\Omega \quad (4.2.12)$$

$$= \sum_{m=1}^{l} \left[\int_{\Omega^{(m)}} f_i^{(m)}(v_i^{(m)} - u_i^{(m)}) d\Omega + \int_{\Gamma_F^{(m)}} F_i^{(m)}(v_i^{(m)} - u_i^{(m)}) d\Gamma \right]$$

$$+ \sum_{q=1}^{k} \left[\int_{\Gamma^{(q)}} S_N^{(q)} ([v_N^{(q)}] - [u_N^{(q)}]) d\Gamma + \int_{\Gamma^{(q)}} S_{T_i}^{(q)} ([v_{T_i}^{(q)}] - [u_{T_i}^{(q)}]) d\Gamma \right] \quad \forall v \in V_{\text{ad}},$$

where $V_{\text{ad}} = \bigcup_{m=1}^{l} V_{\text{ad}}^{(m)}$

In (4.2.12) the integrals along the joints Γ_q, $q = 1, \ldots, k$, have been introduced. The new enumeration of the $\Gamma_S^{(m)}$-boundaries has the advantage that finally the energy of each joint appears. One should take into account that the variation of the energy of each constraint of the form (4.2.3) and (4.2.4) must appear only once in the last terms of (4.2.12). Further we introduce the elastic energy of the m-structure

$$\alpha(u^{(m)}, v^{(m)}) = \int_{\Omega^{(m)}} C_{ijhk}^{(m)} \varepsilon_{ij}(u^{(m)}) \varepsilon_{hk}(v^{(m)}) d\Omega \quad (4.2.13)$$

and by taking into account (4.2.5), (4.2.6) and (4.2.13), we get from (4.2.12) the following hemivariational inequality:
Find $u \in V_{\text{ad}}$ such as to satisfy

$$\sum_{m=1}^{l} \alpha(u^{(m)}, v^{(m)} - u^{(m)}) + \sum_{q=1}^{k} \left[\int_{\Gamma_q} \left[j_{N(q)}^{0}([u_N^{(q)}], [v_N^{(q)}] - [u_N^{(q)}]) \right. \right.$$

$$\left. \left. + j_{T(q)}^{0}([u_T^{(q)}], [v_T^{(q)}] - [u_T^{(q)}]) \right] d\Gamma \right] \geq \sum_{m=1}^{l} \left[\int_{\Omega^{(m)}} f_i^{(m)}(v_i^{(m)} - u_i^{(m)}) d\Omega \right.$$

$$\left. + \int_{\Gamma_F^{(m)}} F_i^{(m)}(v_i^{(m)} - u_i^{(m)}) d\Gamma \right] \quad \forall v \in V_{\text{ad}}. \quad (4.2.14)$$

This hemivariational inequality is the expression of the principle of virtual work in its inequality form for the structure under consideration.

To check in which sense a solution of (4.2.14) fullfils (4.2.7), the boundary conditions on $\Gamma_F^{(m)}$ $m = 1, \ldots, l$ and the interface relations (4.2.3), (4.2.4) we must make the functional setting of the problem more precise. So we assume that $f_i^{(m)} \in L^2(\Omega^{(m)})$, $F_i^{(m)} \in L^2(\Gamma_F^{(m)})$, $C_{ijhk}^{(m)} \in L^\infty(\Omega^{(m)})$, $u_i^{(m)}, v_i^{(m)} \in H^1(\Omega^{(m)})$. Then $u_N^{(m)}, u_{T_i}^{(m)} \in H^{1/2}(\Gamma^{(m)})$ and $S_N^{(m)}, S_{T_i}^{(m)} \in H^{-1/2}(\Gamma^{(m)})$. We set in (4.2.14) $v_i^{(m)} - u_i^{(m)} = \pm \phi_i^{(m)}$ where $\phi_i^{(m)}$ belongs to the space $C_0^\infty(\Omega^{(m)})$ of infinitely differentiable functions with compact support in $\Omega^{(m)}$. Then from (4.2.14) by setting $v_i^{(m)} - u_i^{(m)} = \pm \phi_i^{(m)}$ for $m = n$ and $v_i^{(m)} - u_i^{(m)} = 0$ for $m \neq n$ we obtain

$$a(u^{(n)}, \phi^{(n)}) = \int_{\Omega^{(n)}} f_i^{(n)} \phi_i^{(n)} d\Omega \qquad (4.2.15)$$

since $\phi_i^{(n)} = 0$ on $\Gamma^{(n)}$. Relation (4.2.15) implies that (4.2.7) holds on $\Omega^{(n)}$ in the sense of distributions over $\Omega^{(n)}$. This procedure is repeated for $n = 1, 2, \ldots, l$. Now applying the Green-Gauss theorem to each body we obtain the equality

$$a(u^{(m)}, v^{(m)} - u^{(m)}) = \int_{\Omega^{(m)}} f_i^{(m)}(v_i^{(m)} - u_i^{(m)}) d\Omega \qquad (4.2.16)$$
$$+ \int_{\Gamma_F^{(m)}} S_i^{(m)}(v_i^{(m)} - u_i^{(m)}) d\Gamma$$
$$+ \int_{\Gamma_S^{(m)}} [S_N^{(m)}(v_N^{(m)} - u_N^{(m)}) + S_{T_i}^{(m)}(v_{T_i}^{(m)} - u_{T_i}^{(m)})] d\Gamma.$$

More correctly in (4.2.16) we should have instead of $\int_{\Gamma_F^{(m)}}$ and $\int_{\Gamma_S^{(m)}}$ the corresponding expressions with the duality pairings $\langle \cdot, \cdot \rangle$, $\langle \cdot, \cdot \rangle_{1/2}$, and $\langle \cdot, \cdot \rangle_{H_T}$ (cf. Sect. 2.5) From (4.2.16) and (4.2.14) we obtain the inequality

$$\sum_{q=1}^{k} \left[\int_{\Gamma_q} [j_{N(q)}^0([u_N^{(q)}], [v_N^{(q)}] - [u_N^{(q)}]) + j_{T(q)}^0([u_T^{(q)}], [v_T^{(q)}] - [u_T^{(q)}])] d\Gamma \right]$$
$$+ \sum_{m=1}^{l} \langle S_i^{(m)} - F_i^{(m)}, v_i^{(m)} - u_i^{(m)} \rangle_{\Gamma_F^{(m)}} + \sum_{q=1}^{k} \{ \langle S_N^{(q)}, [v_N^{(q)}] - [u_N^{(q)}] \rangle_{\frac{1}{2}, \Gamma_q}$$
$$+ \langle S_{T_i}^{(q)}, [v_{T_i}^{(q)}] - [u_{T_i}^{(q)}] \rangle_{H_T, \Gamma_q} \} \geq 0 \quad \forall v \in V_{ad}. \qquad (4.2.17)$$

If in (4.2.17) we consider that on $\Gamma_F^{(m)}$, $v_i^{(m)} - u_i^{(m)} = \pm r_i^{(m)} \in H^{1/2}(\Gamma^{(m)})$ for $m = n$, and that $v_i^{(m)} - u_i^{(m)} = 0$ for $m \neq n$ on $\Gamma_F^{(m)}$ and on Γ_q for every q, we obtain $S_i^{(n)} = F_i^{(n)}$ as an equality in $H^{-1/2}(\Gamma^{(n)})$; this can be shown for every n. From (4.2.17) by setting $[v_N^{(q)}] - [u_N^{(q)}] = r_N^{(q)}$ on Γ_q for $q = n$ and the same difference is zero for $q \neq n$, and setting $[v_T^{(q)}] - [u_T^{(q)}] = 0$ on Γ_q for every q we obtain

$$\int_{\Gamma_n} j_{N(n)}^0([u_N^{(n)}], r_N^{(n)}) d\Gamma \geq -\langle S_N^{(n)}, r_N^{(n)} \rangle_{\frac{1}{2}, \Gamma_n} \quad \forall r_N^{(n)} \in H^{1/2}(\Gamma) \qquad (4.2.18)$$

4.2 Hemivariational and Variational-Hemivariational Inequalities

which constitutes a "weak" formulation of (4.2.3) on $H^{-1/2}(\Gamma) \times H^{1/2}(\Gamma)$. Analogously we obtain from (4.2.17) a weak form of (4.2.4).

Suppose now that the substructures $\Omega^{(m)}$ obey a general nonmonotone law of the form

$$\sigma^{(m)} \in \bar{\partial} w_{(m)}(\varepsilon) \qquad (4.2.19)$$

where $W_{(m)}$ is an extended real-valued function nonconvex and noneverywhere differentiable (cf. e.g. Sect. 3.2). Then the variational expression of the problem is the same as in (4.2.14) but now the term $\sum_{m=1}^{l} \alpha(u^{(m)}, v^{(m)} - u^{(m)})$ has to be replaced by $\sum_{m=1}^{l} \int_{\Omega^{(m)}} w_{(m)}^{\uparrow}(\varepsilon(u^{(m)}), \varepsilon(v^{(m)} - u^{(m)})) d\Omega$, where the integral is set equal to ∞, if an integrand is not a $L^1(\Omega)$-function.

If the $w_{(m)}$'s are convex superpotentials, i.e. are convex, l.s.c and proper functionals, then we define

$$W_{(m)}(\varepsilon) = \begin{cases} \int_{\Omega^{(m)}} w_{(m)}(\varepsilon) d\Omega & \text{if } w_{(m)}(\cdot) \in L^1(\Omega^{(m)}) \\ \infty & \text{otherwise} \end{cases} \qquad (4.2.20)$$

and we are led to a variational formulation analogous to (4.2.14); now the elastic energy function $\alpha(\cdot, \cdot)$ is replaced by the difference $\sum_{m=1}^{l}[w_{(m)}(\varepsilon(v^{(m)})) - w_{(m)}(\varepsilon(u^{(m)}))]$. This variational form is called variational-hemivariational inequality and is the expression of the principle of virtual work for the considered problem.

ii) Laminated von Kármán plates

Now we will study the delamination effect for laminated plates undergoing large displacements (von Kármán plates). Delamination [Jone] is one of the main causes of strength-degradation. For a laminated plate the mechanical behaviour of the interlayer binding material, together with the possibility of debonding is described by a nonmonotone, possibly multivalued law connecting the interlaminar bonding forces with the corresponding relative displacements. At the boundary of the plate monotone boundary conditions are assumed to hold, e.g. the Signorini-Fichera boundary condition or the plastic hinge boundary condition (see e.g. [Pan85]). The interlayer law (resp. the boundary law) is expressed through nonconvex (resp. convex) superpotentials leading to hemivariational (resp. variational) inequalities. Thus the whole problem gives rise, as we shall see further, to a hemivariational inequality concerning the bending of the laminae and to a variational inequality concerning the stretching of the plate.

110   4. Hemivariational Inequalities

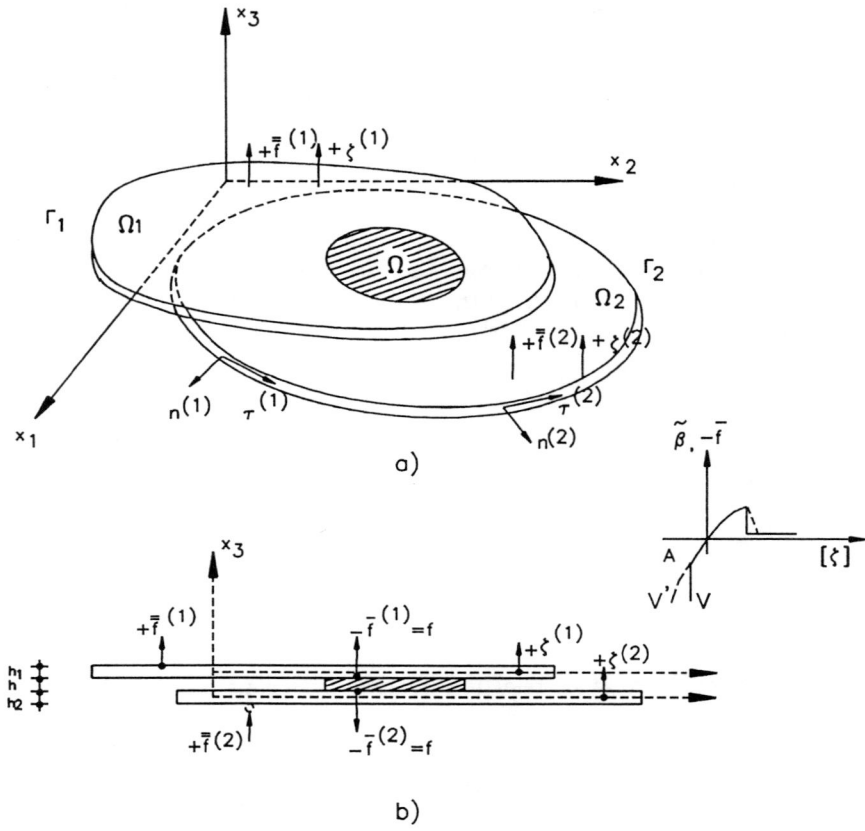

Fig. 4.2.1. Notation and geometry of the laminated plate.

Consider a laminated plate consisting of two laminae and the binding material between them (Fig 4.2.1b). In the undeformed state the middle surface of lamina j occupies an open, bounded and connected subset Ω_j of $\mathbb{R}^2$, referred to a fixed right-handed Cartesian coordinate system $Ox_1x_2x_3$. Let $\Gamma_j, j = 1, 2$, be the boundary of the j-th lamina. Γ_j is assumed to be appropriately regular (in general, a Lipschitz boundary $C^{0,1}$ is sufficient). Let also the interlaminar binding material occupy a subset Ω' such that $\Omega' \subset \Omega_1 \cap \Omega_2$ and $\bar{\Omega}' \cap \Gamma_1 = \emptyset, \bar{\Omega}' \cap \Gamma_2 = \emptyset$. The binding material and the two laminae together form one integral structural element. We denote by $\zeta^{(j)}(x)$ the vertical deflection of the point $x \in \Omega_j$ of the j-th lamina, and by $f^{(j)} = (0, 0, f_3^{(j)}(x))$ the distributed vertical load acting on the j-th lamina.

Further, let $u^{(j)} = \{u_1^{(j)}, u_2^{(j)}\}$ be the in-plane displacement of the j-th lamina. We assume that the j-th lamina has constant thickness h_j, while the interlaminar binding layer has constant thickness h. Moreover, we assume that each lamina obeys the von Kármán plate theory, i.e. it is a thin plate having large deflections. The following system of differential equations holds for von Kármán plates.

$$K_j \Delta\Delta \zeta^{(j)} - h_j(\sigma_{\alpha\beta}^{(j)}\zeta_{,\beta}^{(j)})_{,\alpha} = f^{(j)} \quad \text{in } \Omega_j, \qquad (4.2.21)$$

$$\sigma_{\alpha\beta,\beta}^{(j)} = 0 \quad \text{in } \Omega_j, \qquad (4.2.22)$$

and

$$\sigma_{\alpha\beta}^{(j)} = C_{\alpha\beta\gamma\delta}^{(j)}(\varepsilon_{\gamma\delta}^{(j)}(u^{(j)}) + \frac{1}{2}\zeta_{,\gamma}^{(j)}\zeta_{,\delta}^{(j)}) \quad \text{in } \Omega_j. \qquad (4.2.23)$$

Here the subscripts $\alpha, \beta, \gamma, \delta = 1, 2$ correspond to the coordinate directions; the superscript or the subscript $j = 1, 2$ refers to the j-th lamina; $\{\sigma_{\alpha\beta}^{(j)}\}, \{\varepsilon_{\alpha\beta}^{(j)}\}$, and $\{C_{\alpha\beta\gamma\delta}^{(j)}\}$ denote the stress, strain and elasticity tensors in the plane of the plate. The components of $C^{(j)}$ are elements of $L^\infty(\Omega_j)$ and have the usual symmetry and ellipticity properties. Moreover, $K_j = Eh_j^3/12(1-\nu^2)$ is the bending rigidity of the j-th plate with E the modulus of elasticity and ν the Poisson ratio. For the sake of simplicity, we consider here isotropic homogeneous plates of constant thickness. In laminated and layered plates, the interlaminar normal stress σ_{33} is one of the main causes for delamination effects (see e.g. [Mos] p.318). Note that this is a simplification of the mechanical problem. In order to model the action of σ_{33}, $f^{(j)}$ is split into $\bar{f}^{(j)}$, which describes the interaction of the two plates, and $\bar{\bar{f}}^{(j)} \in L^2(\Omega_j)$, which represents the external loading applied on the j-th plate:

$$f^{(j)} = \bar{f}^{(j)} + \bar{\bar{f}}^{(j)} \quad \text{in } \Omega_j, \quad j = 1, 2. \qquad (4.2.24)$$

If f denotes the stress in the interlaminar binding layer, the following holds (see Fig. 4.2.1)

$$f = \bar{f}^{(1)} = -\bar{f}^{(2)} \quad \text{in } \Omega'. \qquad (4.2.25)$$

We introduce now a phenomenological law connecting f with the corresponding relative deflection of the plates (see Fig. 4.2.1) $[\zeta] = \zeta^{(1)} - \zeta^{(2)}$.

We assume that

$$-f \in \tilde{\beta}([\zeta]) \quad \text{in } \Omega' \qquad (4.2.26)$$

where $\tilde{\beta}$ is a multivalued function defined as in (1.2.53). We note here that cracking as well as crushing effects of either a brittle or semi-brittle nature can be accounted for by means of this law. The impenetrability restriction would imply a vertical branch AV (Fig. 4.2.1b) in the final compression state of the binding layer. However a slightly inclined branch AV' is taken here into account in order to consider the compression of the laminae in the Ox_3-direction. The following relations complete in a natural way the definition of $\bar{f}^{(j)}$:

$$\bar{f}^{(j)} = 0 \quad \text{in } \Omega_j - \Omega', \quad j = 1, 2. \qquad (4.2.27)$$

In order to obtain a variational formulation of the problem, we express relation (4.2.26) in a superpotential form. If $\beta(\xi_{\pm 0})$ exists for every $\xi \in \mathbb{R}$ then (cf. Sect.1.2) a locally Lipschitz (nonconvex) function $J : \mathbb{R} \to \mathbb{R}$ can be determined up to an additive constant, such that

4. Hemivariational Inequalities

$$\tilde{\beta}(\xi) = \bar{\partial} J(\xi). \tag{4.2.28}$$

Moreover, we assume that the following boundary conditions hold on the subset $\tilde{\Gamma}_j$ of the plate boundaries (cf. (2.3.31), (2.3.32))

$$M_{nj}(\zeta^{(j)}) \in \beta_j\left(\frac{\partial \zeta^{(j)}}{\partial n}\right) \quad \text{on } \tilde{\Gamma}_j, \quad j = 1, 2, \tag{4.2.29a}$$

$$-K_{nj}(\zeta^{(j)}) \in \beta'_j(\zeta^{(j)}) \quad \text{on } \tilde{\Gamma}_j, \quad j = 1, 2 \tag{4.2.29b}$$

whereas

$$\zeta^{(j)} = \frac{\partial \zeta^{(j)}}{\partial n} = 0 \quad \text{on } \Gamma_j - \tilde{\Gamma}_j, \quad j = 1, 2. \tag{4.2.29c}$$

Here $\beta_j, \beta'_j, j = 1, 2$ are possibly multivalued maximal monotone operators from $\mathbb{R}$ into $\mathcal{P}(\mathbb{R})$ (cf. Sect.2.3). Accordingly, convex, l.s.c., proper functionals $\phi_j, \phi'_j, j = 1, 2$ can be determined such that

$$\beta_j\left(\frac{\partial \zeta^{(j)}}{\partial n}\right) = \partial \phi_j\left(\frac{\partial \zeta^{(j)}}{\partial n}\right), \quad j = 1, 2, \tag{4.2.30a}$$

and

$$\beta'_j(\zeta^{(j)}) = \partial \phi'_j(\zeta^{(j)}), \quad j = 1, 2. \tag{4.2.30b}$$

For the physical meaning of the boundary conditions (4.2.29a,b), which also include all the classical boundary conditions, the reader is referred to [Pan85]. Let us introduce the Sobolev space $H^2(\Omega_j)$ for the deflections $\zeta^{(j)}$ and define the following convex, l.s.c. and proper functionals on $H^2(\Omega_j), j = 1, 2$:

$$\Phi_j(z^{(j)}) = \begin{cases} \int_{\tilde{\Gamma}_j} \phi_j\left(\frac{\partial z^{(j)}}{\partial n}\right) d\Gamma + \int_{\tilde{\Gamma}_j} \phi'_j(z^{(j)}) d\Gamma, \\ \text{if } \phi_j\left(\frac{\partial z^{(j)}}{\partial n}\right) \in L^1(\tilde{\Gamma}_j) \text{ and } \phi'_j(z^{(j)}) \in L^1(\tilde{\Gamma}_j) \\ \infty \quad \text{otherwise}, \quad j = 1, 2. \end{cases} \tag{4.2.31}$$

For the in-plane displacement we assume the boundary conditions

$$\sigma_{\alpha\beta}^{(j)} n_\beta^{(j)} = 0 \quad \text{on } \Gamma_j. \tag{4.2.32}$$

We can now derive the variational formulation of the problem. From (4.2.21), by assuming sufficiently regular functions, multiplying by $z^{(j)} - \zeta^{(j)}$, integrating and applying the Green-Gauss theorem, we obtain the expressions:

$$a(\zeta^{(j)}, z^{(j)} - \zeta^{(j)}) + \int_{\Omega_j} h_j \sigma_{\alpha\beta}^{(j)} \zeta_{,\alpha}^{(j)} (z^{(j)} - \zeta^{(j)})_{,\beta} d\Omega \tag{4.2.33}$$

$$= \int_{\Gamma_j} h_j \sigma_{\alpha\beta,\beta}^{(j)} \zeta^{(j)} n_\alpha^{(j)} (z^{(j)} - \zeta^{(j)}) d\Gamma + \int_{\Omega_j} \bar{\bar{f}}^{(j)}(z^{(j)} - \zeta^{(j)}) d\Omega$$

$$+ \int_{\Gamma_j} K_{nj}(\zeta^{(j)})(z^{(j)} - \zeta^{(j)}) d\Gamma - \int_{\Gamma_j} M_{nj}(\zeta^{(j)}) \frac{\partial(z^{(j)} - \zeta^{(j)})}{\partial n^{(j)}} d\Gamma,$$

$$j = 1, 2, \quad \alpha, \beta = 1, 2.$$

4.2 Hemivariational and Variational-Hemivariational Inequalities

Here $n^{(j)}$ denotes the outward normal unit vector to Γ_j

$$\alpha(\zeta, z) = K \int_\Omega [(1-\nu)\zeta_{,\alpha\beta} z_{,\alpha\beta} + \nu \Delta\zeta \Delta z] d\Omega, \quad \alpha,\beta = 1,2, 0 < \nu < 0.5, \quad (4.2.34)$$

$$M_n(\zeta) = -K[\nu \Delta\zeta + (1-\nu)(2n_1 n_2 \zeta_{,12} + n_1^2 \zeta_{,11} + n_2^2 \zeta_{,22}], \quad (4.2.35)$$

$$K_n(\zeta) = -K\left[\frac{\partial \Delta\zeta}{\partial n} + (1-\nu)\frac{\partial}{\partial \tau}[n_1 n_2 (\zeta_{,22} - \zeta_{,11}) + (n_1^2 - n_2^2)\zeta_{,12}]\right]. \quad (4.2.36)$$

Here ν is the Poisson ratio and τ is the unit vector tangential to Γ, such that ν, τ and the Ox_3-axis form a right-handed system. Applying the same technique to (4.2.22), implies the expression

$$\int_{\Omega_j} \sigma_{\alpha\beta}^{(j)} \varepsilon_{\alpha\beta}^{(j)} (v^{(j)} - u^{(j)}) d\Omega \quad (4.2.37)$$

$$= \int_{\Gamma_j} \sigma_{\alpha\beta}^{(j)} n_\beta^{(j)} (v_\alpha^{(j)} - u_\alpha^{(j)}) d\Gamma, \quad j = 1, 2, \ \alpha, \beta = 1, 2.$$

Further, the following notations are introduced:

$$R(m, k) = \int_\Omega C_{\alpha\beta\gamma\delta} m_{\alpha\beta} k_{\gamma\delta} d\Omega, \quad \alpha, \beta, \gamma, \delta = 1, 2 \quad (4.2.38)$$

and

$$P(\zeta, z) = \{\zeta_{,\alpha} z_{,\beta}\}, \quad P(\zeta, \zeta) = P(\zeta), \quad (4.2.39)$$

where $m = \{m_{\alpha\beta}\}$ and $k = \{k_{\alpha\beta}\}, \alpha, \beta = 1, 2$ are 2×2 tensors.

Let us also introduce a functional framework for the B.V.P. We assume that $u^{(j)}, v^{(j)} \in [H^1(\Omega_j)]^2$ and that $\zeta^{(j)}, z^{(j)} \in Z_j$, where

$$Z_j = \left\{z \Big| z \in H^2(\Omega_j), \ z = 0 \text{ on } \Gamma_j - \tilde{\Gamma}_j, \ \frac{\partial \zeta}{\partial n} = 0 \text{ a.e. } \Gamma_j - \tilde{\Gamma}_j\right\}. \quad (4.2.40)$$

Taking into account the notation introduced in (4.2.38), (4.2.39) and the variational equalities (4.2.33), (4.2.37), the boundary conditions (4.2.29a,b,c) and (4.2.32) the interface conditions (4.2.26), (4.2.27), the definition (4.2.31) and the inequalities defining the multivalued operators ∂ and $\bar{\partial}$ we obtain the following problem :
Find $u^{(j)} \in [H^1(\Omega_j)]^2$ and $\zeta^{(j)} \in Z_j, j = 1, 2$, such as to satisfy the variational-hemivariational inequality

$$\sum_{j=1}^{2} \alpha_j(\zeta^{(j)}, z^{(j)} - \zeta^{(j)}) + \sum_{j=1}^{2} h_j R(\varepsilon(u^{(j)}) + \frac{1}{2}P(\zeta^{(j)}), P(\zeta^{(j)}, z^{(j)} - \zeta^{(j)}))$$

$$+ \int_{\Omega'} J^0([\zeta], [z] - [\zeta]) d\Omega + \sum_{j=1}^{2} \{\Phi_j(z^{(j)}) - \Phi_j(\zeta^{(j)})\}$$

$$\geq \sum_{j=1}^{2} \int_{\Omega_j} \bar{\bar{f}}^{(j)} (z^{(j)} - \zeta^{(j)}) d\Omega, \quad \forall z^{(1)} \in Z_1, \ \forall z^{(2)} \in Z_2, \quad (4.2.41)$$

and the variational equalities

$$R(\varepsilon(u^{(j)}) + \frac{1}{2}P(\zeta^{(j)}), \varepsilon(v^{(j)} - u^{(j)})) = 0, \quad \forall v^{(j)} \in [H^1(\Omega_j)]^2, \ j = 1, 2. \quad (4.2.42)$$

We call the inequality (4.2.41) a variational-hemivariational inequality, because of the presence of the convex (resp. nonconvex) terms $\Phi(z) - \Phi(\zeta)$ (resp. $J^0(\cdot,\cdot)$) as in variational (resp. hemivariational) inequalities.

Analogously we may derive the variational formulation for r-laminae. Then in (4.2.41) the summation $\sum_{j=1}^{2}$ is replaced by $\sum_{j=1}^{r}$ and the term $\int_{\Omega'} J^0(\cdot,\cdot)d\Omega$ is replaced by the term $\sum_{m=1}^{m'} \int_{\Omega'} J_m^0(\cdot,\cdot)d\Omega$, where m' is the total number of interfaces, i.e. $m' = r - 1$. Then (4.2.41) must hold for every $z^{(j)} \in Z$, $j = 1, \ldots, r$. Several other types of boundary conditions may be considered. For instance, if (4.2.29a) holds on $\tilde{\Gamma}_j$ and (4.2.29b) on $\tilde{\Gamma}'_j$, the integrals in (4.2.31) are extended over $\tilde{\Gamma}_j$ and $\tilde{\Gamma}'_j$, respectively; moreover, β_j and β'_j in (4.2.29a,b) may have different forms on different boundary parts and (4.2.29c) may be replaced by any other classical boundary conditions. Moreover, the boundary conditions (4.2.32) may be replaced by the condition $u_1 = u_2 = 0$ or any other combination of classical boundary conditions which, together with the boundary conditions in bending, will guarantee the vanishing of the term $\int_{\Gamma_j} h_j \sigma_{\alpha\beta}^{(j)} \zeta_{,\beta}^{(j)} n_\alpha^{(j)} (z^{(j)} - \zeta^{(j)}) d\Gamma$ in (4.2.33).

Nonclassical monotone or nonmonotone boundary conditions concerning the stretching (cf. [Pan85] p.225) would lead to a variational or hemivariational inequality in place of (4.2.42); these are still open problems. Also instead of the homogeneous boundary conditions (4.2.29c) we can have nonhomogeneous ones. They lead to the same variational formulations through appropriate translations. We assume now that the classical boundary conditions of each plate in bending define the kinematically admissible sets Z_j which are assumed to be (linear) subspaces of $H^2(\Omega_j)$. The following problem is formulated:
Find $\zeta^{(j)} \in Z_j$ and $u^{(j)} \in [H^1(\Omega_j)], j = 1, \ldots, r$, such as to satisfy the variational-hemivariational inequality

$$\sum_{j=1}^{r} \alpha_j(\zeta^{(j)}, z^{(j)} - \zeta^{(j)}) + \sum_{j=1}^{r} h_j R(\varepsilon(u^{(j)}) + \frac{1}{2}P(\zeta^{(j)}), P(\zeta^{(j)}, z^{(j)} - \zeta^{(j)}))$$

$$+ \sum_{m=1}^{m'} \int_{\Omega'_m} J_m^0([\zeta]^{(m)}, [z]^{(m)} - [\zeta]^{(m)})d\Omega + \sum_{j=1}^{r} \{\Phi_j(z^{(j)}) - \Phi_j(\zeta^{(j)})\}$$

$$\geq \sum_{j=1}^{r} \int_{\Omega_j} \bar{\bar{f}}^{(j)} (z^{(j)} - \zeta^{(j)})d\Omega, \quad \forall z^{(j)} \in Z_j, \ j = 1, \ldots, r, \quad (4.2.43)$$

and the variational equalities

$$R(\varepsilon(u^{(j)}) + \frac{1}{2}P(\zeta^{(j)}), \varepsilon(v^{(j)} - u^{(j)})) = 0, \quad (4.2.44)$$
$$\forall v^{(j)} \in [H^1(\Omega_j)]^2, \ j = 1, \ldots, r.$$

4.2 Hemivariational and Variational-Hemivariational Inequalities

Further we shall eliminate the in-plane displacements of the plate. To this end we note first that $R(\cdot,\cdot)$ as defined in (4.2.38) is a continuous symmetric,coercive bilinear form on $[L^2(\Omega)]^4$, and that $P : [H^2(\Omega)]^2 \to [L^2(\Omega)]^4$ of (4.2.39) is a completely continuous operator (cf. e.g. [Ber67,68] and [Pan85] p.219). Thus (4.2.42) and the Lax-Milgram theorem imply that to every deflection $\zeta^{(j)} \in Z_j, j = 1, 2, \ldots, r$, there corresponds a plane displacement $u^{(j)}(\zeta^{(j)}) \in [H^2(\Omega_j)]^2$. Indeed, due to Korn's inequality [Fich72] $R(\varepsilon(u), \varepsilon(v))$ is a bilinear coercive form on the quotient space $[H^1(\Omega)]^2/\bar{R}$, where $\bar{R}$ is the space of in-plane rigid-plate displacements defined by

$$\bar{R} = \{\bar{r} | \bar{r} \in [H^1(\Omega)]^2, \quad \bar{r}_1 = \alpha_1 + bx_2, \bar{r}_2 = \alpha_2 - bx_2, \alpha_1, \alpha_2, b \in \mathbb{R}\}. \quad (4.2.45)$$

From (4.2.44) it results (see e.g. [Pan85]) that

$$\varepsilon^{(j)}(u^{(j)}(\zeta^{(j)})) : Z_j \to [L^2(\Omega_j)]^4 \quad (4.2.46)$$

is uniquely determined and is a completely continuous quadratic function of $\zeta^{(j)}, j = 1, 2, \ldots, r$, since $\varepsilon^{(j)}(u^{(j)}(\zeta^{(j)}))$ is a linear continuous function of $P(\zeta^{(j)})$. We also introduce the completely continuous, quadratic functions $G_j : Z_j \to [L^2(\Omega_j)]^4$ which are defined by

$$\zeta^{(j)} \to G_j(\zeta^{(j)}) = \varepsilon^{(j)}(u^{(j)}(\zeta^{(j)})) + \frac{1}{2}P(\zeta^{(j)}) \quad (4.2.47)$$

and satisfy the equations (cf. (4.2.42))

$$R(G_j(\zeta^{(j)}), \varepsilon^{(j)}(u^{(j)}(\zeta^{(j)}))) = 0. \quad (4.2.48)$$

We now define the operators $A_j : Z_j \to Z'_j$ and $C_j : Z_j \to Z'_j$ (where Z'_j is the dual space to Z_j)such that

$$\alpha(\zeta^{(j)}, z^{(j)}) = \langle A_j \zeta^{(j)}, z^{(j)} \rangle. \quad (4.2.49)$$

and

$$h_j R(G_j(\zeta^{(j)}), P(\zeta^{(j)}, z^{(j)})) = \langle C_j(\zeta^{(j)}), z^{(j)} \rangle. \quad (4.2.50)$$

Let

$$T_j = A_j + C_j. \quad (4.2.51)$$

The A_j's are continuous monotone linear operators. The C_j's are completely continuous operators; and $\langle \cdot, \cdot \rangle$ denotes the duality pairing between Z_j and Z'_j. The following two properties (4.2.52) and (4.2.53) will be used. By means of (4.2.48) we obtain that.

$$\langle C_j(\zeta^{(j)}), \zeta^{(j)} \rangle = h_j R(G_j(\zeta^{(j)}), 2G_j(\zeta^{(j)})) \geq 0,$$

$$\forall \zeta^{(j)} \in Z_j, \quad j = 1, 2, \ldots, r. \quad (4.2.52)$$

Moreover if $\alpha_j(\zeta^{(j)}, \zeta^{(j)})$ are coercive on Z_j, i.e. if the boundary conditions in bending prevent rigid-plate deflections (e.g. a plate is partially clamped, or has a curved boundary partly fixed with free rotation) then

$$\langle (A_j + C_j)(\zeta^{(j)}), \zeta^{(j)} \rangle = \langle T_j(\zeta^{(j)}), \zeta^{(j)} \rangle \geq c ||\zeta^{(j)}||^2,$$

$$\forall \zeta^{(j)} \in Z_j, \quad j = 1, 2, \ldots, r, c \text{ const.} > 0 \quad (4.2.53)$$

where $||\cdot||$ denotes the Z_j-norm. Thus (4.2.43)(4.2.44) yield the following variational problem:
Find $\zeta^{(j)} \in Z_j$, $j = 1, 2, \ldots, r$, so as to satisfy the variational-hemivariational inequality

$$\sum_{j=1}^{r} \langle T_j(\zeta^{(j)}), z^{(j)} - \zeta^{(j)} \rangle + \sum_{m=1}^{m'} \int_{\Omega'_m} J_m^0([\zeta]^{(m)}, [z]^{(m)} - [\zeta]^{(m)}) d\Omega \quad (4.2.54)$$

$$+ \sum_{j=1}^{r} \{\Phi_j(z^{(j)}) - \Phi_j(\zeta^{(j)})\} \geq \sum_{j=1}^{r} \int_{\Omega_j} \overline{f}^{(j)} (z^{(j)} - \zeta^{(j)}) d\Omega \quad \forall z^{(j)} \in Z_j.$$

The last variational hemivariational inequality describes a large class of nonlinear B.V.Ps for the laminated von Kármán plates.

4.3 Substationarity Problems for the Potential or the Complementary Energy

Let us consider again the adhesive contact problem of the previous section and the corresponding hemivariational inequality (4.2.14). We recall that (4.2.14) expresses the "principle" of virtual work in inequality form. Now with respect to this hemivariational inequality we formulate the potential energy and the corresponding substationarity problem (see in Sect. 2.2. (2.2.10), (2.2.11)) which reads:
Find $u \in V_{\text{ad}}$ such that the potential energy of the structure

$$\Pi(v) = \frac{1}{2} \sum_{m=1}^{l} a(v^{(m)}, v^{(m)}) + \sum_{q=1}^{k} \int_{\Gamma_q} \left[j_{N_{(q)}}([v_N^{(q)}]) + j_{T_{(q)}}([v_T^{(q)}]) \right] d\Gamma \quad (4.3.1)$$

$$- \sum_{m=1}^{l} \int_{\Omega^{(m)}} f_i^{(m)} v_i^{(m)} d\Omega - \sum_{m=1}^{l} \int_{\Gamma^{(m)}} F_i^{(m)} v_i^{(m)} d\Gamma$$

is substationary at $v = u$, where $v \in V_{\text{ad}}$.

In other words $u \in V_{\text{ad}}$ is a solution of the inclusion

$$0 \in \bar{\partial} \Pi(u) \quad \text{for} \quad v \in V_{\text{ad}}. \quad (4.3.2)$$

Proposition 4.3.1 Every solution of the substationarity problem (4.3.2) fulfills the hemivariational inequality (4.2.14) on the assumption that $j_{N_{(q)}}$ and $j_{T_{(q)}}$ are locally Lipschitz such that the constants $c_{N_{(q)}}$ and $c_{T_{(q)}}$ of the inequalities of the type (1.2.22) holding by definition for locally Lipschitz functions have the property

4.3 Substationarity Problems for the Potential or the Complementary Energy

$$c_{N_{(q)}} = c_{N_{(q)}}(x) \in L^2(\Gamma_q) \tag{4.3.3}$$

$$c_{T_{(q)}} = c_{T_{(q)}}(x) \in L^2(\Gamma_q) \tag{4.3.4}$$

Proof. In the functional framework introduced in the previous section we can write (4.3.2) equivalently as

$$l \in \bar{\partial}\Pi_1(u) \quad \text{for} \quad u \in V_{\text{ad}}, \tag{4.3.5}$$

where

$$(l,v) = \sum_{m=1}^{l} \int_{\Omega^{(m)}} f_i^{(m)} v_i^{(m)} d\Omega + \sum_{m=1}^{l} \int_{\Gamma^{(m)}} F_i^{(m)} v_i^{(m)} d\Gamma \tag{4.3.6}$$

and

$$\Pi(v) = \Pi_1(v) - (l,v). \tag{4.3.7}$$

Now let us calculate $\bar{\partial}\Pi_1(u)$ directly, using the definition of $\Pi_1^0(u,v)$. We have according to (1.2.41) that

$$\bar{\partial}\Pi_1(u) \subset \frac{1}{2} \sum_{m=1}^{l} \bar{\partial} a(u^{(m)}, u^{(m)}) \tag{4.3.8}$$

$$+ \sum_{q=1}^{k} \bar{\partial} \int_{\Gamma_q} j_{N_{(q)}}([u_N^{(q)}]) d\Gamma + \sum_{q=1}^{k} \bar{\partial} \int_{\Gamma_q} j_{T_{(q)}}([u_T^{(q)}]) d\Gamma.$$

We shall prove now that

$$\bar{\partial} \int_{\Gamma_q} j_{N_{(q)}}([u_N^{(q)}]) d\Gamma \subset \int_{\Gamma_q} \bar{\partial} j_{N_{(q)}}([u_N^{(q)}]) d\Gamma. \tag{4.3.9}$$

Let us set

$$J_{N_{(q)}}([u_N^{(q)}]) = \int_{\Gamma_q} j_{N_{(q)}}([u_N^{(q)}]) d\Gamma. \tag{4.3.10}$$

We have that (let us omit all indices $N, (q)$ and the brackets for the sake of simplicity)

$$J^0(u,v) = \limsup_{\substack{h \to 0 \\ \lambda \to 0_+}} \frac{J(u+h+\lambda v) - J(u+h)}{\lambda} \tag{4.3.11}$$

$$= \limsup_{\substack{h \to 0 \\ \lambda \to 0_+}} \frac{1}{\lambda} \int_{\Gamma} (j(u+h+\lambda v) - j(u+h)) d\Gamma.$$

Let us denote by $g_{\lambda,h}$ the quotient

$$g_{\lambda,h}(u,v) = \frac{j(u+h+\lambda v) - j(u+h)}{\lambda}. \tag{4.3.12}$$

But $\xi \to j(\xi)$ is locally Lipschitz and thus on Γ

$$|g_{\lambda,h}(u,v)| \leq c|v|, \qquad (4.3.13)$$

where $|\cdot|$ denotes the absolute value and c depends on the neighborhood in the L^2-topology of $(u+h)(x)$. But due to (4.3.3) and (4.3.4) the relation (4.3.13) holds with c depending on x only. Since $u, v \in L^2(\Gamma)$ and $\xi \to j(\xi)$ is continuous, $x \to g_{\lambda,h}(u(x), v(x))$ is measurable. We may apply now Fatou's lemma (cf. e.g. [Evan] p. 19) and we have, by changing the signs that

$$\int_\Gamma \limsup_{\substack{\lambda \to 0_+ \\ h \to 0}} (g_{\lambda,h} - c(x)|v|) d\Gamma \geq \limsup_{\substack{\lambda \to 0_+ \\ h \to 0}} \int_\Gamma (g_{\lambda,h} - c(x)|v|) d\Gamma. \qquad (4.3.14)$$

But since $c_{N_{(q)}}(x) \in L^2(\Gamma_q)$ and $[u_N^{(q)}] \in L^2(\Gamma_q)$ we may eliminate $\int_\Gamma c(x)|v|d\Gamma$ from both sides in (4.3.14) (indeed by Schwartz inequality $\int_\Gamma c(x)|v| < \infty$). From (4.3.14) and (4.3.12) we obtain that

$$\int_{\Gamma_q} j^0_{N_{(q)}}([u_N^{(q)}], [v_N^{(q)}] - [u_N^{(q)}]) d\Gamma \geq J^0_{N_{(q)}}([u_N^{(q)}], [v_N^{(q)}] - [u_N^{(q)}]) \qquad (4.3.15)$$

which implies (4.3.9). From (4.3.8) applying (4.3.15), the analogous inequality for $j_{T_{(q)}}$, and the fact that $\alpha(u^{(m)}, u^{(m)})$ is convex and therefore

$$\bar{\partial}\alpha(u^{(m)}, u^{(m)}) = \partial\alpha(u^{(m)}, u^{(m)}) = \{\operatorname{grad} \alpha(u^{(m)}, u^{(m)})\}, \qquad (4.3.16)$$

we obtain the hemivariational inequality (4.2.14) q.e.d.

In Chapt. 6 we shall give a proposition concerning the full equivalence of the substationarity problem with the hemivariational inequality.

Proposition 4.3.2 Suppose that $\Omega \subset \mathbb{R}^2$ and that $j_{N_{(q)}}$ and $j_{T_{(q)}}$ are obtained from functions $\beta_{N_{(q)}} \in L^\infty_{\text{loc}}(\mathbb{R})$, $\beta_{T_{(q)}} \in L^\infty_{\text{loc}}(\mathbb{R})$ as it is shown in relations $(1.2.51) \div (1.2.55)$. Then if for each β, say for $\beta_{N_{(q)}}$, there exists for every $v \in L^2(\Gamma_q)$, a $c_{N_{(q)}} \in L^2(\Gamma_q)$ and $\varrho > 0$ such that for any $h \in B_{L^2(\Gamma_q)}(0, \varrho)$ and λ sufficiently small

$$\operatorname*{ess\,sup}_{u_{N_{(q)}}(x)+h(x) \leq \xi \leq u_{N_{(q)}}(x)+h(x)+\lambda v(x)} |\beta_{N_{(q)}}(\xi)| \in L^2(\Gamma_q) \qquad (4.3.17)$$

and the same holds for all $\beta_{T_{(q)}}$, then every solution of (4.3.2) satisfies the hemivariational inequality (4.2.14). (Here $B(\cdot, \cdot)$ denotes the ball in $L^2(\Gamma_q)$).

Proof. We have due to (1.2.54) and to (4.3.17) that (omitting again q, N, T and the brackets)

$$\limsup_{\substack{\lambda \to 0_+ \\ h \to 0}} (g_{\lambda,h}(u,v)) = \limsup_{\substack{\lambda \to 0_+ \\ h \to 0}} \left(\frac{1}{\lambda} \int_{(u+h)(x)}^{(u+h+\lambda v)(x)} \beta(t) dt\right) \qquad (4.3.18)$$

$$\leq \limsup_{\substack{\lambda \to 0_+ \\ h \to 0}} \operatorname*{ess\,sup}_{(u+h)(x) \leq t \leq (u+h+\lambda v)(x)} |\beta(t)| |v(x)| \leq c(x)|u(x)|.$$

4.3 Substationarity Problems for the Potential or the Complementary Energy 119

for every h such that $||h||_{L^2(\Gamma_q)} < \varrho$ and λ sufficiently small and for every u and v. Therefore we can write that

$$\limsup_{\substack{\lambda \to 0_+ \\ h \to 0}} (g_{\lambda,h}(u,v)) \leq c(x)|v(x)| \tag{4.3.19}$$

which together with the assumption (4.3.17) implies that in (4.3.14) the term

$$\int_\Gamma c(x)|v|d\Gamma = \int_\Gamma c(x)|v|d\Gamma < \infty \tag{4.3.20}$$

and therefore it may disappear from both sides. Thus (4.3.15) is proved and we proceed as in the proof of Prop. 4.3.1. q.e.d. Note that if instead of (4.2.3) and (4.2.4) relations of the form

$$-[u_N^{(m)}] \in \bar{\partial} j^\star_{N_{(m)}}(S_N^{(m)}) \tag{4.3.21}$$

$$-[u_T^{(m)}] \in \bar{\partial} j^\star_{T_{(m)}}(S_T^{(m)}) \tag{4.3.22}$$

hold, then (4.2.5) and (4.2.6) are replaced by the inequalities

$$j^{\star 0}_{N_{(m)}}(S_N^{(m)}, T_N - S_N^{(m)}) \geq -[u_N^{(m)}](T_N - S_N^{(m)}) \quad \forall T_N \in \mathbb{R} \tag{4.3.23}$$

and

$$j^{\star 0}_{T_{(m)}}(S_T^{(m)}, T_T - S_T^{(m)}) \geq -[u_{T_i}^{(m)}](T_{T_i} - S_{T_i}^{(m)}) \quad \forall T_{T_i} \in \mathbb{R} \quad i = 1,2,3. \tag{4.3.24}$$

For the problem defined by (4.2.1),(4.2.2),(4.2.7)÷(4.2.9) and (4.3.21), (4.3.22) we can formulate a hemivariational inequalitiy by considering stress variations $\tau_{ij}^{(m)} - \sigma_{ij}^{(m)}$ and the statically admissible set

$$\Sigma_{ad}^{(m)} = \{\tau^{(m)}|\tau^{(m)} = \{\tau_{ij}^{(m)}\}, \tau_{ij}^{(m)} \in L^2(\Omega^{(m)}), \tau_{ij,j}^{(m)} + f_i^{(m)} = 0 \text{ in } \Omega^{(m)},$$

$$T_i^{(m)} = F_i^{(m)} \text{ on } \Gamma_F^{(m)}\} \tag{4.3.25}$$

and by formulating the principle of complementary virtual work for the m-body. We obtain for the whole structure through addition that

$$\sum_{m=1}^l \int_{\Omega^{(m)}} \varepsilon_{ij}^{(m)}(\tau_{ij}^{(m)} - \sigma_{ij}^{(m)})d\Omega = \sum_{m=1}^l \int_{\Gamma_U^{(m)}} U_i^{(m)}(T_i^{(m)} - S_i^{(m)})d\Gamma \tag{4.3.26}$$

$$+ \sum_{q=1}^k [\int_{\Gamma_q} [u_N^{(q)}](T_N^{(q)} - S_N^{(q)}) + \int_{\Gamma_q} [u_{T_i}^{(q)}](T_{T_i}^{(q)} - S_{T_i}^{(q)})]d\Gamma \quad \forall \tau \in \Sigma_{ad}.$$

Here $T_i = \tau_{ij}n_j$, T_N and T_T correspond to S_N and S_T, $\Sigma_{ad} = \cup_{m=0}^l \Sigma_{ad}^{(m)}$, and $\tau = \{\tau^{(1)}, \ldots, \tau^{(l)}\}$. Moreover let us define the elastic energy of the m-structure expressed in terms of the stresses

$$A(\sigma^{(m)}, \sigma^{(m)}) = \int_{\Omega^{(m)}} c_{ijhk}^{(m)} \sigma_{ij}^{(m)} \sigma_{hk}^{(m)} d\Omega, \qquad (4.3.27)$$

where $c^{(m)} = \{c_{ijhk}^{(m)}\}$ is the inverse tensor of $C^{(m)}$. From (4.3.23), (4.3.24), (4.3.26) and (4.3.27) we obtain the following hemivariational inequality which expresses the principle of complementary virtual work for the whole structure: Find $\sigma \in \Sigma_{\mathrm{ad}}$ such as to satisfy

$$\sum_{m=1}^{l} A(\sigma^{(m)}, \tau^{(m)} - \sigma^{(m)}) + \sum_{q=1}^{k} \int_{\Gamma_q} [j_{N_{(q)}}^{*0}(S_N^{(q)}, T_N^{(q)} - S_N^{(q)}) \qquad (4.3.28)$$

$$+ j_{T_{(q)}}^{*0}(S_T^{(q)}, T_T^{(q)} - S_T^{(q)})]d\Gamma \geq \sum_{m=1}^{l} \int_{\Gamma_U^{(m)}} U_i^{(m)}(T_i^{(m)} - S_i^{(m)}) d\Gamma \quad \forall \tau \in \Sigma_{\mathrm{ad}}.$$

Let Π^c denote the complementary energy of the structure, which reads

$$\Pi^c(\tau) = \frac{1}{2} \sum_{m=1}^{l} A(\tau^{(m)}, \tau^{(m)}) \qquad (4.3.29)$$

$$+ \sum_{q=1}^{k} \int_{\Gamma_q} [j_{N_{(q)}}^{*}(T_N^{(q)}) + j_{T_{(q)}}^{*}(T_T^{(q)})]d\Gamma - \sum_{m=1}^{l} \int_{\Gamma_U^{(m)}} U_i^{(m)}(T_i^{(m)}) d\Gamma.$$

Then the following substationarity problem is considered:
Find $\sigma \in \Sigma_{\mathrm{ad}}$ such that

$$0 \in \bar{\partial} \Pi^c(\tau) \quad \text{for } \tau \in \Sigma_{\mathrm{ad}}. \qquad (4.3.30)$$

A solution of the substationarity problem (4.3.30) is related to a solution of the hemivariational inequality (4.3.28) by a proposition analogous to Prop. 4.3.1.

If each of the substructures $\Omega^{(m)}$ obeys a general nonmonotone law of the form $\varepsilon^{(m)} \in \bar{\partial} w_{(m)}^{\star}(\sigma^{(m)})$ then the variational formulation of the problem results from (4.3.28) by replacing the quadratic terms by $\sum_{m=1}^{l} \int_{\Omega^{(m)}} w_{(m)}^{\star \uparrow}(\sigma^{(m)}, \tau^{(m)} - \sigma^{(m)}) d\Omega$ on the assumption that the integrals make sense; If moreover, monotone laws $\sigma^{(m)} \in \partial w_{(m)}(\varepsilon^{(m)})$ hold, where the $w_{(m)}$'s are convex, l.s.c. and proper functionals, then using the conjugate functionals $w_{(m)}^c$ we can write the above relation in the form $\varepsilon^{(m)} \in \partial w_{(m)}^c(\sigma^{(m)})$. Then we define

$$W_{(m)}^c(\sigma) = \begin{cases} \int_{\Omega^{(m)}} w_{(m)}^c(\sigma) d\Omega & \text{if } w_{(m)}^c(\sigma) \in L^1(\Omega^{(m)}) \\ \infty & \text{otherwise} \end{cases} \qquad (4.3.31)$$

and we obtain a variational formulation analogous to (4.3.28) but with the elastic energy replaced by $\sum_{m=1}^{l}[W_{(m)}^c(\tau^{(m)}) - W_{(m)}^c(\sigma^{(m)})]$. This is a variational-hemivariational inequality with respect to the stress variations.

4.4 Loading and Unloading Problems, Eigenvalue Problems for Hemivariational Inequalities and Dynamic Problems

The loading-unloading problems as well as the hysteresis problems, due to the appearing nonmonotone possibly multivalued branches, are typical examples for the theory of hemivariational inequalities especially in solid mechanics, where we may consider hysteresis laws between stress and strains. As it can be easily understood (cf. Prop. 3.4.2), the global behaviour of a system obeying a given hysteresis law is governed by a sequence of hemivariational inequalities, one for each branch. The passing from the one branch to the other branch is closely connected with the eigenvalue problem for hemivariational inequalities. This problem is a nonclassical problem due to the lack of monotonicity.

In solid mechanics and especially in structural analysis hysteresis loops appear in loading-unloading processes. In a simple bar submitted to forces acting at the two ends along its axis the loading-unloading process may be easily understood and gives rise to onedimensional stress-strain $\{\sigma, \varepsilon\}$ diagrams like the one depicted in Fig. 4.4.1. Here we accept that the stress-strain law is generally nonmonotone and multivalued, i.e. it contains filled-in parts parallel to the σ-axis. In a more complicated structure (resp. in a solid body) we have a finite or infinite number of elements obeying a law analogous to the one of Fig. 4.4.1 (resp. a law analogous to the one of Fig. 4.4.1 in $R^6 \times R^6$ has to be satisfied at every point of the solid).

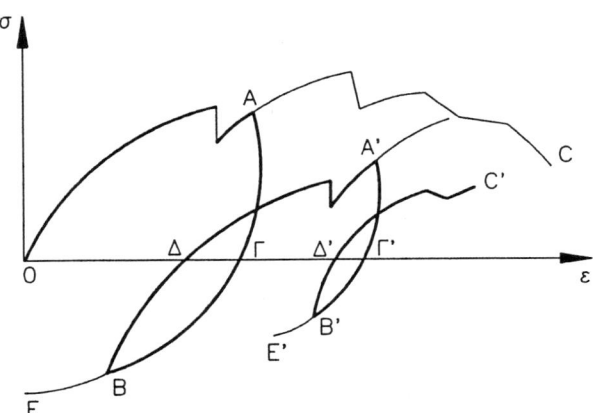

Fig. 4.4.1. Loading-unloading hysteresis loops.

A loading $t \to p(t)$ is given and one has to determine the stress, strain and displacement fields $\sigma = \{\sigma_{ij}\}$, $\varepsilon = \{\varepsilon_{ij}\}$ and $u = \{u_i\}$ at every point of the body. We know the stress-strain law in loading (path OAC) and we also know that in the case of unloading (resp. loading) at A the path AΓBE (resp. AC) will be realized. Analogously, if at a given moment an element has a stress and strain state corresponding to B (resp. to B'), then in the case of loading the path BΔA' (resp. B'Δ'C') will be realized, whereas in the case of unloading the path BE (resp. B'E'). Note that we do not know a priori the points A, A' etc or B, B' etc, where the branching of the solution takes place and we have to determine these points for varying external loading. It is well-known that a nonmonotone possibly multivalued, (i.e. with vertical parts, but without branching) stress-strain or reaction-displacement law gives rise to a hemivariational inequality with respect to the displacements.

In the case of smooth $\sigma - \varepsilon$ laws the hemivariational inequality reduces to a variational equality which is the variational form of a partial differential equation. Then the branching at A leads (cf. e.g. [Budi][Thom]) to an eigenvalue problem for a variational equality. In the case of a nonsmooth $\sigma - \varepsilon$ law with multivalued parts the branching at A leads analogously to an eigenvalue problem for a hemivariational inequality. Let us explain this fact: The potential energy Π is a functional of the "generalized" displacement u. It also depends on a scalar variable λ which determines the magnitude of the external loading on the system. The variable u must satisfy certain boundary or support conditions of classical (nonunilateral) type, e.g. $u = 0$, on a boundary part Γ_1. Conditions resulting from nonconvex superpotential conditions between the stress vector S and the displacements u or relative displacements $[u]$ are not taken into account on the remaining part of the boundary Γ-Γ_1 for the moment, i.e. we assume that the corresponding stress vector (interface forces, or reaction forces) S is given. Since we have a bilateral structure the variational equality of equilibrium reads (δ is the symbol of classical variation, $\langle,\rangle$ is a duality pairing – physically a work expression)

$$\delta \Pi(u, \lambda) = \langle \Pi'_u(u, \lambda), \delta u \rangle = 0. \tag{4.4.1}$$

for every admissible variation δu. Here $\Pi'_u(u, \lambda)$ denotes the Gâteaux differential of Π with respect to u. We assume now that the structure is linear elastic and that a nonmonotone reaction-displacement relation $(-S, u)$ holds on the boundary part Γ-Γ_1, which has a graph analogous to the one-dimensional (σ, ε) graph of Fig. 4.1.1. Moreover the branches OAC, EBΓAC, EBΔA' etc. are described by relations of the type

$$-S \in \bar{\partial} j_\alpha(u) \quad \alpha = 1, 2, \ldots, \tag{4.4.2}$$

where j_α is a nonconvex function and $\bar{\partial}$ is Clarke's generalized gradient.

Suppose now that there exists a fundamental solution u_0 varying with λ which corresponds to the OAC curve of Fig. 4.4.1, and let another solution exists, say the one corresponding to the curve AΓB in Fig. 4.4.1 that intersects the fundamental solution at $\lambda = \lambda_c$ and coincides with it for $\lambda < \lambda_c$. Let the other solution be $u = u_0(\lambda) + v(\lambda)$. Then

4.4 Loading-Unloading. Eigenvalue and Dynamic Problems

$$\lim_{\lambda \to \lambda_c} v(\lambda) = 0. \tag{4.4.3}$$

Moreover we have that

$$\langle \Pi'_u(u_0(\lambda); \lambda), \delta u \rangle = 0 \text{ and } \langle \Pi'_u(u_0(\lambda) + v(\lambda); \lambda), \delta u \rangle = 0. \tag{4.4.4}$$

Assuming now that Π'_u is expandable in Taylor-series as a function of u in the neighborhood of $u_0(\lambda_c)$, implies together with the two previous relations that

$$\langle \Pi''_u(u_0(\lambda); \lambda)v(\lambda), \delta u \rangle + \frac{1}{2}\langle \Pi'''_u(u_0(\lambda); \lambda)v^2(\lambda), \delta u \rangle + \ldots = 0. \tag{4.4.5}$$

We denote by $u_1 = \lim(v(\lambda)/||u||)$ as $\lambda \to \lambda_c$ the "hysteresis bifurcation mode". Dividing the above relation by $||v||$ and letting $\lambda \to \lambda_c$ we obtain that

$$\langle \Pi''_u(u_0(\lambda_c); \lambda_c)u_1(\lambda_c), \delta u \rangle = 0. \tag{4.4.6}$$

which constitutes a variational statement of an eigenvalue problem. We write now the total potential energy functional Π as

$$\Pi(u) = \frac{1}{2}a(u,u) + \langle S, u \rangle - \langle \lambda, \Delta u \rangle \tag{4.4.7}$$

where $a(\cdot, \cdot)$ denotes the quadratic elastic strain energy, S is assumed as given and Δ denotes the load-shortening function ([Budi] p.26) assumed to be at least of the type $\frac{1}{2}u^2 + cu + d$, a fact justified by the physical nonlinearity.

Here $\langle \cdot, \cdot \rangle$ denotes expressions of work; the first is extended over Γ-Γ_1 and the second over the body part, on which the external loading is applied. Recall also that in the usual Sobolev space framework S is a linear bounded mapping of the displacements and that $a(u,v) = (A(u), v)$, where A is a linear bounded operator. Then the last two relations imply that (set $\langle S(\delta u), u_1(\lambda_c) \rangle \simeq 0$; this may be justified from the mechanical model)

$$a(u_1(\lambda_c), \delta u) + \langle S(u_1(\lambda_c)), \delta u \rangle - \langle \lambda_c u_1(\lambda_c), \delta u \rangle = 0. \tag{4.4.8}$$

Setting $\delta u = w - u_1(\lambda_c)$ and using the definition of the generalized gradient

$$\int_{\Gamma-\Gamma_1} j^0(u_1(\lambda_c), w - u_1(\lambda_c))d\Gamma \geq -\langle S(u_1(\lambda_c)), w - u_1(\lambda_c) \rangle \quad \forall w \in V_{\text{ad}}, \tag{4.4.9}$$

where V_{ad} denotes the kinematically admissible set yields an eigenvalue problem for a hemivariational inequality of the type:
Find λ_c and $u_1(\lambda_c) \in V_{\text{ad}}$ such as to satisfy

$$a(u_1(\lambda_c), w - u_1(\lambda_c)) + \int_{\Gamma-\Gamma_1} j^0(u_1(\lambda_c), w - u_1(\lambda_c))d\Gamma \tag{4.4.10}$$
$$- \langle \lambda_c u_1(\lambda_c), w - u_1(\lambda_c) \rangle \geq 0 \quad \forall w \in V_{\text{ad}}.$$

4. Hemivariational Inequalities

Eigenvalue problems for hemivariational inequalities result also in the common case of buckling problems involving nonconvex superpotentials. To give an example we consider an elastic von Kármán plate Ω_1 with constant thickness h, which is connected with an adhesive material of negligible thickness to the rigid plane body Ω' on Ω, $\Omega \subset \Omega_1 \cap \Omega'$ (Fig. 4.4.2). We suppose that $\bar{\Omega} \cap \Gamma_1 = \emptyset$ where Γ_1 is the boundary of Ω_1 which is appropriately smooth (Lipschitzian boundary is sufficient). The points of the plate are referred to a fixed right-handed Cartesian coordinate system $Ox_1x_2x_3$ and the middle plane of the underformed plate coincides with the Ox_1x_2-plane. The set Ω_1 is an open, bounded and connected subset of $\mathbb{R}^2$.

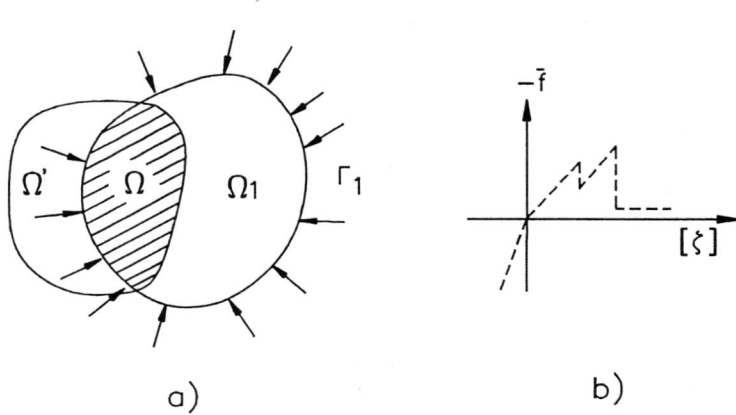

Fig. 4.4.2. On the buckling of a von Kármán plate in adhesive contact.

We suppose that the plate has a buckling because of the boundary loading in the plane of the plate. The theory of von Kármán plates leads to the differential equations (4.2.21)÷(4.2.23a) for $j = 1$. We put $f^{(1)} = f$ The boundary loading which is responsible for the buckling has the form

$$\sigma_{\alpha\beta}n_\beta = \lambda g_\alpha \qquad \alpha = 1,2 \text{ on } \Gamma_1 \qquad (4.4.11)$$

where $g = g(x)$, $x \in \Gamma_1$ is a self-equilibrating compressive load distribution on the plate boundary Γ_1 and λ is a real number. When λ exceeds a certain value called the critical value, the plate leaves its position of equilibrium and buckling occurs. In this case, the small displacement theory is not adequate for precise description of buckling phenomena, and thus the von Kármán theory must be applied.

We assume now that $f = \bar{\bar{f}} + \bar{f}$ where generally $\bar{\bar{f}} \in L^2(\Omega_1)$ and $\bar{f}$ is a given function of ζ and describes the action of the given support on the plate. Let $\bar{f}$ be a multivalued nonmonotone function of ζ in Ω and

$$-\bar{f} \in \bar{b}(\zeta) \text{ on } \Omega \subset \Omega_1, \quad \bar{f} = 0 \text{ on } \Omega_1 - \Omega \qquad (4.4.12)$$

where $\tilde{b}$ is denoted a nonmonotone multivalued function on $\mathbb{R}^1$ which results from $b \in L^\infty_{loc}(\mathbb{R})$ by applying the procedure defined by the relations (1.2.51)÷(1.2.55).

At the same time let for the function b hold that

$$b(0) = 0. \tag{4.4.13}$$

We denote by j the nonconvex superpotential corresponding to $\tilde{b}$ through (1.2.55). The expressions for the virtual work for the bending and the streching respectively of a free plate are obtained from (4.2.33), (4.2.37); the corresponding relations are

$$\begin{aligned} a(\zeta, z - \zeta) &+ \int_{\Omega_1} h\sigma_{\alpha\beta}\zeta_{,\alpha}(z-\zeta)_{,\beta}d\Omega = \int_{\Gamma_1} h\sigma_{\alpha\beta}\zeta_{,\alpha}n_\beta(z-\zeta)d\Gamma \\ &+ \int_{\Gamma_1} K_n(\zeta)(z-\zeta)d\Gamma - \int_{\Gamma_1} M_n(\zeta)\frac{\partial(z-\zeta)}{\partial n}d\Gamma \\ &+ \int_{\Omega_1} \overline{\overline{f}}(z-\zeta)d\Omega \end{aligned} \tag{4.4.14}$$

and

$$\int_{\Omega_1} \sigma_{\alpha\beta}\varepsilon_{\alpha\beta}(v - u)d\Omega = \int_{\Gamma_1} \sigma_{\alpha\beta}n_\beta(v_\alpha - u_\alpha)d\Gamma \tag{4.4.15}$$

where $\alpha, \beta = 1, 2$ $n = (n_1, n_2)$ denotes the outward normal unit vector on Γ_1, $M_n = M_n(\zeta)$, $K_n = K_n(\zeta)$ are known functions of ζ (cf. e.g. (4.2.34), (4.2.35)) expressing the bending moment and the total shearing force at the plate boundary Γ_1. Now, we define the space

$$Z = \left\{ z \Big| z \in H^2(\Omega_1), z = 0 \text{ on } \Gamma_1 \text{ and } \frac{\partial z}{\partial n} = 0 \text{ a.e. on } \Gamma_1 \right\} \tag{4.4.16}$$

where $H^2(\Omega_1)$ is the Sobolev space for the plate deflections. Then, setting in (4.4.14) $\overline{\overline{f}} = 0$, and assuming that in (4.4.11) $g \in L^2(\Gamma_1)$, (4.4.16) and the definition of the generalized gradient yield the inequality

$$\begin{aligned} a(\zeta, z - \zeta) &+ \int_{\Omega_1} h\sigma_{\alpha\beta}\zeta_{,\alpha}(z-\zeta)_{,\beta}d\Omega + \int_\Omega j^0(\zeta, z-\zeta)d\Omega \\ &\geq \lambda \int_{\Gamma_1} hg_\alpha\zeta_{,\alpha}(z-\zeta)d\Gamma \quad \forall z \in Z. \end{aligned} \tag{4.4.17}$$

Here, following the method which is presented in Sect. 4.2 and provided that the symbols and the notions are the same as there, we obtain the following problem after eliminating the in-plane displacements:
Find $\lambda \in \mathbb{R}$ and $\zeta \in Z$ such as to satisfy the hemivariational inequality eigenvalue problem

$$\langle T(\zeta), z - \zeta \rangle + \int_\Omega j^0(\zeta, z - \zeta)d\Omega \geq \lambda \langle B\zeta, z - \zeta \rangle. \tag{4.4.18}$$

Analogous eigenvalue problems for hemivariational inequalities are obtained if one has two or more plates glued together.

We shall close this section by deriving a dynamic hemivariational inequality. Let us place ourselves in the framework of the adhesive contact problem of Sect. 4.2. Analogously to the derivation of the hemivariational inequality (4.2.14) we proceed in dynamic problems on the assumption of small strains and small displacements. Then $f_i^{(m)}$ has to be replaced by $f_i^{(m)} - \rho^{(m)}\frac{\partial^2 u_i^{(m)}}{\partial t^2}$, where $\rho^{(m)}$ is the density of the m-body; moreover initial conditions for the displacements $u_i^{(m)}$ and the velocities $\partial u_i^{(m)}/\partial t$ have to be considered. The resulting hemivariational inequality is analogous to (4.2.14) and expresses d'Alembert's principle in inequality form. In the dynamic case the holonomic interface relations (4.2.3), (4.2.4) may be replaced by the relations

$$-S_N^{(m)} \in \bar{\partial} j_{N_{(m)}}(S^{(m)}; \left[\frac{\partial u_N^{(m)}}{\partial t}\right]), \qquad (4.4.19)$$

$$-S_T^{(m)} \in \bar{\partial} j_{T_{(m)}}(S^{(m)}; \left[\frac{\partial u_T^{(m)}}{\partial t}\right]). \qquad (4.4.20)$$

In this case we write again (4.2.10) by considering instead of displacement variations, velocity variations. Thus we are led to a hemivariational inequality similar to (4.2.14), where now instead of $v^{(m)} - u^{(m)}$ and $[v_N^{(m)}] - [u_N^{(m)}]$ we have the variations $v^{(m)} - \frac{\partial u^{(m)}}{\partial t}$ and $[v_N^{(m)}] - [\frac{\partial u_N^{(m)}}{\partial t}]$. Analogously, dynamic variational-hemivariational inequalities are derived.

4.5 On the F-superpotential and the V-superpotential. Quasidifferentiability in Mechanics

In Sect. 3.5 we introduced material laws and boundary conditions derived by fans. We mean here the relations (3.5.1), (3.5.2) and the corresponding fan-variational inequalities (3.5.3) and (3.5.4). Let us consider a solid mechanics problem where both (3.5.1) and (3.5.2) hold. Then from (4.1.1) or (4.1.16) and by means of (3.5.3) and (3.5.4) we obtain the following variational problem: Find $u \in [H^1(\Omega)]^3$ such as to satisfy

$$G_1(\varepsilon(u^\star - u), \varepsilon(u)) + G_2(u^\star - u, u) \qquad (4.5.1)$$
$$- \int_\Omega f_i(u_i^\star - u_i)d\Omega \geq 0 \quad \forall u_i^\star \in [H^1(\Omega)]^3$$

where

$$G_1(\varepsilon(u^\star), \varepsilon(u)) = \begin{cases} \int_\Omega g_1(\varepsilon(u^\star), \varepsilon(u))d\Omega & \text{if } g_1(\varepsilon(u^\star(\cdot)), \varepsilon(u(\cdot))) \in L^1(\Omega) \\ \infty & \text{otherwise} \end{cases} \qquad (4.5.2)$$

$$G_2(u^\star, u) = \begin{cases} \int_\Gamma g_2(u^\star, u)d\Gamma & \text{if } g_2(u^\star, u) \in L^1(\Omega) \\ \infty & \text{otherwise} \end{cases} \qquad (4.5.3)$$

4.5 On the F and V-superpotential. Quasidifferentiability

The variational inequality (4.5.1) is called a "fan-variational inequality" or a "F-variational inequality". The correspondig superpotentials, for instance F_1 and F_2 in (3.5.1), (3.5.2), are called F-superpotentials. In Sect. 3.5 we have given several boundary conditions and the material laws expressed in terms of F-superpotentials. Especially the cases depicted in Fig. 3.5.1 and 3.5.2 are very interesting. Note that only by means of F-superpotentials we can formulate variational "principles" for these two cases.

Every generalization of the notion of derivative leads to a new variational problem and permits variational formulations for problems in mechanics which cannot be achieved by means of the classical calculus of variations. Thus, the subdifferential leads to the variational inequalities, the generalized gradient in the sense of Clarke (resp. Rockafellar) to hemivariational inequalities containing the directional differential of Clarke $\Phi^0(\cdot,\cdot)$ (resp. $\Phi^\uparrow(\cdot,\cdot)$) and the Ioffe fan to F-variational inequalities. The resulting variational problems express the principle of virtual work (or power). Thus, e.g. $\Phi^\uparrow(\varepsilon, \varepsilon^\star - \varepsilon)$, $\Phi^\uparrow(u, u^\star - u)$ etc. express a virtual work (or power) for the virtual generalized displacement $\varepsilon^\star - \varepsilon$, $u^\star - u$ etc. Thus we could describe a (σ, ε) or a $(-f, u)$ relatioship simply by specifying the expressions of the virtual work $A(\varepsilon, \varepsilon^\star - \varepsilon)$ and $B(u, u^\star - u)$ and we would obtain a variational expression similar to a hemivariational inequality but involving instead of $\Phi^0(u, u^\star - u)$ etc the virtual work (or power) expressions $A(\varepsilon, \varepsilon^\star - \varepsilon)$ and $B(u, u^\star - u)$.

Indeed as it occured several times until now in mechanics, the physical meaning of a mechanical expression permits the mechanical theory to free itself from the mathematics and in some cases to have a better and more attractive development leading to new interesting results. In the present case of the superpotentials (convex, nonconvex, etc) we have a very important "mechanical" tool in our hands: the principle of virtual work (or power).

Accordingly we propose the following: On the mechanical system Σ we consider a multivalued mechanical law (or constraint) of the form

$$-f \in W(u) \tag{4.5.4}$$

equivalent to the expression

$$\begin{aligned} W(u) &= \{f \in F \big| G(u, u^\star - u, \{B\}\{\beta\}, \text{path,history, time derivatives.}) \\ &\geq \langle f, u^\star - u \rangle \quad \forall u^\star \in U\}. \end{aligned} \tag{4.5.5}$$

Here F (resp. U) is the space of generalized forces f (resp. displacement u) and G is an appropriately defined function generally of a nonlocal nature which expresses the virtual work (or power) of the system under consideration. It may depend on several observable variables $\{B\}$ and on several hidden variables $\{\beta\}$, and may depend on the "path" of the variation $u^\star - u$, on the history of the mechanical system, etc. Obviously (4.5.4) is much more general than the convex, nonconvex and F-superpotential laws since it enters into the "material behaviour" of the system. By definition the law (4.5.4)-which we shall call V-superpotential (virtual work - superpotential) law-is appropriate for the description of dissipative phenomena.

The foregoing considerations permit the introduction of the following "Hypothesis of Dissipation" which constitutes a generalization of the "Hypothesis of Normal Dissipation" (cf. Sect. 3.3).

(H) For every real thermodynamic process a V-superpotential W exists such that the "flux" $v \in U$ associated with the "force" $f \in F$ satisfies the relation (4.5.4).
By the second principle of thermodynamics the inequality (3.3.30)

$$\langle f, v \rangle \geq 0 \tag{4.5.6}$$

must always hold. As obvious if

$$G(v, -v, \{B\}, \{\beta\}, \text{path}, \ldots) \leq 0 \tag{4.5.7}$$

then (4.5.6) is satisfied. We call "V-substationarity point" a point v_0 such that

$$0 \in W(v_0) \iff G(v_0, v^\star - v_0, \{B\}, \{\beta\}, \text{path,history}, \ldots) \geq 0 \tag{4.5.8}$$
$$\forall v^\star \in U.$$

We can easily prove the following proposition:

Proposition 4.5.1 Suppose that the spaces U, F are finite dimensional and that

$$G(v, v^\star, \ldots) > G(v, 0, \ldots) \quad \forall v^\star \in U, v^\star \neq 0 \tag{4.5.9}$$

Then to any flux $v \in U$ there corresponds at least one force $f \in F$, if and only if at v

$$G(v, 0, \{B\}, \{\beta\}, \text{path}, \ldots) \geq 0. \tag{4.5.10}$$

Proof. Relations (4.5.9), (4.5.10) imply that $W(v) \neq \emptyset$ since F and U are finite dimensional. Conversely, if $W(u) \neq \emptyset$ there exists f such as to satisfy (4.5.5) and thus for $u^\star = u$ we obtain (4.5.10) q.e.d.

Analogously to (4.5.4) we may consider dissipation mechanisms given by a relation of the form

$$-u \in \tilde{W}(f) \tag{4.5.11}$$

where $\tilde{W}$ is a given V-superpotential. With the notations of Sect. 3.3, (cf. relation (3.3.38)) we define a V-generalized standard material (i.e. compatible to a given expression of the virtual work) as a material for which (4.5.11) holds with

$$u = -((\tilde{d}F_p/dt)F_p^{-1}, -\tilde{d}\{a_j\}/dt), \ f = \{R, \{A_j\}\}. \tag{4.5.12}$$

In order to show the importance of the V-superpotentials we shall explain that this notion is inherent in the study of elastoplastic materials with hardening, whose elastic properties change with plastic deformation. We shall derive for this situation a material law of the form

4.5 On the F and V-superpotential. Quasidifferentiability

$$\dot{\varepsilon}^p \in W(\sigma), \quad \sigma \in K \tag{4.5.13}$$

where $\dot{\varepsilon}^p dt$ denotes a plastic strain increment. Let us consider a subset K of the stress space defined by the yield criterion $F(\sigma) \leq k^2$ such that $0 \in K$. We modify Drucker's "proof" with which the convexity of the yield surface is justified. Let us consider a stress state $\sigma^\star = \{\sigma_{ij}^\star\}$, on, or inside the yield surface, and an external agency causing a change of the stresses until a stress state $\sigma = \{\sigma_{ij}\}$ is reached at the boundary of the yield surface. Suppose further that the agency acting within the infinitesimal time interval dt produces a plastic-strain increment $\dot{\varepsilon}^p dt$ and a corresponding stress increment $\dot{\sigma} dt$ (BC in Fig. 4.5.1) directed outwards from K. Further, the loading releases $\dot{\sigma} dt$ and returns the stresses to the initial stress state $\sigma^\star$ along the elastic path BDO. If the path OABDO lies entirely inside K or on the boundary of K, then, due to the reversibility of elastic deformation, the work done over the cycle OABDO is zero, if the elastic properties are constant. However, we assume here, that, due to the plastic deformation, the elastic properties change, and we denote by G the nonrecoverable elastic work per unit time. We assume that G depends on σ, on $\sigma^\star - \sigma$ and on the path followed and that for $\sigma^\star = \sigma, G = 0$. The work produced over the cycle OABCDO is assumed to be nonnegative as it is the case in classical plasticity and is given by

$$(\sigma_{ij} - \sigma_{ij}^\star)\dot{\varepsilon}_{ij}^p dt + \dot{\varepsilon}_{ij}^p \dot{\sigma}_{ij} dt dt + G(\sigma, \sigma^\star - \sigma, \text{path}, \ldots) dt \geq 0. \tag{4.5.14}$$

This inequality holds for every $\sigma^\star \in K$ and for every path OABDO which does not pass outside K. (4.5.14) is also valid for $\sigma^\star = \sigma$. Accordingly, $\dot{\varepsilon}_{ij}^p \dot{\sigma}_{ij}(dt)^2 \geq 0$. But this expression can be ignored due to $(dt)^2$ and we get for $\sigma \in K$ the inequality

$$(\sigma_{ij} - \sigma_{ij}^\star)\dot{\varepsilon}_{ij}^p + G(\sigma, \sigma^\star - \sigma, \text{path}, \ldots) \geq 0 \quad \forall \sigma^\star \in K, \tag{4.5.15}$$

which implies the validity of (4.5.13). Combining the inequality which defines the V-superpotential (cf. (4.5.5)) with (4.1.1) or (4.1.16) we obtain variational inequality expressions analogous to (4.5.1) involving the virtual work function G of (4.5.5). These variational expressions are called V-variational inequalities. For instance an important class of V-variational inequalities are the variational-hemivariational inequalities. They are obtained if the V-superpotential W is given by

$$W(\cdot) = \bar{\partial}\Phi(\cdot) + \text{grad}\Psi(\cdot) \tag{4.5.16}$$

$$\text{or} \quad W(\cdot) = \bar{\partial}\Phi(\cdot) + \partial\Psi(\cdot) \tag{4.5.17}$$

where Φ is a locally Lipschitz functional and Ψ is a convex, l.s.c. proper functional.

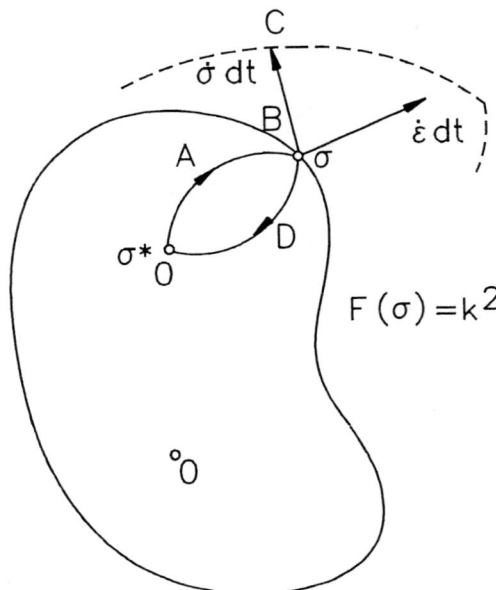

Fig. 4.5.1. Justification of the material law (4.5.13).

We shall close this section by deriving some variational formulations for material laws and boundary conditions expressed in terms of quasidifferentials. Let us place ourselves in the framework of (4.2.12) and let us assume that all the structures are twodimensional. We write then (4.2.12) in the form

$$\sum_{m=1}^{l} \{a(u^{(m)}, v^{(m)} - u^{(m)}) - l^{(m)}(v^{(m)} - u^{(m)})\} \tag{4.5.18}$$

$$- \sum_{m=1}^{l} \int_{\Gamma_F^{(m)}} F_i^{(m)}(v_i^{(m)} - u_i^{(m)}) d\Gamma$$

$$= \sum_{q=1}^{k} \left[\langle S_N^{(q)}, ([v_N^{(q)}] - [u_N^{(q)}]) \rangle_{\frac{1}{2}, \Gamma_q} + \langle S_T^{(q)}, ([v_T^{(q)}] - [u_T^{(q)}]) \rangle_{H_T, \Gamma_q} \right] \quad \forall v \in V_{\text{ad}}.$$

Here $i = 1, 2$, $a(u^{(m)}, v^{(m)})$ is given by (4.2.13), V_{ad} by (4.2.11) and

$$l^{(m)}(v^{(m)}) = \int_{\Omega^{(m)}} f_i^{(m)} v_i^{(m)} d\Omega. \tag{4.5.19}$$

The problem written in (4.5.18) is understood within the following functional framework: $u^{(m)}, v^{(m)} \in [H^1(\Omega^{(m)})]^2, f^{(m)} \in [L^2(\Omega^{(m)})]^2, F^{(m)} \in [L^2(\Gamma_F^{(m)})]^2$, and in (4.2.11) $V(\Omega^{(m)}) = H^1(\Omega^{(m)})$.

Introducing of the quasidifferential interface law (3.5.14), (3.5.15) in the form (3.5.16) into (4.5.18) implies the following formulation of the problem:

4.5 On the F and V-superpotential. Quasidifferentiability 131

$$\sum_{m=1}^{l}\{a(u^{(m)},v^{(m)}-u^{(m)})-l^{(m)}(v^{(m)}-u^{(m)})-\int_{\Gamma_F^{(m)}}F_i^{(m)}(v_i^{(m)}-u_i^{(m)})d\Gamma\}$$

$$+\sum_{q=1}^{k}\Big\{\{\max_{w_{1,N}^*}\langle w_{1,N}^*,([v_N]-[u_N])\rangle_{\frac{1}{2},\Gamma_q}\big|w_{1,N}^*\in \underline{\partial}'F_N([u_N])\}_{(q)} \quad (4.5.20)$$

$$+\{\max_{w_{1,T}^*}\langle w_{1,T}^*,([v_T]-[u_T])\rangle_{H_T,\Gamma_q}\big|w_{1,T}^*\in \underline{\partial}'F_T([u_T])\}_{(q)}$$

$$+\{\min_{w_{2,N}^*}\langle w_{2,N}^*,([v_N]-[u_N])\rangle_{\frac{1}{2},\Gamma_q}\big|w_{2,N}^*\in \bar{\partial}'F_N([u_N])\}_{(q)}$$

$$+\{\min_{w_{2,T}^*}\langle w_{2,T}^*,([v_T]-[u_T])\rangle_{H_T,\Gamma_q}\big|w_{2,T}^*\in \bar{\partial}'F_T([u_T])\}_{(q)}\Big\}=0 \quad \forall v\in V_{\text{ad}}.$$

Here we have put the subscript (q) to denote that all the quantities refer to Γ_q. Applying (3.5.17) and (3.5.18) we may write the corresponding formulations of the problem. The min-max form reads:
Find $u\in V_{\text{ad}}$ such as to satisfy the relation:

$$\sum_{m=1}^{l}\{a(u^{(m)},v^{(m)}-u^{(m)})-l^{(m)}(v^{(m)}-u^{(m)})\}-\int_{\Gamma_F^{(m)}}F_i^{(m)}(v_i^{(m)}-u_i^{(m)})d\Gamma$$

$$+\sum_{q=1}^{k}\Big\{\min_{w_{2,N}^*}\max_{w_{1,N}^*}\{\langle w_{1,N}^*+w_{2,N}^*,([v_N]-[u_N])\rangle_{\frac{1}{2},\Gamma_q}\big|w_{1,N}^*\in \underline{\partial}'F_N([u_N]),$$

$$w_{2,N}^*\in \bar{\partial}'F_N([u_N])\}_{(q)}+\min_{w_{2,T}^*}\max_{w_{1,T}^*}\{\langle w_{1,T}^*+w_{2,T}^*,([v_T]-[u_T])\rangle_{H_T,\Gamma_q}\big|$$

$$w_{1,T}^*\in \underline{\partial}'F_T([u_T]),w_{2,T}^*\in \bar{\partial}'F_T([u_T])\}_{(q)}\Big\}=0 \quad \forall v\in V_{\text{ad}}. \quad (4.5.21)$$

Let us introduce the abbreviating notation $A(u,v-u)-L(v-u)$ for the first three terms in (4.5.21), i.e., the terms after the summation over m. Applying now (3.5.16) we can write (4.5.21) as the variational equality

$$A(u,v-u)-L(v-u) \quad (4.5.22)$$

$$+\sum_{q=1}^{k}\Big\{\tilde{F}_N'([u_N],[v_N]-[u_N])_{(q)}+\tilde{F}_T'([u_T],[v_T]-[u_T])_{(q)}\Big\}=0 \quad \forall v\in V_{\text{ad}}.$$

From (4.5.21) we obtain the following variational inequality

$$A(u,v-u)-L(v-u) \quad (4.5.23)$$

$$+\sum_{q=1}^{k}\Big\{\max_{w_{1,N}^*}\{\langle w_{1,N}^*+w_{2,N}^*,([v_N]-[u_N])\rangle_{\frac{1}{2},\Gamma_q}\big|w_{1,N}^*\in \underline{\partial}'F_N([u_N])\}_{(q)}$$

$$+\max_{w_{1,T}^*}\{\langle w_{1,T}^*+w_{2,T}^*,([v_T]-[u_T])\rangle_{H_T,\Gamma_q}\big|w_{1,T}^*\in \underline{\partial}'F_T([u_T])\}_{(q)}\Big\}\geq 0$$

$$\forall w_{2,N}^*\in \bar{\partial}'F_N([u_N])_{(q)},\quad \forall w_{2,T}^*\in \bar{\partial}'F_T([u_T])$$

$$\text{on}\quad \Gamma_q, q=1,\ldots,k,\quad \forall v\in V_{\text{ad}}.$$

Obviously if we leave the min-expressions in (4.5.21) and not the max-expressions we obtain a reverse variational inequality. Indeed from (3.5.17) and (4.5.22) we obtain the following variational form

$$0 \geq A(u, v-u) - L(v-u) \tag{4.5.24}$$
$$+ \sum_{q=1}^{k} \left\{ \min_{w_{2,N}^{\star}} \{ \langle w_{1,N}^{\star} + w_{2,N}^{\star}, ([v_N] - [u_N]) \rangle_{\frac{1}{2}, \Gamma_q} \big| w_{2,N}^{\star} \in \bar{\partial}' F_N([u_N]) \}_{(q)} \right.$$
$$\left. + \min_{w_{2,T}^{\star}} \{ \langle w_{1,T}^{\star} + w_{2,T}^{\star}, ([v_T] - [u_T]) \rangle_{H_T} \big| w_{2,T}^{\star} \in \bar{\partial}' F_T([u_T]) \}_{(q)} \right\}$$
$$\forall w_{1,N}^{\star} \in \underline{\partial}' F_N([u_N])_{(q)}, \quad \forall w_{1,T}^{\star} \in \underline{\partial}' F_T([u_T])_{(q)}$$
$$\text{on} \quad \Gamma_q, q = 1, \ldots, k, \quad \forall v \in V_{\text{ad}}.$$

From (4.5.23) we obtain the following expression since the maximum is attained: There exists $w_{1,N} \in \underline{\partial}' F_N([u_N])_{(q)}$ and $w_{1,T} \in \underline{\partial}' F_T([u_T])_{(q)}$ on Γ_q for $q = 1, \ldots, k$ such that

$$A(u, v-u) - L(v-u) + \sum_{q=1}^{k} \left\{ \langle w_{1,N} + w_{2,N}^{\star}, ([v_N] - [u_N]) \rangle_{\frac{1}{2}, \Gamma_q} \right.$$
$$\left. + \langle w_{1,T} + w_{2,T}^{\star}, ([v_T] - [u_T]) \rangle_{H_T, \Gamma_q} \right\}_{(q)} \geq 0$$
$$\forall w_{2,N}^{\star} \in \bar{\partial}' F_N([u_N])_{(q)}, \quad \forall w_{2,T}^{\star} \in \bar{\partial}' F_T([u_T])_{(q)}$$
$$\text{on} \quad \Gamma_q, q = 1, \ldots, k, \quad \forall v \in V_{\text{ad}}. \tag{4.5.25}$$

One obtains from (4.5.24) an analogous reverse inequality since the min is also attained: There exists $w_{2,N} \in \bar{\partial}' F_N([u_N])_{(q)}$ and $w_{2,T} \in \bar{\partial}' F_T([u_T])_{(q)}$ on Γ_q for $q = 1, \ldots, k$ such that

$$0 \geq A(u, v-u) - L(v-u) + \sum_{q=1}^{k} \left\{ \langle w_{1,N}^{\star} + w_{2,N}, ([v_N] - [u_N]) \rangle_{\frac{1}{2}, \Gamma_q} \right.$$
$$\left. + \langle w_{1,T}^{\star} + w_{2,T}, ([v_T] - [u_T]) \rangle_{H_T, \Gamma_q} \right\}_{(q)}$$
$$\forall w_{1,N}^{\star} \in \underline{\partial}' F_N([u_N])_{(q)}, \quad \forall w_{1,T}^{\star} \in \underline{\partial}' F_T([u_T])_{(q)}$$
$$\text{on} \quad \Gamma_q, q = 1, \ldots, k, \quad \forall v \in V_{\text{ad}}. \tag{4.5.26}$$

Let us assume now that the relations

$$-S = w_1 + w_2 \quad \text{with } \{w_1, w_2\} \in DF([u]) = \{\underline{\partial}' F([u]), \bar{\partial}' F([u])\} \tag{4.5.27}$$

hold, where S stays for S_N or S_T and $[u]$ for $[u_N]$ or $[u_T]$ respectively. Suppose further that F_N and F_T can be expressed as the difference of two convex functions (d.c.f.). Then (1.4.27) holds, i.e. if $F_N = \Phi_{N_1} - \Phi_{N_2}$, with Φ_{N_1} and Φ_{N_2} convex, then

$$\underline{\partial}' F_N = \partial \Phi_{N_1}, \quad \bar{\partial}' F_N = -\partial \Phi_{N_2}, \tag{4.5.28}$$

where ∂ is the subdifferential of the convex functions. In this case (4.5.25) holds and gives rise to the following system of variational inequalities as it easily

results by using the definition of the subdifferential:
Find $u \in V_{\mathrm{ad}}$, such as to satisfy

$$A(u, v-u) - L(v-u) \tag{4.5.29a}$$

$$+ \sum_{q=1}^{k} \left\{ \langle w_{2,N}^{\star}, ([v_N] - [u_N]) \rangle_{\frac{1}{2}, \Gamma_q} + \langle w_{2,T}^{\star}, ([v_T] - [u_T]) \rangle_{H_T, \Gamma_q} \right\}_{(q)}$$

$$+ \sum_{q=1}^{k} \left\{ \Phi_{N_1}([v_N]) - \Phi_{N_1}([u_N]) + \Phi_{T_1}([v_T]) - \Phi_{T_1}([u_T]) \right\}_{(q)} \geq 0 \quad \forall v \in V_{\mathrm{ad}},$$

$\forall w_{2,N}^{\star} \in H^{1/2}(\Gamma_q), \forall w_{2,T}^{\star} \in H^{T}(\Gamma_q) \quad q = 1, \ldots, k$ such that

$$-\langle w_{2,N}^{\star}, ([v_N] - [u_N]) \rangle_{\frac{1}{2}, \Gamma_q} \leq \left\{ \Phi_{N_2}([v_N]) - \Phi_{N_2}([u_N]) \right\}_{(q)}$$

$$q = 1, \ldots, k, \quad \forall v \in V_{\mathrm{ad}} \tag{4.5.29b}$$

$$-\langle w_{2,T}^{\star}, ([v_N] - [u_N]) \rangle_{H_T, \Gamma_q} \leq \left\{ \Phi_{T_2}([v_T]) - \Phi_{T_2}([u_T]) \right\}_{(q)}$$

$$q = 1, \ldots, k, \quad \forall v \in V_{\mathrm{ad}}. \tag{4.5.29c}$$

Prop. 2.5.1 (see [Dem86b] p. 119-121 the theorem of von Strassen) implies that pointwise boundary conditions of the type (4.5.27), (4.5.28), can be extended to function spaces (actually to $L^2(\Gamma_q)$-spaces) by integrating the directional differential. Suppose, for instance, that (4.5.27), (4.5.28) hold pointwise on Γ_q for $S_N, [u_N]$. Then we define

$$\bar{F}_N([u_N]) = \begin{cases} \int_{\Gamma_q} F_N([u_N]) d\Gamma & \text{if } F_N([u_N]) \in L^1(\Gamma_q) \\ \infty & \text{otherwise} \end{cases} \tag{4.5.30}$$

and analogously we define $\bar{\Phi}_{N_1}$ and $\bar{\Phi}_{N_2}$ from Φ_{N_1} and Φ_{N_2} respectively. Then we may write for the directional differential $\widetilde{\bar{F}}_N'([u_N], [v_N])$ a relation analogous to (3.5.16). Thus in the case of pointwise quasidifferential relations, where the quasidifferentiable function is of the d.c.f. type, we obtain relations analogous to (4.5.29a, b, c) with the difference that instead of $\Phi_{N_1}, \Phi_{N_2}, \Phi_{T_1}, \Phi_{T_2}$ we have the $\bar{\Phi}_{N_1}, \bar{\Phi}_{N_2}, \bar{\Phi}_{T_1}, \bar{\Phi}_{T_2}$. Then we can easily verify that the solution of the variational formulation (4.5.29a,b,c) satisfies the conditions of equilibrium in the sense of distributions over $\Omega^{(m)}$ and thus in the sense of $L^2(\Omega^{(m)})$, since $f_i^{(m)} \in L^2(\Omega^{(m)})$, the boundary condition on $\Gamma_F^{(m)}$ in the sense of $L^2(\Gamma_F^{(m)})$ and the interface conditions in the sense of $H^{1/2}(\Gamma_q) \times H^{-1/2}(\Gamma_q)$ (resp. $H_T(\Gamma_q) \times H_T'(\Gamma_q)$) in the normal (resp. the tangential) direction to Γ_q for $q = 1, \ldots, k$.

We shall close this section by considering material laws of the quasidifferential type, for instance the law (3.5.20). Let us consider for simplicity a plane body without interfaces and with homogeneous boundary conditions (i.e. $u_i = 0$ on $\Gamma, i = 1, 2$). For this structure the principle of virtual work reads for $f_i \in L^2(\Omega)$, $i = 1, 2$:

134   4. Hemivariational Inequalities

$$\int_\Omega \sigma_{ij}(u)\varepsilon_{ij}(v-u)d\Omega - \int_\Omega f_i(v_i - u_i)d\Omega = 0, \quad \forall v \in V_{ad} \qquad (4.5.31)$$

where V_{ad} is the set of kinematically abmissible displacements i.e. $V_{ad} = \{v = \{v_i\}|v_i \in H^1(\Omega), i = 1, 2, v = 0 \text{ on } \Gamma\}$. Assuming the validity of (3.5.20) (resp. (3.5.21)) yields the following problem:
Find $u \in V_{ad}$ such as to satisfy the variational equality (resp. the hemivariational inequality):

$$\int_\Omega (\tilde{w}'(\varepsilon(u)), \varepsilon(v-u))d\Omega - \int_\Omega f_i(v_i - u_i)d\Omega = (\text{resp.}, \geq)0, \quad \forall v \in V_{ad} \qquad (4.5.32)$$

If w is quasidifferentiable, then two convex compact sets $A = \underline{\partial}w(\varepsilon)$ and $B = \bar{\partial}'w(\varepsilon)$ can be determined such that (3.5.22) or (3.5.23) or (3.5.24) hold. Thus (4.5.32) yields the following variational formulations (only the case of equality in (4.5.32) is considered; we denote $\int \tau_{ij}\varepsilon_{ij}d\Omega$ by $\int(\tau, \varepsilon)d\Omega$.

i) Find $u \in V_{ad}$ such as to satisfy the relations: there exists $\bar{\bar{\tau}} \in \bar{\partial}'w(\varepsilon(u))$ such that

$$\int_\Omega f_i(v_i - u_i)d\Omega \geq \int_\Omega (\bar{\tau} + \bar{\bar{\tau}}, \varepsilon(v-u))d\Omega, \qquad (4.5.26)$$

$$\forall v \in V_{ad} \quad \text{and} \quad \forall \bar{\tau} \in \underline{\partial}'w(\varepsilon(u)).$$

ii) Find $u \in V_{ad}$ such as to satisfy the relation

$$\int_\Omega [\min_{\bar{\bar{\tau}} \in B} \max_{\bar{\tau} \in A}((\bar{\tau} + \bar{\bar{\tau}}, \varepsilon(v-u)) - f_i(v_i - u_i)]d\Omega = 0, \quad \forall v \in V_{ad} \qquad (4.5.34)$$

iii) Find $u \in V_{ad}, \bar{\tau} \in A, \bar{\bar{\tau}} \in B$ such that:

$$\int_\Omega [\max_{\bar{\tau}^* \in A}((\bar{\tau}^* + \bar{\bar{\tau}}^*, \varepsilon(v-u)) - f_i(v_i - u_i)]d\Omega \geq 0, \qquad (4.5.35a)$$

$$\forall v \in V_{ad} \quad \forall \bar{\bar{\tau}}^* \in B$$

and

$$\int_\Omega [\min_{\bar{\bar{\tau}}^* \in B}((\bar{\tau}^* + \bar{\bar{\tau}}^*, \varepsilon(v-u)) - f_i(v_i - u_i)]d\Omega \leq 0, \qquad (4.5.35b)$$

$$\forall v \in V_{ad} \quad \forall \bar{\tau}^* \in A.$$

Note that if instead of the material law (3.5.20), the material law (3.5.21) holds, then (4.5.34) holds as an inequality, i.e. the left hand side is ≥ 0. If instead of quasidifferential, codifferentiable energy functions are given, analogous variational expressions are obtained.

5. Multivalued Boundary Integral Equations

Multivalued nonmonotone boundary conditions lead to hemivariational formulations on the boundary of the body which are equivalent to multivalued integral boundary equations. Analogous results hold for interface problems with the difference that, in this case the multivalued integral equations are extended over the interface. As the present chapter is rather brief, we refer the reader to the monograph by Antes and the author [Ant92] and to the monograph by the author [Pan85] p. 160-162 for further information on the material.

5.1 The Indirect and the Direct Method for Nonmonotone Boundary Conditions

i) The Indirect Method.

We consider a three-dimensional linear elastic body subjected to nonmonotone multivalued boundary conditions which are obtained from nonconvex superpotentials (cf. Sect. 2.4). The method which we give here remains valid also for shells, plates, beams etc. Let Ω be an open bounded subset of the three-dimensional Euclidean space $\mathbb{R}^3$ with a Lipschitz boundary Γ. Ω is occupied by a linear elastic body in its undeformed state. We refer to a Cartesian orthogonal coordinate system $Ox_1x_2x_3$. Γ is decomposed into nonoverlapping parts Γ_U, Γ_F and Γ_S nonempty and open in Γ. It is assumed that on Γ_U (resp. Γ_F) the displacements (resp. the tractions) are given and that on Γ_S the boundary conditions causing the inequality formulation of the problem hold. We assume that on Γ_U

$$u_i = U_i, \qquad U_i = U_i(x), \tag{5.1.1}$$

and on Γ_F

$$S_i = F_i, \qquad F_i = F_i(x). \tag{5.1.2}$$

The nonconvex superpotential boundary conditions have the general form (2.4.2) or the "inverse" form

$$u \in \bar{\partial}\tilde{j}(-S) \text{ on } \Gamma_S, \tag{5.1.3}$$

where $\tilde{j}, j$ are nonconvex superpotentials which are locally Lipschitz on Γ_S. If S_N (resp. S_T) are the normal (resp. the tangential) components of S with respect to Γ and u_N and u_T are the corresponding components of the displacement u, the method presented here remains valid if (5.1.3) is replaced by

$$u_N \in \bar{\partial}\tilde{j}_N(-S_N) \quad \text{and} \quad u_T \in \bar{\partial}\tilde{j}_T(-S_T) \text{ on } \Gamma_S \qquad (5.1.4)$$

or (2.4.2) by

$$-S_N \in \bar{\partial}j_N(u_N) \quad \text{and} \quad -S_T \in \bar{\partial}j_T(u_T) \text{ on } \Gamma_S. \qquad (5.1.5)$$

The equations of the B.V.P. read

$$\sigma_{ij,j} + f_i = 0 \quad \text{in } \Omega, \qquad (5.1.6)$$

$$\varepsilon_{ij} = \varepsilon_{ij}(u) = \frac{1}{2}(u_{i,j} + u_{j,i}) \text{ in } \Omega, \qquad (5.1.7)$$

$$\sigma_{ij} = C_{ijhk}\varepsilon_{hk} \quad \text{in } \Omega, \qquad (5.1.8)$$

where the comma denotes the partial derivation and $f = \{f_i\}$ is the volume force vector. Let us denote by $\tilde{V}$ the linear space of the displacements v_i and by V the set of the kinematically admissible displacement fields

$$V = \{v|v = \{v_i\}, v \in \tilde{V}, v_i = U_i, i = 1, 2, 3 \text{ on } \Gamma_U\} \qquad (5.1.9)$$

without taking yet into account the constraints on Γ_S. The work of the force $f = \{f_i\}$ (resp $F = \{F_i\}$) for the displacement $v = \{v_i\}$ on Ω (resp. on Γ_F) is written as (f,v) (resp. as $[F,v]_{\Gamma_F}$) etc. Note that if $\tilde{V} = [H^1(\Omega)]^3$ and $f_i \in L^2(\Omega)$, $C_{ijhk} \in L^\infty(\Omega)$, $F_i \in L^2(\Gamma)$ and $U_i \in H^{1/2}(\Gamma)|_{\Gamma_1}$ then $[S,v]_{\Gamma_S} = \langle S,v \rangle_{\Gamma_S}$, $[S_N, v_N]_{\Gamma_S} = \langle S_N, v_N \rangle_{1/2,\Gamma_S}$ and $[S_T, v_T]_{\Gamma_S} = \langle S_T, v_T \rangle_{H_T,\Gamma_S}$ etc. The bilinear form of elasticity is again denoted by $a(\cdot,\cdot)$ and the relations (4.1.5), (4.1.6), (4.1.7) hold. Moreover

$$l(v) = (f,v) + \int_{\Gamma_F} F_i v_i d\Gamma. \qquad (5.1.10)$$

In order to make the problem homogeneous on Γ_U we introduce a kinematically admissible displacement field u_0 such that $u_{0i} = U_i$ on Γ_U, and let

$$\bar{u} = u - u_0, \quad \bar{v} = v - u_0 \qquad (5.1.11)$$

where

$$\bar{u}, \bar{v} \in V_0 = \{v|v = \{v_i\}, v \in V, v_i = 0 \text{ on } \Gamma_U\}. \qquad (5.1.12)$$

We denote also by L the admissible vector space of the tractions S on Γ_S, i.e. L is the restriction of $[H^{-1/2}(\Gamma)]^3$ to Γ_S.

In order to formulate the indirect Boundary Integral Equation Method (B.I.E.M.) for the present problem we imbed our problem in an infinite elastic medium and we determine layers of singular solutions for the equations of elastostatics for the infinite elastic medium such that the boundary conditions are satisfied (cf. e.g. [Ant92] for references). Suppose for instance that q is the vector of the unknown force and/or dislocation layers on the boundary Γ in the infinite

5.1 The Indirect and the Direct Method for Nonmonotone Boundary Conditions

elastic medium. We can express $u, S, u_N, S_N, u_{T_i}, S_{T_i}$, in (5.1.2) and (5.1.3) or (5.1.4) in terms of q through appropriate operators and we have to solve the resulting system of equations and multivalued equations defined through the boundary conditions. To be more precise let us treat a plane stress elasticity problem. The body is subjected to the forces $f_i, i = 1, 2$. Our aim is to replace in the boundary conditions on Γ_U, Γ_F and Γ_S, u and S with their expressions as functions of the unknown force distribution $R_i, i = 1, 2$ on Γ. The displacement $u_i(x)$ at $x = (x_1, x_2)$ due to a unit force $R(\xi)$ at $\xi = (\xi_1, \xi_2)$ is given by the formula

$$u_i(x) = [u_{,R}(x, \xi)]_{ij} R_j(\xi) \tag{5.1.13}$$

where (cf. e.g. [Bane])

$$[u_{,R}(x, \xi)]_{ij} = -\frac{1}{8\pi G(1 - \bar{\nu})} \left((3 - 4\bar{\nu})\delta_{ij} \ln r - \frac{(x_i - \xi_i)(x_j - \xi_j)}{r^2} \right) + C_{ij}. \tag{5.1.14}$$

Here $G = \frac{E}{2(1+\nu)}$, with E the modulus of elasticity and ν the Poisson ratio, $\bar{\nu} = \nu/(1 + \nu)$, $r^2 = (x_i - \xi_i)(x_i - \xi_i)$ and C_{ij} is a constant tensor which introduces rigid-body displacements into the problem. The traction $S_i(x)$ due to a unit force $R(\xi)$ reads

$$S_i(x) = [S_{,R}(x, \xi)]_{ij} R_j(\xi), \tag{5.1.15}$$

where

$$[S_{,R}(x, \xi)]_{ij} = -\frac{1}{4\pi(1 - \bar{\nu})r^2} \Big[(1 - 2\bar{\nu})(n_j(x_i - \xi_i) - n_i(x_j - \xi_j)) \tag{5.1.16}$$
$$+ ((1 - 2\bar{\nu})\delta_{ij} + \frac{2(x_i - \xi_i)(x_j - \xi_j)}{r^2})(x_j - \xi_j)n_j \Big].$$

Note that the above formulas hold for plane strain problems, if one replaces $\bar{\nu}$ by the Poisson ratio ν. The singularity in (5.1.14) is a $\ln r$-singularity, i.e. a weak singularity, whereas in (5.1.15) is a $1/r$-singularity, i.e. a strong singularity. Due to the unknown force distribution R_i, $i = 1, 2$ over Γ and due to the given forces f_i, $i = 1, 2$, over Ω we obtain the following displacements $u_i(x_0)$ and tractions $S_i(x_0)$ at a point $x_0 \in \Gamma$

$$u_i(x_0) = \int_\Gamma [u_{,R}(x_0, \xi)]_{ij} R_j(\xi) d\Gamma + \int_\Omega [u_{,R}(x_0, \xi)]_{ij} f_j(\xi) d\Omega + C_i \tag{5.1.17}$$

$$S_i(x_0) = \frac{1}{2} \delta_{ij} R_i(x_0) \tag{5.1.18}$$
$$+ \int_\Gamma [S_{,R}(x_0, \xi)]_{ij} R_j(\xi) d\Gamma + \int_\Omega [S_{,R}(x_0, \xi)]_{ij} R_j(\xi) d\Omega.$$

Here the $\ln r$-singularity does not cause any serious problem even for $x_i = \xi_i$ but the $1/r$-singularity compels us to consider the line integral as a Cauchy principal-value integral (i.e. it makes sense as the limiting value of the integral as point x approaches the loading point on Γ (cf. e.g. [Kress])). Moreover we assume that x tends to the point x_0 on Γ from inside Γ (if it tends from outside Γ to x_0 then the first term in (5.1.18) must have the minus sign) and that Γ does not have corners. For the case of corners we refer to [Ant92]. Moreover C_i denotes a rigid body displacement which can be suppressed by the boundary conditions. Further we write (5.1.17) and (5.1.18) symbolically as

$$u_i(x_0) = [D_u(R)]_i, \quad S_i(x_0) = [D_S(R)]_i \tag{5.1.19}$$

and we obtain for instance from (5.1.1), (5.1.2) and (5.1.3) the system of equations and multivalued equations whose solution gives the unknown force distributions R. Of course there are still many open questions concerning the solution of the arising multivalued integral equations. We may also consider as unknowns of the problem layers of dislocations, or both layers of forces and layers of dislocations on complementary parts of the boundary. Every case leads to other operators D_u and D_S, and therefore to different forms of integral equations, regular or singular (cf. also [Pan83b]).

ii) The Direct Method.

For convex superpotential boundary conditions the direct B.I.E.M. is formulated by using the duality of convex functionals and cetrain Lagrangian formulations of the minimum energy problems for the potential and the complementary energy. In this context we refer to [Ant92] and the references given there and to [Pana87,91]. Here due to the lack of convexity we will apply a method based on Betti's theorem of elasticity, developed in [Pana89, Ant92] which will lead to multivalued B.I.Es.

Now we assume that $S \in L$ is given on Γ_S and is equal to $\mu = \{\mu_i\}$. Then the solution of the arising classical problem satisfies the following problem: Find $\bar{u} = \bar{u}(\mu) \in V_0$ such that

$$a(\bar{u}, \bar{v}) + a(u_0, \bar{v}) - [\mu, \bar{v}]_{\Gamma_S} - (f, \bar{v}) - [F, \bar{v}]_{\Gamma_F} = 0 \quad \forall \bar{v} \in V_0. \tag{5.1.20}$$

Obviously (5.1.20) expresses the principle of virtual work for a structure resulting from the initial one by eliminating the superpotential constraints on Γ_S and by applying the forces $\mu = \{\mu_i\}$ on Γ_S. Because of the linearity of (5.1.20) the solution $\bar{u}$ of it can be written as the sum $\bar{u}_{(1)} \in V_0$ and $\bar{u}_{(2)} \in V_0$ where $\bar{u}_{(1)}$ and $\bar{u}_{(2)}$ are solutions of the two variational equalities

$$a(\bar{u}_{(1)}, \bar{v}) - l(\bar{v}) + a(u_0, \bar{v}) = 0, \quad \forall \bar{v} \in V_0 \tag{5.1.21}$$

and

$$a(\bar{u}_{(2)}, \bar{v}) - [\mu, \bar{v}]_{\Gamma_S} = 0 \quad \forall \bar{v} \in V_0 \tag{5.1.22}$$

respectively. Here $\bar{u}_{(1)}$ and $\bar{u}_{(2)}$ are equilibrium configurations of two bilateral structures resulting from the initial one by ignoring the superpotential boundary

5.1 The Indirect and the Direct Method for Nonmonotone Boundary Conditions

conditions on Γ_S, and assuming that on certain parts of the boundary the load is zero; thus in the case of (5.1.21) the structure is fixed on Γ_U and is loaded by the forces f in Ω and F on Γ_F, whereas on Γ_S the loading is zero. Moreover the structure is subjected to an initial displacement field u_0. In the case of (5.1.22) the structure is loaded by a force $\mu = \{\mu_i\}$ only on Γ_S and is fixed along Γ_U; the loading in Ω and on Γ_F is zero. The solutions $\bar{u}_{(1)}$ and $\bar{u}_{(2)}$ are uniquely determined, as it is well known from the classical (bilateral) elasticity theory. For the bilateral structures the solutions $\bar{u}_{(1)}$ and $\bar{u}_{(2)}$ can be written in terms of Green's operator G, which is the same for both structures due to the same type of boundary conditions holding in each structure. Accordingly, we can write that

$$\bar{u}_{(1)} = G(l), \quad \bar{u}_{(2)} = G(\mu), \quad u = \bar{u}_{(1)} + \bar{u}_{(2)}, \quad l = \{f, F, u_0\}. \tag{5.1.23}$$

We have to determine the unknown force distribution $\mu = \{\mu_i\} \in L$ on Γ_S. With respect to the linear elasticity problem described by (5.1.22) we apply Betti's theorem: Assume that $\lambda = \{\lambda_i\} \in L$ on Γ_S is a force distribution corresponding to a displacement field $\bar{v}_{(2)} \in V_0$ if $f = 0$, $F = 0$ on Γ_F and $u_0 = 0$. Then we have that for every $\lambda \in L$

$$[\lambda, \bar{u}_{(2)}]_{\Gamma_S} = [\mu, \bar{v}_{(2)}]_{\Gamma_S}. \tag{5.1.24}$$

Obviously we may write that

$$\bar{v}_{(2)} = G(\lambda). \tag{5.1.25}$$

Now (5.1.24) implies with (5.1.25) that

$$[\lambda, \bar{u}]_{\Gamma_S} = [\lambda, \bar{u}_{(1)}]_{\Gamma_S} + [\lambda, \bar{u}_{(2)}]_{\Gamma_S} = [\lambda, \bar{u}_{(1)}]_{\Gamma_S} + [\mu, \bar{v}_{(2)}]_{\Gamma_S}, \quad \forall \lambda \in L \tag{5.1.26}$$

which becomes further

$$[\lambda, \bar{u}]_{\Gamma_S} = [\lambda, [G(l)]]_{\Gamma_S} + [\mu, [G(\lambda)]]_{\Gamma_S} \quad \forall \lambda \in L. \tag{5.1.27}$$

Now we introduce the bilinear symmetric (by Betti's theorem) form

$$\beta(\lambda, \mu) = [\mu, [G(\lambda)]]_{\Gamma_S}, \tag{5.1.28}$$

and the linear form

$$\bar{\gamma}(\lambda) = -[\lambda, [G(l)]]_{\Gamma_S}. \tag{5.1.29}$$

Assuming now that the tractions μ on Γ_S are related to the displacement field u through the relation (5.1.3) we may write that

$$\bar{j}^0(-\mu, -\lambda^*) \geq -\lambda_i^*(\bar{u}_i + u_{0i}) \quad \forall \lambda^* \in L, \tag{5.1.30}$$

where $\bar{j}^0(\cdot, \cdot)$ denotes the directional differential of Clarke. From (5.1.26), (5.1.27) and (5.1.30) we obtain for $\lambda^* = \lambda$ that

$$\int_{\Gamma_S} \tilde{j}^0(-\mu, -\lambda)d\Gamma \geq [-\lambda, (\bar{u}+u_0)]_{\Gamma_S} = \bar{\gamma}(\lambda) - \beta(\lambda, \mu) - [\lambda, u_0]_{\Gamma_S} = \gamma(\lambda) - \beta(\lambda, \mu),$$
(5.1.31)

where
$$\gamma(\lambda) = \bar{\gamma}(\lambda) - [\lambda, u_0]_{\Gamma_S}.$$
(5.1.32)

Relation (5.1.31) holds for all $\lambda \in L$ and thus we are led to the following hemivariational inequality:
Find $\mu \in L$ such as to satisfy

$$\beta(\mu, \lambda) - \gamma(\lambda) + \int_{\Gamma_S} \tilde{j}^0(-\mu, -\lambda)d\Gamma \geq 0 \quad \forall \lambda \in L.$$
(5.1.33)

Let us consider now the "substationarity" problem

$$\mu \in L, \quad 0 \in \bar{\partial}\Pi(\mu), \quad \Pi(\mu) = \frac{1}{2}\beta(\mu, \mu) - \gamma(\mu) + \int_{\Gamma_S} \tilde{j}(-\mu)d\Gamma,$$
(5.1.34)

provided that the integral on Γ_S makes sense. Then every solution of (5.1.34) satisfies (5.1.33) but not conversely, if relations analogous to (4.3.3) and (4.3.4) hold, to guarantee the validity of (4.3.9). Obviously (5.1.34) implies the multivalued boundary integral equation

$$\gamma - \frac{1}{2}\text{grad}\beta(\mu, \mu) \in \bar{\partial}\left(\int_{\Gamma_S} \tilde{j}(-\mu)d\Gamma\right)$$
(5.1.35)

which is explicitly written as

$$\frac{\partial}{\partial \mu}\left\{[-\mu, [G(\bar{l})] + u_0]_{\Gamma_S} - \frac{1}{2}[\mu, [G(\mu)]]_{\Gamma_S}\right\} \in \bar{\partial}\left(\int_{\Gamma_S} \tilde{j}(-\mu)d\Gamma\right).$$
(5.1.36)

We recall here that every local minimum and every saddle point of the energy Π is a substationarity problem. Also a local maximum, say $\mu_0 \in L$, is a substationarity point if Π is Lipschitzian around μ_0 (cf. Sect. 1.2)

Until now we have derived a multivalued B.I.E. with respect to the boundary stresses on Γ_S. Here a multivalued boundary integral formulation with respect to the displacements on Γ_S will be derived. We assume that on Γ_S the nonmonotone possibly multivalued boundary conditions are expressed in the form (2.4.2). Note that in the case of monotonicity we do not need to distinguish between (5.1.3) and (2.4.2) since then j is convex and $\tilde{j}$ is the conjugate functional of j. But if convexity does not hold an appropriate definition of the "conjugacy operation" does not exist which could make possible to invert (2.4.2) in order to get (5.1.3).

Let us assume first that the displacements u on Γ_S are given. Then we denote by Σ the set of all symmetric stress-tensors and let

5.1 The Indirect and the Direct Method for Nonmonotone Boundary Conditions

$$\Sigma_1 = \{\tau | \tau = \{\tau_{ij}\}, \tau_{ij} = \tau_{ji} \in L^2(\Omega), \tau_{ij,j} + f_i = 0 \quad (5.1.37)$$
$$\text{a.e. in } \Omega, T_i = F_i \text{ a.e. on } \Gamma_F\}$$

be the statically admissible set. In (5.1.37) $\{T_i\}$ denotes the traction on Γ corresponding to the stress field $\{\tau_{ij}\}$. Let also $c = \{c_{ijhk}\}$ be the inverse tensor to $C = \{C_{ijhk}\}$, i.e.

$$\varepsilon_{ij} = c_{ijhk}\sigma_{hk} \tag{5.1.38}$$

and let

$$A(\sigma, \tau) = (c\sigma, \tau) = \int_\Omega c_{ijhk}\sigma_{ij}\tau_{hk}d\Omega. \tag{5.1.39}$$

For given displacements v on Γ_S we can write the "principle" of complementary virtual work for the structure in the form:
Find $\sigma = \sigma(v) \in \Sigma_1$ such that

$$A(\sigma, \tau) = [\bar{U}, T]_{\Gamma_U} + [v, T]_{\Gamma_S} \quad \forall \tau \in \Sigma_1. \tag{5.1.40}$$

Let us now introduce a strain-field $\sigma_0 \in \Sigma_1$, i.e. a stress field satisfying the equations of equilibrium and the static boundary conditions on Γ_F and let us introduce the new variables

$$\bar{\sigma} = \sigma - \sigma_0 \text{ and } \bar{\tau} = \tau - \tau_0 \tag{5.1.41}$$

where $\bar{\sigma}, \bar{\tau} \in \Sigma_0$ and

$$\Sigma_0 = \{\tau | \tau = \{\tau_{ij}\}, \tau_{ij} = \tau_{ji} \in L^2(\Omega), \tau_{ij,j} = 0 \quad (5.1.42)$$
$$\text{a.e. in } \Omega, T_i = 0 \text{ a.e. on } \Gamma_F\}.$$

Thus (5.1.40) becomes:
Find $\bar{\sigma} = \bar{\sigma}(v) \in \Sigma_0$ such as to satisfy

$$A(\bar{\sigma}, \bar{\tau}) - [\bar{U}, \bar{T}]_{\Gamma_U} - [v, \bar{T}]_{\Gamma_S} + A(\sigma_0, \bar{\tau}) = 0, \quad \forall \bar{\tau} \in \Sigma_0. \tag{5.1.43}$$

Here we choose a σ_0 satisfying

$$A(\sigma_0, \bar{\tau}) = 0 \quad \forall \bar{\tau} \in \Sigma_0 \tag{5.1.44}$$

in order to simplify all the arising expressions, i.e. we assume that σ_0 is the unique solution of a classical (bilateral or equality) problem having on Γ_U and Γ_S zero displacements.

The stress $\bar{\sigma}$ in (5.1.43) can be written as the sum $\bar{\sigma}_{(1)} + \bar{\sigma}_{(2)}$ where $\bar{\sigma}_{(1)}$ and $\bar{\sigma}_{(2)}$ are solutions of the variational equalities

$$A(\bar{\sigma}_{(1)}, \bar{\tau}) - [v, \bar{T}]_{\Gamma_S} = 0 \quad \forall \bar{\tau} \in \Sigma_0, \tag{5.1.45}$$

$$A(\bar{\sigma}_{(2)}, \bar{\tau}) - [\bar{U}, \bar{T}]_{\Gamma_U} = 0 \quad \forall \bar{\tau} \in \Sigma_0, \tag{5.1.46}$$

respectively. Both (5.1.45) and (5.1.46) respectively express the "principle" of complementary virtual work for bilateral structures resulting from the initial one in the following way: for (5.1.45) (resp. (5.1.46)) we consider the structure Ω under the action of "given" displacements v (resp. zero) on Γ_S, zero forces in Ω and Γ_F and zero (resp. $\bar{U}$) displacements on Γ_U. Since these structures are linear elastic, $\bar{\sigma}_{(1)}$ and $\bar{\sigma}_{(2)}$ are uniquely determined. Therefore (5.1.45) and (5.1.46) imply that

$$\bar{\sigma}_{(1)} = H(v), \quad \bar{\sigma}_{(2)} = H(\bar{U}), \quad \bar{\sigma} = \bar{\sigma}_{(1)} + \bar{\sigma}_{(2)}, \quad \bar{\sigma} \in \Sigma_0, \qquad (5.1.47)$$

where H is the Green's stress-displacement operator for the two problems (5.1.45) and (5.1.46). Both fictive bilateral structures corresponding to (5.1.45) and (5.1.46) have the same H-operator because of the same type of boundary conditions. Moreover we denote by $\tilde{H}$, the operator transforming the displacement at the boundary into the traction $S = \{S_i\}$. Thus we may write that

$$\bar{S}_{(1)} = \tilde{H}(v), \quad \bar{S}_{(2)} = \tilde{H}(\bar{U}). \qquad (5.1.48)$$

We have to determine the unknown displacement distribution $v = \{v_i\} \in N$ on Γ_S. Note that in the functional framework introduced previously N = restriction of $[H^{1/2}(\Gamma)]^3$ to Γ_S. Let $w = \{w_i\} \in N$ be another displacement distribution on Γ_S corresponding to the stress field $\bar{\tau}_{(1)} \in \Sigma_0$ through (5.1.45). Moreover $v = \{v_i\} \in N$ corresponds to $\bar{\sigma}_1 \in \Sigma_0$. Applying Betti's theorem we can write that

$$[\bar{T}_{(1)}, v]_{\Gamma_S} = [\bar{S}_{(1)}, w]_{\Gamma_S} \quad \forall w \in N. \qquad (5.1.49)$$

Moreover we can write analogously to (5.1.48) that

$$\bar{T}_{(1)} = \tilde{H}(w). \qquad (5.1.50)$$

Relation (5.1.49) implies with (5.1.50), and for $\bar{S} = \bar{S}_{(1)} + \bar{S}_{(2)}$ that

$$\begin{aligned}
[w, \bar{S}]_{\Gamma_S} &= [w, \bar{S}_{(1)}]_{\Gamma_S} + [w, \bar{S}_{(2)}]_{\Gamma_S} = [v, \bar{T}_{(1)}]_{\Gamma_S} + [w, \bar{S}_{(2)}]_{\Gamma_S} \qquad (5.1.51)\\
&= [v, [\tilde{H}(w)]]_{\Gamma_S} + [w, \bar{S}_{(2)}]_{\Gamma_S} = [v, [\tilde{H}(w)]]_{\Gamma_S} + [w, [\tilde{H}(\bar{U})]]_{\Gamma_S}.
\end{aligned}$$

Now the bilinear symmetric (due to Betti's theorem) form

$$\delta(v, w) = [[\tilde{H}(v)], w]_{\Gamma_S}. \qquad (5.1.52)$$

is introduced and the linear form

$$\bar{\zeta}(w) = [[\tilde{H}(\bar{U})], w]_{\Gamma_S} \qquad (5.1.53)$$

and thus (5.1.51) implies that

$$[w, \bar{S}]_{\Gamma_S} = \delta(v, w) - \bar{\zeta}(w). \qquad (5.1.54)$$

But (2.4.2) implies by definition that

$$j^0(v, w^*) \geq -S_i w_i^* = -(\bar{S}_i + S_{0i}) w_i^* \quad \forall w_i^* \in N. \qquad (5.1.55)$$

From (5.1.54) and (5.1.55) we obtain for $w^* = w$ that

$$\int_{\Gamma_S} j^0(v,w) d\Gamma \geq [\bar{S}, w]_{\Gamma_S} - [S_0, w]_{\Gamma_S} = -\delta(v,w) + \zeta(w), \quad (5.1.56)$$

where

$$\zeta(w) = \bar{\zeta}(w) - [S_0, w]_{\Gamma_S}. \quad (5.1.57)$$

Relation (5.1.56) holds for all $w \in N$ and thus we are led to the following hemivariational inequality:
Find $v \in N$ such as to satisfy

$$\delta(v,w) - \zeta(w) + \int_{\Gamma_S} j^0(v,w) d\Gamma \geq 0 \quad \forall w \in N. \quad (5.1.58)$$

We formulate further the corresponding "substationarity" problem

$$v \in N, \quad 0 \in \bar{\partial}\tilde{\Pi}(v), \quad \tilde{\Pi}(v) = \frac{1}{2}\delta(v,v) - \zeta(v) + \int_{\Gamma_S} j(v) d\Gamma \quad (5.1.59)$$

and the corresponding multivalued boundary integral equation

$$\zeta - \frac{1}{2}\mathrm{grad}\,\delta(v,v) \in \bar{\partial}\left(\int_{\Gamma_S} j(v) d\Gamma\right), \quad (5.1.60)$$

which may be put in the form

$$\frac{\partial}{\partial v}\left\{[[\tilde{H}(\bar{U})] + S_0, v]_{\Gamma_S} - \frac{1}{2}[[\tilde{H}(v)], v]_{\Gamma_S}\right\} \in \bar{\partial}\left(\int_{\Gamma_S} j(v) d\Gamma\right). \quad (5.1.61)$$

Analogously we proceed if instead of (2.4.2) or (5.1.3) the relations (5.1.4) or (5.1.5) hold.

5.2 Complement for Adhesively Bonded Cracks

The theory developed in the previous section can be appropriately modified to deal with the influence of adhesives on the behaviour of cracks in two-dimensional linear elastic bodies. The crack is considered to be "repaired" through an adhesive material which exhibits a nonlinear, nonmonotone, possibly multivalued behaviour in the tangential and/or the normal direction with respect to the cracks [The91]. Of course, now the singularities caused by the crack should be taken into account. We consider that the crack has a given length, that the adhesive between the two sides of the crack has a negligible thickness and that debonding may take place both in the normal and in the tangential direction. Moreover, we consider only the case of loading; for unloading or cyclic loading we have to repeat the procedure for the loading (cf., e.g. Sect. 3.4). Here we develop the direct boundary integral equation method (B.I.E.M)

144   5. Multivalued Boundary Integral Equations

for the present problem. For the indirect B.I.E.M we refer the reader to the previous section.

It should also be mentioned that the indirect method is not adequate for the numerical treatment of the present problem because it leads to multivalued singular integral equations of nonclassical type.

i) Formulation with respect to the crack stresses

Let us consider a two-dimensional linear elastic body Ω. We assume that Ω is a subset of $\mathbb{R}^2$ and that it has a regular boundary Γ. We refer Ω to a Cartesian orthogonal coordinate system Ox_1x_2. The boundary Γ is decomposed into two mutually disjoint parts Γ_U and Γ_F. It is assumed that on Γ_U (resp. Γ_F) the displacements (resp. the tractions) are given. Let $n = \{n_i\}$ be the outward unit normal vector to Γ and $S = \{S_i\} = \{\sigma_{ij}n_j\}$ the traction vector on the boundary where $\sigma = \{\sigma_{ij}\}$ is the stress tensor. By $u = \{u_i\}$, we denote the displacement vector, by $\varepsilon = \{\varepsilon_{ij}\}$ the strain tensor (small strain assumption) and by $C = \{C_{ijhk}\}$ $i,j,h,k = 1,2$, Hooke's tensor of elasticity obeying the well-known symmetry and ellipticity conditions.

On Γ_U we have, for the sake of simplicity that

$$u_i = 0, \tag{5.2.1}$$

otherwise we have to make the problem homogeneous through a translation, and on Γ_F we have

$$S_i = F_i, \quad F_i = F_i(x). \tag{5.2.2}$$

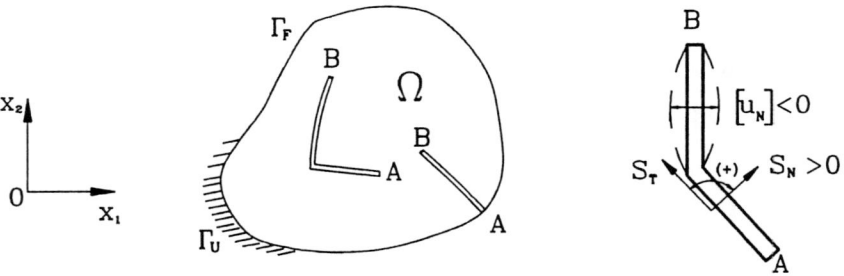

Fig. 5.2.1. Geometry of the cracked body.

Further, we consider some cracks in the body and we denote them by AB. They have been repaired by means of an adhesive material. Let us assume for the present that its behaviour normally to the crack is independent of its behaviour in the tangential direction. In order to describe this behaviour we introduce the components S_N and S_T of the tractions $S = \{S_i\}$, $i = 1,2$, in the interior side of the crack AB, where S_N is normal to the crack and S_T is tangential to it (Fig. 5.2.1). Similarly, we define the normal and tangential components of the relative displacement $[u] = \{[u_i]\}$ of the two crack sides. These components are denoted by $[u_N]$ and $[u_T]$.

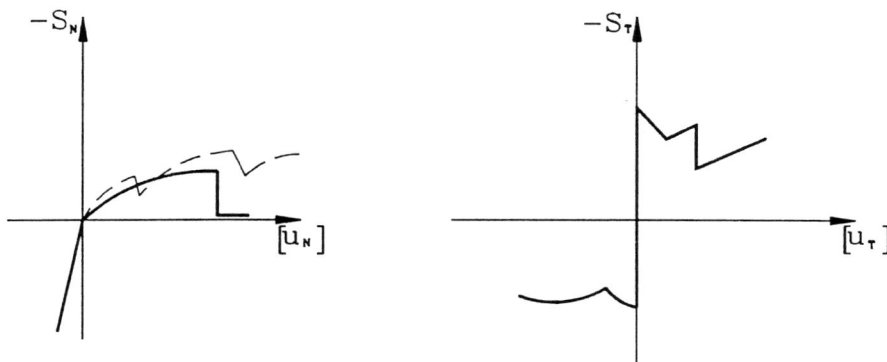

Fig. 5.2.2. Possible laws of the adhesive bonding material.

In Fig. 5.2.2 we give possible diagrams depicting the behaviour of the adhesive in the normal and in the tangential direction. All these non-monotone, multivalued laws can be put in the general form

$$[u_N] \in \bar{\partial} j_N(-S_N) \quad -S_N \in \bar{\partial} \tilde{j}_N([u_N]) \text{ on } \cup (\text{AB}), \qquad (5.2.3a,b)$$

$$[u_T] \in \bar{\partial} j_T(-S_T) \quad -S_T \in \bar{\partial} \tilde{j}_T([u_T]) \text{ on } \cup (\text{AB}), \qquad (5.2.3c,d)$$

where $\cup$ (AB) denotes the union of all cracks denoted further by γ and $j_N, j_T, \tilde{j}_N, \tilde{j}_T$ are non-convex superpotentials.

The equations of the boundary value problem (B.V.P) read, $i,j = 1,2$,

$$\sigma_{ij,j} + f_i = 0 \text{ in } \Omega, \qquad (5.2.4)$$

$$\varepsilon_{ij} = \varepsilon_{ij}(u) = \frac{1}{2}(u_{i,j} + u_{j,i}) \text{ in } \Omega, \qquad (5.2.5)$$

$$\sigma_{ij} = C_{ijhk}\varepsilon_{hk} \text{ in } \Omega. \qquad (5.2.6)$$

Here the comma denotes the partial derivative and $f = \{f_i\}$ is the volume force vector.

Now, let V be the linear space of the displacements v_i and let V_0 be the set of the kinematically admissible displacements, i.e.

$$V_0 = \{v | v = \{v_i\}, v_i \in V, \ v_i = 0 \text{ on } \Gamma_U \text{ for } i = 1,2\}. \qquad (5.2.7)$$

We denote the bilinear form of elasticity by $a(\cdot,\cdot)$ and we apply the notation (5.1.10).

Let L be the admissible space for the traction S on each crack AB. First we assume that S is given on γ and is equal to $\mu = \{\mu_i\}$. Then the solution of the arising classical problem satisfies the following problem:
Find $u = u(\mu) \in V_0$ such that

$$a(u,v) - \int_\gamma \mu_i[v_i]d\Gamma - (f,v) - \int_{\Gamma_F} F_i v_i d\Gamma = 0 \quad \forall v \in V_0. \tag{5.2.8}$$

Obviously (5.2.8) expresses the principle of virtual work for a structure resulting from the initial one by eliminating the superpotential constraints on γ and by applying the coresponding forces $\mu = \{\mu_i\}$. The linearity of (5.2.8) implies that the solution u of it can be written as the sum $\tilde{u}_{(1)} \in V_0$ and $\tilde{u}_{(2)} \in V_0$ where $\tilde{u}_{(1)}$ and $\tilde{u}_{(2)}$ are solutions of the two variational equalities

$$a(\tilde{u}_{(1)}, v) - l(v) = 0 \quad \forall v \in V_0 \tag{5.2.9}$$

$$a(\tilde{u}_{(2)}, v) - \int_\gamma \mu_i[v_i]d\Gamma = 0 \quad \forall v \in V_0, \tag{5.2.10}$$

respectively. Here $\tilde{u}_{(1)}$ and $\tilde{u}_{(2)}$ are equilibrium configurations of two classical (bilateral) structures resulting from the initial one by ignoring the superpotential conditions on γ and taking that on certain part of the boundary the load is zero. Thus, in (5.2.9) the structure is loaded by the forces f in Ω and F on Γ_F, whereas on γ the loading is zero. Moreover, the structure is fixed on Γ_U. In (5.2.10) the structure is loaded by a force $\mu = \{\mu_i\}$ on γ only and is fixed along Γ_U; the loading in Ω and on Γ_F is zero. The solutions $\tilde{u}_{(1)}$ and $\tilde{u}_{(2)}$ exist and are unique, as it is well-known from the classical theory of elasticity. For the aforementioned bilateral structures the solution $\tilde{u}_{(1)}$ and $\tilde{u}_{(2)}$ can be written in terms of Green's operator G which is the same for both structures, because in each case the same type of boundary conditions holds. Thus we have

$$\tilde{u}_{(1)} = G(\bar{l}), \quad \tilde{u}_{(2)} = G(\mu), \quad u = \tilde{u}_{(1)} + \tilde{u}_{(2)}, \quad \bar{l} = \{f, F\}. \tag{5.2.11}$$

It remains to determine the unknown force distribution $\mu = \{\mu_i\}$ on γ. With respect to the linear elasticity problem corresponding to (5.2.10) we apply Betti's theorem. Assume that $\lambda = [\lambda_i]$ on γ is a force distribution corresponding to a displacement field $v_{(2)} \in V_0$ if $f = 0$ and $F = 0$ on Γ_F. Then we have for all $\lambda = \{\lambda_i\} \in L$ that

$$\int_\gamma \lambda_i[\tilde{u}_{(2)i}]d\Gamma = \int_\gamma \mu_i[v_{(2)i}]d\Gamma. \tag{5.2.12}$$

Obviously we may write that

$$v_{(2)} = G(\lambda). \tag{5.2.13}$$

Now (5.2.12), together with (5.2.11) and (5.2.13), implies that

$$\int_\gamma \lambda_i[\tilde{u}_i]d\Gamma = \int_\gamma \lambda_i[\tilde{u}_{(1)i}]d\Gamma + \int_\gamma \lambda_i[\tilde{u}_{(2)i}]d\Gamma \qquad (5.2.14)$$

$$= \int_\gamma \lambda_i[\tilde{u}_{(1)i}]d\Gamma + \int_\gamma \mu_i[v_{(2)i}]d\Gamma$$

$$= \int_\gamma \lambda_i[G(\bar{l})_i]d\Gamma + \int_\gamma \mu_i[G(\lambda)_i]d\Gamma.$$

Let us introduce now the bilinear form

$$\beta(\lambda,\mu) = \int_\gamma \mu_i[G(\lambda)_i]d\Gamma \qquad (5.2.15)$$

which is symmetric by Betti's theorem and the linear form

$$b(\lambda) = -\int_\gamma \lambda_i[G(\bar{l})_i]d\Gamma. \qquad (5.2.16)$$

Assuming now that the tractions $\mu = [\mu_N, \mu_T]$ on Γ_S are related to the displacement field u through the relations (5.2.3a), (5.2.3c) we may write, using the definition of the generalized gradient that

$$j_N^0(-\mu_N, -\lambda_N^\star) + j_T^0(-\mu_T, -\lambda_T^\star) \geq -[u_N]\lambda_N^\star - [u_T]\lambda_T^\star \quad \forall \lambda^\star \in L. \qquad (5.2.17)$$

From (5.2.17) and (5.2.14) we get for $\lambda^\star = \lambda$ that

$$\int_\gamma [j_N^0(-\mu_N, -\lambda_N) + j_T^0(-\mu_T, -\lambda_T)]d\Gamma \geq b(\lambda) - \beta(\lambda,\mu). \qquad (5.2.18)$$

Since this relation holds for all $\lambda \in L$ we are led to the following hemivariational inequality:
Find $\mu \in L$ such as to satisfy

$$\beta(\mu,\lambda) - b(\lambda) + \int_\gamma [j_N^0(-\mu_N, -\lambda_N) + j_T^0(-\mu_T, -\lambda_T)]d\Gamma \geq 0 \quad \forall \lambda \in L. \qquad (5.2.19)$$

The corresponding substationarity problem reads

$$\mu \in L, \quad 0 \in \bar{\partial}\Pi(\mu), \quad \Pi(\mu) = \frac{1}{2}\beta(\mu,\mu) - b(\mu) \qquad (5.2.20)$$
$$+ \int_\gamma [j_N(-\mu_N) + j_T(-\mu_T)]d\Gamma$$

and the corresponding multivalued boundary integral equation is

$$b - \frac{1}{2}\text{grad}\,\beta(\mu,\mu) \in \bar{\partial}\left\{\int_\gamma [j_N(-\mu_N) + j_T(-\mu_T)]d\Gamma\right\} \qquad (5.2.21)$$

or explicitly written

$$\frac{\partial}{\partial \mu}\left[-\int_\gamma \mu_i([G(\bar{l})_i])d\Gamma - \frac{1}{2}\int_\gamma \mu_i([G(\mu)_i])d\Gamma\right] \tag{5.2.22}$$

$$\in \bar{\partial}\left\{\int_\gamma [j_N(-\mu_N) + j_T(-\mu_T)]d\Gamma\right\}.$$

ii) Formulation with respect to the relative displacements of the cracks

Let us assume first that $u_i = U_i$ on Γ_U; subsequently we are free to take $U_i = 0$. We suppose now that the relative displacements $[u]$ on γ are prescribed. Let Σ be the set of all symmetric stress-tensors and let

$$\begin{aligned}\Sigma_1 &= \{\tau | \tau = \{\tau_{ij}\}, \tau_{ij} = \tau_{ji}, \tau_{ij,j} + f_i = 0 \text{ in } \Omega, \\ &\quad T_i = F_i \text{ on } \Gamma_F, \quad i,j = 1,2\} \end{aligned} \tag{5.2.23}$$

be the statically admissible set. Here $\{T_i\}$ denotes the traction on Γ corresponding to the stress $\tau = \{\tau_{ij}\}$. Also, let $c = \{c_{ijhk}\}$ be the inverse tensor to $C = \{C_{ijhk}\}$. For given relative displacements $[v]$ on γ the principle of complementary virtual work for the structure has the form:
Find $\sigma \in \Sigma_1$ such that

$$A(\sigma, \tau) = \int_{\Gamma_U} U_i T_i d\Gamma + \int_\gamma [v_i] T_i d\Gamma \quad \forall \tau \in \Sigma_1. \tag{5.2.24}$$

Let us now introduce a stress-field $\sigma_0 \in \Sigma_1$, i.e. a stress field satisfying the equations of equilibrium and the static boundary conditions on Γ_F and let us introduce the new variables

$$\bar{\sigma} = \sigma - \sigma_0, \quad \bar{\tau} = \tau - \sigma_0, \tag{5.2.25}$$

where $\sigma, \tau \in \Sigma_0$ and

$$\Sigma_0 = \{\tau | \tau = \{\tau_{ij}\}, \tau_{ij} = \tau_{ji}, \tau_{ij,j} = 0 \text{ in } \Omega, T_i = 0 \text{ on } \Gamma_F\}. \tag{5.2.26}$$

Thus (5.2.24) becomes:
Find $\sigma \in \Sigma_0$ such as to satisfy

$$A(\bar{\sigma}, \bar{\tau}) - \int_{\Gamma_U} U_i \bar{T}_i d\Gamma - \int_\gamma [v_i]\bar{T}_i d\Gamma - A(\sigma_0, \tau) = 0 \quad \forall \bar{\tau} \in \Sigma_0. \tag{5.2.27}$$

Note also that by the Green-Gauss theorem

$$\begin{aligned}A(\bar{\sigma}_0, \bar{\tau}) &= \int_\Omega \varepsilon_{0ij}\bar{\tau}_{ij}d\Omega = -\int_\Omega \tilde{u}_{0i}\bar{\tau}_{ij,j}d\Omega + \int_\gamma [\tilde{u}_{0i}]\bar{T}_i d\Gamma + \int_{\Gamma_F} \tilde{u}_{0i}\bar{T}_i d\Gamma \\ &\quad + \int_{\Gamma_U} \tilde{u}_{0i}\bar{T}_i d\Gamma = \int_\gamma [\tilde{u}_{0i}]T_i d\Gamma + \int_{\Gamma_U} \tilde{u}_{0i}T_i d\Gamma \quad \forall \tau \in \Sigma_0. \end{aligned} \tag{5.2.28}$$

Here $\varepsilon_{0ij} = c_{ijhk}\sigma_{0hk}$ and $\tilde{u}_0$ is a displacement field corresponding to σ_0 which must be such that $\sigma_0 \in \Sigma_1$; we are free to consider any type of kinematic or static boundary conditions on Γ_U and on γ. Thus e.g., σ_0 can be the unique solution of a bilateral problem having on Γ_U and on γ zero displacements and subjected to forces f_i in Ω and F_i on Γ_F. Then (5.2.28) takes the form

$$A(\bar{\sigma}_0, \bar{\tau}) = 0 \quad \forall \bar{\tau} \in \Sigma_0. \tag{5.2.29}$$

The stress σ in (5.2.27) is now written as the sum $\bar{\sigma}_{(1)} + \bar{\sigma}_{(2)}$ where $\bar{\sigma}_{(1)}$ and $\bar{\sigma}_{(2)}$ are solutions of the variational equalities

$$A(\bar{\sigma}_{(1)}, \bar{\tau}) - \int_\gamma [v_i]\bar{T}_i d\Gamma = 0 \quad \forall \bar{\tau} \in \Sigma_0, \tag{5.2.30}$$

$$A(\bar{\sigma}_{(2)}, \bar{\tau}) - \int_{\Gamma_U} U_i \bar{T}_i d\Gamma = 0 \quad \forall \bar{\tau} \in \Sigma_0, \tag{5.2.31}$$

respectively. Both (5.2.30) and (5.2.31) express the principle of complementary virtual work for bilateral structures resulting from the initial one in the following way. For the first (resp. the second) we consider the structure Ω under the action of given relative displacements $[v]$ (resp. zero) on γ, zero forces in Ω and Γ_F and zero (resp. U) displacements on Γ_U. Since these structures are linear elastic, $\bar{\sigma}_{(1)}$ and $\bar{\sigma}_{(2)}$ are uniquely determined. Therefore, from (5.2.30) and (5.2.31) we obtain that

$$\bar{\sigma}_{(1)} = H([v]), \quad \bar{\sigma}_{(2)} = H([U]), \quad \bar{\sigma} = \bar{\sigma}_{(1)} + \bar{\sigma}_{(2)}, \quad \sigma \in \Sigma_0 \tag{5.2.32}$$

where H is the stress-displacement operator. The bilateral structures corresponding to (5.2.30) and (5.2.31) have the same H-operator because of the same type of boundary conditions.

Moreover, let $\tilde{H}$ be the operator $J \circ H$ where J transforms σ into the boundary traction $S = \{S_i\} = \{\sigma_{ij} n_j\}$. Thus we may write

$$\bar{S}_{(1)} = \tilde{H}([v]), \quad \bar{S}_{(2)} = \tilde{H}([U]). \tag{5.2.33}$$

We have to determine the unknown relative displacement distribution $[v] = \{[v_i]\} \in N$ on γ, where N denotes their space. Let $[w] = \{[w_i]\} \in N$ be another relative displacement distribution on γ corresponding to the stress field $\bar{\tau}_{(1)} \in \Sigma_0$ through (5.2.30). Applying Betti's theorem we can write for every $[v] \in N$ that

$$\int_\gamma \bar{T}_{(1)i}[v_i] d\Gamma = \int_\gamma \bar{S}_{(1)i}[w_i] d\Gamma, \tag{5.2.34}$$

where $\bar{T}_{(1)}$ corresponds to $\bar{\tau}_{(1)}$. Analogously to (5.2.33) we may write that

$$\bar{T}_{(1)} = \tilde{H}([w]). \tag{5.2.35}$$

Relation (5.2.34) implies for $\bar{S} = \bar{S}_{(1)} + \bar{S}_{(2)}$ with (5.2.35)

5. Multivalued Boundary Integral Equations

$$\int_\gamma [w_i]\bar{S}_i d\Gamma = \int_\gamma [w_i]\bar{S}_{(1)i} d\Gamma + \int_\gamma [w_i]\bar{S}_{(2)i} d\Gamma \qquad (5.2.36)$$

$$= \int_\gamma [v_i]\bar{T}_{(1)i} d\Gamma + \int_\gamma [w_i]\bar{S}_{(2)i} d\Gamma$$

$$= \int_\gamma [v_i](\tilde{H}([w]))_i d\Gamma + \int_\gamma [w_i]\bar{S}_{(2)i} d\Gamma$$

$$= \int_\gamma [v_i](\tilde{H}([w]))_i d\Gamma + \int_\gamma [w_i](\tilde{H}([U]))_i d\Gamma.$$

The bilinear symmetric (due to Betti's theorem) form

$$\delta([v],[w]) = \int_\gamma (\tilde{H}([v]))_i [w_i] d\Gamma \qquad (5.2.37)$$

and the linear form

$$\bar{d}([w]) = -\int_\gamma (\tilde{H}([U]))_i [w_i] d\Gamma \qquad (5.2.38)$$

are introduced, and thus (5.2.36) implies that

$$\int_\gamma ([w_i])\bar{S}_i d\Gamma = \delta([v],[w]) - \bar{d}([w]). \qquad (5.2.39)$$

But (5.2.3b), (5.2.3d), imply by definition that

$$\tilde{j}_N^0([v]_N,[w]_N^\star) + \tilde{j}_T^0([v]_T,[w]_T^\star) \geq -S_N[w_N^\star] - S_T[w_T^\star] \qquad (5.2.40)$$
$$= -S_i[w_i^\star] = -(\bar{S}_i + S_{0i})[w_i^\star] \quad \forall [w]^\star \in N.$$

Setting w instead of $w^\star$ implies that

$$\int_\gamma [\tilde{j}_N^0([v]_N,[w]_N) + \tilde{j}_T^0([v]_T,[w]_T)] d\Gamma \qquad (5.2.41)$$
$$\geq -\delta([v],[w]) + d([w]) \forall [w] \in N$$

where

$$d([w]) = \bar{d}([w]) - \int_\gamma S_{0i}[w_i] d\Gamma. \qquad (5.2.42)$$

Thus we are led to the following hemivariational inequality:
Find $[v] \in N$ such as to satisfy

$$\delta([v],[w]) - d([w]) + \int_\gamma \{\tilde{j}_N^0([v]_N,[w]_N) \qquad (5.2.43)$$
$$+ \tilde{j}_T^0([v]_T,[w]_T)\} d\Gamma \geq 0 \forall [w] \in N$$

The corresponding substationarity problem reads

5.2 Complement for Adhesively Bonded Cracks

$$[v] \in N, \quad 0 \in \bar{\partial}\tilde{\Pi}([v]), \quad \tilde{\Pi}([v]) = \frac{1}{2}\delta([v],[v]) - d([v]) \quad (5.2.44)$$
$$+ \int_\gamma [\tilde{j}_N([v_N]) + \tilde{j}_T([v_T])] d\Gamma,$$

whereas the corresponding multivalued boundary integral equation (on the crack interfaces γ) becomes

$$d - \frac{1}{2}\text{grad}\,\delta([v],[v]) \in \bar{\partial}\int_\gamma [\tilde{j}_N([v_N]) + \tilde{j}_T([v_T])] d\Gamma. \quad (5.2.45)$$

It is explicitly written in the form

$$\frac{\partial}{\partial[v]}\left[-\int_\gamma [v_i](\tilde{H}(U))_{,i} d\Gamma - \int_\gamma S_{0i}[v_i] d\Gamma - \frac{1}{2}\int_\gamma (\tilde{H}([v]))_{,i}[v_i] d\Gamma\right] \quad (5.2.46)$$
$$\in \bar{\partial}\int_\gamma [\tilde{j}_N([v_N]) + \tilde{j}_T([v_T])] d\Gamma.$$

Analogous multivalued B.I.Es can be formulated, as it is obvious, if instead of the nonconvex superpotential laws (2.4.2) or (5.1.3), reaction-displacement boundary laws resulting from F-superpotentials or V-superpotentials hold. In this case in (5.2.22) and in (5.2.46), for instance, only the right hand side is replaced by the corresponding multivalued expression.

Concerning the mathematical study of the formulated B.I.Es, both for the coersive and the semicoersive case, we refer to [Ant92] [Pan92a], where certain existence and approximation results are proved. We refer also the reader to the next chapter in this context and to Props. 4.3.1, 4.3.2 and 6.3.1 which concern the rigorous formulation of the substationarity problem. One can find there the conditions under which the integrals on Γ_S, or on γ in the expressions of the potential and the complementarity energy make sense.

Part III
MATHEMATICAL THEORY

6. Static Hemivariational Inequalities

In the present chapter we study static hemivariational inequalities concerning the existence of their solutions. Some approximation results are also given. We distinguish the coercive and the more difficult semicoercive case where the rigid body displacements play an important role. After the study of hemivariational inequalities we deal with variational hemivariational inequalities and we derive some existence and approximation results. Finally the mathematical results obtained are applied to concrete engineering problems. The present chapter is mainly based on [Pan88a,89a,90,91,92c]. For other types of existence results we refer to Naniewicz [Nan88,89a,89b].

6.1 Coercive Hemivariational Inequalities

The theory of variational inequalities is a well-developed theory in mathematics which, as it is well known, is closely connected with the convexity of the energy functionals involved. Indeed the existence theory of variational inequalities is based on monotonicity arguments.

As pointed out in Chapt. 2, if the corresponding energy functionals involved are nonconvex, another type of inequality expressions arises as variational formulation of the problem which are called hemivariational inequalities. Their derivation is based on the mathematical notion of the generalized gradient and in contrast to the variational inequalities, the hemivariational inequalities are not equivalent to minimum problems, but they give rise to substationarity problems. Several types of hemivariational inequalities have already been studied (see, for example, [Pana83,85a,b,88a,b,c, Pan85,88a,89a,90,92c], and the references given there) with respect to certain mechanical problems, e.g. in nonmonotone semipermeability problems, in the theory of simple and multilayered plates, in the theory of composite structures, etc. It should be noted that the hemivariational inequalities have been proved very efficient for the treatment of certain as yet unsolved or partially solved problems, e.g. the delamination problem of multilayered plates, the partial debonding of adhesive joints, etc.

The aim of this section is the formulation of a mathematical theory for a simple type of coercive hemivariational inequality, i.e. for the problem: find $u \in V$ such as to satisfy the inequality

6. Static Hemivariational Inequalities

$$a(u, v - u) + \int_\Omega j^0(u, v - u) d\Omega \geq (l, v - u) \qquad \forall v \in V. \tag{6.1.1}$$

Let V be a real Hilbert space with the property that

$$V \subset L^2(\Omega) \subset V', \tag{6.1.2}$$

where V' denotes the dual space of V, Ω is an open bounded subset of $\mathbb{R}^n$, and the injections are continuous and dense. We denote by $(\cdot, \cdot)$ the $L^2(\Omega)$ product and the duality pairing, by $\|\cdot\|$ the norm of V and by $|\cdot|_2$ the $L^2(\Omega)$-norm. We recall [Aub79a] that the form $(\cdot, \cdot)$ extends uniquely from $V \times L^2(\Omega)$ to $V \times V'$. Further, assume that

$$V \subset L^2(\Omega) \quad \text{is compact} \tag{6.1.3}$$

and that

$$V \cap L^\infty(\Omega) \quad \text{is dense in } V \text{ for the } V\text{-norm}, \tag{6.1.4}$$

and has a Galerkin base. It is also assumed that $a(\cdot, \cdot) \colon V \times V \to \mathbb{R}$ is a bilinear symmetric continuous form which is coercive, i.e. there exists $c > 0$ constant such that

$$a(v, v) \geq c \|v\|^2 \qquad \forall v \in V. \tag{6.1.5}$$

We denote by $j \colon \mathbb{R} \to \mathbb{R}$, a locally Lipschitz function defined as in (1.2.55): let $\beta \in L^\infty_{\text{loc}}(\mathbb{R})$ and consider the functions $\bar{\beta}_\mu$ and $\bar{\bar{\beta}}_\mu$ defined by (1.2.51) (where ρ is replaced by μ) i.e.

$$\bar{\beta}_\mu(\xi) = \operatorname*{ess\,sup}_{|\xi_1 - \xi| \leq \mu} \beta(\xi_1) \quad \text{and} \quad \bar{\bar{\beta}}_\mu(\xi) = \operatorname*{ess\,inf}_{|\xi_1 - \xi| \leq \mu} \beta(\xi_1). \tag{6.1.6}$$

They are increasing and decreasing functions of μ, respectively; therefore the limits for $\mu \to 0_+$ exist. We denote them by $\bar{\beta}(\xi)$ and $\bar{\bar{\beta}}(\xi)$ respectively; the multivalued function is defined by

$$\hat{\beta}(\xi) = [\bar{\beta}(\xi), \bar{\bar{\beta}}(\xi)]. \tag{6.1.7}$$

If $\beta(\xi_{\pm 0})$ exists for every $\xi \in \mathbb{R}$ then a locally Lipschitz function $j \colon \mathbb{R} \to \mathbb{R}$ can be determined (up to an additive constant) such that [Ch]

$$\hat{\beta}(\xi) = \bar{\partial} j(\xi). \tag{6.1.8}$$

Here j is obtained from β by simple integration, (cf. (1.2.54)).

On the assumption that $l \in V'$ we formulate the following coercive hemivariational inequality (problem P^C): Find $u \in V$ such that

$$a(u, v - u) + \int_\Omega j^0(u, v - u) d\Omega \geq (l, v - u) \qquad \forall v \in V. \tag{6.1.9}$$

In order to define the regularized problem P^C_ε we consider the mollifier

6.1 Coercive Hemivariational Inequalities

$$p \in C_0^\infty(-1,+1),\ p \geq 0, \quad \text{with} \int_{-\infty}^{+\infty} p(\xi)d\xi = 1 \tag{6.1.10}$$

and let

$$\beta_\varepsilon = p_\varepsilon \star \beta \text{ with } p_\varepsilon(\xi) = \frac{1}{\varepsilon}p\left(\frac{\xi}{\varepsilon}\right) \quad 0 < \varepsilon < 1 \tag{6.1.11}$$

where $(\star)$ denotes the convolution product. The regularized problem P_ε^C reads: Find $u_\varepsilon \in V$ such as to satisfy the variational equality

$$a(u_\varepsilon, v) + \int_\Omega \beta_\varepsilon(u_\varepsilon)v d\Omega = (l, v) \quad \forall v \in V. \tag{6.1.12}$$

In order to define the corresponding finite dimensional problem $P_{\varepsilon n}^C$ we consider a Galerkin basis of $V \cap L^\infty(\Omega)$ and let V_n be the resulting n-dimensional subspace. This problem reads:
Find $u_{\varepsilon n} \in V_n$ such as to satisfy the variational equality

$$a(u_{\varepsilon n}, v) + \int_\Omega \beta_\varepsilon(u_{\varepsilon n})v d\Omega = (l, v) \quad \forall v \in V_n. \tag{6.1.13}$$

Now it is assumed that the graph $\{\xi, \beta(\xi)\}$ ultimately increases, i.e. that there exists $\xi \in \mathbb{R}$ such that

$$\underset{(-\infty,-\xi)}{\text{ess sup}}\ \beta(\xi_1) \leq 0 \leq \underset{(\xi,\infty)}{\text{ess inf}}\ \beta(\xi_1). \tag{6.1.14}$$

The existence proof is based on the following propositions.

Proposition 6.1.1 Suppose that (6.1.14) holds. Then we can determine $\rho_1 > 0$, $\rho_2 > 0$ such that for every $u_{\varepsilon n}$

$$\int_\Omega \beta_\varepsilon(u_{\varepsilon n})u_{\varepsilon n} d\Omega \geq -\rho_1 \rho_2 \text{mes } \Omega. \tag{6.1.15}$$

Proof. From (6.1.14) we obtain that

$$\beta_\varepsilon(\xi) = (p_\varepsilon \star \beta)(\xi) = \int_{-\varepsilon}^{+\varepsilon} \beta(\xi - t)p_\varepsilon(t)dt \leq \underset{|t|\leq\varepsilon}{\text{ess sup}}\ \beta(\xi - t). \tag{6.1.16}$$

and analogously

$$\underset{|t|\leq\varepsilon}{\text{ess inf}}\ \beta(\xi - t) \leq \beta_\varepsilon(\xi). \tag{6.1.17}$$

In the above two inequalities we set $x = \xi - t$, $|x - \xi| \leq \varepsilon$ and enlarge the bounds for $-\infty < x \leq \varepsilon + \xi$ and $\xi - \varepsilon \leq x < \infty$, respectively. Then the supremum and the infimum for $\xi \in (-\infty, -\xi_1)$ and $\xi \in (\xi_1, \infty)$, respectively are formed and the bounds are enlarged by replacing $\varepsilon + \xi$ by $1 - \xi_1$ and $\xi - \varepsilon$ by $\xi_1 - 1$ ($\varepsilon < 1$); we obtain from (6.1.14) that there exists $\xi \in \mathbb{R}$ such that

$$\sup_{(-\infty,-\xi)} \beta_e(\xi_1) \leq 0 \leq \inf_{(\xi,\infty)} \beta_e(\xi_1). \qquad (6.1.18)$$

Thus we can determine $\rho_1 > 0$ and $\rho_2 > 0$ such that $\beta_e(\xi) \geq 0$ if $\xi > \rho_1$, $\beta_e(\xi) \leq 0$ if $\xi < -\rho_1$, and $|\beta_e(\xi)| \leq \rho_2$ if $|\xi| \leq \rho_1$ and may write

$$\int_\Omega \beta_e(u_{en}) u_{en} d\Omega = \int_{|u_{en}(x)|>\rho_1} \ldots d\Omega \qquad (6.1.19)$$

$$+ \int_{|u_{en}(x)|\leq\rho_1} \ldots d\Omega \geq 0 - \rho_1 \rho_2 \mathrm{mes}\Omega. \quad \text{q.e.d}$$

Proposition 6.1.2 The problem P_{en}^C has at least one solution $u_{en} \in V_n$.

Proof. Equation (6.1.13) is written in the form

$$(\Lambda(u_{en}), v) = 0, \qquad \forall v \in V_n \qquad (6.1.20)$$

and because of (6.1.5) and (6.1.15) we have the estimate

$$(\Lambda(u_{en}), u_{en}) \geq c|u_{en}|^2 - \rho_1\rho_2 \mathrm{mes}\Omega - c_1|u_{en}| \qquad c, c_1 > 0. \qquad (6.1.21)$$

By applying Brouwer's fixed point theorem (cf. [Lio69] p.53) we obtain that (6.1.20) admits a bounded solution u_{en}. q.e.d.

Proposition 6.1.3 The sequence $\{\beta_e(u_{en})\}$ is weakly precompact in $L^1(\Omega)$.

Proof. The Dunford-Pettis theorem (cf. [Eke], p.239) implies that it suffices to show that for each $\mu > 0$ a $\delta(\mu) > 0$ can be determined such that for $\omega \subset \Omega$ with $\mathrm{mes}\,\omega < \delta$

$$\int_\omega |\beta_e(u_{en})| d\Omega < \mu. \qquad (6.1.22)$$

The inequality

$$\xi_0 |\beta_e(\xi)| \leq |\beta_e(\xi)\xi| + \xi_0 \sup_{|\xi|\leq\xi_0} |\beta_e(\xi)| \qquad (6.1.23)$$

implies that

$$\int_\omega |\beta_e(u_{en})| d\Omega \leq \frac{1}{\xi_0} \int_\Omega |\beta_e(u_{en})u_{en}| d\Omega + \int_\omega \sup_{|u_{en}(x)|\leq\xi_0} |\beta_e(u_{en})| d\Omega. \qquad (6.1.24)$$

But

$$\int_\Omega |\beta_e(u_{en})u_{en}| d\Omega = \int_{|u_{en}(x)|>\rho_1} |\beta_e(u_{en})u_{en}| d\Omega + \int_{|u_{en}(x)|\leq\rho_1} |\beta_e(u_{en})u_{en}| d\Omega \quad (6.1.25)$$

$$= \int_{|u_{en}(x)|>\rho_1} |\beta_e(u_{en})u_{en}| d\Omega - \int_{|u_{en}(x)|\leq\rho_1} |\beta_e(u_{en})u_{en}| d\Omega$$

$$+2\int_{|u_{\varepsilon n}(x)|\leq\rho_1}|\beta_\varepsilon(u_{\varepsilon n})u_{\varepsilon n}|d\Omega$$

$$\leq \int_{|u_{\varepsilon n}(x)|>\rho_1}|\beta_\varepsilon(u_{\varepsilon n})u_{\varepsilon n}|d\Omega + \int_{|u_{\varepsilon n}(x)|\leq\rho_1}\beta_\varepsilon(u_{\varepsilon n})u_{\varepsilon n}d\Omega$$

$$+2\int_{|u_{\varepsilon n}(x)|\leq\rho_1}|\beta_\varepsilon(u_{\varepsilon n})u_{\varepsilon n}|d\Omega$$

$$= \int_\Omega \beta_\varepsilon(u_{\varepsilon n})u_{\varepsilon n}d\Omega + 2\int_{|u_{\varepsilon n}(x)|\leq\rho_1}|\beta_\varepsilon(u_{\varepsilon n})u_{\varepsilon n}|d\Omega$$

$$= (l,u_{\varepsilon n}) - a(u_{\varepsilon n},u_{\varepsilon n}) + 2\int_{|u_{\varepsilon n}(x)|\leq\rho_1}|\beta_\varepsilon(u_{\varepsilon n})u_{\varepsilon n}|d\Omega$$

$$\leq c + 2\rho_1\rho_2\text{mes}\,\Omega, \quad c \text{ constant.}$$

In the last two inequalities we have used the boundedness of $|u_{\varepsilon n}|$ and assumptions (6.1.5) and (6.1.14). Further, the relation

$$\sup_{|\xi|\leq\xi_0}|\beta_\varepsilon(\xi)| \leq \operatorname*{ess\,sup}_{|\xi|\leq\xi_0+1}|\beta(\xi)|, \tag{6.1.26}$$

is applied which can be easily verified using (6.1.11). Now choose ξ_0 such that for all ε and n

$$\frac{1}{\xi_0}\int_\omega|\beta_\varepsilon(u_{\varepsilon n})u_{\varepsilon n}|d\Omega \leq \frac{1}{\xi_0}(c+2\rho_1\rho_2\text{mes}\,\Omega) \leq \frac{\mu}{2} \tag{6.1.27}$$

and δ such that for mes $\omega < \delta$

$$\operatorname*{ess\,sup}_{|\xi|\leq\xi_0+1}|\beta(\xi)| \leq \frac{\mu}{2\delta}. \tag{6.1.28}$$

Relation (6.1.26) implies with (6.1.27) that

$$\int_\omega \sup_{|u_{\varepsilon n}(x)|\leq\xi_0}|\beta_\varepsilon(u_{\varepsilon n})|d\Omega \leq \operatorname*{ess\,sup}_{|u_{\varepsilon n}(x)|\leq\xi_0+1}|\beta(u_{\varepsilon n})|\text{mes}\,\omega \leq \frac{\mu}{2\delta}\cdot\delta = \frac{\mu}{2}. \tag{6.1.29}$$

From the relations (6.1.24), (6.1.27) and (6.1.29), the relation (6.1.22) results, i.e. that $\{\beta_\varepsilon(u_{\varepsilon n})\}$ is weakly precompact in $L^1(\Omega)$. q.e.d.

Now the proof of the following theorem can be given.

Theorem 6.1.1 Problem P^C has at least one solution.

Proof. From Prop. 6.1.2 we have that $|u_{\varepsilon n}| < c$ where c is independent of ε and n. Thus as $\varepsilon \to 0$, $n \to \infty$ and by considering subsequences if necessary we may write that

$$u_{\varepsilon n} \to u \quad \text{weakly in } V \tag{6.1.30}$$

and because of (6.1.3)

$$u_{\varepsilon n} \to u \quad \text{strongly in } L^2(\Omega) \tag{6.1.31}$$

and thus
$$u_{\varepsilon n} \to u \quad \text{a.e. on } \Omega. \tag{6.1.32}$$

Moreover due to Prop. 6.1.3 we can write that
$$\beta_\varepsilon(u_{\varepsilon n}) \to \chi \quad \text{weakly in } L^1(\Omega). \tag{6.1.33}$$

Using assumption (6.1.4) and the properties of the Galerkin basis we can pass to the limit $\varepsilon \to 0$ $n \to \infty$ in (6.1.13) and obtain
$$a(u,v) + \int_\Omega \chi v d\Omega = (l,v) \quad \forall v \in V. \tag{6.1.34}$$

In order to complete the proof it will be shown that
$$\chi \in \hat{\beta}(u) = \bar{\partial}j(u) \quad \text{a.e. on } \Omega. \tag{6.1.35}$$

From (6.1.32) by applying Egoroff's theorem we can find that for any $\alpha > 0$ we can determine $\omega \subset \Omega$ with mes $\omega < \alpha$ such that
$$u_{\varepsilon n} \to u \quad \text{uniformly on } \Omega - \omega \tag{6.1.36}$$

with $u \in L^\infty(\Omega - \omega)$. Thus for any $\alpha > 0$ we can find $\omega \subset \Omega$ with mes $\omega < \alpha$ such that for any $\mu > 0$ and for $\varepsilon < \varepsilon_0 < \mu/2$ and $n > n_0 > 2/\mu$ we have
$$|u_{\varepsilon n} - u| < \frac{\mu}{2}, \quad \forall x \in \Omega - \omega. \tag{6.1.37}$$

From (6.1.13) (6.1.14) we obtain that
$$\beta_\varepsilon(u_{\varepsilon n}) \leq \operatorname*{ess\,sup}_{|u_{\varepsilon n}-\xi|\leq\varepsilon} \beta(\xi) \leq \operatorname*{ess\,sup}_{|u_{\varepsilon n}-\xi|<\frac{\mu}{2}} \beta(\xi) \leq \operatorname*{ess\,sup}_{|u-\xi|<\mu} \beta(\xi) = \bar{\bar{\beta}}_\mu(u), \tag{6.1.38}$$

where $\bar{\bar{\beta}}_\mu$ is defined in (6.1.6). Analogously we prove the inequality
$$\bar{\beta}_\mu(u) = \operatorname*{ess\,inf}_{|u-\xi|<\mu} \beta(\xi) \leq \beta_\varepsilon(u_{\varepsilon n}). \tag{6.1.39}$$

We choose $e \geq 0$ a.e. on $\Omega - \omega$ with $e \in L^\infty(\Omega - \omega)$, and we obtain from (6.1.38) and (6.1.39) the inequality
$$\int_{\Omega-\omega} \bar{\beta}_\mu(u) e d\Omega \leq \int_{\Omega-\omega} \beta_\varepsilon(u_{\varepsilon n}) e d\Omega \leq \int_{\Omega-\omega} \bar{\bar{\beta}}_\mu(u) e d\Omega. \tag{6.1.40}$$

Taking the limits as $\varepsilon \to 0$ and $n \to \infty$ we obtain that
$$\int_{\Omega-\omega} \bar{\beta}_\mu(u) e d\Omega \leq \int_{\Omega-\omega} \chi e d\Omega \leq \int_{\Omega-\omega} \bar{\bar{\beta}}_\mu(u) e d\Omega \tag{6.1.41}$$

and as $\mu \to 0$ that

$$\int_{\Omega-\omega} \bar{\beta}(u) e \, d\Omega \leq \int_{\Omega-\omega} \chi e \, d\Omega \leq \int_{\Omega-\omega} \bar{\bar{\beta}}(u) e \, d\Omega . \tag{6.1.42}$$

Since e is arbitrary we have that

$$\chi \in [\bar{\beta}(u), \bar{\bar{\beta}}(u)] = \hat{\beta}(u), \quad \text{a.e. on } \Omega - \omega , \tag{6.1.43}$$

where mes $\omega < \alpha$. Taking α as small as possible, implies (6.1.35).   q.e.d.

The method applied for the proof of this theorem is similar to the method developed in [Rauch] for semilinear differential equations.

Further a proposition concerning the strong convergence of $u_{\varepsilon n}$ to u is proved.

Proposition 6.1.4 Suppose that there is c const > 0 such that

$$|\beta(\xi)| \leq c(1 + |\xi|) \quad \forall \xi \in R . \tag{6.1.44}$$

Then $u_{\varepsilon n} \to u$ strongly in V as $\varepsilon \to 0$, $n \to \infty$.

Proof. Let us put $v = u_{\varepsilon n}$ in (6.1.9). It results that

$$a(u, u) \leq a(u, u_{\varepsilon n}) + \int_{\Omega} j^0(u, u_{\varepsilon n} - u) d\Omega - (l, u_{\varepsilon n} - u). \tag{6.1.45}$$

Similarly we get from (6.1.13) that

$$a(u_{\varepsilon n}, u_{\varepsilon n}) = a(u_{\varepsilon n}, v_n) + \int_{\Omega} \beta_{\varepsilon}(u_{\varepsilon n})(v_n - u_{\varepsilon n}) d\Omega - (l, v_n - u_{\varepsilon n}), \quad \forall v_n \in V_n. \tag{6.1.46}$$

The coercivity of $a(\cdot, \cdot)$, and (6.1.45) and (6.1.46) imply that

$$\begin{aligned} c\|u - u_{\varepsilon n}\|^2 &\leq a(u - u_{\varepsilon n}, u - u_{\varepsilon n}) \\ &\leq a(u_{\varepsilon n}, v_n - u) + \int_{\Omega} j^0(u, u_{\varepsilon n} - u) d\Omega \\ &\quad + \int_{\Omega} \beta_{\varepsilon}(u_{\varepsilon n})(v_n - u_{\varepsilon n}) d\Omega - (l, v_n - u). \end{aligned} \tag{6.1.47}$$

The definition of j^0 and (6.1.44) imply that

$$\begin{aligned} \int_{\Omega} j^0(u, u_{\varepsilon n} - u) d\Omega &= \int_{\Omega} \left(\limsup_{\substack{\lambda \to 0_+ \\ h \to 0}} \frac{1}{\lambda} \int_{u+h}^{u+h+\lambda(u_{\varepsilon n}-u)} \beta(\xi) d\xi \right) d\Omega \tag{6.1.48} \\ &\leq c \int_{\Omega} \left[\limsup_{\substack{\lambda \to 0_+ \\ h \to 0}} \frac{1}{\lambda} \int_{u+h}^{u+h+\lambda(u_{\varepsilon n}-u)} (1 + |\xi|) d\xi \right] d\Omega \\ &= c \int_{\Omega} (1 + |u|)(u_{\varepsilon n} - u) d\Omega . \end{aligned}$$

Due to (6.1.31) the right-hand side of (6.1.48) tends to zero for $\varepsilon \to 0$, $n \to \infty$. Moreover

$$\left| \int_\Omega \beta_\varepsilon(u_{\varepsilon n})(v_n - u_{\varepsilon n})d\Omega \right| \leq \|\beta_\varepsilon(u_{\varepsilon n})\|_{L^2} \|v_n - u_{\varepsilon n}\|_{L^2} \tag{6.1.49}$$

$$\leq \|\beta_\varepsilon(u_{\varepsilon n})\|_{L^2}(\|v_n - u\|_{L^2} + \|u - u_{\varepsilon n}\|_{L^2})$$

and $\|\beta_\varepsilon(u_{\varepsilon n})\|_{L^2} < c$ (independently of ε, n) as it results from (6.1.11) and (6.1.44). Now set into (6.1.47) a v_n such that

$$v_n \to u \quad \text{strongly in } V. \tag{6.1.50}$$

Thus as $\varepsilon \to 0$ $n \to \infty$, the right-hand side of (6.1.49) tends to zero. From the above limits we conclude that the right-hand side of (6.1.47) tends to zero as $\varepsilon \to 0$, $n \to \infty$. This proves the strong convergence of $u_{\varepsilon n}$ to u in V. q.e.d.

Note that a more abstract existence proof for problem P^C can be given by using the fixed point theory of multivalued mappings. Here we prefer the present proof i.e. the use of problems P_ε^C and $P_{\varepsilon n}^C$ because it is directly applicable to most engineering and mechanical problems and because it permits the treatment of the semicoercive case under slightly more general assumptions.

Note that in the problem P^C, i.e. in (6.1.9), we have assumed, as it is obvious from the construction of j and from the assumptions (6.1.2)(6.1.3), that $u(x) \in \mathbb{R}$. However, the foregoing proof holds for the more general case, in which $u(x) \in \mathbb{R}^n$, if the term $\int_\Omega j^0(u, v-u)d\Omega$ is replaced by the term $\int_\Omega j^0(\hat{u}, \hat{v} - \hat{u})d\Omega$, where $\hat{u}(x), \hat{v}(x) \in \mathbb{R}$ and the mapping L, $\hat{u} = Lu, \hat{u} \in L^2(\Omega)$ is linear continuous. The problem P^C becomes now:
Find $u \in V$ such that

$$a(u, v-u) + \int_\Omega j^0(\hat{u}, \hat{v} - \hat{u})d\Omega \geq (l, v-u) \quad \forall v \in V. \tag{6.1.51}$$

Here V is a Hilbert space and $l \in V'$. Relation (6.1.2) is replaced by

$$V \subset [L^2(\Omega)]^n \subset V' \tag{6.1.52}$$

(6.1.3) by

$$L : V \to L^2(\Omega) \quad \text{is compact} \tag{6.1.53}$$

and (6.1.4) by

$$\{v \in V | \hat{v} \in L^\infty(\Omega)\} \quad \text{is dense in } V \text{ for the } V\text{-norm} \tag{6.1.54}$$

and has a Galerkin basis. We assume now that V_n is a finite dimensional subspace of $\{v \in V | \hat{v} \in L^\infty(\Omega)\}$. Then Props. 6.1.1, 6.1.2 and 6.1.3 hold again (with $\beta_\varepsilon(u_{\varepsilon n})$ replaced by $\beta_\varepsilon(\hat{u}_{\varepsilon n})$). Similarly Theorem 6.1.1 holds also; its proof will be slightly modified: (6.1.30) still holds but in (6.1.31) till the end of the proof $u_{\varepsilon n}$ and u are replaced by $\hat{u}_{\varepsilon n}$ and $\hat{u}$ respectively. Prop. 6.1.4 holds with analogous modifications in its proof. Due to the lack of convexity no uniqueness result can be obtained (cf. also Prop. 7.2.2).

6.2 Semicoercive Hemivariational Inequalities

In this section we will study the hemivariational inequality (6.1.51) on the assumption that $a(\cdot,\cdot)$ is no longer coercive but semicoercive, i.e. $a(\cdot,\cdot)$ is continuous and symmetric but it has a nonzero kernel, i.e.

$$\ker a(u,u) = \{q | a(q,q) = 0\} \neq \emptyset. \quad (6.2.1)$$

Moreover let the

$$\ker a \text{ be finite dimensional.} \quad (6.2.2)$$

The norm $||v||$ on V is assumed to be equivalent to $||v|| = p(\tilde{v}) + |q|_2$, where $v = \tilde{v} + q$, $q \in \ker a$, $\tilde{v} \in \ker a^\perp$ (i.e. $(\tilde{v}, q) = 0 \; \forall q \in \ker a$) and $p(\tilde{v})$ is a seminorm on V such that $p(v) = p(v+q) \; \forall v \in V$, $q \in \ker a$ and let

$$a(v,v) \geq c(p(v))^2, \quad \forall v \in V, \; c \text{ const} > 0. \quad (6.2.3)$$

This semicoercivity inequality replaces (6.1.5). Further, keep the assumptions (6.1.52), (6.1.53), (6.1.54), $l \in V'$ and the assumption $\beta \in L^\infty_{\text{loc}}(R)$, which leads to (6.1.8). Now the semicoercive problem P^S, reads:
Find $u \in V$ such that

$$a(u, v-u) + \int_\Omega j^0(\hat{u}, \hat{v} - \hat{u}) d\Omega \geq (l, v-u) \quad \forall v \in V. \quad (6.2.4)$$

We denote by q_+ and q_- the positive and the negative parts of $\hat{q}$, where $\hat{q} = Lq$, i.e. $q_+ = \sup\{0, \hat{q}\}$, $q_- = \sup\{0, -\hat{q}\}$ and the notation below is introduced

$$\beta(-\infty) = \limsup_{\xi \to -\infty} \beta(\xi) \quad \text{and} \quad \beta(\infty) = \liminf_{\xi \to \infty} \beta(\xi). \quad (6.2.5)$$

The following proposition gives a necessary condition for the existence of the solution.

Proposition 6.2.1 Let

$$\beta(-\infty) \leq \beta(\xi) \leq \beta(\infty), \quad \forall \xi \in \mathbb{R}. \quad (6.2.6)$$

Then a necessary condition for the existence of a solution $u \in V$ of problem P^S is the inequality

$$\int_\Omega [\beta(-\infty)q_+ - \beta(\infty)q_-]d\Omega \leq (l,q) \quad (6.2.7)$$

$$\leq \int_\Omega [\beta(\infty)q_+ - \beta(-\infty)q_-]d\Omega \quad \forall q \in \ker a.$$

If (6.2.6) holds strictly (with $<$ instead of $\leq$) then (6.2.7) also holds strictly.

Proof. We set in (6.2.4) $v - u = \pm q \in \ker a$, $q \neq 0$. We obtain

164   6. Static Hemivariational Inequalities

$$\int_\Omega j^0(\hat{u}, \pm\hat{q})d\Omega \geq \pm(l, q) \quad \forall q \in \ker a,\ q \neq 0. \tag{6.2.8}$$

Then (6.2.8) is written as

$$\int_\Omega j^0(\hat{u}, \hat{q})d\Omega \geq (l, q) \geq -\int_\Omega j^0(\hat{u}, -\hat{q})d\Omega, \quad \forall q \in \ker a,\ q \neq 0 \tag{6.2.9}$$

because $\hat{q} \to j^0(\hat{u}, \hat{q})$ is positively homogeneous. From the definition of j and j^0 we obtain by means of (6.2.6) that

$$\int_\Omega j^0(\hat{u}, \hat{q})d\Omega \leq \int_\Omega [\beta(\infty)q_+ - \beta(-\infty)q_-]d\Omega, \quad \forall q \in \ker a,\ q \neq 0 \tag{6.2.10}$$

and analogously $\int_\Omega j^0(\hat{u}, -\hat{q})d\Omega$. Thus (6.2.7) is proved. Analogously results the rest of the proposition.   q.e.d.

Theorem 6.2.1 Suppose that

$$\beta(-\infty) < \beta(\infty). \tag{6.2.11}$$

Then if

$$\int_\Omega [\beta(-\infty)q_+ - \beta(\infty)q_-]d\Omega < (l, q) \tag{6.2.12}$$

$$< \int_\Omega [\beta(\infty)q_+ - \beta(-\infty)q_-]d\Omega \quad \forall q \in \ker a,\ q \neq 0$$

problem P^S has a solution.

Proof. The proof follows the same steps as the proof of theorem 6.1.1. Estimate (6.1.15) is used; it holds due to (6.2.11), but a more sharp estimate for the same quantity is needed. As in the coercive case the regularized problem P^S_ε is defined and the finite dimensional problem $P^S_{\varepsilon n}$ may be put in the form (6.1.20). From (6.1.20), (6.2.3) and (6.1.15) it is found that

$$(\Lambda(u_{\varepsilon n}), u_{\varepsilon n}) \geq c[p(\tilde{u}_{\varepsilon n})]^2 - c|||u_{\varepsilon n}||| - \rho_1\rho_2 \mathrm{mes}\ \Omega, \quad c\ \mathrm{const} > 0. \tag{6.2.13}$$

Now we apply Brouwer's theorem to prove that (6.1.20) has at least one bounded solution $u_{\varepsilon n}$. According to this theorem one has to show that $r > 0$ exists such that

$$||u_{\varepsilon n}|| = r \Longrightarrow (\Lambda(u_{\varepsilon n}), u_{\varepsilon n}) \geq 0. \tag{6.2.14}$$

Here it is proved that a number $M > 0$ can be determined such that

$$||u_{\varepsilon n}|| > M \Longrightarrow (\Lambda(u_{\varepsilon n}), u_{\varepsilon n}) > 0 \tag{6.2.15}$$

and thus one may take $||u_{\varepsilon n}|| = r > M$. Instead of (6.2.15) we shall prove equivalently that

$$(A(u_{en}), u_{en}) \leq 0 \implies ||u_{en}|| \leq c. \tag{6.2.16}$$

From (6.2.13) if $(A(u_{en}), u_{en}) \leq 0$, then a constant $c > 0$ exists such that

$$p(\tilde{u}_{en}) \leq c(\sqrt{(|q_{en}|_2)} + 1) \tag{6.2.17}$$

where $u_{en} = \tilde{u}_{en} + q_{en}$. Thus it is enough to prove that $(A(u_{en}), u_{en}) \leq 0$ implies $|q_{en}|_2 \leq c$, or equivalently, that a number $R > 0$ can be determined such that

$$|q_{en}|_2 > R \text{ and } (6.2.17) \implies (A(u_{en}), u_{en}) > 0. \tag{6.2.18}$$

This last relation will be proved.

The definition of β_ε implies that

$$\beta_\varepsilon(\infty) = \lim_{\xi \to \infty} \beta_\varepsilon(\xi) = \lim_{\xi \to \infty} \int_{-\varepsilon}^{+\varepsilon} \beta(\xi - t) p_\varepsilon(t) dt \geq \lim_{\xi \to \infty} \operatorname*{ess\,inf}_{|x-\xi| \leq \varepsilon} \beta(x) \tag{6.2.19}$$

$$\geq \lim_{\xi \to \infty} \operatorname*{ess\,inf}_{\xi - \varepsilon \leq x < \infty} \beta(x) = \liminf_{\xi \to \infty} \beta(x) = \beta(\infty).$$

Similarly $\beta_\varepsilon(-\infty) \leq \beta(-\infty)$. Thus (6.2.12) implies that

$$\int_\Omega [\beta_\varepsilon(-\infty) q_+ - \beta_\varepsilon(\infty) q_-] d\Omega < (l, q) \tag{6.2.20}$$

$$< \int_\Omega [\beta_\varepsilon(\infty) q_+ - \beta_\varepsilon(-\infty) q_-] d\Omega \, \forall q \in \ker a, q \neq 0.$$

We note now that (6.2.11) implies the exterior inequality of (6.1.14) as a strict inequality (i.e. esssup < essinf); a number $\tilde{M} > \rho_1$ (cf. (6.1.15)) can be chosen such that for any function $u \in V$ with $|\hat{u}(x)| > \tilde{M}$ and sign $\hat{u}(x) = $ sign $\hat{q}(x)$ for almost every $x \in \Omega$, we have from (6.2.20) that for $\hat{q} > 0$, $\hat{q} = q_+$, $\hat{u}(x) > \tilde{M}$ and thus

$$\int_{\{x|\hat{q}(x)>0\}} \beta_\varepsilon(\hat{u}(x)) \hat{q}(x) d\Omega - (l, q) > 0 \tag{6.2.21}$$

$$- \int_{\{x|\hat{q}(x)>0\}} \beta_\varepsilon(-\hat{u}(x)) \hat{q}(x) d\Omega + (l, q) > 0. \tag{6.2.22}$$

For $\hat{q} < 0$ we have $\hat{q} = -q_-$, $\hat{u}(x) < -M$ and thus (6.2.20) implies that

$$\int_{\{x|\hat{q}(x)<0\}} \beta_\varepsilon(\hat{u}(x)) \hat{q}(x) d\Omega - (l, q) > 0 \tag{6.2.23}$$

$$- \int_{\{x|\hat{q}(x)<0\}} \beta_\varepsilon(-\hat{u}(x)) \hat{q}(x) d\Omega + (l, q) > 0. \tag{6.2.24}$$

166   6. Static Hemivariational Inequalities

By an appropriate choice of the numbers $\delta \in (0,1], N > 1, \eta > 0$, and $\alpha > 0$, and by taking into consideration that $\beta_\epsilon(\hat{u}(x))\hat{u}(x) \geq 0$ and that sign $\hat{u}(x) =$ sign $\hat{q}(x)$ these inequalities imply for every u as above, the relations

$$\left(1 - \frac{1}{N}\right) \int_{\{x \mid |\hat{q}(x)| > \delta\alpha\}} \beta_\epsilon(\hat{u}(x))\hat{q}(x)d\Omega - (l,q) > \eta|q|_2 \qquad (6.2.25)$$

$$-\left(1 - \frac{1}{N}\right) \int_{\{x \mid |\hat{q}(x)| > \delta\alpha\}} \beta_\epsilon(-\hat{u}(x))\hat{q}(x)d\Omega + (l,q) > \eta|q|_2 \qquad (6.2.26)$$

as is obvious for $\delta \to 0_+$. Now we write $u_{en} = \tilde{u}_{en} + q_{en}$, and let us take N as in (6.2.25), (6.2.26). Then for $\alpha > \alpha_0 = \tilde{M}\delta^{-1}(1-1/N)^{-1}$ it results

$$\int_\Omega \beta_\epsilon(\hat{u}_{en})\hat{u}_{en}d\Omega = \int_{\substack{|\hat{u}_{en}(x)|<\delta\alpha/N \\ |\hat{q}_{en}(x)|>\delta\alpha}} \cdots + \int_{\substack{|\hat{u}_{en}(x)|\geq\delta\alpha/N \\ |\hat{q}_{en}(x)|>\delta\alpha}} \cdots + \int_{|\hat{q}_{en}(x)|\leq\delta\alpha} \cdots$$

$$\geq \int_{\substack{|\hat{u}_{en}(x)|<\delta\alpha/N \\ |\hat{q}_{en}(x)|>\delta\alpha}} \beta_\epsilon(\hat{u}_{en})\hat{u}_{en}d\Omega - \rho_1\rho_2 \text{ mes } \Omega \qquad (6.2.27)$$

$$\geq \left(1 - \frac{1}{N}\right) \int_{\substack{|\hat{u}_{en}(x)|<\delta\alpha/N \\ |\hat{q}_{en}(x)|>\delta\alpha}} \hat{q}_{en}\beta_\epsilon(\hat{u}_{en})d\Omega - \rho_1\rho_2 \text{ mes } \Omega.$$

Indeed for $|\hat{\tilde{u}}_{en}(x)| < \delta\alpha/N$ and $|\hat{q}_{en}(x)| > \delta\alpha$ one has that for $\alpha > \alpha_0$

$$|\hat{u}_{en}| = |\hat{\tilde{u}}_{en} + \hat{q}_{en}| > \left(1 - \frac{1}{N}\right)\delta\alpha > \tilde{M} \qquad (6.2.28)$$

and thus $\beta_\epsilon(\hat{u}_{en})\hat{u}_{en} \geq 0$, and $\beta_\epsilon(\hat{u}_{en})\hat{q}_{en} \geq 0$. Further it can be shown that for $\hat{q}_{en} > \delta\alpha$

$$\hat{u}_{en}(x) = \hat{\tilde{u}}_{en}(x) + \hat{q}_{en}(x) > -\frac{\delta\alpha}{N} + \hat{q}_{en}(x) > \hat{q}_{en}(x)\left(1 - \frac{1}{N}\right) \qquad (6.2.29)$$

and thus

$$\beta_\epsilon(\hat{u}_{en})\hat{u}_{en} \geq \left(1 - \frac{1}{N}\right)\hat{q}_{en}\beta_\epsilon(\hat{u}_{en}). \qquad (6.2.30)$$

Similarly for $\hat{q}_{en} < -\delta\alpha$. Thus we obtain that

$$(\Lambda(u_{en}), u_{en}) \geq \left(1 - \frac{1}{N}\right) \int_{\substack{|\hat{u}_{en}(x)|<\delta\alpha/N \\ |\hat{q}_{en}(x)|>\delta\alpha}} \hat{q}_{en}\beta_\epsilon(\hat{u}_{en})d\Omega - \rho_1\rho_2\text{mes }\Omega - (l, q_{en}) - (l, \tilde{u}_{en})$$

$$(6.2.31)$$

is obtained. For $\alpha > \alpha_0$ sufficiently large (6.2.31) and (6.2.25) imply that

$$\begin{aligned}(\Lambda(u_{en}), u_{en}) &> \eta|q_{en}|_2 - \rho_1\rho_2 \text{ mes } \Omega - (l, \tilde{u}_{en}) \qquad (6.2.32) \\ &\geq \eta|q_{en}|_2 - c_1 - c_2||\tilde{u}_{en}|| \geq \eta|q_{en}|_2 - c_1 - c_2'|||\tilde{u}_{en}||| \\ &= \eta|q_{en}|_2 - c_1 - c_2'p(\tilde{u}_{en}), \quad c_1, c_2, c_2' \text{ const} > 0.\end{aligned}$$

From (6.2.32), assuming that (6.2.17) holds and that $\alpha > \alpha_0$, we get the estimate

$$(\Lambda(u_{en}), u_{en}) > \eta|q_{en}|_2 - c|q_{en}|_2^{1/2} - c', \quad c, c' \text{ const} > 0. \tag{6.2.33}$$

The right-hand side of (6.2.33) is positive if $R > 0$ is such that

$$|q_{en}|_2 > R > \delta\alpha_0(\text{mes } \Omega)^{1/2}. \tag{6.2.34}$$

Thus we have proved (6.2.18) and therefore Brouwer's fixed point theorem implies that problem P_{en}^S has a solution u_{en} with $\|u_{en}\| < c$. The rest of the proof is the same as the proof of Theorem 6.1.1.   q.e.d.

6.3 On the Substationarity of the Energy

In this section we shall discuss the relation between the hemivariational formulation and the corresponding substationarity problem. Let us consider the following problem P_1:
Find $u \in V$ such that the "energy" functional

$$\Pi(v) = \frac{1}{2}a(v,v) + \int_\Omega j(\hat{v})d\Omega - (l,v) \tag{6.3.1}$$

is substationary at $v = u$.

Problem P_1 is by definition equivalent to the problem: $u \in V$ is a solution of the inclusion

$$0 \in \bar{\partial}\Pi(v). \tag{6.3.2}$$

The following proposition will be proved now.

Proposition 6.3.1 Suppose that j is locally Lipschitz and $\bar{\partial}$-regular and (6.1.44) holds. Then every solution of (6.3.2) is solution of the (coercive or semicoercive) hemivariational inequality and conversely.

Proof. Equation (6.3.2) can be written equivalently as

$$l \in \bar{\partial}\Pi_1(v) \quad \text{for } v \in V. \tag{6.3.3}$$

where

$$\Pi(v) = \Pi_1(v) - (l,v). \tag{6.3.4}$$

Now compute directly $\bar{\partial}\Pi_1(u)$ by using the definition of $\Pi^0(u,v)$. Note that $\frac{1}{2}a(u,u)$ is $\bar{\partial}$-regular and that

$$\Pi_1^0(u,v) \leq a(u,v) + J^0(\hat{u},\hat{v}), \tag{6.3.5}$$

where J is the (finite) integral

$$J(\hat{u}) = \int_{\Omega} j(\hat{u})d\Omega. \qquad (6.3.6)$$

We will show first that

$$J^0(\hat{u}, \hat{v}) = \int_{\Omega} j^0(\hat{u}, \hat{v})d\Omega. \qquad (6.3.7)$$

Let us denote by $g_{\lambda,h}$ the difference quotient

$$g_{\lambda,h}(\hat{u}, \hat{v}) = \frac{j(\hat{u} + h + \lambda\hat{v}) - j(\hat{u} + h)}{\lambda}. \qquad (6.3.8)$$

Function $\xi \to j(\xi)$ is locally Lipschitz and therefore

$$|g_{\lambda,h}(\hat{u}, \hat{v})| \le c|\hat{v}| \qquad (6.3.9)$$

where c depends on the neighbourhood of $(\hat{u}+h)(x)$ and $|\cdot|$ denotes the absolute value. Note that $\hat{u}, \hat{v} \in L^2(\Omega)$ and that $\xi \to j(\xi)$ is continuous. Thus $x \to g_{\lambda,h}(\hat{u}(x), \hat{v}(x))$ is measurable. Now let us apply Fatou's lemma for not integrable functions ([Dunf]). We get that

$$\int_{\Omega} \limsup_{\substack{\lambda \to 0_+ \\ h \to 0}} (g_{\lambda,h} - c|\hat{v}|)d\Omega \ge \limsup_{\substack{\lambda \to 0_+ \\ h \to 0}} \int_{\Omega} (g_{\lambda,h} - c|\hat{v}|)d\Omega. \qquad (6.3.10)$$

Due to the growth assumption (6.1.44) c in (6.3.10) is a function of $L^2(\Omega)$. Accordingly $\int c|\hat{v}|d\Omega$ is finite and may "disappear" from both sides in (6.3.10). From (6.3.10) we obtain

$$\int_{\Omega} j^0(\hat{u}, \hat{v})d\Omega \ge J^0(\hat{u}, \hat{v}) \qquad (6.3.11)$$

where the integrals are finite. Using the definition of lim sup, Fatou's lemma, the $\bar{\partial}$-regularity of j and (6.3.11) imply that

$$\begin{aligned}
J^0(\hat{u}, \hat{v}) &\ge \liminf_{\lambda \to 0_+} \frac{J(\hat{u} + \lambda\hat{v}) - J(\hat{u})}{\lambda} \ge \int_{\Omega} \liminf_{\lambda \to 0_+} \frac{j(\hat{u} + \lambda\hat{v}) - j(\hat{u})}{\lambda} d\Omega \\
&= \int_{\Omega} \lim_{\lambda \to 0_+} \frac{j(\hat{u} + \lambda\hat{v}) - j(\hat{u})}{\lambda} d\Omega = \int_{\Omega} \tilde{j}'(\hat{u}, \hat{v})d\Omega \\
&= \int_{\Omega} j^0(\hat{u}, \hat{v})d\Omega \ge J^0(\hat{u}, \hat{v}).
\end{aligned} \qquad (6.3.12)$$

From (6.3.11) and (6.3.12) we get (6.3.7). Thus (6.3.5) and (6.3.7) imply the hemivariational inequality (6.1.51).

Now the converse will be shown, i.e. that any solution of P^C or P^S is a solution of the substationarity problem (6.3.2). First we show that J is $\bar{\partial}$-regular: indeed as in (6.3.12) Fatou's lemma implies that

$$
\begin{aligned}
J^0(\hat{u}, \hat{v}) &\geq \liminf_{\lambda \to 0_+} \frac{J(\hat{u} + \lambda \hat{v}) - J(\hat{u})}{\lambda} \geq \int_\Omega \liminf_{\lambda \to 0_+} \frac{j(\hat{u} + \lambda \hat{v}) - j(\hat{u})}{\lambda} d\Omega \\
&= \int_\Omega \tilde{j}'(\hat{u}, \hat{v}) d\Omega = \int_\Omega \limsup_{\lambda \to 0_+} \frac{j(\hat{u} + \lambda \hat{v}) - j(\hat{u})}{\lambda} \\
&\geq \limsup_{\lambda \to 0_+} \frac{J(\hat{u} + \lambda \hat{v}) - J(\hat{u})}{\lambda} \geq \liminf_{\lambda \to 0_+} \frac{J(\hat{u} + \lambda \hat{v}) - J(\hat{u})}{\lambda}.
\end{aligned}
$$

Thus

$$\tilde{J}'(\hat{u}, \hat{v}) = \int_\Omega \tilde{j}'(\hat{u}, \hat{v}) d\Omega = \int_\Omega j^0(\hat{u}, \hat{v}) d\Omega = J^0(\hat{u}, \hat{v}). \tag{6.3.13}$$

Because of the $\bar{\partial}$-regularity of $J(\cdot)$ and $\frac{1}{2}a(\cdot, \cdot)$ in $\Pi_1(\cdot)$ it can be shown that Π_1 is $\bar{\partial}$-regular and therefore

$$\Pi_1^0(u, v) = \tilde{\Pi}_1'(u, v). \tag{6.3.14}$$

Thus one may write P^C or P^S in the form (cf. (6.3.5), (6.3.13) and (6.3.14))

$$u \in V, \tilde{\Pi}_1'(u, v - u) = \Pi_1^0(u, v - u) \geq (l, v - u), \quad \forall v \in V \tag{6.3.15}$$

which yields the substationarity ptoblem (6.3.2)   q.e.d.

In this section we have actually proved Prop. 2.5.3. We refer the reader also to Prop. 4.3.1 and Prop. 4.3.2.

6.4 Variational Hemivariational Inequalities

Let us consider in this section a functional $\Phi: V \to (-\infty, +\infty]$, $\Phi \not\equiv \infty$, which is a convex, l.s.c and proper functional on V. The following problem $\tilde{P}^C$ is posed: Find $u \in V$ such as to satisfy

$$a(u, v - u) + \int_\Omega j^0(\hat{u}, \hat{v} - \hat{u}) d\Omega + \Phi(v) - \Phi(u) \geq (l, v - u), \quad \forall v \in V. \tag{6.4.1}$$

Here a, j, and V have the same properties ((6.1.5), (6.1.8), (6.1.52), (6.1.54)) as in the case of problem (6.1.51) in Sect. 6.1, i.e. as for coercive hemivariational inequalities. Moreover $l \in V'$ we shall first consider the "differentiable" case in which $\operatorname{grad} \Phi(\cdot)$ exists everywhere and then the "nondifferentiable" case.

(a) *The "differentiable" problem* $\tilde{P}^C$. The following proposition holds:

Proposition 6.4.1 The inequality (6.4.1) is equivalent to the inequality

$$\begin{aligned} u \in V, \; a(u, v - u) &+ \int_\Omega j^0(\hat{u}, \hat{v} - \hat{u}) d\Omega + (\operatorname{grad} \Phi(u), v - u) \\ &\geq (l, v - u), \quad \forall v \in V. \end{aligned} \tag{6.4.2}$$

170   6. Static Hemivariational Inequalities

Proof. From (6.4.1), we obtain (6.4.2) by setting $v = u + \lambda(w - u)$, $\lambda \in (0,1)$, letting $\lambda \to 0_+$, and using the fact that $\xi \to j^0(\hat{u}, \xi)$ is positively homogeneous. Conversely (6.4.2) implies (6.4.1) because of the inequality

$$\Phi(v) - \Phi(u) \geq (\mathrm{grad}\Phi(u), v - u), \qquad \forall v \in V \tag{6.4.3}$$

which holds due to the convexity of Φ.   q.e.d.

Now the corresponding regularized problem $\tilde{P}^C_e$ and the corresponding finite dimensional problem $\tilde{P}^C_{en}$ are defined respectively:
Find $u_e \in V$ such as to satisfy the variational equality

$$a(u_e, v) + \int_\Omega \beta_e(\hat{u}_e)\hat{v}d\Omega + (\mathrm{grad}\Phi(u_e), v) = (l, v), \qquad \forall v \in V. \tag{6.4.4}$$

Find $u_{en} \in V_n$ such that

$$a(u_{en}, v) + \int_\Omega \beta_e(\hat{u}_{en})\hat{v}d\Omega + (\mathrm{grad}\Phi(u_{en}), v) = (l, v), \qquad \forall v \in V_n. \tag{6.4.5}$$

The following theorem can be proved.

Theorem 6.4.1 Suppose that (6.1.14) holds and that

$$\mathrm{grad}\Phi(0) = 0. \tag{6.4.6}$$

Moreover, let the linear continuous operator L, mapping u to $\hat{u}$, have the property

$$L : V \to L^\infty(\Omega) \text{ be compact.} \tag{6.4.7}$$

Then problem $\tilde{P}^C$ has at least one solution.

Proof. The monotonicity of $\mathrm{grad}\Phi$ and the assumption (6.4.6) imply that

$$(\mathrm{grad}\Phi(u_{en}), u_{en}) \geq 0, \qquad \forall u_{en} \in V_n. \tag{6.4.8}$$

Thus using (6.1.5), (6.1.15) and (6.4.8) we may write that

$$\begin{aligned}(\Lambda(u_{en}), u_{en}) &= a(u_{en}, u_{en}) + (\mathrm{grad}\Phi(u_{en}), u_{en}) \\ &+ \int_\Omega \beta_e(\hat{u}_{en})\hat{u}_{en}d\Omega - (l, u_{en}) \\ &\geq c_1|u_{en}|^2 - \rho_1\rho_2\mathrm{mes}\,\Omega - c_2|u_{en}|, c_1, c_2, \mathrm{const} > 0.\end{aligned} \tag{6.4.9}$$

From (6.4.9) by Brouwer's fixed point theorem, problem $\tilde{P}^C_{en}$ has a solution u_{en} with $||u_{en}|| < c$. Thus as $\varepsilon \to 0$, $n \to \infty$ we may extract a subsequence denoted also by $\{u_{en}\}$ such that (6.1.30), (6.1.31) and (6.1.32) hold (due to (6.4.7)).

A slight modification of the proof of Prop. 6.1.3 implies (6.1.22), i.e. the weak convergence of (6.1.33). The modification consists in the fact that in the last inequality of (6.1.25) the term $(\mathrm{grad}\Phi(u_{en}), u_{en})$ appears, which, because of (6.4.8) does not change the result of (6.1.25). From the above estimates and from (6.4.5) we obtain that

$$|\operatorname{grad} \Phi(u_{en})|_{V'} < c \qquad (6.4.10)$$

and thus as $\varepsilon \to 0$, $n \to \infty$

$$\operatorname{grad} \Phi(u_{en}) \to \psi \qquad \text{weakly in } V'. \qquad (6.4.11)$$

Because of (6.1.54) and the properties of Galerkin basis we may pass to the limit. It results from (6.4.5) the equality

$$a(u,v) + \int_\Omega \chi \hat{v} d\Omega + (\psi, v) = (l, v), \qquad \forall v \in V. \qquad (6.4.12)$$

In order to complete the proof we have to show that

$$\psi = \operatorname{grad} \Phi(u) \text{ in } V' \qquad (6.4.13)$$

and that

$$\chi \in \hat{\beta}(\hat{u}), \quad \text{a.e. on } \Omega. \qquad (6.4.14)$$

The monotonicity of $\operatorname{grad} \Phi$ implies that

$$X_n = (\operatorname{grad} \Phi(u_{en}) - (\operatorname{grad} \Phi(\theta), u_{en} - \theta) \geq 0, \qquad \forall \theta \in V, \qquad (6.4.15)$$

and by means of (6.4.5) that

$$X_n = -a(u_{en}, u_{en}) - \int_\Omega \beta_\varepsilon(\hat{u}_{en})\hat{u}_{en} d\Omega + (l, u_{en}) \qquad (6.4.16)$$
$$-(\operatorname{grad} \Phi(u_{en}), \theta) - (\operatorname{grad} \Phi(\theta), u_{en} - \theta) \geq 0 \qquad \forall \theta \in V.$$

Due to (6.4.7) we may show that

$$\lim_{\substack{\varepsilon \to 0 \\ n \to \infty}} \int_\Omega \beta_\varepsilon(\hat{u}_{en})\hat{u}_{en} d\Omega = \int_\Omega \chi \hat{u} d\Omega. \qquad (6.4.17)$$

Indeed we have that

$$\int_\Omega \{\beta_\varepsilon(\hat{u}_{en})\hat{u}_{en} - \chi\hat{u}\} d\Omega = \int_\Omega \beta_\varepsilon(\hat{u}_{en})(\hat{u}_{en} - \hat{u}) d\Omega + \int_\Omega \hat{u}(\beta_\varepsilon(\hat{u}_{en}) - \chi) d\Omega = A + B. \qquad (6.4.18)$$

Since V is compactly imbedded into $L^\infty(\Omega)$, $\lim B = 0$. Again (6.1.22) implies that

$$\|\beta_\varepsilon(\hat{u}_{en})\|_{L^1(\Omega)} < c. \qquad (6.4.19)$$

From (6.4.19) and (6.4.7) we get that $\lim A = 0$.

For $\varepsilon \to 0$ and $n \to \infty$ we obtain from (6.4.16) using (6.4.17) and the inequality

$$\liminf a(u_{en}, u_{en}) \geq a(u, u) \qquad (6.4.20)$$

that

$$0 \leq \limsup X_n \leq -a(u,u) - \int_\Omega \chi \hat{u} d\Omega + (l,u) - (\psi,\theta) - (\operatorname{grad} \Phi(\theta), u-\theta), \forall \theta \in V.$$
(6.4.21)

From (6.4.21) and (6.4.12) the inequality

$$(\psi - \operatorname{grad} \Phi(\theta), u - \theta) \geq 0, \qquad \forall \theta \in V \qquad (6.4.22)$$

results. Now Minty's monotonicity argument is applied: setting in (6.4.22) $u - \theta = \lambda w$, $\lambda > 0$, we get the expression

$$(\psi - \operatorname{grad} \Phi(u - \lambda w), w) \geq 0, \qquad \forall w \in V \qquad (6.4.23)$$

from which by passing to the limit $\lambda \to 0_+$, because of the monotonicity of the function $\lambda \to (\operatorname{grad} \Phi(u - \lambda w), w)$, we obtain that

$$(\psi - \operatorname{grad} \Phi(u), w) \geq 0, \qquad \forall w \in V. \qquad (6.4.24)$$

Then (6.4.13) results from (6.4.24) by setting $\pm w$. The proof of (6.4.14) is the same as in Theorem 6.1.1   q.e.d.

(b) *The "nondifferentiable" problem* $\tilde{P}^C$. In this case $\operatorname{grad} \Phi$ does not exist everywhere. Through a regularization of Φ the problem can be transformed into a sequence of differentiable problems. The following assumption is made: there exist convex Gâteaux differentiable functionals Φ_ρ, $\rho > 0$, which have the following properties:

i) $\Phi_\rho(v) \to \Phi(v)$, $\quad \forall v \in V \quad$ as $\rho \to 0$ $\hfill (6.4.25)$

ii) $\operatorname{grad} \Phi_\rho(0) = 0 \quad$ for every ρ $\hfill (6.4.26)$

iii) if $v_\rho \to v$ weakly in V for $\rho \to 0$ and $\Phi_\rho(v_\rho) < M$, where M is a constant, then

$$\liminf_{\rho \to 0} \Phi_\rho(v_\rho) \geq \Phi(v). \qquad (6.4.27)$$

The regularized problem $\tilde{P}^C_{\epsilon\rho}$ is given below:
Find $u \in V$ such as to satisfy the variational equality

$$a(u_{\epsilon\rho}, v) + \int_\Omega \beta_\epsilon(u_{\epsilon\rho}) v d\Omega + (\operatorname{grad} \Phi(u_{\epsilon\rho}), v) = (l,v), \qquad \forall v \in V. \quad (6.4.28)$$

From (6.4.28) the finite dimensional problem $\tilde{P}^C_{\epsilon\rho n}$ is obtained analogously to (6.4.5). We denote by $u_{\epsilon\rho n} \in V_n$ its solution, if any exists. The following theorem can be proved.

Theorem 6.4.2 Suppose that (6.1.14) and (6.4.7) hold and that Φ satisfies (6.4.25), (6.4.26), (6.4.27). Then the nondifferentiable problem $\tilde{P}^C$ has at least one solution.

Proof. For problem $\tilde{P}^C_{\varepsilon\rho n}$ it can be shown using Brouwer's fixed point theorem (as in Theorem 6.4.1) that a solution exists and that

$$|u_{\varepsilon\rho n}| \leq c, \tag{6.4.29}$$

where c is a constant independent of ε, ρ and n. Thus for $\varepsilon \to 0$, $n \to \infty$

$$u_{\varepsilon\rho n} \to u_\rho \quad \text{weakly in } V. \tag{6.4.30}$$

As in Theorem 6.4.1 an estimate similar to (6.1.22) is proved and thus

$$\beta_\varepsilon(\hat{u}_{\varepsilon\rho n}) \to \chi_\rho \quad \text{weakly in } L^1(\Omega). \tag{6.4.31}$$

Moreover

$$|\text{grad}\, \Phi_\rho(u_{\varepsilon\rho n})|_{V'} \leq c \tag{6.4.32}$$

where c is independent of ε, n and ρ. Thus for $\varepsilon \to 0$, $n \to \infty$

$$\text{grad}\, \Phi_\rho(u_{\varepsilon\rho n}) \to \psi_\rho \quad \text{weakly in } V'. \tag{6.4.33}$$

Since the above estimates are independent of ε, n and ρ, for $\rho \to 0$

$$u_\rho \to u \quad \text{weakly in } V \tag{6.4.34}$$

$$\chi_\rho \to \chi \quad \text{weakly in } L^1(\Omega) \tag{6.4.35}$$

and

$$\psi_\rho \to \psi \quad \text{weakly in } V'. \tag{6.4.36}$$

Analogously to (6.4.17) it is shown that

$$\lim_{\substack{\varepsilon \to 0 \\ n \to \infty}} \int_\Omega \beta_\varepsilon(\hat{u}_{\varepsilon\rho n})\hat{u}_{\varepsilon\rho n} d\Omega = \int_\Omega \chi_\rho \hat{u}_\rho d\Omega. \tag{6.4.37}$$

The monotonicity argument applied as in the previous theorem implies that

$$\psi_\rho = \text{grad}\, \Phi_\rho(u_\rho). \tag{6.4.38}$$

Finally, for $\varepsilon \to 0$ and $n \to \infty$ the variational equality

$$a(u_\rho, v) + \int_\Omega \chi_\rho \hat{v} d\Omega + (\text{grad}\, \Phi_\rho(u_\rho), v) = (l, v), \quad \forall v \in V \tag{6.4.39}$$

is obtained from $P^C_{\varepsilon\rho n}$. The next step in the proof is to pass to the limit in (6.4.39) for $\rho \to 0$. From (6.4.38) and (6.4.39) by applying (6.4.3) to Φ_ρ we obtain the inequality

$$\Phi_\rho(v) - \Phi_\rho(u_\rho) + a(u_\rho, v - u_\rho) + \int_\Omega \chi_\rho(\hat{v} - \hat{u}_\rho)d\Omega \geq (l, v - u_\rho), \quad \forall v \in V. \tag{6.4.40}$$

If in (6.4.40) a v is chosen such that $\Phi(v) < \infty$, then (6.4.25) implies that a constant M_1 exists such that $\Phi_\rho(v) < M_1$, and from (6.4.40) that $\Phi_\rho(u_\rho) < M'_1$, where M_1' is another constant. From this last inequality and from (6.4.39), relation (6.4.27) results. From (6.4.40) we obtain

$$D = \Phi_\rho(v) + a(u_\rho, v) + \int_\Omega \chi_\rho \hat{v} d\Omega \qquad (6.4.41)$$

$$\geq \Phi_\rho(u_\rho) + a(u_\rho, u_\rho) + \int_\Omega \chi_\rho \hat{u}_\rho d\Omega + (l, v - u_\rho) = F, \quad \forall v \in V,$$

which for $\rho \to 0$ implies that

$$\liminf_{\rho \to 0} D = \lim_{\rho \to 0} \left\{ \Phi_\rho(v) + a(u_\rho, v) + \int_\Omega \chi_\rho \hat{v} d\Omega \right\} \qquad (6.4.42)$$

$$= \Phi(v) + a(u, v) + \int_\Omega \chi \hat{v} d\Omega \geq \liminf_{\rho \to 0} F$$

$$= \liminf_{\rho \to 0} \left\{ \Phi_\rho(u_\rho) + a(u_\rho, u_\rho) + \int_\Omega \chi_\rho \hat{u}_\rho d\Omega + (l, v - u_\rho) \right\}$$

$$\geq \Phi(u) + a(u, u) + \int_\Omega \chi \hat{u} d\Omega + (l, v - u), \quad \forall v \in V.$$

Here we have used (6.4.27), the property (6.4.20) of the bilinear form $a(u_\rho, u_\rho)$, and analogously to (6.4.17) that

$$\lim_{\rho \to 0} \int_\Omega \chi_\rho \hat{u}_\rho d\Omega = \int_\Omega \chi \hat{u} d\Omega. \qquad (6.4.43)$$

In the final part of the proof we have to show (6.4.14). Its proof is the same as in Theorem 6.1.1. q.e.d.

Note that due to assumption (6.4.7) the proof of Prop. 6.1.3 and of (6.4.14) can be simplified. Note here that (6.4.7) is needed only for the proof of (6.4.17) and (6.4.43). The semicoercive case of (6.4.1) is still an open problem.

6.5 Applications to Engineering Problems

The results of the previous sections can be applied directly to the study of B.V.Ps in mechanics and engineering involving nonmonotone multivalued boundary conditions, material laws, or interface laws (cf. e.g. Sect. 2.4 and Sect. 3.2). For such problems the derivation of variational formulations and the existence proof for their solution has been made possible only with the theory of hemivariational inequalities.

i) Unilateral contact of a linear elastic body with a granular support. In this case we assume that on a subset Γ_S of the boundary Γ of Ω the nonmonotone multivalued law of Fig. 2.4.1b holds between normal displacements u_N and reactions $-S_N$. Moreover on Γ_S the tangential forces S_T are prescribed. Obviously we can write that

6.5 Applications to Engineering Problems

$$-S_N \in \bar{\partial} j(u_N) \quad \text{on } \Gamma_S \tag{6.5.1}$$

where j is a locally Lipschitz functional. Similarly the law of Fig. 2.4.1c represents the bilateral contact with a fibre-reinforced support can be also put in the form (6.5.1). Let on $\Gamma_U = \Gamma - \Gamma_S$ the displacements be zero. Accordingly we get a hemivariational inequality of the form (6.1.51) where $a(\cdot, \cdot)$ is the elastic energy of the body, l represents the volume forces $f \in V'$ in Ω and the given tangential forces $S_T = C_T$, $C_T \in L^2(\Gamma)$ (i.e. $(l, v) = (f, v) + \int_{\Gamma_S} C_{T_i} v_{T_i} d\Gamma$ where $v_T = \{v_{T_i}\}$ are the tangential displacements on Γ_S) and instead of $\int_\Omega j^0(\hat{u}, \hat{v} - \hat{u}) d\Omega$ we have $\int_{\Gamma_S} j^0(u_N, v_N - u_N) d\Gamma$. For this problem $V = \{v | v \in [H^1(\Omega)]^3, v = 0 \text{ on } \Gamma_U\}$ and $a(\cdot, \cdot)$ is coercive on V. Recalling the trace theorem $v \to v_N : [H^1(\Omega)]^3 \to H^{\frac{1}{2}}(\Gamma)$ and because $H^1(\Omega) \subset L^2(\Gamma)$ is compact, Theorem 6.1.1 can be proved for the present problem. Moreover assumption (6.1.54) is replaced by the fact that $\{v \in V | v_N \in L^\infty(\Gamma_S)\}$ is dense in V for the H^1-norm. Note that if $\Gamma_U = \emptyset$ then $a(\cdot, \cdot)$ is semicoercive. In this case Theorem 6.2.1 may be proved, where $\hat{q} = q_N$, (q is a rigid body displacement), and in (6.2.7) and (6.2.12) the integrals are extended on Γ.

ii) Analogously to the above cases Theorems 6.1.1 and 6.2.1 apply to the case of a plane linear elastic body, which on $\Gamma_S \subset \Gamma$ has an adhesive support described under loading conditions in the normal direction of the law of Fig. 2.4.1a and in the tangential direction of the stick-slip law of Fig. 2.4.1f. If mes $\Gamma_U > 0$ the problem is described by the hemivariational inequality (6.1.51) where V and $a(\cdot, \cdot)$ are as in (i), $l \in V'$ are the volume forces in Ω and $\int_\Omega j^0(\hat{u}, \hat{v} - \hat{u}) d\Omega$ is replaced by

$$\int_{\Gamma_S} j_N^0(u_N, v_N - u_N) d\Gamma + \int_{\Gamma_S} j_T^0(u_T, v_T - u_T) d\Gamma,$$

where the j_N and j_T are the nonconvex superpotentials corresponding to the laws of the aforementioned figures. Then the Theorems 6.1.1 and 6.2.1 may be proved for the present problem as well, with the difference that (6.2.12) takes the form

$$\int_{\Gamma_S} [(\beta_N(-\infty)q_{N+} + \beta_T(-\infty)q_{T+}) - (\beta_N(\infty)q_{N-} + \beta_T(\infty)q_{T-})] d\Gamma \tag{6.5.2}$$

$$< (l, q) < \int_{\Gamma_S} [(\beta_N(\infty)q_{N+} + \beta_T(\infty)q_{T+}) - (\beta_N(-\infty)q_{N-} + \beta_T(-\infty)q_{T-})]$$

$$\forall q \in \ker a, q \neq 0,$$

where q_{N+} and q_{N-} (respectively q_{T+} and q_{T-}) are the positive and negative parts of q_N (respectively of q_T), i.e. of the normal (respectively tangential) rigid displacement at the boundary.

iii) The interface problem of massonry structures. We consider m-linear elastic twodimensional bodies connected in q-interfaces and assume that the laws of

Fig. 2.4.1c,d between $-S_N$ (respectively $-S_T$) and the relative displacements $[u_N]$ (respectively $[u_T]$) hold on the interfaces. The following hemivariational inequality holds (cf. also Sect. 4.2)

$$\{u^{(1)}, u^{(2)}, \ldots, u^{(m)}\} \in [H^1(\Omega^{(1)})]^2 \times \ldots \times [H^1(\Omega^{(m)})]^2 = V$$

such that

$$\sum_{i=1}^{m} a(u^{(i)}, v^{(i)} - u^{(i)}) + \sum_{k=1}^{q} \int_{\Gamma^{(k)}} j_{N,k}^0([u_N^{(k)}], [v_N^{(k)}] - [u_N^{(k)}])d\Gamma \quad (6.5.3)$$

$$+ \sum_{k=1}^{q} \int_{\Gamma^{(k)}} j_{T,k}^0([u_T^{(k)}], [v_T^{(k)}] - [u_T^{(k)}])d\Gamma \geq \sum_{i=1}^{m}(l^{(i)}, v^{(i)} - u^{(i)})$$

$$\forall \{v^{(1)}, \ldots, v^{(m)}\} \in V.$$

Here index (i) refers to the bodies and index (k) to the interfaces. Again Theorems 6.1.1 and 6.2.1 can be proved. Now the sufficient condition reads

$$\sum_{k=1}^{q} \int_{\Gamma^{(k)}} [(\beta_{N,k}(-\infty)[q_N^{(k)}]_+ + \beta_{T,k}(-\infty)[q_T^{(k)}]_+) \quad (6.5.4)$$

$$-(\beta_{N,k}(\infty)[q_N^{(k)}]_- + \beta_{T,k}(\infty)[q_T^{(k)}]_-)]d\Gamma$$

$$< \sum_{i=1}^{m}(l^{(i)}, v^{(i)}) < \sum_{k=1}^{q} \int_{\Gamma^{(k)}} [(\beta_{N,k}(\infty)[q_N^{(k)}]_+ + \beta_{T,k}(\infty)[q_T^{(k)}]_+)$$

$$-(\beta_{N,k}(-\infty)[q_N^{(k)}]_- - \beta_{T,k}(-\infty)[q_T^{(k)}]_-)]d\Gamma.$$

Here $[q_N^{(k)}]_+$ denotes the positive part of the relative rigid normal displacement $[q_N^{(k)}]$, etc.

iv) The delamination problem of adhesively connected n-sandwich plates. Here an interface law of the form of Fig. 4.2.1b (with the inclined branch AV′) between the adjacent plates is considered, and $V = H^2(\Omega^{(1)}) \times \ldots \times H^2(\Omega^{(n)})$, $\Omega^{(i)} \subset R^2$. The problem leads to a hemivariational inequality of the form

$$\sum_{i=1}^{n} a(u^{(i)}, v^{(i)} - u^{(i)}) + \sum_{k=1}^{n-1} \int_{\Omega_{k,k+1}} j^0([u^{k,k+1}], [v^{k,k+1}] - [u^{k,k+1}])d\Omega \quad (6.5.5)$$

$$\geq \sum_{i=1}^{n}(l^{(i)}, v^{(i)} - u^{(i)}), \quad \forall \{v^{(1)}, \ldots, v^{(n)}\} \in V.$$

Here $[u^{k,k+1}]$ denotes the relative deflexion of the plates k and $k+1$ at their interface $\Omega_{k,k+1}$. The sufficient condition is analogous to (6.5.4).

v) Due to the compact imbedding $H^2(\Omega^{(i)}) \subset C^0(\bar{\Omega}^{(i)})$ and the imbeddings $C^0(\bar{\Omega}^{(i)}) \subset C^0(\Gamma^{(i)}) \subset L^\infty(\Gamma^{(i)})$ we have the compact imbedding $H^2(\Omega^{(i)}) \subset L^\infty(\Gamma^{(i)})$ and thus in the framework of the problem of laminated plates we may

6.5 Applications to Engineering Problems

assume monotone multivalued boundary conditions (e.g. the conditions (2.3.31), (2.3.32), for instance plastic hinge boundary conditions or contact with a rigid support etc.). In this case a variational-hemivariational inequality of the type (6.4.1) results and Theorems 6.4.1 and 6.4.2 hold.

vi) A variational-hemivariational inequality of the type (6.4.1) is obtained if a nonlinear elastic body is studied which obeys a monotone law of the form (3.1.1) and is subjected to nonmonotone boundary conditions of the type (2.4.1). In this context cf. [Pana88c].

Note that all the problems expressed via hemivariational or variational-hemivariational inequalities are free boundary problems; for instance in the case of the delamination problem it is not *a priori* known the free boundary between the interface points which remain adhesively bonded and which not.

The sufficient conditions derived are similar to the sufficient conditions of the Landesman-Lazer theory [Land]. Concerning the number of solutions of hemivariational inequalities most questions are still open. Perhaps some new results of the Landesman-Lazer approach to the theory of semilinear differential equations could be extended to the theory of hemivariational inequalities to study the problem of the multiplicity of their solutions.

7. Eigenvalue and Dynamic Problems

The present chapter is devoted to the study of the eigenvalue problem for hemivariational inequalities and to the study of dynamic hemivariational inequalities. In Chapt. 4 we described certain mechanical problems leading to the eigenvalue problem for hemivariational inequalities and to dynamic hemivariational inequalities. Here we shall give some existence results. For the eigenvalue problem the minimax technique is applied [Mot93] and for the dynamic hemivariational inequalities the regularization method. The corresponding results for dynamic hemivariational inequalities appear here for a first time.

7.1 On the Eigenvalue Problem for Hemivariational Inequalities

Let us consider the following eigenvalue problem for a hemivariational inequality (problem P):
Find $u \in V \subset L^2(\Omega)$ and $\lambda \in \mathbb{R}$ satisfying the inequality

$$a(u,v) + \int_{\Omega'} j^0(u;v)dx \geq \lambda(u,v)_V \quad \forall v \in V. \tag{7.1.1}$$

Here Ω is an open bounded connected subset of $\mathbb{R}^n$, Ω' is a subdomain of Ω, V is a real Hilbert space endowed with the scalar product $(.,.)_V$, continuously imbedded into $L^2(\Omega)$, $a : V \times V \to \mathbb{R}$ denotes a continuous symmetric bilinear form on V and the function $j : \mathbb{R} \to \mathbb{R}$ is defined by means of

$$j(t) = \int_0^t \beta(s)ds, \quad t \in \mathbb{R}, \tag{7.1.2}$$

where $\beta \in L^\infty_{loc}(\mathbb{R})$. A possible method to treat problem P is to regularize the nonsmooth integral term and then to pass to limit (cf. e.g. [Kar91,92]). Here, however a different approach based on the critical point theory is applied, because it permits us to obtain certain qualitative informations for the solution.

Thus we write problem P in the substationarity form (problem P')

$$0 \in \bar{\partial} I_\lambda(u), \quad u \in V, \tag{7.1.3}$$

where $I_\lambda : V \to \mathbb{R}$ denotes a suitably defined locally Lipschitz functional depending on λ, which is associated to problem P, and $\bar{\partial} I_\lambda(u)$ represents Clarke's generalized gradient of I_λ at $u \in V$. Problem P' expresses that $u \in V$ is a critical point of the locally Lipschitz functional I_λ in the sense of K.C.Chang [Ch] or a substationarity point according to R.T.Rockafellar [Rock79]. As we know from Prop. 4.3.1 and Prop. 6.3.1, problem P may have a larger number of solutions that problem P', unless the conditions of Prop. 6.3.1 hold.

Since we are looking here for at least one solution of P it is sufficient to prove that P' has at least one solution. Thus the study of problem P is reduced to finding a (the) critical point(s) of the corresponding functional I_λ on V. To prove their existence use is made of two minimax existence results due to K.C.Chang [Ch]. In order to verify the assumptions of Chang's results we employ an argument inspired from the work of Rabinowitz ([Rabi], Chapt. 2). The main tool is the nonsmooth version of the Palais-Smale condition given in [Ch] (cf. also [Mot86], where the Palais-Smale condition is weakened). We will show the existence of nontrivial solutions of problem P in the form of a global minimum or a minimax point of an appropriately defined functional I_λ. Moreover we prove that there exist solutions of problem P in spaces $L^{2+d}(\Omega)$ with $d > 0$ in the case where the function β satisfies an appropriate growth condition.

Let X be a Banach space with norm $\|\cdot\|$ and let X' be its dual space. The locally Lipschitz functional $f : X \to \mathbb{R}$ is said to satisfy the Palais-Smale condition if every sequence $\{x_n\}$ in X for which $\{f(x_n)\}$ is bounded in $\mathbb{R}$ and

$$\min_{w \in \bar{\partial} f(x_n)} \|w\|_{X'} \to 0 \text{ as } n \to \infty, \tag{7.1.4}$$

contains a convergent subsequence in X.

The next result represents a nonsmooth version of Palais-Smale Minimization Theorem, due to Chang [Ch].

Proposition 7.1.1 Suppose that the Banach space X is reflexive and the locally Lipschitz functional $f : X \to \mathbb{R}$ is bounded from below and satisfies the Palais-Smale condition (7.1.4). Then $\inf_X f$ is a critical value of f, i.e., there exists a point $x \in X$ such that

$$\inf_X f = f(x) \text{ and thus } 0 \in \bar{\partial} f(x). \tag{7.1.5}$$

The next result is a different formulation of a result of Chang [Ch] and it has a similarity with the classical Mountain Pass Theorem (see [Rabi], p.7).

Proposition 7.1.2 Let the locally Lipschitz functional $f : X \to \mathbb{R}$ on the reflexive Banach space X satisfy the Palais-Smale condition together with the following assumptions

(i) $f(0) = 0$ and there exist positive constants ρ, d such that

$$f(x) \geq d \text{ for each } x \in X \text{ with } \|x\| = \rho; \tag{7.1.6}$$

7.1 On the Eigenvalue Problem for Hemivariational Inequalities

(ii) there is a point $e \in X$ such that

$$\|e\| > \rho \quad \text{and} \quad f(e) \leq 0. \tag{7.1.7}$$

Then there exists a critical point $x_0 \in X$ of f satisfying

$$f(x_0) = \inf_{\gamma \in \Gamma} \max_{t \in [0,1]} f(\gamma(t)) \tag{7.1.8}$$

where

$$\Gamma = \{\gamma \in C([0,1]), X) |\ \gamma(0) = 0, \ \gamma(1) = e\}. \tag{7.1.9}$$

Proof. We proceed as in the proof of the Mountain Pass Theorem of Ambrosetti and Rabinowitz (cf. [Rabi]). Since the sphere $\|x\| = \rho$ separates in X the points 0 and e, it is clear that

$$c = \inf_{\gamma \in \Gamma} \max_{t \in [0,1]} f(\gamma(t)) \geq d. \tag{7.1.10}$$

It remains only to show that the number c in (7.1.10) is a critical value of f, i.e. that there is a critical point x_0 of f with $f(x_0) = c$. Arguing by contradiction let us assume that c is not a critical value of f. Then the nonsmooth Deformation Lemma of K.C.Chang ([Ch], Theorem 3.1) yields the existence of a real number $\varepsilon \in (0, d/2)$ and of a homeomorphism $h : X \to X$ such that

$$h(x) = x \text{ for every } x \notin \{y \in X |\ |f(x) - c| < d/2\} \tag{7.1.11}$$

and

$$f(h(x)) \leq c - \varepsilon \text{ for every } x \in X \text{ with } f(x) \leq c + \varepsilon. \tag{7.1.12}$$

From (7.1.10) one can find a $\gamma \in \Gamma$ such that

$$f(\gamma(t)) \leq c + \varepsilon \quad \forall t \in [0,1]. \tag{7.1.13}$$

Let

$$g(t) = h(\gamma(t)), \quad t \in [0,1]. \tag{7.1.14}$$

From (7.1.11) and the assumptions (i) and (ii) it follows that $g \in \Gamma$, while inequalities (7.1.12), (7.1.13) imply that

$$f(g(t)) \leq c - \varepsilon, \quad t \in [0,1]. \tag{7.1.15}$$

Thus we arrive at the contradiction

$$c \leq \max_{t \in [0,1]} f(g(t)) \leq c - \varepsilon. \tag{7.1.16}$$

This completes the proof of Prop. 7.1.2   q.e.d.

Concerning problem P we deal with functionals $J : L^{\sigma+1}(\Omega) \to \mathbb{R}$ of the type

$$J(u) = \int_{\Omega'} j(u(x)) dx, \quad u \in L^{\sigma+1}(\Omega), \tag{7.1.17}$$

where Ω' is a subdomain of Ω and $j : \mathbb{R} \to \mathbb{R}$ is the function

$$j(t) = \int_0^t \beta(s)ds, \ t \in \mathbb{R}, \quad (7.1.18)$$

which corresponds to some locally bounded measurable function $\beta : \mathbb{R} \to \mathbb{R}$. To have J defined on $L^{\sigma+1}(\Omega)$ we must impose upon β the following growth condition

$$|\beta(t)| \leq c_1 + c_2 |t|^\sigma, \ t \in \mathbb{R}, \quad (7.1.19)$$

for some positive constants c_1, c_2 independent of $t \in \mathbb{R}$. We can easily verify that under assumption (7.1.19) the functional J in (7.1.17) is locally Lipschitz. Then using the same argument as in [Ch], (p.108-110), we obtain the following characterization of the generalized gradient $\bar{\partial} J(u)$ (cf. also (1.2.55) and Prop. 2.5.3).

Proposition 7.1.3 If $w \in \bar{\partial} J(u)$, then $w \in (L^{\sigma+1}(\Omega))'$ is such that $w(x) \in [\bar{\beta}(u(x)), \bar{\bar{\beta}}(u(x))]$, for a.e. $x \in \Omega'$. The functions $\bar{\beta}$ and $\bar{\bar{\beta}}$ are defined as follows (cf. (1.2.51), (1.2.52)) for $t \in \mathbb{R}$

$$\bar{\beta}(t) = \lim_{\delta \to 0} \operatorname*{ess\,inf}_{|s-t|<\delta} \beta(s), \quad (7.1.20)$$

$$\bar{\bar{\beta}}(t) = \lim_{\delta \to 0} \operatorname*{ess\,sup}_{|s-t|<\delta} \beta(s).$$

If β admits the one-sided limits $\beta(t_\pm)$ at $t \in \mathbb{R}$, the following inequalities hold

$$\bar{\beta}(t) = \min(\beta(t_-), \beta(t_+)) \quad (7.1.21)$$

$$\bar{\bar{\beta}}(t) = \max(\beta(t_-), \beta(t_+)).$$

Further we shall investigate the existence of solutions of the eigenvalue problem in $L^2(\Omega)$. Let us make now the following assumptions for problem P:

(H_1) the function $\beta \in L^\infty_{\text{loc}}(\mathbb{R})$ satisfies the growth condition

$$|\beta(t)| \leq c_1 + c_2|t| \ \forall t \in R, \quad (7.1.22)$$

where c_1 and c_2 are positive constants independent of t.

(H_2) Ω is an open bounded connected subset of some Euclidean space $\mathbb{R}^n$, V is a real Hilbert space with the scalar product $(\cdot, \cdot)_V$ and the induced norm $\| \ \|_V$ such that V is continuously and densely imbedded as a linear subspace of $L^2(V)$, and $a : V \times V \to \mathbb{R}$ is a continuous symmetric bilinear form on V. Let us denote by $A : V \to V'$ the linear operator which corresponds to the bilinear form $a(\cdot, \cdot)$ (cf. (4.2.29))

(H_3) Ω' is a subdomain of Ω such that $W = \{u|_{\Omega'} \ |u \in V\}$ is a reflexive Banach space such that the imbedding $W \subset L^2(\Omega')$ is compact and the restriction mapping $r : V \to W$ given by $r(u) = u|_{\Omega'} \ \forall u \in V$ is continuous.

7.1 On the Eigenvalue Problem for Hemivariational Inequalities

For each number $\lambda \in \mathbb{R}$ let us consider the functional $I_\lambda : V \to \mathbb{R}$

$$I_\lambda(u) = \frac{1}{2}(a(u,u) - \lambda \|u\|_V^2) + J(u), \quad u \in V, \tag{7.1.23}$$

where $J : L^2(\Omega) \to \mathbb{R}$ is defined in (7.1.17) by means of the function $j : \mathbb{R} \to \mathbb{R}$ with β satisfying (H_1). Assumptions (H_1) and (H_2), imply that the functional I_λ is locally Lipschitz on V.

Proposition 7.1.4 Suppose that assumptions $(H_1), (H_2), (H_3)$ hold. If $u_0 \in V$ is a critical point of the locally Lipschitz functional $I_\lambda : V \to \mathbb{R}$, for some $\lambda \in \mathbb{R}$, then the pair (u_0, λ) solves problem P.

Proof. It follows from (7.1.23) and (H_2) that the generalized gradient $\partial I_\lambda(u)$ is given by

$$\bar{\partial} I_\lambda(u) = Au - \lambda \Lambda u + \bar{\partial}(J|_V)(u), \tag{7.1.24}$$

where $\Lambda : V \to V'$ denotes the duality mapping

$$(\Lambda u)(v) = (u,v)_V \quad \forall v \in V. \tag{7.1.25}$$

Using hypothesis (H_1) and relations (7.1.20) one obtains

$$|w| \leq \max(|\bar{\beta}(t)|, |\bar{\bar{\beta}}(t)|) \leq c_1 + c_2|t| \quad \forall t \in \mathbb{R} \text{ and } w \in \bar{\partial}j(t). \tag{7.1.26}$$

The estimate (7.1.26) implies by means of Prop. 2.5.3 (cf. also [Clar83] p.83) that

$$\bar{\partial}(J|_{L^2(\Omega')})(u) \subset \int_{\Omega'} \bar{\partial}j(u(x))dx \quad \forall u \in L^2(\Omega'). \tag{7.1.27}$$

The restriction map $r : V \to W$ is continuously differentiable and also onto. Thus the following formula holds

$$\bar{\partial}(J|_V)(u) = \bar{\partial}(J|_W)(u|_{\Omega'}) \circ r \quad \forall u \in V. \tag{7.1.28}$$

Due to (H_2) W is dense in $L^2(\Omega')$. This property and hypothesis (H_3) allow us to apply a theorem of K.C.Chang ([Ch], p.110, Theorem 2.2), which implies that

$$\bar{\partial}(J|_W)(u) \subset \bar{\partial}(J|_{L^2(\Omega')})(u) \quad \forall u \in W. \tag{7.1.29}$$

Let us consider the critical point $u_0 \in V$ of I_λ for some $\lambda \in \mathbb{R}$; thus u_0 is a solution of the problem P'. Relations (7.1.24), (7.1.27), (7.1.28) and (7.1.29) lead to the relation

$$\lambda(u_0, v)_V - a(u_0, v) \leq \int_{\Omega'} (\max_{z \in \bar{\partial}j(u_0(x))} (z, v(x)))dx \tag{7.1.30}$$

$$= \int_{\Omega'} j^0(u_0(x), v(x))dx \quad \forall v \in V$$

due to (1.2.29). Relation (7.1.30) shows that $u_0 \in V$ is a solution of problem P which corresponds to the real number λ appearing in P' q.e.d.

The following technical result is useful to check the Palais-Smale condition for the functional I_λ in (7.1.23).

Proposition 7.1.5 Assume that conditions (H_1), (H_2), (H_3) are verified. Let us choose $\lambda \in \mathbb{R}$ such that

$$\lambda < \inf\{a(u,u) \big| \|u\|_V^2 \mid u \in V, \ u \neq 0\}. \tag{7.1.31}$$

If a sequence $\{u_n\}$ is bounded in V and there is a sequence $w_n \in \bar{\partial}(J|_V)(u_n)$ such that

$$Au_n - \lambda \Lambda u_n + w_n \to 0 \text{ in } V' \text{ as } n \to \infty, \tag{7.1.32}$$

then $\{u_n\}$ contains a subsequence which is convergent in V.

Proof. The boundedness of $\{u_n\}$ in V and the compactness of the imbedding $W \subset L^2(\Omega')$ imply the existence of a convergent subsequence in $L^2(\Omega')$ that we denote again by $\{u_n\}$. From (7.1.28) and (7.1.29) we have that

$$w_n \in \partial(J|_{L^2(\Omega')})(u_n). \tag{7.1.33}$$

Since $J|_{L^2(\Omega')}$ is locally Lipschitz and $\{u_n\}$ converges in $L^2(\Omega')$ one deduces from (7.1.33) that $\{w_n\}$ is a bounded sequence in $L^2(\Omega')$. By hypothesis (H_3) it turns out that the imbedding $L^2(\Omega') \subset W'$ is compact, so $\{w_n\}$ contains a convergent subsequence in W', which is denoted again by $\{w_n\}$. The continuity of the linear mapping $r : V \to W$ and relations (7.1.33), (7.1.28) ensure the convergence of the subsequence $\{w_n\}$ in V'. Because the duality mapping $\Lambda : V \to V'$ is a topological linear isomorphism, we obtain from (7.1.32) that

$$\Lambda^{-1}Au_n - \lambda u_n \text{ converges in } V, \tag{7.1.34}$$

where $\{u_n\}$ denotes the subsequence of $\{u_n\}$ corresponding by (7.1.33) to the subsequence $\{w_n\}$. Let us fix now some $\lambda \in \mathbb{R}$ satisfying (7.1.31). Then, by Schwarz inequality we can write that for all positive integers m, n

$$\Big(\inf_{\substack{u \in V \\ u \neq 0}} \frac{a(u,u)}{\|u\|_V^2} - \lambda\Big)\|u_n - u_m\|_V^2 \leq a(u_n - u_m, u_n - u_m) \tag{7.1.35}$$

$$-\lambda\|u_n - u_m\|_V^2 \leq \|\Lambda^{-1}A(u_n - u_m)$$

$$-\lambda(u_n - u_m)\|_V \|u_n - u_m\|_V.$$

Relations (7.1.31),(7.1.34) and (7.1.35) assure that $\{u_n\}$ converges in V. q.e.d.

We are now in position to prove the first main result of this Section.

Theorem 7.1.1 Suppose that hypotheses (H_1), (H_2), (H_3) are verified. Let $C(\Omega)$ be a positive constant such that

$$\|u\|_{L^2(\Omega)} \leq C(\Omega)\|u\|_V \quad \forall u \in V. \tag{7.1.36}$$

Then, for every $\lambda \in \mathbb{R}$ satisfying

7.1 On the Eigenvalue Problem for Hemivariational Inequalities

$$\lambda < \inf_{\substack{u \in V \\ u \neq 0}} \frac{a(u,u)}{\|u\|_V^2} - c_2(C(\Omega))^2, \qquad (7.1.37)$$

with the positive real number c_2 from (H_1), there exists a point $u_0 \in V$ such that

$$\inf_{u \in V} I_\lambda(u) = I_\lambda(u_0). \qquad (7.1.38)$$

The point $u_0 \in V$ is thus a critical point of the functional $I_\lambda : V \to \mathbb{R}$, and the pair (u_0, λ) is a solution of problem P.

Proof. The constant $C(\Omega)$ exists because of hypothesis (H_2). First we show that the functional $I_\lambda : V \to \mathbb{R}$ is bounded from below. Let λ be a real number verifying (7.1.37). Then one has

$$\gamma = \inf_{\substack{u \in V \\ u \neq 0}} \frac{a(u,u)}{\|u\|_V^2} - c_2(C(\Omega))^2 - \lambda > 0. \qquad (7.1.39)$$

Relations (7.1.23), (7.1.36), (7.1.39), assumption (H_1) and the Schwarz inequality imply that

$$I_\lambda(u) \geq \frac{1}{2}(\gamma + c_2(C(\Omega))^2)\|u\|_V^2 - \int_\Omega |\int_0^{u(x)} \beta(t)dt|dx \geq \frac{1}{2}\gamma\|u\|_V^2 \qquad (7.1.40)$$
$$-c_1(\text{mes}(\Omega))^{1/2}C(\Omega)\|u\|_V \quad \forall u \in V.$$

The estimate (7.1.40) justifies the lower boundedness of the functional $I_\lambda : V \to \mathbb{R}$. Let us now show that if λ satisfies (7.1.37) the corresponding locally Lipschitz functional $I_\lambda : V \to \mathbb{R}$ verifies the Palais-Smale condition. To prove this let $\{u_n\}$ be a sequence in V such that $\{I_\lambda(u_n)\}$ is bounded and

$$\min_{w \in \bar{\partial} I_\lambda(u_n)} \|w\|_{V^*} \to 0 \quad \text{as } n \to \infty. \qquad (7.1.41)$$

We may verify that the boundedness of $\{I_\lambda(u_n)\}$ in $\mathbb{R}$ ensures the boundedness of $\{u_n\}$ in V: indeed, let $M > 0$ denote a constant such that

$$I_\lambda(u_n) \leq M \quad \forall n. \qquad (7.1.42)$$

Then from (7.1.40) and (7.1.42) we obtain the existence of a positive constant b such that the following inequality holds

$$M \geq \frac{1}{2}\gamma\|u_n\|_V^2 - b\|u_n\|_V \quad \forall n. \qquad (7.1.43)$$

Because of (7.1.39) inequality (7.1.43) implies the boundedness of the sequence $\{u_n\}$ in V. In addition, from (7.1.41) we see that (7.1.32) holds. Therefore one can apply Prop. 7.1.5 from which we derive the existence of a subsequence of $\{u_n\}$ converging in V. Thus the functional $I_\lambda : V \to \mathbb{R}$ satisfies the Palais-Smale condition.

We proved that, if $\lambda \in \mathbb{R}$ verifies (7.1.37), the corresponding functional $I_\lambda : V \to \mathbb{R}$ satisfies all the hypotheses of Prop. 7.1.1. This Proposition implies the existence of a point $u_0 \in V$ satisfying (7.1.38); thus u_0 is a critical point of I_λ. Prop. 7.1.4 then implies that the pair (u_0, λ) is a solution of problem P, q.e.d.

With a change in the assumptions (H_1), (H_2) and (H_3) we can guarantee the existence of minimax solutions of the eigenvalue problem in $L^{2+d}(\Omega)$. Now we assume that:

(H_1') the function $\beta \in L^\infty_{loc}(\mathbb{R})$ satisfies the growth condition

$$|\beta(t)| \leq c_1 + c_2 |t|^{1+d} \quad \forall t \in \mathbb{R}, \tag{7.1.44}$$

where $d > 0$ and $c_1, c_2 \geq 0$ are constants independent of $t \in \mathbb{R}$.

(H_2') Ω is a domain in some Euclidean space $\mathbb{R}^n$, V is a real Hilbert space which is continuously imbedded as a dense linear subspace of $L^{2+d}(\Omega)$, and $a : V \times V \to \mathbb{R}$ is a continuous symmetric bilinear form on V.

(H_3') Ω' is a subdomain of Ω such that $W = \{u|_{\Omega'} \, |u \in V\}$ is a reflexive Banach space with the property that the imbedding $W \subset L^{2+d}(\Omega')$ is compact and the restriction map $r : V \to W$ $r(u) = u|_{\Omega'}$ $\forall u \in V$ is continuous.

Notice that under hypothesis (H_1') the functional J introduced in (7.1.17) is defined on $L^{2+d}(\Omega)$. Furthermore, taking $\sigma = 1 + d$ in Prop. 7.1.3 one finds that, if $w \in \bar{\partial} J(u)$ with $u \in L^{2+d}(\Omega)$ then

$$w(x) \in [\bar{\beta}(u(x)), \bar{\bar{\beta}}(u(x))] \text{ a.e. for } x \in \Omega' \tag{7.1.45}$$

where $\bar{\beta}$ and $\bar{\bar{\beta}}$ are defined in (7.1.20). Here we have to deal with problem P in the space V on the assumptions (H_2'), (H_3'). Since we work in the more general setting of hypothesis (H_1'), the previous approach does not hold. Therefore we proceed by using a minimax method whose smooth pattern can be found in Rabinowitz ([Rabi], Chapt. 2).

Assumption (H_3') implies that the functional $I_\lambda : V \to \mathbb{R}, \lambda \in \mathbb{R}$, defined in (7.1.23) is locally Lipschitz. The result below shows that Prop. 7.1.4 remains true.

Proposition 7.1.6 Under conditions (H_1'), (H_2'), (H_3'), if $u_0 \in V$ is a critical point of $I_\lambda : V \to \mathbb{R}$ in (7.1.23) for some $\lambda \in \mathbb{R}$, then the pair $(u_0, \lambda) \in V \times R$ is a solution of problem P.

Proof. Hypothesis (H_1') guarantees that in place of (7.1.26) the following inequality holds

$$|w| \leq \max(|\bar{\beta}(t)|, |\bar{\bar{\beta}}(t)|) \leq c_1 + c_2|t|^{1+d} \quad \forall t \in \mathbb{R} \text{ and } w \in \bar{\partial} j(t), \tag{7.1.46}$$

7.1 On the Eigenvalue Problem for Hemivariational Inequalities

where $j : \mathbb{R} \to \mathbb{R}$ is the function defined in (7.1.18). Applying Prop. 2.5.3 (cf. also [Clar83] p.83, Theorem 2.7.5) we may write that

$$\bar{\partial}(J|_{L^{2+d}(\Omega')})(u) \subset \int_{\Omega'} \bar{\partial}j(u(x))dx \quad \forall u \in L^{2+d}(\Omega'). \tag{7.1.47}$$

Arguing as in the proof of Prop. 7.1.4, by means of (H'_2), (H'_3), we can verify (7.1.28) and the inclusion

$$\bar{\partial}(J|_W)(u) \subset \bar{\partial}(J|_{L^{2+d}(\Omega')})(u) \quad \forall u \in V. \tag{7.1.48}$$

Since $u_0 \in V$ is a critical point of I_λ, we deduce from the above relations that for the critical point $u_0 \in V$ of the functional I_λ the inequality (7.1.30) holds for each $v \in V$,   q.e.d.

Further, a proposition analogous to Prop. 7.1.5 holds in the new framework of the assumptions (H'_1), (H'_2), (H'_3).

Proposition 7.1.7 Assume that (H'_1), (H'_2), (H'_3) hold, and let λ be a real number satisfying (7.1.31). If $\{u_n\}$ is a bounded sequence in V such that there exists a sequence $w_n \in \bar{\partial}(J|_V)(u_n)$ in V' with the property (7.1.32), then $\{u_n\}$ has a convergent subsequence in V.

Proof. The proof is analogous to the proof of Prop. 7.1.5 with the difference that the imbedding $V \subset L^2(\Omega)$ is replaced by $V \subset L^{2+d}(\Omega)$ and we use (H'_1), (H'_2), (H'_3) instead of (H_1), (H_2), (H_3).   q.e.d.

The following theorem now holds. It is the second main result of this Section.

Theorem 7.1.2 In addition to the assumptions (H'_1), (H'_2), (H'_3) we suppose that the following conditions hold

(i) $\lim_{t \to 0} \beta(t)t^{-1} = 0;$ (7.1.49)

(ii) there exist the real numbers $c_1 > 0$ and c_2 such that

$$\int_0^t \beta(s)ds \leq -c_1|t|^{2+d} + c_2 \quad \text{for } t < 0, \tag{7.1.50}$$

where $d > 0$ is the same constant as in (H'_1);

(iii) there exist the real numbers $\mu > 2$ and $n > 0$ such that

$$\int_0^t \beta(s)ds \geq \mu^{-1} \max(\bar{\beta}(t)t, \bar{\bar{\beta}}(t)t) \quad \forall |t| > n. \tag{7.1.51}$$

Then for every $\lambda \in \mathbb{R}$ satisfying (7.1.31) and every constant c with

$$0 < c < 2^{-1}(\inf_{\substack{u \in V \\ u \neq 0}} (a(u,u)\|u\|_V^2) - \lambda), \tag{7.1.52}$$

the locally Lipschitz functional $I_\lambda : V \to \mathbb{R}$ has a critical point $u_\lambda \in V$ so that $I_\lambda(u_\lambda) \geq c$. Therefore problem P admits a nontrivial solution u_λ in V for every $\lambda \in \mathbb{R}$ verifying (7.1.31).

Proof. Fix some number $\varepsilon > 0$. By condition (i), there is $\delta = \delta(\varepsilon) > 0$ such that $|\beta(t)| \leq \varepsilon |t|$ for each $|t| \leq \delta$. Then the function $j : \mathbb{R} \to \mathbb{R}$ defined in (7.1.18) satisfies

$$|j(t)| \leq \frac{1}{2}\varepsilon |t|^2 \quad \text{for } |t| \leq \delta. \tag{7.1.53}$$

Due to (H'_1) one can determine a constant $A(\delta) > 0$ with

$$|j(t)| \leq A(\delta)|t|^{2+d} \quad \text{for } |t| > \delta. \tag{7.1.54}$$

Combining (7.1.53) and (7.1.54) it results that

$$|j(t)| \leq (\frac{1}{2}\varepsilon + A(\delta)|t|^d)t^2 \quad \forall t \in \mathbb{R}. \tag{7.1.55}$$

Inequality (7.1.55) implies the following estimate

$$\begin{aligned}|J(u)| &\leq \frac{\varepsilon}{2}\|u\|^2_{L^2(\Omega)} + A(\delta)\|u\|^{2+d}_{L^{2+d}(\Omega)} \\ &\leq C(\frac{\varepsilon}{2} + A(\delta)\|u\|^d_V)\|u\|^2_V \quad \forall u \in V,\end{aligned} \tag{7.1.56}$$

where $C > 0$ denotes a constant depending only on the continuous imbedding $V \subset L^{2+d}(\Omega) \subset L^2(\Omega)$. If we take in (7.1.56)

$$\|u\|_V < (\frac{\varepsilon}{2A(\delta)})^{1/d},$$

we find

$$|J(u)| \leq \varepsilon C \|u\|^2_V,$$

and thus

$$\lim_{u \to 0}(J(u)/\|u\|^2_V) = 0. \tag{7.1.57}$$

Due to (7.1.23), one can write that

$$I_\lambda(u) \geq \frac{1}{2}(\inf_{\substack{u \in V \\ u \neq 0}} \frac{a(u,u)}{\|u\|^2_V} - \lambda)\|u\|^2_V + (J|_V)(u). \tag{7.1.58}$$

Relation (7.1.57) implies that

$$I_\lambda(u) \geq c \quad \text{for } \|u\|_V \text{ sufficiently small}. \tag{7.1.59}$$

Hence condition (i) in Prop. 7.1.2 is satisfied in the case of the functional $I_\lambda : V \to \mathbb{R}$. In order to check the condition (ii) of the same proposition let us fix some function $u \in V$, such that $u(x) > 0$ a.e. in Ω. Hypothesis (H'_2) assures the existence of such function u. Then, by assumption (7.1.50), one obtains

7.1 On the Eigenvalue Problem for Hemivariational Inequalities

$$I_\lambda(tu) \leq \frac{1}{2}(a(u,u) - \lambda\|u\|_V^2)t^2 \quad (7.1.60)$$
$$- c_1(\int_\Omega u(x)^{2+d}dx|t|^{2+d} + c_2(\text{meas}(\Omega)), \quad \forall t < 0.$$

Because $c_1 > 0$, it turns out from (7.1.60) that $I(tu) \to -\infty$ as $t \to -\infty$. This property shows that condition (ii) of Prop. 7.1.2 is verified in the case of the functional $f = I_\lambda : V \to \mathbb{R}$.

Let us verify that for every $\lambda \in \mathbb{R}$ as in (7.1.31) the corresponding functional $I_\lambda : V \to \mathbb{R}$ satisfies the Palais-Smale condition: let $\{u_n\}$ be a sequence in V provided there is a constant $M > 0$ such that

$$I_\lambda(u_n) \leq M \quad \text{for every } n \quad (7.1.61)$$

and a sequence $w_n \in \bar{\partial}(J|_V)(u_n)$ satisfying (7.1.32). From (7.1.32) and (7.1.61) one can verify that for each integer n sufficiently large the following inequality holds

$$I_\lambda(u_n) - \mu^{-1}(a(u_n, u_n) - \lambda\|u_n\|_V^2 + w_n(u_n)) \leq M + \mu^{-1}\|u_n\|_V. \quad (7.1.62)$$

Let us replace in the left-hand side of (7.1.62) $I_\lambda(u_n)$ by its expression in (7.1.23). Then let us use (7.1.45). It results that there exists a constant K such that

$$(\frac{1}{2} - \frac{1}{\mu})(a(u_n, u_n) - \lambda\|u_n\|_V^2) + \int_{|u_n(x)| \geq n|} (\int_0^{u_n(x)} \beta(t)dt \quad (7.1.63)$$
$$-\frac{1}{\mu} \max(\bar{\beta}(u_n(x))u_n(x), \bar{\bar{\beta}}(u_n(x))u_n(x)))dx$$
$$\leq M + \frac{1}{\mu}\|u_n\|_V + K \quad \text{for all positive integers } n.$$

From (7.1.63), (7.1.51), and (7.1.31) we obtain the boundedness of $\{u_n\}$ in V. Then Prop. 7.1.7 implies that one can extract from $\{u_n\}$ a convergent subsequence in V. The Palais-Smale condition holds for the functional $I_\lambda : V \to \mathbb{R}$ with λ satisfying (7.1.31). Consequently, if $\lambda \in \mathbb{R}$ verifies (7.1.31), the locally Lipschitz functional $I_\lambda : V \to \mathbb{R}$ satisfies the hypotheses of Prop. 7.1.2. Thus, there exists a critical point $u_\lambda \in V$ of $I_\lambda : V \to \mathbb{R}$ such that

$$I_\lambda(u_\lambda) = \inf_{\gamma \in \Gamma} \max_{t \in [0,1]} I_\lambda(\gamma(t)) \quad (7.1.64)$$

where

$$\Gamma = \{\gamma \in C([0,1], V) \,\big|\, \gamma(0) = 0 \text{ and } \gamma(1) = e\} \quad (7.1.65)$$

for some fixed point $e \in V$ far enough from the origin. The above minimax characterization of $I_\lambda(u_\lambda)$ and relation (7.1.59) show that $I_\lambda(u_\lambda) \geq c$. Finally, Prop. 7.1.6 ensures that $u_\lambda \in V$ is a nontrivial solution of problem P, q.e.d.

We give now an example where Theorem 7.1.2 applies: Let $\beta(t) = -t^3 + h(t), t \in \mathbb{R}$, where the function $h : \mathbb{R} \to \mathbb{R}$ satisfies

$$|h(t)| \leq c_1|t|^{2+\sigma} + c_2, \ t \in R, \tag{7.1.66}$$

and $\lim_{t \to 0} h(t)t^{-1} = 0$, where c_1, c_2, σ are nonnegative real numbers with $\sigma < 1$. Since the function β is dominated at infinity by $-t^3$, it is clear that conditions (i)-(iii) in Theorem 7.1.2 are verified as well as (H_1'). For a space V verifying (H_2') and (H_3') Theorem 7.1.2 holds.

The proofs of the above results are due to D.Motreanu [Mot93]. Theorem 7.1.1. holds only with minor modifications if in (7.1.1) $\int_{\Omega'} j^0(u,v) d\Omega$ is replaced by $\int_{\Gamma} j^0(u_N, v_N) d\Gamma$ where u_N, v_N are the normal boundary displacements in a linear elastic body. In this case $V = [H^1(\Omega)]^3$, $W = H^{1/2}(\Gamma)$, and r is the trace mapping. Theorem 7.1.2 holds for plate problems where $V = H^2(\Omega)$. The above results do not need for their proofs the coercivity of the bilinear form $a : V \times V \to \mathbb{R}$ but they need the growth assumptions (7.1.22) and (7.1.44). The regularization approach to the eigenvalue problem for hemivariational inequalities [Kar91,92] is based on the assumption (6.1.14) and not on the more restrictive growth assumptions, but it leads only to the existence of at least one solution and not to minimax properties. In the regularization approach we make use of the results of Browder [Brow] to show the existence of a solution $(u_{\varepsilon n}, \lambda_{\varepsilon n})$ of the regularized finite dimensional hemivariational inequality (cf. also [Pan85] Chapt. 7, Eigenvalue Problems for Variational Inequalities). Then we pass to the limit as $\varepsilon \to 0$ $n \to \infty$ analogously to the same procedures in Sect. 6.1.1. The regularization approach to the eigenvalue problem for hemivariational inequalities applies directly to the buckling problem of laminated von Kármán plates (see e.g. Sect. 4.2 and Sect. 4.4). Note finally that the Theorem 7.1.2 applies to the eigenvalue problem describing the buckling of adhesively bonded von Kármán plates. In this case instead of the bilinear form $a(\cdot, \cdot) : V \times V \to \mathbb{R}$ we have the form $a(u, u) + (C(v), v)$ where C is a completely continuous operator and $(C(v), v)$ is nonnegative for every $v \in V$.

7.2 Dynamic Hemivariational Inequalities

In this section we shall study as a pilot problem the dynamic hemivariational inequality arising in a plane linear elastic body when nonmonotone skin effects are taken into account [Panagi]. We consider a body which in its undeformed state occupies an open, bounded connected subset $\Omega \subset \mathbb{R}^2$. The boundary Γ of Ω is assumed to be appropriately "regular" (a Lipschitz boundary is sufficient) and the points x of Ω and Γ are referred to a fixed Cartesian coordinate system $0x_1x_2$. The boundary Γ consists of two open disjoint parts Γ_U and Γ_F. On Γ_U (resp. Γ_F) the displacements (resp. the tractions) are prescribed. On $\Omega' \subset \Omega$ nonmonotone skin effects appear. In order to describe skin effects, e.g. nonmonotone skin friction, adhesive forces etc., we assume that the volume forces consist of two parts: $\overline{\overline{f}}_i$ which is given and $\bar{f}_i$ which is the reaction of the constraint introducing the skin effects. Thus we may write

$$f_i = \bar{f}_i + \bar{\bar{f}}_i, \quad i = 1, 2. \tag{7.2.1}$$

Here $\bar{\bar{f}}_i$ is the given external loading and $\bar{f}_i$ is a possibly multivalued function of $\frac{\partial u_i}{\partial t} = u'_i$ the time derivative of u_i. In Figs. 7.2.1a,b,c,d we give certain relations corresponding to skin frictional effects of monotone and nonmonotone nature. In Fig. 7.2.1e,f certain adhesive skin laws are depicted. All the aforementioned laws between $\bar{f}_i$ and u'_i, $i = 1, 2$, are one-dimensional. However they can be generalized if we make the additional assumption that the vectors u and f are at every point collinear.

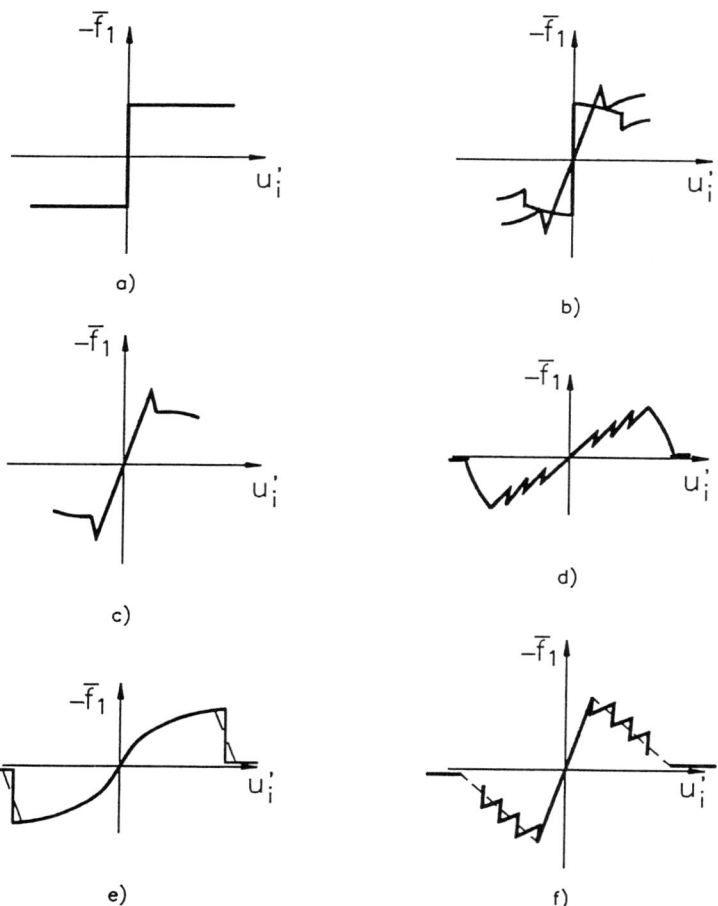

Fig. 7.2.1. Graphs of the skin effects laws.

We consider the multivalued reaction-velocity law (cf.(1.2.53))

$$-\bar{f}_i \in \tilde{\beta}_i(u'_i) = \bar{\partial} j_i(u'_i) \quad i = 1, 2 \text{ on } \Omega', \tag{7.2.2}$$

where Ω' is the part of the body where frictional or adhesive effects take place. We assume that $\Omega' \subset \Omega$, that

$$\bar{f}_i = 0 \quad \text{on} \quad \Omega - \Omega' \tag{7.2.3}$$

and that $\bar{\Omega}' \cap \Gamma = \emptyset$. The dynamic behaviour of the body, is described, on the assumptions of small displacements and small deformations by the relations $(i, j, h, k = 1, 2)$

$$\sigma_{ij,j} + f_i = \rho u_i'' \quad \text{in} \quad \Omega \times (0, T), \quad f_i = \bar{\bar{f}}_i + \bar{f}_i \tag{7.2.4}$$

$$\varepsilon_{ij}(u) = \frac{1}{2}(u_{i,j} + u_{j,i}) \quad \text{in} \quad \Omega \times (0, T), \tag{7.2.5}$$

$$\sigma_{ij} = C_{ijhk}\varepsilon_{hk} \quad \text{in} \quad \Omega \times (0, T), \tag{7.2.6}$$

$$u_i = U_i \quad \text{on} \quad \Gamma_U \times (0, T), \quad \text{mes } \Gamma_U > 0 \tag{7.2.7}$$

$$S_i = F_i \quad \text{on} \quad \Gamma_F \times (0, T), \tag{7.2.8}$$

$$-\bar{f}_i \in \bar{\partial} j_i(u_i') \quad \text{on} \quad \Omega' \times (0, T), \tag{7.2.9}$$

and (7.2.3) holds on $\Omega - \Omega'$. Here, the time ranges over the interval $[0, T]$, the density of the body $\rho = \rho(x), x \in \Omega$, is assumed to be a positive function of $L^\infty(\Omega)$ (with essinf $\rho(x) > 0$). The prescribed forces and displacements f, F and U are functions of x and t. All the functions in (7.2.3)÷(7.2.9) are assumed initially to be appropriately smooth. To the above relations the initial conditions

$$u_i = u_{0i}, \quad u_i' = u_{1i}, \quad \text{for} \quad x \in \Omega \quad \text{and} \quad t = 0 \tag{7.2.10}$$

are added. Here u_{0i} and u_{1i} are given functions of x expressing the displacement and velocity field at $t = 0$. Obviously, u_0 and u_1 must be compatible with the kinematical boundary condition on Γ_U. The appearance of u_i' instead of u_i in (7.2.9), necessitates consideration of variational expressions with respect to the velocities. Then the kinematically admissible set

$$U_{\text{ad}} = \{v | v \in [H^1(\Omega)]^2, v_i = U_i' \text{ on } \Gamma_U\} \tag{7.2.11}$$

is introduced and the following variational formulation of the B.V.P. is considered:

Find $u : [0, T] \to [H^1(\Omega)]^2$ with $u'(t) \in U_{\text{ad}}$ and $u''(t) \in [L^2(\Omega)]^2$ to satisfy the hemivariational inequality

$$(\rho u'', v - u') + a(u, v - u') + \sum_{i=1,2} \int_\Omega j_i^0(u_i', v_i - u_i') d\Omega \tag{7.2.12}$$

$$\geq (\bar{\bar{f}}, v - u') + \int_{\Gamma_F} F_i(v_i - u_i') d\Gamma, \quad \forall v \in U_{\text{ad}}$$

and the initial conditions (7.2.10).

7.2 Dynamic Hemivariational Inequalities

This hemivariational inequality results from (7.2.4)÷(7.2.6) by multiplying (7.2.4) by $v - u'$, integrating over Ω, and applying formally the Green-Gauss theorem. Then the boundary conditions and the skin conditions (7.2.7)÷(7.2.9) are taken into account, as in the static problem. At this point, we would like to call attention to the fact that the final functional framework of the problem has not yet been stated; further, after considering a more precise dependence of the data on time, the functional framework will be completed. The hemivariational inequality (7.2.12) expresses physically the principle of virtual power in its inequality form. For the dynamic problem, if $\Gamma_U = \emptyset$, one must guarantee that no rigid-body motion incompatible with the boundary conditions will occur. Further the subspace

$$V_0 = \{v | v \in [H^1(\Omega)]^2, v = 0 \text{ on } \Gamma_U\} \quad (7.2.13)$$

is introduced. By means of the substitutions $\tilde{v} = v - w$ and $\tilde{u} = u - w$, where $w : [0, T] \to [H^1(\Omega)]^2$ is a function such that $w(t)|_{\Gamma_U} = U(t)$ we obtain the following form of the problem (problem P):
Find a function $\tilde{u} : [0, T] \to V_0$ with $\tilde{u}'(t) \in V_0$ and $\tilde{u}''(t) \in [L^2(\Omega)]^2$ such that

$$(\rho\tilde{u}'', \tilde{v} - \tilde{u}') + a(\tilde{u}, \tilde{v} - \tilde{u}') + \sum_{i=1,2} \int_{\Omega'} j^0(\tilde{u}'_i + w'_i, \tilde{v}_i - \tilde{u}'_i) d\Omega \quad (7.2.14)$$

$$\geq (\psi, \tilde{v} - \tilde{u}'), \quad \forall \tilde{v} \in V_0,$$

and for $t = 0$

$$\tilde{u} = \tilde{u}_0 = u_0 - w(0), \quad \tilde{u}' = \tilde{u}_1 = u_1 - w'(0). \quad (7.2.15)$$

Here

$$(\psi, \tilde{v} - \tilde{u}') = (\bar{\bar{f}}, \tilde{v} - \tilde{u}') + \int_{\Gamma_F} F_i(\tilde{v}_i - \tilde{u}'_i) d\Gamma - (\rho w'', \tilde{v} - \tilde{u}') - a(w, \tilde{v} - \tilde{u}'). \quad (7.2.16)$$

Let us consider now that $U = 0$ on Γ_U for the sake of simplicity. We notice further that $H^1(\Omega) \to L^2(\Omega')$ is compact and that $V = V_0 \cap \{v|_{\Omega'} \in [L^\infty(\Omega')]^2\}$ is dense in V_0 for the H^1-norm. In order to prove the existence of the solution, we consider, as in Ch. 6 the regularized finite dimensional problem. The following problem (problem $P_{\varepsilon n}$) is considered now. Here $\beta_{i\varepsilon}$ is defined by (6.1.11) and V_n is a finite dimensional Galerkin subspace of V. Problem $P_{\varepsilon n}$ reads:
Find a function $u_{\varepsilon n} : [0, T] \to V_n$ with $u'_{\varepsilon n}(t) \in V_0$ and $u''_{\varepsilon n}(t) \in [L^2(\Omega)]^2$ which satisfies the variational equality

$$(\rho u''_{\varepsilon n}, v) + a(u_{\varepsilon n}, v) + \sum_{i=1,2} \int_{\Omega'} \beta_{i\varepsilon}(u'_{i\varepsilon n}) v_i d\Omega = (\psi, v), \quad \forall v \in V_n \quad (7.2.17)$$

and the initial conditions (7.2.10).
The solution of problem (7.2.14), (7.2.15) is "constructed" as the limit of the solution $\tilde{u}_{\varepsilon n}$ of (7.2.17) as $\varepsilon \to 0, n \to \infty$.

Proposition 7.2.1 Suppose that (6.1.14) holds for each β_i, $i = 1, 2$ and that

$$\bar{\bar{f}}, \bar{\bar{f}}', \bar{\bar{f}}'' \in L^2(0T, [L^2(\Omega)]^2), \tag{7.2.18}$$

$$F, F', F'' \in L^2(0T, [L^2(\Gamma_F)]^2), \tag{7.2.19}$$

for $i = 1, 2$, $\dfrac{\beta_i(\xi + h) - \beta_i(\xi)}{h} \geq -c \quad \forall \xi \in \mathbb{R} \quad \forall h > 0, \quad c \text{ const} > 0 \tag{7.2.20}$

$$u_0 \in [H^1(\Omega)]^2, \quad u_1 \in [H^1(\Omega)]^2, \tag{7.2.21}$$

and

$$v \to \sum_{i=1,2} \int_{\Omega'} \beta_{i\varepsilon}(u_{i1}) v\, d\Omega - \int_\Omega [C_{ijhk}\varepsilon_{hk}(u_0)]_{,j} v_i\, d\Omega \tag{7.2.22}$$

is a continuous linear functional on V_0 for the $[L^2(\Omega)]^2$-topology. Under these conditions the solution $\tilde{u}_{en}$ of the regularized finite dimensional problem exists and has the following properties: for $\varepsilon \to 0 \quad n \to \infty$

$$u_{en} \to u, \quad u'_{en} \to u' \text{ in } L^\infty(0T, V_0) \text{ weakly}_*, \tag{7.2.23}$$

$$u''_{en} \to u'' \text{ in } L^\infty(0T, [L^2(\Omega)]^3) \text{ weakly}_*; \tag{7.2.24}$$

u is a solution of problem (7.2.14)(7.2.15) (with $U = 0$ on Γ_U).

Proof. a) Let us put in (7.2.17) $v = u'_{en}$. Using (6.1.15) we obtain that

$$-\sum_{i=1,2} \rho_{1i}\rho_{2i}\text{mes}\Omega' + (\rho u''_{en}, u'_{en}) + a(u_{en}, u'_{en}) \leq (\psi, u'_{en}) \tag{7.2.25}$$

which reduces to

$$\frac{1}{2}\frac{d}{dt}\left(\int_\Omega \rho(u'_{en})^2 d\Omega + a(u_{en}, u_{en})\right) \leq (\psi, u'_{en}) + \sum_{i=1,2} \rho_{1i}\rho_{2i}\text{mes}\,\Omega'. \tag{7.2.26}$$

Relation (7.2.26) combined with the obvious inequality

$$a(v, v) + c_1|v|_2^2 \geq c_2||v||^2, \quad c_1 \text{ const} \geq 0 \quad c_2 \text{ const} > 0 \quad \forall v \in V_0 \tag{7.2.27}$$

implies through integration from 0 to t that

$$\frac{1}{2}\rho|u'_{en}|_2^2 + c_2||u_{en}||^2 - c_1|u_{en}|_2^2 \leq \frac{1}{2}\rho|u_1|_2^2 + c_2||u_0||^2 - c_1|u_0|_2^2 \tag{7.2.28}$$

$$+ 2\int_0^t (\psi, u'_{en})dt + 2\int_0^t c\,dt \quad c \text{ const} > 0$$

because $\rho_{1i}\rho_{2i}$ are positive constants independent of t. Due to (7.2.18)(7.2.19) $\psi, \psi' \in L^2(0T, V'_0)$. Thus it follows from (7.2.28), by means of the inequality (further c denotes the various positive constants)

$$|u_{en}|_2^2 \leq c\int_0^t |u'_{en}|_2^2 dt + |u_0|_2^2, \tag{7.2.29}$$

the relation

$$|u'_{en}|_2^2 + ||\tilde{u}_{en}||^2 \le c\left(1 + \int_0^t (|u'_{en}|_2^2 + ||u_{en}||^2)dt\right) + \int_0^t cdt$$

$$+2\int_0^t (\psi, u'_{en})dt \le c\left(1 + \int_0^t (|u'_{en}|_2^2 + ||u_{en}||^2)dt\right) + 2(\psi, u_{en})$$

$$-2(\psi(0), u_0) + 2\int_0^t ||\psi'||_{V'_0} ||u_{en}||dt. \qquad (7.2.30)$$

Thus, the relation

$$|u'_{en}|_2^2 + ||u_{en}||_1^2 \le c\left(1 + \int_0^t (|u'_{en}|_2^2 + ||u_{en}||_1^2)dt\right) \qquad (7.2.31)$$

is obtained, from which it results, by applying Gronwall's inequality (cf. e.g. [Walt]) that

$$|u'_{en}|_2 \le c, \qquad ||u_{en}|| \le c \qquad (7.2.32)$$

(b) From (7.2.17), after "differentiating" with respect to t and by setting $v = u''_e$, we get the relation

$$(\rho u'''_{en}, u''_{en}) + a(u'_{en}, u''_{en}) + \sum_{i=1,2} \int_{\Omega'} [\beta_{ie}(u'_{ien})]' u''_{ien} d\Omega = (\psi', u''_e). \qquad (7.2.33)$$

Note that the time differentiation is formal. The procedure which follows remains valid if the difference quotients are considered instead of the time derivatives. Because of the assumptions (7.2.20) it results by means of (6.1.11) that the inequality

$$\int_{\Omega'} [\beta_{ie}(u'_{ien})]' u''_{ien} d\Omega = \int_{\Omega'} \frac{d\beta_{ie}(u'_{ien})}{d(u'_{ien})} (u''_{ien})^2 d\Omega \qquad (7.2.34)$$

$$\ge -c\int_{\Omega'} (u''_{ien})^2 d\Omega \ge -c|u''_{en}|_2^2$$

holds. This can be easily verified by formulating the convolution product defining β_{ie}. Due to (7.2.34) the variational equality (7.2.33) gives rise to the inequality

$$\frac{1}{2}\frac{d}{dt}\left(\int_\Omega \rho(u''_{en})^2 d\Omega + a(u'_{en}, u'_{en})\right) \le (\psi', u''_{en}) + \bar{c}|u''_{en}|_2^2 \qquad (7.2.35)$$

From (7.2.35), after integrating from 0 to t, by means of (7.2.27) and of

$$|u''_{en}|_2^2 \le c\int_0^t |u''_{en}|_2^2 dt + |u_1|_2^2, \qquad (7.2.36)$$

we obtain the inequality

$$|u''_{en}|_2^2 + ||u'_{en}||^2 \leq c \left(||u_1||^2 + |u''_{en}(0)|_2^2 + \int_0^t |u''_{en}|_2^2 dt \right) \quad (7.2.37)$$

$$+ 2\int_0^t (\psi', u''_{en})dt + \bar{c}\int_0^t |u''_{en}|_2^2 dt.$$

Because of (7.2.18),(7.2.19), $\psi'' \in L^2(0T, V'_0)$, and thus we can write the relation

$$\int_0^t (\psi', u''_{en})dt = (\psi', u''_{en}) - (\psi'(0), u_1) - \int_0^t (\psi'', u'_{en})dt \quad (7.2.38)$$

$$\leq c(||\psi'||_{V'_0}||u'_{en}|| + 1) + \int_0^t ||\psi''||_{V'_0}||u'_{en}||dt.$$

From (7.2.37), (7.2.38) we get the inequality

$$|u''_{en}|_2^2 + ||u'_{en}||^2 \leq c(||u_1||^2 + |u''_{en}(0)|_2^2) + c\int_0^t (||u'_{en}||^2 + |u''_{en}|_2^2)dt. \quad (7.2.39)$$

(c) By setting $t = 0$ in (7.2.17), and using (7.2.22), we obtain the relation

$$(\rho u''_{en}(0), v) + a(u_0, v) + \sum_{i=1,2} \int_{\Omega'} \beta_{i\varepsilon}(u_{i1})v d\Omega = (\psi(0), v) \quad (7.2.40)$$

$$= (\bar{\bar{f}}(0), v) + \int_{\Gamma_F} F_i(0) v_i d\Gamma.$$

But

$$a(u_0, v) = \int_\Omega C_{ijhk}\varepsilon_{ij}(u_0)\varepsilon_{hk}(v)d\Omega \quad (7.2.41)$$

$$= -\int_\Omega (C_{ijhk}\varepsilon_{hk}(u_0))_{,j} v_i d\Omega + \int_{\Gamma_F} F_i(0) v_i d\Gamma$$

and, therefore,

$$(\rho u''_{en}(0), \tilde{v}) - \int_\Omega (C_{ijhk}\varepsilon_{hk}(u_0))_{,j} v_i d\Omega = (\bar{\bar{f}}(0), \tilde{v}). \quad (7.2.42)$$

Using (7.2.21) we obtain from (7.2.42)

$$|u''_{en}(0)|_2 \leq c. \quad (7.2.43)$$

Estimates (7.2.39) and (7.2.43) imply that

$$|u''_{en}|_2^2 + ||u'_{en}||^2 \leq c\left(1 + \int_0^t (|u''_{en}|_2^2 + ||u'_{en}||^2)dt\right), \tag{7.2.44}$$

from which, through Gronwall's inequality, we find that

$$|u''_{en}|_2 \leq c, \quad ||u'_{en}|| \leq c. \tag{7.2.45}$$

(d) From (7.2.32) and (7.2.45), it is concluded that the sequences $\{u_{en}\}, \{u'_{en}\}$, and $\{u''_{en}\}$ are bounded, the first two in the space $L^\infty(0T, V_0)$ and the last one in $L^\infty(0T, [L^2(\Omega)]^2)$ independently of ε and n. Thus we can select subsequences, again denoted by $\{u_{en}\}, \{u'_{en}\}$, and $\{u''_{en}\}$, which are weakly$_*$ convergent. We denote the limits by u, u' and u'' respectively. In the next steps we will show that u, u' and u'' fulfill the hemivariational inequality (7.2.14) and the initial conditions (7.2.15) (with $w = 0$).

(e) The above estimates permit us to prove as in Prop. 6.1.3 that the sequences $\{\beta_{i\varepsilon}(u'_{ien})\}$ are weakly precompact in $L^1((0T) \times \Omega')$. The proof uses the estimates (6.1.25) but with the additional time integration from 0 to t. Then we may take $t = T$. Accordingly

$$\beta_{i\varepsilon}(u'_{ien}) \to \chi_i \text{ weakly in } L^1((0T) \times \Omega') \quad i = 1, 2. \tag{7.2.46}$$

(f) Let us consider the finite dimensional problem (7.2.17) for $n = m$ fixed. Without loss of generality we can take $u_0, u_1 \in V_{m_0}$ for a fixed m_0. The following initial value problem is now considered for the functions $g_{m_i}(i = 1, \ldots, m, m \geq m_0)$ satisfying the relation

$$u_{em}(t) = \sum_{t=1}^m g_{m_i}(t)v_i \in V_m \quad \text{a.e. on } (0, T). \tag{7.2.47}$$

$$(\rho u''_{em}, v_p) + a(u_{em}, v_p) + \sum_{i=1,2} \int_{\Omega'} \beta_{i\varepsilon}(u'_{iem})v_{ip}d\Omega = (\psi, v_p), \quad 1 \leq p \leq m \tag{7.2.48}$$

$$u_{em} = u_0, \quad u'_{em} = u_1 \quad \text{for } t = 0. \tag{7.2.49}$$

As is known from the theory of ordinary differential equations, there exist for every $m \geq m_0$ a $t_m \in [0, T]$ and continuous functions g_{m_i} on $[0, t_m], i = 1, \ldots, m$, which fulfill (7.2.48), i.e. u_{em} exists. Actually, $t_m = T$; indeed from the previous estimates we obtain that the solution u_{en} can be extended by continuation to the whole time interval $[0, T]$.

(g) Passing to the limit $\varepsilon \to 0$ $n \to \infty$ we obtain that

$$\int_0^T [(\rho u'', v) + a(u, v) + \sum_{i=1,2} \int_{\Omega'} \chi_i v_i d\Omega] dt = \int_0^T (\psi, v) dt \quad \forall v \in L^2(0T, V_0). \tag{7.2.50}$$

Because of the $H^1(\Omega) \to L^2(\Omega')$ compactness we may apply Egoroff's theorem to show as in the proof of Theorem 6.1.1 that

198   7. Eigenvalue and Dynamic Problems

$$\chi_i \in \bar{\partial}j(u_i') \quad \text{a.e. on} \quad \Omega' \times [0,T] \quad i=1,2. \tag{7.2.51}$$

(h) From (7.2.50) and (7.2.51) we obtain the hemivariational inequality

$$\int_0^T [(\rho u'', v - u') + a(u, v - u') + \sum_{i=1,2} \int_{\Omega'} j_i^0(u_i', v_i - u_i') d\Omega \tag{7.2.52}$$
$$- (\psi, v - u')] dt \geq 0 \quad \forall v \in L^2(0T, V_0)$$

where $u' \in L^\infty(0T, V_0)$, by applying the definition of the generalized gradient. Note that (7.2.52) can be considered as the expression of Hamilton's principle for the problem under consideration. Moreover u and u' satisfy the initial conditions (7.2.15) (with $w = 0$) because of (7.2.23), (7.2.24) (cf. e.g. [Lio69] p.223-224).

(i) From (7.2.50), (7.2.51), the pointwise inequality (7.2.14) (with $w = 0$) will be derived. To this end, we consider the sequence $\{\theta_K\}$ of the intervals of the form $\theta_K = (t_0 - 1/K, t_0 + 1/K) \subset (0, T)$, and we put in (7.2.52) $v = u'$, $\forall t \notin \theta_K$ and $v = \tilde{v}$, $\forall t \in \theta_K$. Here $\tilde{v}$ denotes a fixed element of V_0. Then (7.2.52) reduces to

$$\frac{1}{|\theta_K|} \int_{\theta_K} [(\rho u'', \tilde{v} - u') + a(u, \tilde{v} - u') + \sum_{i=1,2} \int_{\Omega'} j_i^0(u_i', \tilde{v}_i - u_i') d\Omega \tag{7.2.53}$$
$$- (\psi, \tilde{v} - u')] dt \geq 0.$$

Further, we let $K \to \infty$ and we apply Lebesgue's theorem (cf. e.g. [Evan] p.43 or [Pan85] eq.(1.4.10)). Then

$$\frac{1}{|\theta_K|} \int_{\theta_K} [(\rho u'', \tilde{v}) + a(u, \tilde{v}) - (\psi, \tilde{v})] dt \tag{7.2.54}$$
$$\to (\rho u''(t_0), \tilde{v}) + a(u(t_0), \tilde{v}) - (\psi(t_0), \tilde{v})$$

for every $t_0 \notin \theta' \subset (0, T)$ with mes $\theta' = 0$, and

$$\frac{1}{|\theta_K|} \int_{\theta_K} [(\rho u'', u') + a(u, u') + \sum_{i=1,2} \int_{\Omega'} j_i^0(u_i', \tilde{v}_i - u_i') d\Omega + (\psi, u')] dt$$
$$\to (\rho u''(t_0), u'(t_0)) + a(u(t_0), u'(t_0))$$
$$+ \sum_{i=1,2} \int_{\Omega'} j_i^0(u_i'(t_0), \tilde{v}_i - u_i'(t_0)) d\Omega + (\psi(t_0), u') \tag{7.2.55}$$

for every $t_0 \notin \theta'' \subset (0, T)$ with mes $\theta'' = 0$. From (7.2.53)÷(7.2.55), it results that for $t_0 \notin \theta' \cup \theta''$ we can take in (7.2.53) the limit for $K \to \infty$. Thus the hemivariational inequality (7.2.14) with $w = 0$ results, q.e.d.

Because $u \in L^\infty(0T, V_0)$ and $u' \in L^\infty(0T, V_0)$, it follows by the continuity theorem (cf. e.g. [Pan85] Sect. 1.4.1) that $u : [0, T] \to V_0$ is continuous, and thus the initial condition $u = u_0$ for $t = 0$ makes sense. The same applies to the other boundary conditions as well. In the above existence proof we cannot prove that all Galerkin sequences converge to the solution. Moreover there is generally nonuniqueness of solution due to the lack of convexity. The proof of Prop. 7.2.1 may be performed by using the method of time discretization in order to reduce the dynamic hemivariational inequality to a sequence of static once. We leave this proof as an exercise for the reader (cf. for the method [Rekt]).

Concerning the uniqueness of the solution we can have such a result if around the solution under consideration the nonconvex superpotential are locally convex. Analogous results hold obviously for all hemivariational inequalities. These results do not exclude that outside the region of local convexity another solution may exist. We can state such a result for the present problem.

Proposition 7.2.2 Suppose that a displacement field $u \in L^\infty(0T, V_0)$ satisfies problem P and that around $u(t)$ an open set A in V_0 can be determined on which for almost every $t \in [0, T]$, $u_i'(t) \to j_i(u_i'(t))$ is convex for $i = 1, 2$. Then in A the solution is uniquely determined for almost every $t \in [0, T]$.

Proof. Arguing by contradiction, we assume that u_1 and u_2 are two solutions of problem P in A. Since j is convex on A the hemivariational inequality is reduced to a variational inequality, i.e. the terms $\int_{\Omega'} j_i^0(\tilde{u}_i' + w_i, \tilde{v}_i - \tilde{u}_i') d\Omega$ are replaced by $\int_{\Omega'} [j_i(\tilde{v}_i + w_i') - j(\tilde{u}_i' + w_i)] d\Omega$, due to (1.2.37). Then through the substitutions $\tilde{u} = u_1$, $\tilde{v} = u_2$ and $\tilde{u} = u_2$, $\tilde{v} = u_1$ and through addition it results that

$$-(\rho(u_1'' - u_2''), u_1' - u_2') - a(u_1 - u_2, u_1' - u_2') \geq 0 \qquad (7.2.56)$$

or

$$\frac{d}{dt}\left[\int_\Omega \rho(u_1' - u_2')^2 d\Omega + a(u_1 - u_2, u_1 - u_2)\right] \leq 0. \qquad (7.2.57)$$

The initial conditions imply that $u_1 - u_2 = 0$ and $u_1' - u_2' = 0$ for $t = 0$, which, combined with (7.2.57), yields the uniqueness of the solution in A, q.e.d.

Further let us deal in this section with another type of dynamic hemivariational inequality for which the existence proof presents some difficulties. The dynamic hemivariational inequality arises in the nonmonotone friction problem of a plane linear elastic body with given normal forces. We consider a body which in its undeformed state occupies an open, bounded connected subset $\Omega \subset \mathbb{R}^2$. The boundary Γ of Ω is assumed to be appropriately "regular" (a Lipschitz boundary v sufficient) end the point x of Ω and Γ are referred to a fixed Cartesian coordinate system $0x_1x_2$. The boundary Γ consists of three open disjoint parts Γ_U, Γ_F and Γ_S. On Γ_U (resp. Γ_F) the displacements (resp. the tractions) are prescribed and on Γ_S the nonmonotone possibly multivalued friction condition holds with given normal forces (cf. e.g. Fig. 2.4.1f). The dynamic behaviour of the body, is described, on the assumptions of small displacements and small deformations by the relations ($i, j, h, k = 1, 2$)

7. Eigenvalue and Dynamic Problems

$$\sigma_{ij,j} + f_i = \rho u_i^{''} \quad \text{in} \quad \Omega \times (0,T), \tag{7.2.58}$$

$$\varepsilon_{ij}(u) = \frac{1}{2}(u_{i,j} + u_{j,i}) \quad \text{in} \quad \Omega \times (0,T), \tag{7.2.59}$$

$$\sigma_{ij} = C_{ijhk}\varepsilon_{hk} \quad \text{in} \quad \Omega \times (0,T), \tag{7.2.60}$$

$$u_i = U_i \quad \text{on} \quad \Gamma_U \times (0,T), \quad \text{mes}\,\Gamma_U > 0 \tag{7.2.61}$$

$$S_i = F_i \quad \text{on} \quad \Gamma_F \times (0,T), \tag{7.2.62}$$

$$-S_T \in \bar{\partial}j(u_T^{'}) \quad \text{on} \quad \Gamma_S \times (0,T) \tag{7.2.63}$$

and

$$S_N = C_N \quad \text{on} \quad \Gamma_S \times (0,T). \tag{7.2.64}$$

As in the previous B.V.P. the density of the body $\rho = \rho(x), x \in \Omega$, is assumed to be a positive function of $L^\infty(\Omega)$ (with essinf $\rho(x) > 0$) and the prescribed forces and displacements f, F, C_N and U are functions of x and t. All the functions in (7.2.58)÷(7.2.64) are assumed initially to be appropriately smooth. Also the initial conditions

$$u_i = u_{0i}, \quad u_i^{'} = u_{1i}, \quad \text{for} \quad x \in \Omega \quad \text{and} \quad t = 0 \tag{7.2.65}$$

hold, where u_0 and u_1 must be compatible with the kinematical boundary condition on Γ_U. Then the set of kinematically admissible velocities

$$U_{\text{ad}} = \{v | v \in [H^1(\Omega)]^2, v_i = U_i^{'} \quad \text{on} \quad \Gamma_U\} \tag{7.2.66}$$

is introduced and the following variational formulation of the B.V.P. is considered:

Find $u : [0,T] \to [H^1(\Omega)]^2$ with $u^{'}(t) \in U_{\text{ad}}$ and $u^{''}(t) \in [L^2(\Omega)]^2$ to satisfy the variational inequality

$$(\rho u^{''}, v - u^{'}) + a(u, v - u^{'}) + \int_{\Gamma_S} j^0(u_T^{'}, v_T - u_T^{'})d\Gamma \geq (f, v - u^{'}) \tag{7.2.67}$$
$$+ \int_{\Gamma_F} F_i(v_i - u_i^{'})d\Gamma + \int_{\Gamma_S} C_N(v_N - u_N^{'})d\Gamma, \quad \forall v \in U_{\text{ad}}$$

and the initial conditions (7.2.65).

This variational inequality results from (7.2.58)÷(7.2.64) by multiplying (7.2.58) by $v - u^{'}$, integrating over Ω, and applying formally the Green-Gauss theorem. Then the boundary conditions are taken into account, as in the static problem. The variational inequality (7.2.67) expresses physically the principle of virtual power in its inequality form. As in the previous B.V.P. let us introduce the subspace

$$V_0 = \{v | v \in [H^1(\Omega)]^2, v = 0 \quad \text{on} \quad \Gamma_U\}. \tag{7.2.68}$$

The substitutions $\tilde{u} = v - w$ and $\tilde{u} = u - w$, where $w : [0,T] \to [H^1(\Omega)]^2$ is a function such that $w(t)|_{\Gamma_U} = U(t)$ lead to the following form of the problem

(problem P):
Find a function $\tilde{u} : [0, T] \to V_0$ with $\tilde{u}'(t) \in V_0$ and $\tilde{u}''(t) \in [L^2(\Omega)]^2$ such that

$$(\rho\tilde{u}'', \tilde{v} - \tilde{u}') + a(\tilde{u}, \tilde{v} - \tilde{u}') + \int_{\Gamma_S} j^0(\tilde{u}'_T + w'_T, \tilde{v}_T - \tilde{u}'_T) d\Gamma \quad (7.2.69)$$

$$\geq (\psi, \tilde{v} - \tilde{u}') \quad \forall \tilde{v} \in V_0.$$

For $t = 0$
$$\tilde{u} = \tilde{u}_0 = u_0 - w(0), \qquad \tilde{u}' = \tilde{u}_1 = u_1 - w'(0), \quad (7.2.70)$$
and
$$(\psi, \tilde{v} - \tilde{u}') = (f, \tilde{v} - \tilde{u}') + \int_{\Gamma_F} F_i(\tilde{v}_i - \tilde{u}') d\Gamma \quad (7.2.71)$$

$$+ \int_{\Gamma_S} C_N(\tilde{v}_N - \tilde{u}'_N) d\Gamma - (\rho w'', \tilde{v} - \tilde{u}') - a(w, \tilde{v} - \tilde{u}').$$

We take now $U = 0$ on Γ_U for the sake of simplicity. Note that $H^1(\Omega) \to L^2(\Gamma)$ is compact and that $V = V_0 \cap \{v|v_T \in L^\infty(\Gamma_S)\}$ is dense in V for the H-norm. The following regularized finite dimensional problem (problem P_{en}) is considered now. Here β_ε is defined by (6.1.11) and V_n is a finite dimensional Galerkin subspace of V. Problem P_{en} reads:
Find a function $u_{en} : [0, T] \to V_n$ with $u'_{en}(t) \in V_0$ and $u''_{en}(t) \in [L^2(\Omega)]^2$ which satisfies the variational equality

$$(\rho u''_{en}, v) + a(u_{en}, v) + \int_{\Gamma_S} \beta_\varepsilon(u'_{Ten}) v_T d\Gamma = (\psi, v), \quad \forall v \in V_n \quad (7.2.72)$$

and the initial conditions (7.2.65).
We shall sketch further the basic steps of the existence proof, which also poses an open question.

Proposition 7.2.3 Suppose that (6.1.14) holds and that

$$f, f', f'' \in L^2(0T, [L^2(\Omega)]^2), \quad (7.2.73)$$

$$C_N, C'_N, C''_N \in L^2(0T, [L^2(\Gamma_S)]^2), \quad (7.2.74)$$

$$F, F', F'' \in L^2(0T, [L^2(\Gamma_F)]^2), \quad (7.2.75)$$

let β_ε have the regularity

$$\int_0^t \int_{\Gamma_S} \frac{d\beta_\varepsilon(v)}{dv_T}(v'_T)^2 d\Gamma dt \quad (7.2.76)$$

$$\geq -c \int_0^t |v'|_2^2 dt \quad \forall v(t) \in V_0, \ v'(t) \in [L^2(\Omega)]^2, 0 \leq t \leq T$$

where c const > 0.

$$u_0 \in [H^2(\Omega)]^2, \tag{7.2.77}$$

$$u_1 \in [H^1(\Omega)]^2, \tag{7.2.78}$$

and

$$v \to -\langle S_T(0), v_T\rangle_{\Gamma_S} + \int_\Omega [C_{ijhk}\varepsilon_{hk}(u_0)]_{,j} v_i d\Omega - \int_{\Gamma_S} \beta_e(u_{T1}) v_T d\Gamma \tag{7.2.79}$$

is a continuous linear functional on V_0 for the $[L^2(\Omega)]^2$-topology. Under these conditions the solution $\tilde{u}_{en}$ of the problem P_{en} exists and has the following properties: for $\varepsilon \to 0 \; n \to \infty$

$$u_{en} \to u \text{ in } L^\infty(0T, V_0) \text{ weakly}_*, \tag{7.2.80}$$

$$u'_{en} \to u' \text{ in } L^\infty(0T, V_0) \text{ weakly}_*, \tag{7.2.81}$$

$$u''_{en} \to u'' \text{ in } L^\infty(0T, [L^2(\Omega)]^3) \text{ weakly}_*; \tag{7.2.82}$$

u is a solution of problem (7.2.69)(7.2.70) (with $U = 0$ on Γ_U).

Proof. a) Let us put in (7.2.72) $v = u'_{en}$. By means of (6.1.15) we obtain that

$$-\rho_1 \rho_2 \text{mes } \Gamma_S + (\rho u''_{en}, u'_{en}) + a(u_{en}, u'_{en}) \leq (\psi, u'_{en}) \tag{7.2.83}$$

and we proceed exactly as in Prop. 7.2.1.

Finally the relation

$$|u'_{en}|_2^2 + \|u_{en}\|_1^2 \leq c\left(1 + \int_0^t (|u'_{en}|_2^2 + \|u_{en}\|_1^2) dt\right) \tag{7.2.84}$$

is obtained, from which it results, by applying Gronwall's inequality that

$$|u'_{en}|_2 \leq c, \quad \|u_{en}\| \leq c \tag{7.2.85}$$

(b) From (7.2.72), after "differentiating" with respect to t and by setting $u = u''_e$, we get the relation

$$(\rho u'''_{en}, u''_{en}) + a(u'_{en}, u''_{en}) + \int_{\Gamma_S} [\beta_e(u'_{Ten})]' u''_{Ten} d\Gamma = (\psi', u''_e). \tag{7.2.86}$$

Because of the assumption (7.2.76) it results that

$$\int_0^t \int_{\Gamma_S} [\beta_e(u'_{Ten})]' u''_{Ten} d\Gamma dt = \int_0^t \int_{\Gamma_S} \frac{d\beta_e(u'_{Ten})}{d(u'_{Ten})} (u''_{Ten})^2 d\Gamma dt \geq -\int_0^t c|u''_{en}|_2^2 dt \tag{7.2.87}$$

Then we proceed as in the proof of Prop. 7.2.1 and we get the inequality

$$|u''_{en}|_2^2 + ||u'_{en}||^2 \le c(||u_1||^2 + |u''_{en}(0)|_2^2) + \int_0^t (||u'_{en}||^2 + |u''_{en}|_2^2)dt. \quad (7.2.88)$$

(c) By setting $t = 0$ in (7.2.72) we obtain the relation

$$(\rho u''_{en}(0), v) + a(u_0, v) + \int_{\Gamma_S} \beta_e(u_{T1}) v_T d\Gamma = (\psi(0), v) \quad (7.2.89)$$

$$= (f(0), v) + \int_{\Gamma_F} F_i(0) v_i d\Gamma + \int_{\Gamma_S} C_N(0) v_N d\Gamma$$

But

$$a(u_0, v) = \int_\Omega C_{ijhk} \varepsilon_{ij}(u_0) \varepsilon_{hk}(v) d\Omega \quad (7.2.90)$$

$$= -\int_\Omega (C_{ijhk} \varepsilon_{hk}(u_0))_{,j} v_i d\Omega$$

$$+ \int_{\Gamma_F} F_i(0) v_i d\Gamma + \int_{\Gamma_S} C_N(0) v_N d\Gamma + \langle S_T(0), v_T \rangle_{\Gamma_S},$$

and, therefore,

$$(\rho u''_{en}(0), v) - \int_\Omega (C_{ijhk} \varepsilon_{hk}(u_0))_{,j} v_i d\Omega \quad (7.2.91)$$

$$+ \langle S_T(0), v_T \rangle_{\Gamma_S} + \int_{\Gamma_S} \beta_e(u_{T1}) v_T d\Gamma = (f(0), v).$$

From (7.2.79), (7.2.91) we find that

$$|u''_{en}(0)|_2 \le c. \quad (7.2.92)$$

Estimates (7.2.88) and (7.2.92) imply through Gronwall's inequality, that

$$|u''_{en}|_2 \le c, \quad ||u'_{en}|| \le c. \quad (7.2.93)$$

(d) From (7.2.85) and (7.2.93), it is concluded that the sequences $\{u_{en}\}$, $\{u'_{en}\}$, and $\{u''_{en}\}$ are bounded, the first two in the space $L^\infty(0T, V_0)$ and the last one in $L^\infty(0T, [L^2(\Omega)]^3)$ independently of ε and n. Thus we can determine subsequences, again denoted by $\{u_{en}\}, \{u'_{en}\}$, and $\{u''_{en}\}$, which are weakly$_*$ convergent. The limits u, u' and u'' are considered. In the next steps we will show that u, u' and u'' fulfill the hemivariational inequality (7.2.69) and the initial conditions (with $w = 0$).

(e) Further we can prove that the sequence $\{\beta_e(u'_{Ten})\}$ is weakly precompact in $L^1((0T) \times \Gamma_S)$. Accordingly

$$\beta_e(u'_{Ten}) \to \chi \quad \text{weakly in} \quad L^1((0T) \times \Gamma_S). \quad (7.2.94)$$

The rest of the proof is the same as in Prop. 7.2.1. For instance because of the $H^1(\Omega) \to L^2(\Gamma)$ compactness we may apply Egoroff's theorem to show as in the proof of Theorem 6.1.1 that

$$\chi \in \bar{\partial} j(u'_T) \quad \text{a.e. on} \quad \Gamma_S \times [0,T]. \tag{7.2.95}$$

From (7.2.72), after passing to the limit $\varepsilon \to 0$ $n \to \infty$, and from (7.2.95) we obtain the hemivariational inequality

$$\int_0^T [(\rho u'', v - u') + a(u, v - u') + \int_{\Gamma_S} j^0(u'_T, v_T - u'_T) d\Gamma \tag{7.2.96}$$
$$- (\psi, v - u')] dt \geq 0 \quad \forall v \in L^2(0T, V_0)$$

where $u' \in L^\infty(0T, V_0)$. Here (7.2.96) can be considered as the expression of Hamilton's principle for the present problem. From (7.2.96), the pointwise inequality (7.2.69) (with $w = 0$) results. q.e.d.

We shall briefly discuss in what sense the variational solution of the problem satisfies the relations (7.2.58)÷(7.2.64). We proceed as in the static case (cf. Sect. 4.1.). Thus, it is easily concluded that a variational solution satisfies (7.2.58) in the sense of distributions over Ω, and (7.2.62)÷(7.2.64) in the sense of traces. In the above existence proof we cannot prove that all Galerkin sequences converge to the solution. Moreover the solution is nonunique due to the lack of convexity.

Concerning the uniqueness of the solution we can formulate an analogous result to the one of Prop. 7.2.2.

Note that the regularity assumption (7.2.76) is quite unnatural but for the present problem it cannot be replaced by a more natural assumption of the type (7.2.20). Indeed if (7.2.20) would hold then one could write by means of (6.1.11) that

$$\int_{\Gamma_S} [\beta_\varepsilon(u'_{Ten})]' u''_{Ten} d\Gamma = \int_{\Gamma_S} \frac{d\beta_\varepsilon(u'_{Ten})}{du'_{Ten}} (u''_{Ten})^2 d\Gamma \tag{7.2.97}$$
$$\geq -c \int_{\Gamma_S} (u''_{Ten})^2 d\Gamma \geq -c\|u''_{en}\|^2.$$

Then in the right hand side of (7.2.88) the additional term $c \int_0^t \|u''_{en}\|^2 dt$ appears, where $\|\cdot\|$ is the $[H^1(\Omega)]^2$-norm, and this fact does not permit the derivation of the estimates (7.2.93). Thus the existence of the solution is still an open problem if an assumption of the type (7.2.20) holds.

7.3 Applications to Engineering Problems: Von Kármán Plates and Thermoelasticity

i) Dynamic hemivariational inequalities in the theory of von Kármán plates.

7.3 Applications to Engineering Problems: Von Kármán Plates and Thermoelasticity

Within the framework of Sect. 4.2, the dynamic behaviour of von Kármán plates is governed by the folowing system of differential equations ($\alpha, \beta = 1, 2$):

$$\rho \zeta'' + K \triangle \triangle \zeta - D \triangle \zeta'' - h(\sigma_{\alpha\beta}\zeta_{,\beta})_{,\alpha} = f_3 \quad \text{in} \quad \Omega \times (0, T), \tag{7.3.1}$$

$$\rho u_\alpha'' - \sigma_{\alpha\beta,\beta} = f_\alpha \quad \text{in} \quad \Omega \times (0, T), \tag{7.3.2}$$

$$\sigma_{\alpha\beta} = C_{\alpha\beta\gamma\delta}(\varepsilon_{\gamma\delta}(u) + \frac{1}{2}\zeta_{,\gamma}\zeta_{,\delta}) \quad \text{in} \quad \Omega \times (0, T), \tag{7.3.3}$$

$$\varepsilon_{\alpha\beta}(u) = \frac{1}{2}(u_{\alpha,\beta} + u_{\beta,\alpha}) \quad \text{in} \quad \Omega \times (0, T). \tag{7.3.4}$$

This system originates formally from (4.2.21)÷(4.2.23) by adding the inertia terms (the rotatory inertia terms corresponding to u_1 and u_2 have been neglected). Here t is again the time variable, $[0, T]$ the time interval, ρ the plate density, ($\rho \in L^\infty(\Omega)$, ess inf $\rho(x) > 0$) and $f = (f_1, f_2, f_3)$, $f_i = f_i(x, t)$, the load vector of the plate. $D = \rho h^3/12$ is the product of the density and the moment of inertia of a unit-width strip cut from the plate. For $t = 0$ the following initial conditions hold:

$$u = u_0, \tag{7.3.5}$$

$$u' = u_1, \tag{7.3.6}$$

$$\zeta = \zeta_0, \tag{7.3.7}$$

and

$$\zeta' = \zeta_1, \tag{7.3.8}$$

where u_0, u_1, ζ_0 and ζ_1 are functions of x. First, the relations analogous to (4.2.33), (4.2.37) are formulated on the assumption of sufficiently regular functions. They have the form

$$\int_\Omega \rho \zeta''(z - \zeta')d\Omega + \bar{\alpha}(\zeta'', z - \zeta') + \alpha(\zeta, z - \zeta') \tag{7.3.9}$$

$$+ \int_\Omega h\sigma_{\alpha\beta}\zeta_{,\alpha}(z - \zeta')_{,\beta}d\Omega$$

$$= \int_\Gamma h\sigma_{\alpha\beta}\zeta_{,\alpha}n_\beta(z - \zeta')d\Gamma + \int_\Gamma K_n(\zeta)(z - \zeta')d\Gamma$$

$$- \int_\Gamma M_n(\zeta)\frac{\partial(z - \zeta')}{\partial n}d\Gamma$$

$$+ \int_\Omega f_3(z - \zeta')d\Omega + D\int_\Gamma \frac{\partial \zeta''}{\partial n}(z - \zeta')d\Gamma,$$

where $\alpha(\zeta, z), M_n(\zeta)$ and $K_n(\zeta)$ are given by (4.2.34)÷(4.2.36) and

$$\bar{\alpha}(\zeta'', z) = D \int_\Omega \zeta''_{,\alpha} z_{,\alpha} d\Omega, \quad \alpha = 1, 2. \tag{7.3.10}$$

7. Eigenvalue and Dynamic Problems

Similarly, from (7.3.2) we get, by means of the Green-Gauss theorem, the relation

$$\int_\Omega \rho u''_\alpha (v_\alpha - u'_\alpha) d\Omega + \int_\Omega \sigma_{\alpha\beta}\varepsilon_{\alpha\beta}(v-u')d\Omega \qquad (7.3.11)$$

$$= \int_\Omega f_\alpha(v_\alpha - u'_\alpha)d\Omega + \int_\Gamma \sigma_{\alpha\beta}n_\beta(v_\alpha - u'_\alpha)d\Gamma.$$

The following boundary conditions are assumed to hold pointwise on Γ for every $t \in [0, T]$:

$$M_n \in \tilde{\beta}_1\left(\frac{\partial \zeta'}{\partial n}\right) = \bar{\partial}j_1\left(\frac{\partial \zeta'}{\partial n}\right); \qquad (7.3.12)$$

$$-K_n \in \tilde{\beta}_2(\zeta') = \bar{\partial}j_2(\zeta'); \qquad (7.3.13)$$

where j_1 and j_2 are locally Lipschitz nonconvex superpotentials resulting from $\beta_1, \beta_2 \in L^\infty_{loc}(\mathbb{R})$ through (1.2.54) and

$$\sigma_{1\alpha}n_\alpha = \sigma_{2\beta}n_\beta = 0. \qquad (7.3.14)$$

Moreover, f_3 is split into $\tilde{f}_3$ and $\bar{\bar{f}}_3$, where $\bar{\bar{f}}_3$ is given and

$$-\tilde{f}_3 \in \tilde{\beta}_3(\zeta') = \bar{\partial}j_3(\zeta') \text{ in } \Omega' \subset \Omega, \qquad (7.3.15)$$
$$\bar{\Omega}' \cap \Gamma = \emptyset, \quad \tilde{f}_3 = 0 \text{ in } \Omega - \Omega'.$$

β_3 and j_3 have the same properties as β_1 and β_2 and j_1, j_2. In order to give a variational formulation of the B.V.P. defined by (7.3.1)÷(7.3.8) and (7.3.12)÷(7.3.15) we assume that $\rho \in L^\infty(\Omega), z \in H^2(\Omega), v \in [H^1(\Omega)]^2$ and $f_i(t) \in L^2(\Omega), i = 1, 2$, and $\bar{\bar{f}}_3(t) \in L^2(\Omega)$ for every $t \in [0, T]$. Taking into account the definition of the generalized gradient we are led to the following problem:

Find $u : [0, T] \to [H^1(\Omega)]^2, \zeta : [0, T] \to H^2(\Omega)$, with $u'(t) \in [H^1(\Omega)]^2$ and $\zeta'(t) \in H^2(\Omega), u''(t) \in [L^2(\Omega)]^2, \zeta''(t) \in L^2(\Omega)$ such as to satisfy the variational system (for the notation see Sect. 4.2)

$$(\rho\zeta'', z - \zeta') + \alpha(\zeta, z - \zeta') + \bar{\alpha}(\zeta'', z - \zeta')$$
$$+ hR(\varepsilon(u) + \frac{1}{2}P(\zeta), P(\zeta, z - \zeta')) + \int_\Gamma j_1^0\left(\frac{\partial \zeta'}{\partial n}, \frac{\partial (z - \zeta')}{\partial n}\right) d\Gamma$$
$$+ \int_\Gamma j_2^0(\zeta', z - \zeta')d\Gamma + \int_{\Omega'} j_3^0(\zeta', z - \zeta')d\Omega \geq (\bar{\bar{f}}_3, z - \zeta')$$
$$+ D\int_\Gamma \frac{\partial \zeta''}{\partial n}(z - \zeta')d\Gamma, \quad \forall z \in H^2(\Omega) \qquad (7.3.16)$$

$$(\rho u'', v - u') + R(\varepsilon(u) + \frac{1}{2}P(\zeta), \varepsilon(v - u')) \qquad (7.3.17)$$
$$= (f_e, v - u'), \quad \forall v \in [H^1(\Omega)]^2$$

7.3 Applications to Engineering Problems: Von Kármán Plates and Thermoelasticity

and the initial conditions (7.3.5)÷(7.3.8). Here,

$$f_e = (f_1, f_2). \tag{7.3.18}$$

It is also possible to consider nonmonotone boundary conditions with respect to $S = \{S_\alpha\} = \{\sigma_{\alpha\beta} n_\beta\}$ and u'; for instance, we can assume that on Γ (7.3.12) holds, as do the boundary conditions

$$-K_n \in \bar{\partial} j_2(\zeta') \text{ on } \Gamma_1, \qquad \zeta = 0 \text{ on } \Gamma - \Gamma_1 \tag{7.3.19}$$

and

$$S = 0 \text{ on } \Gamma_1, \qquad -S \in \bar{\partial} j_4(u') \text{ on } \Gamma - \Gamma_1. \tag{7.3.20}$$

In addition, let (7.3.15) hold as well. Here j_4 is again a locally Lipschitz functional on $\mathbb{R}^2$. Then the spaces $Z = \{z | z \in H^2(\Omega), z = 0 \text{ on } \Gamma - \Gamma_1\}$ and $V = \{v | v \in [H^1(\Omega)]^2\}$ are introduced. A pair (u, ζ) is a variational solution of this B.V.P. if and only if it satisfies the following problem:
Find $u : [0, T] \to V$ and $\zeta : [0, T] \to Z$ with $u'(t) \in V$, $\zeta'(t) \in Z$ and $u''(t) \in [L^2(\Omega)]^2$, $\zeta''(t) \in L^2(\Omega)$, such as to satisfy

$$\begin{aligned}(\rho\zeta'', z - \zeta') &+ a(\zeta, z - \zeta') + \bar{a}(\zeta'', z - \zeta') \\ &+ hR(\varepsilon(u) + \frac{1}{2}P(\zeta), P(\zeta, z - \zeta')) + \int_\Gamma j_1^0\left(\frac{\partial \zeta'}{\partial n}, \frac{\partial(z - \zeta')}{\partial n}\right) d\Gamma \\ &+ \int_{\Gamma_1} j_2^0(\zeta', z - \zeta')d\Gamma + \int_{\Omega'} j_3^0(\zeta', z - \zeta')d\Omega \geq (\bar{\bar{f}}_3, z - \zeta') \\ &+ D\int_\Gamma \frac{\partial \zeta''}{\partial n}(z - \zeta')d\Gamma, \quad \forall z \in Z \end{aligned} \tag{7.3.21}$$

$$\begin{aligned}(\rho u'', v - u') &+ R(\varepsilon(u) + \frac{1}{2}P(\zeta), \varepsilon(v - u')) + \int_{\Gamma - \Gamma_1} j_4^0(u', v - u')d\Gamma \\ &\geq (f_e, v - u'), \quad \forall v \in V \end{aligned} \tag{7.3.22}$$

and the initial conditions (7.3.5-8).

The existence of the solution of the system of the hemivariational inequalities (7.3.21) and (7.3.22) has not yet been proved. Here, we shall study the variational system (7.3.16)(7.3.17), where (7.3.13) is replaced by the condition $\zeta = 0$ on Γ and (7.3.12) by the condition $\frac{\partial \zeta}{\partial n} = 0$ on Γ. We call this variational system problem P. Thus we study a system of a hemivariational inequality and a variational equality. The aforementioned boundary conditions and (7.3.14) lead us to the following functional framework. Let

$$V = [H^1(\Omega)]^2 \subset V_0 = [L^2(\Omega)]^2 \equiv V_0' \subset V', \tag{7.3.23}$$

$$Z = \overset{0}{H^2}(\Omega) \subset Z_1 = H^1(\Omega) \subset L^2(\Omega) = Z_0 \equiv Z_0' \subset Z_1' \subset Z', \tag{7.3.24}$$

where the imbeddings are dense. Moreover, the injection $Z \subset L^\infty(\Omega)$ is compact and the bilinear forms $\alpha(z,z)$ and $\bar{\alpha}(z,z)$ are symmetric, continuous and coercive on Z and Z_1 respectively. We define $R(h,k)$ as in (4.2.38) and we may verify that the applications ε, P and R have the properties (cf. [Pan85] p.242)

(i) $\varepsilon : V \to [L^2(\Omega)]^4$ is linear and continuous; (7.3.25)

(ii) $P : Z \times Z \to [L^2(\Omega)]^4$ is bilinear, and completely continuous; (7.3.26)

(iii) $\|P(\zeta,\zeta)\|_{[L^2(\Omega)]^4} \leq c\|\zeta\|_Z \|\zeta\|_{Z_1}$, $\forall \zeta \in Z$; and (7.3.27)

(iv) for $\zeta_i \to \zeta$ in $L^\infty(0T, Z)$ weakly$_*$ and $\zeta_i' \to \zeta'$ in $L^\infty(0T, Z_0)$ weakly$_*$

$$R(P(\zeta_i), P(\zeta_i, z)) \to R(P(\zeta), P(\zeta, z)) \text{ in } D'(0,T), \quad \forall z \in L^2(0T, Z). \quad (7.3.28)$$

Here $D'(0T)$ denotes the space of distribution over $(0,T)$.
In order to prove the existence of the solution we consider as in the static case (Ch.6) the regularized finite dimensional problem $P_{\varepsilon n}$. Here $\beta_{3\varepsilon}$ is defined through the mollifier as in (6.1.11). Moreover we consider Galerkin bases in Z and in V and let us denote the corresponding finite dimensional subspaces by Z_n and V_n respectively. We recall here the compact injection $H^2(\Omega) \subset L^\infty(\Omega)$. Thus problem $P_{\varepsilon n}$ reads:
Find $u_{\varepsilon n} : [0,T] \to V_n$ and $\zeta_{\varepsilon n} : [0,T] \to Z_n$ with $u'_{\varepsilon n}(t) \in V, u''_{\varepsilon n}(t) \in V_0$ and $\zeta'_{\varepsilon n}(t) \in Z, \zeta''_{\varepsilon n}(t) \in Z_1$, which satisfy the equations

$$(\rho \zeta''_{\varepsilon n}, z) + \alpha(\zeta_{\varepsilon n}, z) + \bar{\alpha}(\zeta''_{\varepsilon n}, z) + hR(\varepsilon(u_{\varepsilon n}) + \frac{1}{2}P(\zeta_{\varepsilon n}), P(\zeta_{\varepsilon n}, z))$$
$$+ \int_{\Omega'} \beta_{3\varepsilon}(\zeta'_{\varepsilon n}) z \, d\Omega = (f_3, z) \quad \forall z \in Z_n \quad (7.3.29)$$

$$(\rho u''_{\varepsilon n}, v) + R(\varepsilon(u_{\varepsilon n}) + \frac{1}{2}P(\zeta_{\varepsilon n}), \varepsilon(v)) = (f_e, v), \quad \forall v \in V_n \quad (7.3.30)$$

and the initial conditions at $t = 0$

$$u_{\varepsilon n} = u_0, \quad u'_{\varepsilon n} = u_1, \quad \zeta_{\varepsilon n} = \zeta_0, \quad \zeta'_{\varepsilon n} = \zeta_1. \quad (7.3.31)$$

Now the following propositions holds ($\bar{R}$ is defined in (4.2.45)).

Proposition 7.3.1 Suppose that (6.1.14) holds for β_3 and let

$$f_e, f'_e \in L^2(0T, V_0), (f_e(t), \bar{r}) = 0 \quad \forall \bar{r} \in \bar{R} \quad \forall t \in (0,T), \quad (7.3.32)$$

$$f_3, f'_3 \in L^2(0T, Z_0), \quad (7.3.33)$$

$$u_0, u_1 \in V \quad (7.3.34)$$

$$\zeta_0 \in Z \quad (7.3.35)$$

$$\zeta_1 \in Z. \quad (7.3.36)$$

7.3 Applications to Engineering Problems: Von Kármán Plates and Thermoelasticity

Moreover let

$$\frac{\beta_3(\xi + h) - \beta_3(\xi)}{h} \geq -c_i \quad \forall \xi \in \mathbb{R} \quad \forall h > 0, \quad c \text{ const} > 0. \tag{7.3.37}$$

We assume further that the linear form

$$z \to F_1(z) = -\alpha(\zeta_0, z) - hR(\varepsilon(u_0) + \frac{1}{2}P(\zeta_0), P(\zeta_0, z)) - \int_{\Omega'} \beta_{3\varepsilon}(\zeta_1) z d\Omega, \quad \varepsilon > 0 \tag{7.3.38}$$

is continuous on Z for the Z_1-topology, and finally that the linear form

$$v \to F_2(v) = -R(\varepsilon(u_0) + \frac{1}{2}P(\zeta_0), \varepsilon(v)), \tag{7.3.39}$$

is continuous on V for the V_0-topology. Then a solution $u_{\varepsilon n}$, $\zeta_{\varepsilon n}$ of problem $P_{\varepsilon n}$ exists. Moreover, a $\zeta \in L^\infty(0T, Z)$ and a $u \in L^\infty(0T, V_0)$ exist such that for $\varepsilon \to 0$ $n \to \infty$

$$\zeta_{\varepsilon n} \to \zeta, \quad \zeta'_{\varepsilon n} \to \zeta' \text{ in } L^\infty(0T, Z) \text{ weakly}_*, \tag{7.3.40}$$

$$\zeta''_{\varepsilon n} \to \zeta'' \text{ in } L^\infty(0T, Z_1) \text{ weakly}_*, \tag{7.3.41}$$

$$u_{\varepsilon n} \to u, \quad u'_{\varepsilon n} \to u', \quad u''_{\varepsilon n} \to u'' \text{ in } L^\infty(0T, V_0) \text{ weakly}_*, \tag{7.3.42}$$

and

$$\varepsilon(u_{\varepsilon n}) \to \varepsilon(u), \quad \varepsilon(u'_{\varepsilon n}) \to \varepsilon(u') \text{ in } L^\infty(0T, [L^\infty(\Omega)]^4) \text{ weakly}_*. \tag{7.3.43}$$

The functions u and ζ are solutions of problem P.

Proof. The proof follows the same steps as the proof of Prop. 7.2.1. We refer also to the proof of the analogous variational inequality problem given in [Pan85] p.243-247.

(a) Let us put $z = \zeta'_{\varepsilon n}$ into (7.3.29) and $u = u'_{\varepsilon n}$ into (7.3.30). Thus we obtain the relations

$$(\rho \zeta''_{\varepsilon n}, \zeta'_{\varepsilon n}) + \alpha(\zeta_{\varepsilon n}, \zeta'_{\varepsilon n}) + \bar{\alpha}(\zeta''_{\varepsilon n}, \zeta'_{\varepsilon n}) \tag{7.3.44}$$
$$+ hR(\varepsilon(u_{\varepsilon n}) + \frac{1}{2}P(\zeta_{\varepsilon n}), P(\zeta_{\varepsilon n}, \zeta'_{\varepsilon n})) + \int_{\Omega'} \beta_{3\varepsilon}(\zeta'_{\varepsilon n})\zeta'_{\varepsilon n} d\Omega = (f_3, \zeta'_{\varepsilon n})$$

and

$$(\rho u''_{\varepsilon n}, u'_{\varepsilon n}) + R(\varepsilon(u_{\varepsilon n}) + \frac{1}{2}P(\zeta_{\varepsilon n}), \varepsilon(u'_{\varepsilon n})) = (f_e, u'_{\varepsilon n}). \tag{7.3.45}$$

By means of the identity

$$R(\varepsilon(u_{\varepsilon n}) + \frac{1}{2}P(\zeta_{\varepsilon n}), P(\zeta_{\varepsilon n}, \zeta'_{\varepsilon n})) = \frac{1}{2}\frac{d}{dt}R(\varepsilon(u_{\varepsilon n}) + \frac{1}{2}P(\zeta_{\varepsilon n}), \varepsilon(u_{\varepsilon n})$$
$$+ \frac{1}{2}P(\zeta_{\varepsilon n})) - R(\varepsilon(u_{\varepsilon n}) + \frac{1}{2}P(\zeta_{\varepsilon n}), \varepsilon(u'_{\varepsilon n})), \tag{7.3.46}$$

and applying (6.1.15) to β_{3e}, there results from (7.3.44)(7.3.45), through integration from 0 to t, the relation

$$\frac{1}{2}(||\zeta'_{en}||^2_{Z_0} + c||\zeta_{en}||^2_{Z} + c||\zeta'_{en}||^2_{Z_1}$$
$$+ hR(\varepsilon(u_{en}) + \frac{1}{2}P(\zeta_{en}), \varepsilon(u_{en}) + \frac{1}{2}P(\zeta_{en})) + h||u'_{en}||^2_{V_0})$$
$$- \int_0^t \rho_1 \rho_2 \text{mes}\,\Omega_1 dt \le \int_0^t [(f_3, \zeta'_{en}) + h(f_e, u'_{en})] dt \qquad (7.3.47)$$
$$+ \frac{1}{2}(||\zeta_1||^2_{Z_0} + c||\zeta_0||^2_{Z} + c||\zeta_1||^2_{Z_1}$$
$$+ R(\varepsilon(u_0) + \frac{1}{2}P(\zeta_0), \varepsilon(u_0) + \frac{1}{2}P(\zeta_0)) + h||u_1||^2_{V_0}).$$

To derive (7.3.47), we have used the continuity and the coerciveness of the bilinear forms and the continuity of the linear forms. Moreover ρ_1 and ρ_2 correspond to β_3 according to (6.1.15). From (7.3.47) the inequality

$$||\zeta_{en}||^2_Z + ||\zeta'_{en}||^2_{Z_1} + ||u'_{en}||^2_{V_0} + ||\varepsilon(u_{en}) + \frac{1}{2}P(\zeta_{en})||^2_{[L^2(\Omega)]^4} \qquad (7.3.48)$$
$$\le c_1 \left[\int_0^t (||\zeta_{en}||^2_Z + ||\zeta'_{en}||^2_{Z_1} + ||u'_{en}||^2_{V_0} \right.$$
$$\left. + ||\varepsilon(u_{en}) + \frac{1}{2}P(\zeta_{en})||^2_{[L^2(\Omega)]^4}) dt \right] + c_2$$
$$+ \int_0^t \rho_1 \rho_2 \text{mes}\,\Omega_1 dt \quad c_1, c_2 \text{ const} > 0$$

results. From (7.3.48), by applying Gronwall's inequality, we obtain (ρ_1, ρ_2, are constants independent of t) that

$$||\zeta_{en}||^2_Z \le c, \quad ||\zeta'_{en}||_{Z_1} \le c, \quad ||u'_{en}||_{V_0} \le c, \quad ||\varepsilon(u_{en}) + \frac{1}{2}P(\zeta_{en})||_{[L^2(\Omega)]^4} \le c.$$
$$(7.3.49)$$

Accordingly, the sequences $\{\zeta_{en}\}, \{\zeta'_{en}\}, \{u_{en}\}, \{u'_{en}\}$, and $\{\varepsilon(u_{en})\}$ are bounded in the spaces $L^\infty(0T, Z), L^\infty(0T, Z_1), L^\infty(0T, V_0), L^\infty(0T, V_0)$, and $L^\infty(0T, [L^2(\Omega)]^4)$ respectively.

(b) From (7.3.29), (7.3.30), by differentiating formally with respect to t we obtain the system of variational equalities

$$(\rho\zeta'''_{en}, z) + a(\zeta'_{en}, z) + \bar{a}(\zeta'''_{en}, z) \qquad (7.3.50)$$
$$+ hR(\varepsilon(u'_{en}) + P(\zeta_{en}, \zeta'_{en}), P(\zeta_{en}, z))$$
$$+ hR(\varepsilon(u_{en}) + \frac{1}{2}P(\zeta_{en}), P(\zeta'_{en}, z))$$
$$+ \int_{\Omega'} [\beta_{3e}(\zeta'_{en})]' z d\Omega = (f'_3, z), \quad \forall z \in Z$$

7.3 Applications to Engineering Problems: Von Kármán Plates and Thermoelasticity 211

and

$$(\rho u_{en}''', v) + R(\varepsilon(u_{en}') + P(\zeta_{en}', \zeta_{en}), \varepsilon(v)) = (f_e', v), \quad \forall v \in V. \tag{7.3.51}$$

Because of (7.3.37) we have that (cf. (7.2.34))

$$\int_{\Omega'} [\beta_{3\varepsilon}(\zeta_{en}')]' \zeta_{en}'' d\Omega \geq -c||\zeta_{en}''||^2_{L^2(\Omega')} \geq -\bar{c}||\zeta_{en}''||^2_{Z_0} \tag{7.3.52}$$

Let us put $z = \zeta_{en}''$ and $v = u_{en}''$ into (7.3.50) and into (7.3.51) respectively. Then the inequality

$$\frac{1}{2}\frac{d}{dt}F(t) \leq (f_3', \zeta_{en}'') + \bar{c}||\zeta_{en}''||^2_{Z_0} + h(f_e', u_{en}'') \tag{7.3.53}$$
$$+ hR(\varepsilon(u_{en}') + P(\zeta_{en}, \zeta_{en}'), P(\zeta_{en}'))$$
$$- hR(\varepsilon(u_{en}) + \frac{1}{2}P(\zeta_{en}), \frac{1}{2}\frac{d}{dt}P(\zeta_{en}')),$$

results with

$$F(t) = \int_\Omega \rho(\zeta_{en}'')^2 d\Omega + \alpha(\zeta_{en}', \zeta_{en}') + \bar{\alpha}(\zeta_{en}'', \zeta_{en}'') \tag{7.3.54}$$
$$+ hR(\varepsilon(u_{en}') + P(\zeta_{en}, \zeta_{en}'), \varepsilon(u_{en}') + P(\zeta_{en}, \zeta_{en}')) + h\int_\Omega \rho(u_{en}'')^2 d\Omega.$$

Integrating (7.3.54) from 0 to t yields, by means of (7.3.27), (7.3.49) the inequality

$$F(t) \leq 2F(0) + \int_0^t F(t)dt + c\int_0^t (||f_3'||_{Z_{1'}} + ||f_e'||_{V_0})dt \tag{7.3.55}$$
$$\leq 2F(0) + c\int_0^t F(t)dt + c.$$

The meaning of (7.3.38) and (7.3.39) is that there exist $z_1 \in Z_1'$ and $v_1 \in V_0$ respectively such that

$$F_1(z) = \langle z_1, z \rangle_{Z_1}, \quad \forall z \in Z \tag{7.3.56}$$

and

$$F_2(v) = (v_1, v)_{V_0}, \quad \forall v \in V. \tag{7.3.57}$$

By means of (7.3.56) the variational equality (7.3.29) yields

$$(\rho \zeta_{en}''(0), z) + \bar{\alpha}(\zeta_{en}''(0), z) = -\alpha(\zeta_0, z) - hR(\varepsilon(u_0) + \frac{1}{2}P(\zeta_0), P(\zeta_0, z))$$
$$- \int_{\Omega'} \beta_{3\varepsilon}(\zeta_1) z d\Omega + (f_3(0), z) = \langle f_3(0) + z_1, z \rangle_{Z_1} \tag{7.3.58}$$

for $t = 0$, and thus
$$\|\zeta''_{en}(0)\|_{Z_1} \le c. \tag{7.3.59}$$
Similarly, for $t = 0$ (7.3.30) implies by means of (7.3.57) the inequality
$$\|u''_{en}(0)\|_{V_0} \le c. \tag{7.3.60}$$
Accordingly,
$$F(0) \le c, \tag{7.3.61}$$
and thus Gronwall's inequality can be applied to (7.3.55). It yields that $F(t) \le c$, or equivalently that
$$\|\zeta'_{en}\|_Z \le c, \quad \|\zeta''_{en}\|_{Z_1} \le c, \quad \|u''_{en}\|_{V_0} \le c, \quad \|\varepsilon(u'_{en})\|_{[L^2(\Omega)]^4} \le c. \tag{7.3.62}$$

Hence the sequences $\{\zeta'_{en}\}, \{\zeta''_{en}\}, \{u''_{en}\}$ and $\{\varepsilon(u'_{en})\}$ are bounded in the spaces $L^\infty(0T, Z), L^\infty(0T, Z_1), L^\infty(0T, V_0)$ and $L^\infty(0T, [L^2(\Omega)]^4)$ respectively.

(c) From (7.3.49) and (7.3.62), by considering subsequences if necessary, we obtain that as $\varepsilon \to 0\ n \to \infty$ (7.3.40)÷(7.3.43) hold. The rest of the proof is exactly the same as in the proof of Prop. 7.2.1. q.e.d.

ii) Dynamic Hemivariational Inequalities in Linear Thermoelasticity.

Let us consider a thermoelastic medium imbedded in the Euclidean space $\mathbb{R}^3$. A point is denoted by x and its coordinates in reference to a fixed Cartesian coordinate system $0x_1x_2x_3$ by x_i, $i = 1, 2, 3$. The time variable t takes values over $[0, T] \subset \mathbb{R}$. We further denote by $u = u(x, t)$ the displacement of the material point x at time t with reference to the natural state of the body, which is characterized by zero stresses and constant absolute temperature $\theta_0 > 0$. The density at point x of the natural state is denoted by $\rho = \rho(x)$ and the open, bounded, connected subset of $\mathbb{R}^3$ occupied by the body is denoted by Ω. The boundary Γ of Ω is assumed to be regular. The behaviour of a linear thermoelastic body is governed by the following constitutive equations for the stress tensor $\sigma = \{\sigma_{ij}\}, i = 1, 2, 3$, and the specific entropy deviation $\eta - \eta_0$ (η_0 is the specific entropy of the natural state)

$$\sigma_{ij} = t_{ij} - m_{ij}(\theta - \theta_0) = C_{ijhk}\varepsilon_{hk} - m_{ij}(\theta - \theta_0), \tag{7.3.63}$$

$$\eta - \eta_0 = \frac{1}{\theta_0}c_D(\theta - \theta_0) + \frac{1}{\rho}m_{ij}\varepsilon_{ij}. \tag{7.3.64}$$

Here $\theta = \theta(x, t)$ is the absolute temperature, and $\varepsilon = \{\varepsilon_{ij}\}$ the strain tensor which is related to the displacements by classical formula

$$\varepsilon_{ij}(u) = \frac{1}{2}(u_{i,j} + u_{j,i}) \tag{7.3.65}$$

for a small deformation theory. $C = \{C_{ijhk}\}$, $i, j, h, k = 1, 2, 3$, is the elasticity tensor satisfying the well-known symmetry and ellipticity conditions, $m = \{m_{ij}\}$ is the symmetry tensor of thermal expansion, and $c_D = c_D(x) > 0$ is the specific heat at zero strain of the body. $C(x), m(x)$ and $c_D(x)$ are referred to the natural state of the body. The equations of motion assuming small displacements and the law of conservation of energy read

$$\rho u_i'' = \sigma_{ij,j} + f_i, \quad (7.3.66)$$

and

$$\rho \theta_0 \eta' = -q_{i,i} + Q, \quad (7.3.67)$$

where $f = \{f_i\}, f_i = f_i(x,t)$, is the volume force vector, $q = \{q_i\}, q_i = q_i(x,t)$, is the heat flux vector and $Q = Q(x,t)$ is the radiant heating per unit volume. Further we introduce Fourier's law of heat conduction. It reads

$$q_i = -k_{ij}\theta_{,j}, \quad (7.3.68)$$

where $k = \{k_{ij}\}, k_{ij} = k_{ij}(x)$ is the symmetric tensor of thermal conductivity. k refers to the natural state of the body and satisfies the condition

$$k_{ij}a_i a_j \geq c a_i a_i, \quad \forall a = \{a_i\} \in \mathbb{R}^3, \quad c \text{ const} > 0. \quad (7.3.69)$$

The above relations lead to the following system of differential equations:

$$\rho u_i'' = f_i + (C_{ijhk}\varepsilon_{hk})_{,j} - (m_{ij}(\theta - \theta_0))_{,j} \text{ in } \Omega \times (0,T), \quad (7.3.70)$$

$$\rho c_D \theta' - (k_{ij}\theta_{,j})_{,i} + m_{ij}\theta_0 \varepsilon_{ij}' = Q \text{ in } \Omega \times (0,T). \quad (7.3.71)$$

These last two differential equations describe the linear thermoelastic behaviour of a generally nonhomogeneous and nonisotropic body. For $t = 0$ the following initial conditions hold:

$$u_i = u_{0i} \text{ in } \Omega, \quad (7.3.72)$$

$$u_i' = u_{1i} \text{ in } \Omega \quad (7.3.73)$$

and

$$\theta = \bar{\theta} \text{ in } \Omega. \quad (7.3.74)$$

Here $u_0 = u_0(x), u_1 = u_1(x)$ and $\bar{\theta} = \bar{\theta}(x)$ are given functions on Ω.

The first B.V.P. which we will formulate results if between the boundary temperature and the heat flux the relation

$$q_i n_i = -k_{ij}\theta_{,j}n_i \in \tilde{\beta}(\theta) = \bar{\partial}j(\theta) \text{ on } \Gamma_1 \times (0,T), \quad (7.3.75a)$$

holds, where $\Gamma_1 \subset \Gamma$ and

$$\theta = 0 \text{ on } \Gamma - \Gamma_1. \quad (7.3.75b)$$

For the displacements, we assume simply that

$$u_i = 0 \text{ on } \Gamma \times (0,T). \quad (7.3.76)$$

Here $n = \{n_i\}$ denotes, as usual, the unit normal to Γ directed towards the exterior of Ω, and $\tilde{\beta}$ is a multivalued function obtained from a discontinuous function $\beta \in L^\infty_{loc}(\mathbb{R})$ as in (1.2.53). Then the locally Lipschitz function j satisfies (1.2.54),(1.2.55). The boundary condition (7.3.75) describes a broad class of temperature control problem (cf. Sect. 2.4).

The second B.V.P. which we formulate results if instead of (7.3.75),(7.3.76) we consider the boundary conditions

$$\theta = \theta_0 \quad \text{on} \quad \Gamma \times (0,T), \tag{7.3.77}$$

$$u_i = U_i \quad \text{on} \quad \Gamma_U \times (0,T) \tag{7.3.78}$$

and

$$-S = \{-S_i\} = \{-\sigma_{ij}n_i\} \in \bar{\partial}\psi(u') \quad \text{on} \quad \Gamma_S \times (0,T). \tag{7.3.79}$$

Here $\Gamma = \bar{\Gamma}_U \cup \bar{\Gamma}_S$, where Γ_U and Γ_S are nonempty, disjoint, open sets, $U_i = U_i(x,t)$ is a prescribed displacement vector on Γ_U, assumed to be compatible with the initial conditions (7.3.72)÷(7.3.74) and ψ is a locally Lipschitz generally nonconvex functional on $\mathbb{R}^3$. A more general B.V.P. may result by replacing the boundary condition (7.3.77) with (7.3.75); in this case, we obtain, as it is obvious, a system of two hemivariational inequalities whereas in the two previous B.V.Ps., a system of a hemivariational inequality and a variational equality is obtained.

We introduce now the following notation:

$$a(u,v) = \int_\Omega C_{ijhk}\varepsilon_{ij}(u)\varepsilon_{hk}(v)d\Omega; \tag{7.3.80}$$

$$(u,v) = \int_\Omega u_i v_i d\Omega, \tag{7.3.81}$$

$$M_1(\theta,v) = \int_\Omega (m_{ij}\theta)_{,j} v_i d\Omega; \tag{7.3.82}$$

$$M_2(u,\varphi) = \int_\Omega m_{ij} u_{i,j} \varphi d\Omega; \tag{7.3.83}$$

$$K(\theta,\varphi) = \int_\Omega k_{ij} \theta_{,j} \varphi_{,i} d\Omega; \tag{7.3.84}$$

$$(\theta,\varphi) = \int_\Omega \theta\varphi d\Omega. \tag{7.3.85}$$

Assuming that the variations $v - u'$ and $\varphi - \theta$ are sufficiently smooth, then by multiplying (7.3.70) and (7.3.71) by $v - u'$ and $\varphi - \theta$ respectively, integrating over Ω, and using the Green-Gauss theorem, we obtain the variational equalities

$$(\rho u'', v - u') + a(u, v - u') + M_1(\theta - \theta_0, v - u') \tag{7.3.86}$$
$$= (f, v - u') + \int_\Gamma t_{ij} n_j (v_i - u'_i) d\Gamma \quad \text{in} \quad \Omega \times (0,T)$$

and

7.3 Applications to Engineering Problems: Von Kármán Plates and Thermoelasticity

$$(\rho c_D \theta', \varphi - \theta) + K(\theta, \varphi - \theta) + M_2(\theta_0 u', \varphi - \theta) \qquad (7.3.87)$$
$$= (Q, \varphi - \theta) + \int_\Gamma k_{ij} \theta_{,j} n_i (\varphi - \theta) d\Gamma \quad \text{in} \quad \Omega \times (0, T).$$

Now we assume that $C_{ijhk}, k_{ij}, m_{ij}, \rho > 0$ and $c_D > 0$ are elements of $L^\infty(\Omega)$, and that $f(t) \in [L^2(\Omega)]^3$ and $Q(t) \in L^2(\Omega)$. We introduce the spaces $[\overset{0}{H}{}^1(\Omega)]^3$ for v, u' and $H^1(\Omega)$ for φ, θ. Further, the variational equalities (7.3.86) and (7.3.87) are combined with the boundary conditions (7.3.75) and (7.3.76), and thus we can define the following variational problem (Problem P):

Find functions $u : [0, T] \to [\overset{0}{H}{}^1(\Omega)]^3$ and $\theta : [0, T] \to \Phi = \{H^1(\Omega) | \theta = 0 \text{ on } \Gamma - \Gamma_1\}$, with $u'(t) \in [\overset{0}{H}{}^1(\Omega)]^3$, $u''(t) \in [L^2(\Omega)]^3$, $\theta'(t) \in L^2(\Omega)$, which satisfy the variational equality

$$(\rho u'', v - u') + a(u, v - u') + M_1(\theta - \theta_0, v - u') \qquad (7.3.88)$$
$$= (f, v - u'), \quad \forall v \in [\overset{0}{H}{}^1(\Omega)]^3,$$

the hemivariational inequality

$$(\rho c_D \theta', \varphi - \theta) + K(\theta, \varphi - \theta) + M_2(\theta_0 u', \varphi - \theta) \qquad (7.3.89)$$
$$+ \int_{\Gamma_1} j^0(\theta, \varphi - \theta) d\Gamma \geq (Q, \varphi - \theta), \quad \forall \varphi \in \Phi$$

and the initial conditions (7.3.72)÷(7.3.74).

Let us now consider the B.V.P. which is defined by the relations (7.3.70)÷(7.3.74) and (7.3.77)÷(7.3.79). Let $v, u' \in [H^1(\Omega)]^3$ be such that $v = u' = U'(t)$ on Γ_U and $\varphi, \theta \in H^1(\Omega)$ with $\varphi = \theta = \theta_0$ on Γ. Arguing as before, by using the given boundary conditions, we obtain from (7.3.86), (7.3.87) the relations

$$(\rho u'', v - u') + a(u, v - u') + M_1(\theta - \theta_0, v - u') + \int_{\Gamma_S} \psi^0(u', v - u') d\Gamma$$
$$\geq (f, v - u'), \quad \forall v \in [H^1(\Omega)]^3 \text{ with } v = U'(t) \text{ on } \Gamma_U \quad (7.3.90)$$

and

$$(\rho c_D \theta', \varphi - \theta) + K(\theta, \varphi - \theta) + M_2(\theta_0 u', \varphi - \theta) \qquad (7.3.91)$$
$$= (Q, \varphi - \theta), \quad \forall \varphi \in H^1(\Omega) \text{ with } \varphi = \theta_0 \text{ on } \Gamma.$$

Let us consider a function $w : [0, T] \to [H^1(\Omega)]^3$ such that $w(t) = U(t)$ on Γ_U, and make the substitutions $\tilde{u}' = u' - w'$, $\tilde{v} = v - w'$ (resp. $\tilde{\theta} = \theta - \theta_0$, $\tilde{\varphi} = \varphi - \varphi_0$). Thus we obtain from (7.3.90), (7.3.91) the variational system consisting of the hemivariational inequality

$$(\rho \tilde{u}'', \tilde{v} - \tilde{u}') + a(\tilde{u}, \tilde{v} - \tilde{u}) + M_1(\tilde{\theta}, \tilde{v} - \tilde{u}') + \int_{\Gamma_S} \psi^0(\tilde{u}' + w', \tilde{v} - \tilde{u}') d\Gamma$$
$$\geq (\tilde{f}, \tilde{v} - \tilde{u}'), \quad \forall \tilde{v} \in V_0 \qquad (7.3.92)$$

216   7. Eigenvalue and Dynamic Problems

and the variational equality

$$(\rho c_D \tilde{\theta}', \tilde{\varphi} - \tilde{\theta}) + K(\tilde{\theta}, \tilde{\varphi} - \tilde{\theta}) + M_2(\theta_0 \tilde{u}', \tilde{\varphi} - \tilde{\theta}) \quad (7.3.93)$$
$$= (\tilde{Q}, \tilde{\varphi} - \tilde{\theta}), \quad \forall \tilde{\varphi} \in \overset{0}{H}{}^1(\Omega),$$

where $V_0 = \{v | v \in [H^1(\Omega)]^3, v = 0 \text{ on } \Gamma_U\}$ and $\tilde{f}, \tilde{Q}$ are given by

$$(\tilde{f}, v) = (f, v) - a(w, v) - (\rho w'', v) \quad (7.3.94)$$

and

$$(\tilde{Q}, \varphi) = (Q, \varphi) - M_2(\theta_0 w', \varphi). \quad (7.3.95)$$

The initial conditions (7.3.72)÷(7.3.74) become

$$\tilde{u}_i = u_{0i} - w_i(0), \quad \tilde{u}'_i = u_{1i} - w'_i(0), \quad \tilde{\theta} = \bar{\theta} - \theta_0. \quad (7.3.96)$$

Thus we may consider the following variational system (Problem P'):
Find functions $\tilde{u} : [0, T] \to V_0$ and $\tilde{\theta} : [0, T] \to \overset{0}{H}{}^1(\Omega)$, with $\tilde{u}'(t) \in V_0$, $\tilde{u}''(t) \in [L^2(\Omega)]^3$, $\tilde{\theta}'(t) \in L^2(\Omega)$, which satisfy (7.3.92), (7.3.93), (7.3.96).

Further, a third B.V.P. defined by (7.3.70)÷(7.3.75) and (7.3.78), (7.3.79) will be considered, with $\Gamma_S \neq \emptyset, \Gamma_1 \neq \emptyset$. Arguing as for the preceding problems, and by means of the substitutions used in Problem P', we obtain the following variational formulation (retaining the symbols $\tilde{u}, \tilde{v}$). We call the problem, Problem P'':
Find $\tilde{u} : [0, T] \to V_0$ and $\theta : [0, T] \to \Phi$, with $\tilde{u}'(t) \in V_0$, $\tilde{u}''(t) \in [L^2(\Omega)]^3$, $\theta'(t) \in L^2(\Omega)$, which satisfy the hemivariational inequalities

$$(\rho \tilde{u}'', \tilde{v} - \tilde{u}') + a(\tilde{u}, \tilde{v} - \tilde{u}') + M_1(\theta - \theta_0, \tilde{v} - \tilde{u}') \quad (7.3.97)$$
$$+ \int_{\Gamma_S} \psi^0(\tilde{u}' + w', \tilde{v} - \tilde{u}') d\Gamma \geq (\tilde{f}, v - u'), \quad \forall \tilde{v} \in V_0$$

$$(\rho c_D \theta', \varphi - \theta) + K(\theta, \varphi - \theta) + M_2(\theta_0(\tilde{u}' + w'), \varphi - \theta) \quad (7.3.98)$$
$$+ \int_{\Gamma_1} j^0(\theta + \theta_0, \varphi - \theta) d\Gamma \geq (\tilde{Q}, \varphi - \theta), \quad \forall \varphi \in \Phi,$$

and the initial conditions (7.3.96) (the first two) and (7.3.74).

Further we shall discuss the existence of the solutions of Problem P. We consider a Galerkin base in $V = [\overset{0}{H}{}^1(\Omega)]^3$ and in $W = \Phi \cap \{\theta | \theta \in L^\infty(\Omega)\}$ and let V_n and W_n be the corresponding finite dimensional subspaces. Note that W is dense in Φ for the $H^1(\Omega)$-norm. We consider the finite dimensional regularized problem $P_{\varepsilon n}$ where again β_ε results from β by means of (6.1.11):
Find $u_{\varepsilon n} : [0, T] \to V_n$, with $u'_{\varepsilon n}(t) \in V$, $u''(t) \in [L^2(\Omega)]^3$, and $\theta_{\varepsilon n} : [0, T] \to W_n$, with $\theta'_{\varepsilon n}(t) \in L^2(\Omega)$, to solve the variational system

$$(\rho u''_{\varepsilon n}, v) + a(u_{\varepsilon n}, v) + M_1(\theta_{\varepsilon n} - \theta_0, v) = (f, v), \quad \forall v \in V_n, \quad (7.3.99)$$

7.3 Applications to Engineering Problems: Von Kármán Plates and Thermoelasticity

$$(\rho c_D \theta'_{en}, \varphi) + K(\theta_{en}, \varphi) + M_2(\theta_0 u'_{en}, \varphi) + \int_{\Gamma_1} \beta_\varepsilon(\theta_{en})\varphi d\Gamma \quad (7.3.100)$$
$$= (Q, \varphi), \forall \varphi \in W_n,$$

with the initial conditions

$$u_{en} = u_0, \quad u'_{en} = u_1, \quad \theta_{en} = \bar{\theta} \quad (7.3.101)$$

For the definitive variational formulation of the B.V.P., let us assume that

$$f, f' \in L^2(0T, L^2(\Omega)), \quad (7.3.102)$$

$$Q, Q' \in L^2(0T, [L^2(\Omega)]^3), \quad (7.3.103)$$

$$u_0 \in [H^2(\Omega) \cap \overset{0}{H}{}^1(\Omega)]^3, \quad (7.3.104)$$

$$u_1 \in [\overset{0}{H}{}^1(\Omega)]^3 \quad (7.3.105)$$

and

$$\bar{\theta} \in H^1(\Omega). \quad (7.3.106)$$

The following proposition can now be proved.

Proposition 7.3.2 Assume that (6.1.14) and (7.3.102)÷(7.3.106) hold. Moreover let

$$\frac{\beta(\xi+h)-\beta(\xi)}{h} \geq -c \quad \forall \xi \in \mathbb{R}, \quad \forall h > 0, \quad c \text{ const} > 0 \quad (7.3.107)$$

where

$$c < \frac{\bar{\bar{c}}}{\bar{c}} \quad (7.3.108)$$

with $\bar{c}$ the constant of the $H^1(\Omega) \to L^2(\Gamma)$ imbedding and with $\bar{\bar{c}}$ the coercivity constant of $K(\varphi, \varphi)$ in Φ,

$$m_{ij,i} \in L^\infty(\Omega), \quad i, j = 1, 2, 3, \quad (7.3.109)$$

and

$$F(\varphi) = K(\bar{\theta}, \varphi) + \int_{\Gamma_1} \beta_\varepsilon(\bar{\theta})\varphi d\Gamma + M_2(\theta_0 u_1, \varphi) \quad (7.3.110)$$

is a continuous linear functional with respect to the $L^2(\Omega)$-topology on $H^1(\Omega)$. Then, the solution of the Problem P_{en} exists and as $\varepsilon \to 0 \ n \to \infty$

$$u_{en} \to u \quad \text{weakly}_\star \quad \text{in} \quad L^\infty(0T, [\overset{0}{H}{}^1(\Omega)]^3), \quad (7.3.111)$$

$$u'_{en} \to u' \quad \text{weakly}_\star \quad \text{in} \quad L^\infty(0T, [\overset{0}{H}{}^1(\Omega)]^3), \quad (7.3.112)$$

$$u''_{en} \to u'' \quad \text{weakly}_\star \quad \text{in} \quad L^\infty(0T, [L^2(\Omega)]^3) \quad (7.3.113)$$

and

$$\theta_{en} \to \theta, \quad \theta'_{en} \to \theta \text{ weakly in } L^\infty(0T, H^1(\Omega)) \qquad (7.3.114)$$
$$\text{and weakly}_* \text{ in } L^\infty(0T, L^2(\Omega))$$

The pair $\{u, \theta\}$ thus obtained is a solution of Problem P.

Proof. We shall sketch the basic steps of the proof which is analogous to the proof of Prop. 7.2.1. We put $v = u'_{en}$ in (7.3.99) and $\varphi = \theta_{en}$ in (7.3.100) and we obtain

$$(\rho u''_{en}, u'_{en}) + a(u_{en}, u'_{en}) + M_1(\theta_{en} - \theta_0, u'_{en}) = (f, u'_{en}) \qquad (7.3.115)$$

and (using (6.1.15))

$$\left(\frac{\rho c_D \theta'_{en}}{\theta_0}, \theta_{en}\right) + \frac{1}{\theta_0} K(\theta_{en}, \theta_{en}) + M_2(u'_{en}, \theta_{en}) \leq \left(\frac{Q}{\theta_0}, \theta_{en}\right) + \frac{\rho_1 \rho_2 \text{mes}\,\Gamma_1}{\theta_0} \qquad (7.3.116)$$

By addition, the relation

$$\frac{1}{2}\frac{d}{dt}\left[\int_0^T \rho(u'_{en})^2 d\Omega + \alpha(u_{en}, u_{en}) + \int_\Omega \frac{\rho c_D}{\theta_0}(\theta_{en})^2 d\Omega\right] \qquad (7.3.117)$$
$$+ M_1(\theta_{en} - \theta_0, u'_{en}) + M_2(u'_{en}, \theta_e) + \frac{1}{\theta_0}K(\theta_{en}, \theta_{en})$$
$$= (f, u'_{en}) + \left(\frac{Q}{\theta_0}, \theta_{en}\right) + \frac{\rho_1 \rho_2 \text{mes}\,\Gamma_1}{\theta_0}$$

results. But since $u'_{en}(t) = 0$ on Γ, we find, by means of the Green-Gauss theorem, that

$$M_1(\theta_{en} - \theta_0, u'_{en}) + M_2(u'_{en}, \theta_{en}) = -M_1(\theta_0, u'_{en}). \qquad (7.3.118)$$

Using the inequality

$$K(\varphi, \varphi) + \lambda |\varphi|_2^2 \geq \overline{\overline{c}}\, ||\varphi||^2 \quad \forall \varphi \in \Phi \qquad (7.3.119)$$

where $\lambda > 0$ and $\overline{\overline{c}} > 0$ the coercivity constant of $K(\varphi, \varphi)$, the coerciveness of $\alpha(v, v)$ and the continuity of the linear forms, we find (here $||\varphi||$ denotes the $H^1(\Omega)$-norm, $||u||$ denotes the $[\overset{0}{H}{}^1(\Omega)]^3$-norm, and $|\cdot|_2$ denotes the $L^2(\Omega)$-norms, i.e for the temperature the $L^2(\Omega)$-norm and for the displacements or velocities the $[L^2(\Omega)]^3$-norm) from (7.3.117), (7.3.118) that

$$[|u'_{en}|_2^2 + ||u_{en}||^2 + |\theta_{en}|_2^2] + c\int_0^t ||\theta_{en}||^2 dt \qquad (7.3.120)$$
$$\leq c_1 \int_0^t (|u'_{en}|_2^2 + ||u_{en}||^2 + |\theta_{en}|_2^2) dt + c_2 + \int_0^t \frac{\rho_1 \rho_2 \text{mes}\,\Gamma_1}{\theta_0} dt$$

7.3 Applications to Engineering Problems: Von Kármán Plates and Thermoelasticity

Here c, c_1, c_2 are various positive constants depending on the data of the problem (θ_0, ess $\inf_\Omega \rho$, ess $\sup_\Omega \rho$, etc.) and ρ_1, ρ_2 are independent of t. Thus (7.3.120) implies by Gronwall's inequality that

$$|u'_{en}|_2 \leq c, \quad \|u_{en}\| \leq c, \quad |\theta_{en}|_2 \leq c, \tag{7.3.121}$$

and subsequently, if we take $t = T$, that

$$\int_0^T \|\theta_{en}\|^2 dt \leq c. \tag{7.3.122}$$

We now put $t = 0$ in (7.3.99). It results that

$$(\rho u''_{en}(0), v) + a(u_0, v) + M_1(\bar{\theta} - \theta_0, v) = (f(0), v), \quad \forall v \in V_n, \tag{7.3.123}$$

from which

$$|u''_{en}(0)|_2 \leq c. \tag{7.3.124}$$

Similarly, from (7.3.100), we obtain

$$(\rho c_D \theta'_{en}(0), \varphi) + K(\bar{\theta}, \varphi) + M_2(\theta_0 u_1, \varphi) + \int_{\Gamma_1} \beta_\epsilon(\bar{\theta}) \varphi d\Gamma$$
$$= (Q(0), \varphi), \quad \forall \varphi \in H^1(\Omega). \tag{7.3.125}$$

Due to (7.3.110), there exists $\tau \in L^2(\Omega)$ such that for every $\varphi \in H^1(\Omega)$

$$F(\varphi) = (\tau, \varphi), \tag{7.3.126}$$

and thus it results from (7.3.125) that

$$|\theta'_{en}(0)|_2 \leq c. \tag{7.3.127}$$

Differentiating (7.3.99), (7.3.100) with respect to t, we obtain

$$(\rho u'''_{en}, v) + a(u'_{en}, v) + M_1(\theta'_{en}, v) = (f', v), \quad \forall v \in V_n, \tag{7.3.128}$$

and

$$\left(\frac{\rho c_D \theta''_{en}}{\theta_0}, \varphi\right) + \frac{1}{\theta_0} K(\theta'_{en}, \varphi) + M_2(u''_{en}, \varphi) + \frac{1}{\theta_0}\int_{\Gamma_1} [\beta_\epsilon(\theta_{en})]' \varphi d\Gamma$$
$$= \left(\frac{Q'}{\theta_0}, \varphi\right), \quad \forall \varphi \in W_n. \tag{7.3.129}$$

We set $v = u''_{en}$ in (7.3.128) and $\varphi = \theta'_{en}$ in (7.3.129). We have due to (7.3.107) that

$$\frac{1}{\theta_0}\int_{\Gamma_1}[\beta_e(\theta_{en})]'\theta'_{en}d\Gamma = \frac{1}{\theta_0}\int_{\Gamma_1}\frac{d\beta_e(\theta_{en})}{d\theta_{en}}(\theta'_{en})^2 d\Gamma \qquad (7.3.130)$$

$$= \int_{\Gamma_1}\lim_{h\to 0+}\frac{1}{\theta_0}\int_{-\varepsilon}^{+\varepsilon}p_e(t)\left(\frac{\beta(\xi+h-t)-\beta(\xi-t)}{h}\right)dt\bigg|_{\xi=\theta_{en}(x)}(\theta'_{en})^2 d\Gamma$$

$$\geq -\frac{c}{\theta_0}|\theta'_{en}|^2_{L^2(\Gamma_1)} \geq -\frac{c\bar{c}}{\theta_0}\|\theta'_{en}\|^2.$$

Thus (7.3.128)÷(7.3.130) imply that

$$\frac{1}{2}\frac{d}{dt}\left[\int_\Omega \rho(u''_{en})^2 d\Omega + \alpha(u'_{en}, u'_{en}) + \int_\Omega \frac{\rho c_D}{\theta_0}(\theta'_{en})^2 d\Omega\right] \qquad (7.3.131)$$

$$+ M_1(\theta'_{en}, u''_{en}) + M_2(u''_{en}, \theta'_{en}) + \frac{1}{\theta_0}K(\theta'_{en}, \theta'_{en})$$

$$\leq (f', u''_{en}) + \left(\frac{Q'}{\theta_0}, \theta'_{en}\right) + \frac{c\bar{c}}{\theta_0}\|\theta'_{en}\|^2.$$

From (7.3.131), by applying the Green-Gauss theorem to $M_1(\theta'_{en}, u''_{en}) + M_2(u''_{en}, \theta'_{en})$ (noting that $u''_{en} = 0$ on Γ, by integrating from 0 to t, and by means of (7.3.119) we deduce the relation

$$c_1[|u''_{en}|^2_2 + \|u'_{en}\|^2 + |\theta'_{en}|^2_2] + \frac{\bar{\bar{c}}}{\theta_0}\int_0^t \|\theta'_{en}\|^2 - \frac{\lambda}{\theta_0}\int_0^t |\theta'_{en}|^2_2 dt \qquad (7.3.132)$$

$$\leq c_1[|u''_{en}|^2_2 + \|u'_{en}\|^2 + |\theta'_{en}|^2_2] + \int_0^t \frac{1}{\theta_0}K(\theta'_{en}, \theta'_{en})dt$$

$$\leq c_2\int_0^t [|f'|_2|u''_{en}|_2 + \frac{1}{\theta_0}|Q'|_2|\theta'_{en}|_2]dt + \frac{c\bar{c}}{\theta_0}\int_0^t \|\theta'_{en}\|^2 dt$$

$$+ c_1(|u''_{en}(0)|^2_2 + \|u_1\|^2 + |\theta'_{en}(0)|^2_2),$$

where the c_i terms are positive constants. From (7.3.132) and (7.3.108), by using Gronwall's inequality and (7.3.124) and (7.3.126), we obtain the relations

$$|u''_{en}|_2 \leq c, \quad \|u'_{en}\| \leq c, \quad |\theta'_{en}|_2 \leq c. \qquad (7.3.133)$$

Indeed (7.3.132) implies that

$$c_1[|u''_{en}|^2_2 + \|u'_{en}\|^2 + |\theta'_{en}|^2_2] + \frac{(\bar{\bar{c}}-c\bar{c})}{\theta_0}\int_0^t \|\theta'_{en}\|^2 dt \qquad (7.3.134)$$

$$\leq c_2(|u''_{en}(0)|^2 + \|u_1\|^2 + |\theta'_{en}(0)|^2_2)$$

$$+ c_3\int_0^t (|u''_{en}|^2_2 + \|u'_{en}\|^2 + |\theta'_{en}|^2_2)dt.$$

7.3 Applications to Engineering Problems: Von Kármán Plates and Thermoelasticity

If we set $t = T$ in (7.3.132), we obtain that

$$\int_0^T ||\theta'_\varepsilon||^2 dt \leq c \qquad (7.3.135)$$

Due to (7.3.121), (7.3.122), (7.3.133), (7.3.135) we obtain that we can determine subsequences again denoted by $u_{\varepsilon n}, u'_{\varepsilon n}, u''_{\varepsilon n}$ and $\theta_{\varepsilon n}, \theta'_{\varepsilon n}$, such that for $\varepsilon \to 0$, $n \to \infty$, we obtain that (7.3.111)÷((7.3.114) are true. The rest of the proof is the same as in Prop. 7.2.1. q.e.d.

8. Optimal Control and Identification Problems

The present chapter addresses the optimal control and the parameter indentification problem of systems governed by hemivariational inequalities. This is a nonclassical mathematical problem, because the state of the problem is connected with the control function through a hemivariational inequality. We recall here that optimal control problems governed by state variational inequalities have been already studied (cf. e.g. [Yvon], [Lio71], [Panag77], [Mign76, 84], [Shi], [Hasli86a,b] [Barb]). However the present problem is more complicated, because we have state hemivariational inequalities. Here due to the lack of convexity of the superpotentials involved, compactness arguments will be applied. The mathematical framework is quite general to cover most of the usual engineering structures, as e.g. beams, plates in stretching and bending etc. The chapter is based mainly on [Panag92] and [Hasli93]. We refer also to [Panag84,89,90,91], [Hasli89]. Some application of the theory to engineering problems close the present chapter and illustrate the prospects for applications of the developed theory.

8.1 Formulation of the Problem

Let Ω be an open bounded subset of $\mathbb{R}^n$ and let Γ be its boundary assumed to be Lipschitzian. We define on Ω a Hilbert space V such that

$$V \subset L^2(\Omega) \subset V' \tag{8.1.1}$$

Where V' is the dual space of V and the injections are continuous and dense. We denote by $(\cdot, \cdot)$ the L^2-inner product and the duality pairing on $V \times V'$, by $||\cdot||$ the V-norm and by $|\cdot|_2$ the L^2-norm. We recall that the form $(\cdot, \cdot)$ extends uniquely from $V \times L^2(\Omega)$ to $V \times V'$ [Aub79a]. Let further

$$V \subset L^2(\Omega) \text{ be compact} \tag{8.1.2}$$

and

$$V \cap L^\infty(\Omega) \text{ be separable and dense in } V \text{ for the } V\text{-norm.} \tag{8.1.3}$$

We consider a bilinear form $a : V \times V \to \mathbb{R}$ having the following properties: There exist a M const > 0 such that

224   8. Optimal Control and Identification Problems

$$|a(y,z)| \leq M||y|| \ ||z|| \quad \forall y,z \in V \tag{8.1.4}$$

and there exists a const $\alpha > 0$ such that

$$a(y,y) \geq \alpha ||y||^2 \quad \forall y \in V. \tag{8.1.5}$$

Moreover $f \in V'$ is given. Let also $j : \mathbb{R} \to \mathbb{R}$ be a locally Lipschitz continuous function obtained from $\beta \in L^\infty_{loc}(\mathbb{R})$ as in (1.2.54). It is assumed that $\beta(\xi_\pm)$ exists for each $\xi \in \mathbb{R}$. We shall use the notations (1.2.51), (1.2.52) and the multivalued function

$$\tilde{\beta}(\xi) = [\bar{\beta}(\xi), \bar{\bar{\beta}}(\xi)] \tag{8.1.6}$$

which results from β by "filling in the jumps" and satisfies the relation (up to an additive constant)

$$\tilde{\beta}(\xi) = \bar{\partial} j(\xi). \tag{8.1.7}$$

We consider now the following hemivariational inequality (Problem P):
Find $y \in V$ such as to satisfy

$$a(y, z-y) + \int_\Omega j^0(y, z-y) d\Omega \geq (f, z-y) \quad \forall z \in V. \tag{8.1.8}$$

The following definition is justified by Theorem 6.1.1. By a solution of a hemivariational inequality determined by a, f and j we mean a function $y \in V$ such that a $\chi \in L^1(\Omega) \cap V'$ exists satisfying the relations (Problem $\bar{P}$)

$$a(y, z) + \int_\Omega \chi z d\Omega = (f, z) \quad \forall z \in V, \tag{8.1.9}$$

and

$$\chi(x) \in \bar{\partial} j(y(x)) = \tilde{\beta}(y(x)) \quad \text{a.e. in } \Omega. \tag{8.1.10}$$

In order to define the corresponding finite dimensional problem P_{en} we consider a Galerkin basis of $V \cap L^\infty(\Omega)$ and let V_n be the resulting n-dimensional subspace (e.g. a finite element approximation). Problem P_{en} reads:
Find $y_{en} \in V_n$ such as to satisfy the variational equality

$$a(y_{en}, z) + \int_\Omega \beta_e(y_{en}) z d\Omega = (f, z) \quad \forall z \in V_n. \tag{8.1.11}$$

The following assumption is made for the graph $\{\xi, \beta(\xi)\}$: there exists $\xi \in \mathbb{R}$ such that

$$\underset{(-\infty, -\xi)}{\text{ess sup}}\, \beta(\xi_1) \leq 0 \leq \underset{(\xi, \infty)}{\text{ess inf}}\, \beta(\xi_1). \tag{8.1.12}$$

The existence proof given in Theorem 6.1.1 was based on Prop. 6.1.1. stating that due to (8.1.12) one can determine two constants $\rho_1 > 0, \rho_2 > 0$ such that

$$\int_\Omega \beta_e(y_{en}) y_{en} d\Omega \geq -\rho_1 \rho_2 \text{mes}\, \Omega \quad \forall y_{en} \in V_n, \tag{8.1.13}$$

8.1 Formulation of the Problem

on Prop. 6.1.2 stating that problem P_{en} has at least one solution $y_{en} \in V_n$, and on Prop 6.1.3 stating that the sequence $\{\beta_\epsilon(y_{en})\}$ is weakly precompact in $L^1(\Omega)$. Then if (8.1.12) holds, according to Theorem 6.1.1, problem $\bar{P}$ has at least one solution which satisfies problem P.

Let now U be another Hilbert space (the space of controls) and let U_{ad} be a nonempty, convex, closed subset of U. We denote by B a linear condinuous operator from U into V', i.e. $B \in L(U, V')$. Now we associate with any $u \in U_{ad}$ a mapping $A(u) \in L(V, V')$ generated by a bilinear form $a_u : V \times V \to \mathbb{R}$ through the relation

$$(A(u)y, z) = a_u(y, z) \quad \forall y, z \in V, \quad \forall u \in U_{ad}. \tag{8.1.14}$$

Moreover we assume that $a_u(\cdot, \cdot)$ is bounded and coercive for every $u \in U_{ad}$, i.e. (8.1.4) and (8.1.5) hold for every $u \in U_{ad}$ and let A have the following property:

$$u_n \to u \text{ weakly in } U, u_n, u \in U_{ad} \Longrightarrow A(u_n) \to A(u) \text{ in } L(V, V'). \tag{8.1.15}$$

Now for any $u \in U_{ad}$ the following state-control problem $P(u)$ is considered: Find $y = y(u) \in V$ such as to satisfy the hemivariational inequality

$$a_u(y, z-y) + \int_\Omega j^0(y, z-y)d\Omega \geq (f + Bu, z-y) \quad \forall z \in V. \tag{8.1.16}$$

This inequality corresponds by Theorem 6.1.1 to the following problem $\bar{P}(u)$: Find $y = y(u) \in V$ such that

$$a_u(y, z) + \int_\Omega \chi z d\Omega = (f + Bu, z) \quad \forall z \in V \tag{8.1.17}$$

$$\chi = \chi(u) \in L^1(\Omega) \cap V' \tag{8.1.18}$$

$$\chi(x) \in \tilde{\beta}(y(x)) = \bar{\partial}j(y(x)) \text{ a.e. in } \Omega. \tag{8.1.19}$$

Let us denote further by $X(u)$ the set of all solutions of the problem $\bar{P}(u)$ for a given $u \in U_{ad}$. We introduce further a cost functional $I : V \times U \to \mathbb{R}$ having the following properties:

$$\text{i)} \quad \left.\begin{array}{l} y_n \to y \text{ weakly in } V \\ u_n \to u \text{ weakly in } U \end{array}\right\} \Longrightarrow \liminf I(y_n, u_n) \geq I(y, u) \tag{8.1.20}$$

ii) $\forall \mathbb{R} > 0$ there exists $r > 0$ such that for every u with $\|u\| \geq r$,
$u \in U_{ad}$, and for every $y \in V$, $I(u, y) \geq R$(coercivity). \hfill (8.1.21)

Then for any $u \in U_{ad}$ we may define the following problem $\mathbb{P}(u)$:
Find $\bar{y} \in X(u)$ such that

$$I(\bar{y}, u) \leq I(y, u) \quad \forall y \in X(u), \tag{8.1.22}$$

i.e. $\bar{y}$ minimizes I over $X(u)$. Let us suppose now that $\mathbb{P}(u)$ has a solution. We denote by $E(u)$ the $I(\bar{y}, u)$, where $\bar{y} \in X(u)$ solves $\mathbb{P}(u)$. Then the optimal control problem $\mathbb{P}$ of the hemivariational inequality reads:
Find $u^\star \in U_{\text{ad}}$ such that

$$E(u^\star) \leq E(u) \quad \forall u \in U_{\text{ad}}. \tag{8.1.23}$$

Obviously the above definition of $\mathbb{P}(u)$ and $\mathbb{P}$ have a meaning if $X(u)$ is nonempty.

8.2 Mathematical Study of the Optimal Control Problem Governed by Hemivariational Inequalities

The definition (8.1.23) of the problem $\mathbb{P}$ includes as special cases both the optimal control and the parameter identification problem (see e.g. [Lio71]). The Theorem 6.1.1 implies that $X(u) \neq \emptyset$ for any $u \in U_{\text{ad}}$. A consequence of (8.1.12) is that two positive numbers $\bar{\rho}_1$ and $\bar{\rho}_2$ can be determined such that

$$\begin{aligned} \beta(\xi) &\geq 0 & \text{if } \xi &\geq \bar{\rho}_1 \\ \beta(\xi) &\leq 0 & \text{if } \xi &\leq -\bar{\rho}_1 \\ |\beta(\xi)| &\leq \bar{\rho}_2 & \text{if } |\xi| &\leq \bar{\rho}_1. \end{aligned} \tag{8.2.1}$$

The following result can be proved.

Proposition 8.2.1 The estimate

$$\sup_{y \in X(u)} ||y|| \leq \frac{1}{\alpha}(||f||_{V'} + ||B|| \, ||u||_U) + (\frac{2}{\alpha}\bar{\rho}_1\bar{\rho}_2\text{mes }\Omega)^{1/2} \tag{8.2.2}$$

holds.

Proof. Let $y \in X(u)$. Then due to (8.1.19) and the definition of $\tilde{\beta}$ we get easily that

$$\int_\Omega \chi y d\Omega = \int_{|y(x)| \leq \bar{\rho}_1} \chi y d\Omega \tag{8.2.3}$$

$$+ \int_{|y(x)| \geq \bar{\rho}_1} \chi y d\Omega \geq \int_{|y(x)| \leq \bar{\rho}_1} \chi y d\Omega \geq -\bar{\rho}_1\bar{\rho}_2\text{mes }\Omega.$$

From (8.1.17), (8.2.3) and the coercivity of $a_u(\cdot, \cdot)$ we obtain that

$$\alpha||y||^2 \leq \bar{\rho}_1\bar{\rho}_2\text{mes}\Omega + (||f||_{V'} + ||B|| \, ||u||_U)||y|| \tag{8.2.4}$$

$$\leq \frac{\alpha}{2}||y||^2 + \frac{1}{2\alpha}(||f||_{V'} + ||B|| \, ||u||_U)^2 + \bar{\rho}_1\bar{\rho}_2\text{mes}\Omega$$

from which the estimate (8.2.2) results. q.e.d.

Let as define now as $||X(u)||$ the expression $\sup\{||y|| \,\big|\, y \in X(u)\}$. Then Prop. 8.2.1 implies that

$$||X(u)|| < c \quad \forall u \in U_{\mathrm{ad}} \text{ with } ||u||_U \leq c. \tag{8.2.5}$$

Now we will prove the following result.

Proposition 8.2.2 The solution set $X(u)$ is for any $u \in U_{\mathrm{ad}}$ weakly compact.

Proof. Let $y_n \in V$ denote a solution of $\bar{P}(u)$, i.e. there exist a function $\chi_n \in L^1(\Omega) \cap V'$ such that

$$a_u(y_n, z) + \int_\Omega \chi_n z \, d\Omega = (f + Bu, z) \quad \forall z \in V \tag{8.2.6}$$

$$\chi_n \in \tilde{\beta}(y_n(x)) \text{ a.e. in } \Omega. \tag{8.2.7}$$

This holds for all solutions $y_n \in V$ of $\bar{P}(u)$. From Prop.8.2.1 we have that the set of all solutions $\{y_n\}$ is bounded in V. Therefore a subsequence again denoted by $\{y_n\}$ and a function $y \in V$ exist such that

$$y_n \to y \text{ weakly in } V \text{ as } n \to \infty. \tag{8.2.8}$$

Hence

$$a_u(y_n, z) \to a_u(y, z). \tag{8.2.9}$$

Let us prove now that $y \in X(u)$. To show this we prove first that the sequence $\{\chi_n\}$ is weakly precompact in $L^1(\Omega)\}$. According to the Dunford-Pettis theorem it is sufficient to prove that for each $\mu > 0$, a $\delta(\mu) > 0$ can be determined such that for $\omega \subset \Omega$ with $\mathrm{mes}\,\omega < \delta$

$$\int_\omega |\chi_n(x)| \, d\Omega < \mu. \tag{8.2.10}$$

We apply the inequality

$$\int_\omega \chi_n(y_n(x)) d\Omega \leq \int_\omega \left[\frac{|y_n(x)|\,|\chi_n(y_n(x))|}{\xi_0} + \operatorname*{ess\,sup}_{|y_n(x)| \leq \xi_0} |\chi_n(y_n(x))| \right] d\Omega \tag{8.2.11}$$

which holds for $\xi_0 > 0$ and for $\omega \subset \Omega$ and we write

$$\begin{aligned}
\int_\omega |y_n \chi_n| d\Omega &\leq \int_\Omega |y_n \chi_n| d\Omega = \int_{|y_n(x)| \leq \bar{\rho}_1} |y_n \chi_n| d\Omega + \int_{|y_n(x)| > \bar{\rho}_1} y_n \chi_n d\Omega \\
&= 2 \int_{|y_n(x)| \leq \bar{\rho}_1} |\ldots| d\Omega + \int_{|y_n(x)| > \bar{\rho}_1} \ldots d\Omega - \int_{|y_n(x)| \leq \bar{\rho}_1} |\ldots| d\Omega \\
&\leq 2 \int_{|y_n(x)| \leq \bar{\rho}_1} |\ldots| d\Omega + \int_{|y_n(x)| > \bar{\rho}_1} \ldots d\Omega + \int_{|y_n(x)| \leq \bar{\rho}_1} \ldots d\Omega \\
&= 2 \int_{|y_n(x)| \leq \bar{\rho}_2} |\ldots| d\Omega + \int_\Omega y_n x_n d\Omega. \tag{8.2.12}
\end{aligned}$$

228   8. Optimal Control and Identification Problems

From (8.2.12) and (8.2.1), (8.2.2), (8.2.6) we obtain for a given u the estimate

$$\int_\omega |y_n x_n| d\Omega \leq 2\bar{\rho}_1 \bar{\rho}_2 \operatorname{mes} \Omega + c \qquad (8.2.13)$$

holding for every n. Let us now choose ξ_0 such that

$$\frac{1}{\xi_0} \int_\omega |y_n x_n| d\Omega \leq \frac{2\bar{\rho}_1 \bar{\rho}_2 \operatorname{mes} \Omega + c}{\xi_0} \leq \frac{\mu}{2} \qquad (8.2.14)$$

Also we get from (8.2.7) and (1.2.51), (1.2.52), (1.2.53) that

$$\sup_{|y_n(x)| \leq \xi_0} |\chi_n(y_n(x))| \leq |\lim_{\delta \to 0+} \operatorname{ess\,sup}_{|\xi| \leq \xi_0 + \delta} \beta(\xi)| \leq |\operatorname{ess\,sup}_{|\xi| \leq \xi_0+1} \beta(\xi)|$$
$$= ||\beta||_{L^\infty((-\xi_0-1),(\xi_0+1))} = \tilde{c}. \qquad (8.2.15)$$

Now let $\omega \subset \Omega$ such that $\operatorname{mes} \omega \leq \delta(\mu) = \frac{\mu}{2\tilde{c}}$. Thus from (8.2.15), (8.2.14) and (8.2.11) we obtain the estimate (8.2.10). Accordingly there exists a subsequence of $\{\chi_n\}$ and a $\chi \in L^1(\Omega)$ such that

$$\chi_n \to \chi \text{ weakly in } L^1(\Omega). \qquad (8.2.16)$$

From (8.2.8), (8.2.9) and (8.2.16) we obtain the equality

$$a_u(y, z) + \int_\Omega \chi z d\Omega = (f + Bu, z) \quad \forall z \in L^\infty(\Omega) \cap V. \qquad (8.2.17)$$

At the same time $\chi \in V'$ and (8.2.17) holds for any $z \in V$ because of (8.1.3). It remains to show that

$$\chi(x) \in \tilde{\beta}(y(x)) \text{ a.e. in } \Omega. \qquad (8.2.18)$$

Due to (8.1.2) we obtain from (8.2.8) that

$$y_n \to y \text{ strongly in } L^2(\Omega). \qquad (8.2.19)$$

Then by Egoroff's theorem for any $\tilde{\alpha} > 0$ there exists $\omega \subset \Omega$ with $\operatorname{mes} \omega \leq \tilde{\alpha}$ such that

$$y_n \to y \text{ uniformly in } \Omega - \omega. \qquad (8.2.20)$$

Let $\varepsilon > 0$ be given. Then for any $n > n_0(\varepsilon)$

$$|y_n(x) - y(x)| \leq \frac{\varepsilon}{2} \quad \forall x \in \Omega - \omega. \qquad (8.2.21)$$

Let now $0 < \mu \leq \frac{\varepsilon}{2}$. Then (8.2.7) implies that

$$\chi_n(x) \leq \lim_{\mu \to 0+} (\operatorname{ess\,sup}_{|t - y_n(x)| \leq \mu} \beta(t)) \leq \operatorname{ess\,sup}_{|t - y_n(x)| \leq \mu} \beta(t) \leq \operatorname{ess\,sup}_{|t - y_n(x)| \leq \frac{\varepsilon}{2}} \beta(t)$$
$$\leq \operatorname{ess\,sup}_{|t - y(x)| \leq \varepsilon} \beta(t) = \bar{\bar{\beta}}_\varepsilon(y(x)) \text{ a.e. in } \Omega - \omega \qquad (8.2.22)$$

where $\bar{\bar{\beta}}_\varepsilon(\cdot)$ is defined in (1.2.51) (replace ρ by ε). Analogously

$$\bar{\beta}_\varepsilon(y(x)) \leq \chi_n(x) \text{ a.e. in } \Omega - \omega. \tag{8.2.23}$$

Accordingly we may write for $e \geq 0$ a.e. in $\Omega - \omega$, $e \in L^\infty(\Omega - \omega)$ that

$$\int_{\Omega-\omega} \bar{\beta}_\varepsilon(y(x))e d\Omega \leq \int_{\Omega-\omega} \chi_n(x) e d\Omega \leq \int_{\Omega-\omega} \bar{\bar{\beta}}_\varepsilon(y(x)) e d\Omega \tag{8.2.24}$$

For $n \to \infty$ in (8.2.24), we have, due to (8.2.16), that

$$\int_{\Omega-\omega} \bar{\beta}_\varepsilon(y(x))e d\Omega \leq \int_{\Omega-\omega} \chi(x) e d\Omega \leq \int_{\Omega-\omega} \bar{\bar{\beta}}_\varepsilon(y(x)) e d\Omega \tag{8.2.25}$$

and then letting $\varepsilon \to 0_+$ we obtain

$$\int_{\Omega-\omega} \bar{\beta}(y(x))e d\Omega \leq \int_{\Omega-\omega} \chi(x) e d\Omega \leq \int_{\Omega-\omega} \bar{\bar{\beta}}(y(x)) e d\Omega \tag{8.2.26}$$

i.e.

$$\chi(x) \in \tilde{\beta}(y(x)) \text{ a.e. in } \Omega - \omega. \tag{8.2.27}$$

Since $\tilde{\alpha}$ can be made arbitrarily small, (8.2.18) results. q.e.d.

Proposition 8.2.3 The mapping $u \to X(u)$, $u \in U_{\text{ad}}$, is weakly upper semi-continuous, i.e. if $u_n \to u$ weakly in U, $(u, u_n \in U_{\text{ad}})$, and $y_n \in X(u_n)$ such that $y_n \to y$ weakly in V, then $y \in X(u)$.

Proof. The definition of $X(u_n)$ implies (8.2.6) and (8.2.7). Moreover (8.2.9) holds, as well as that

$$(f + Bu_n, z) \to (f + Bu, z) \quad \forall z \in V. \tag{8.2.28}$$

Then using the same method as in Prop. (8.2.2) we may show that $\{\chi_n\}$ is weakly precompact in $L^1(\Omega)$ and the weak cluster point χ of it satisfies (8.2.18). Passing to the limit $n \to \infty$ in (8.2.6) implies with (8.1.3) the assertion q.e.d.

A consequence of Prop. 8.2.3 is the following: Let $\{u_n\}$, $u_n \in U_{\text{ad}}$, be such that $u_n \to u$ weakly in U, and let $y_n \in X(u_n)$. Then $\{u_n\}$ is bounded and thus $\{y_n\}$ is bounded as well due to (8.2.2). It results that a subsequence can be determined such that $y_n \to y$ weakly in V. Then Prop. 8.2.3 implies that $y \in X(u)$.

Proposition 8.2.4 Problem $\mathbb{P}(u)$ admits for every $u \in U_{\text{ad}}$ at least one solution.

Proof. Follows directly from Prop. (8.2.2) and (8.1.20) q.e.d.
Now the main result of this section can be proved.

Theorem 8.2.1 The optimal control problem $\mathbb{P}$ has at least one solution.

Proof. Let

$$q = \inf\{E(u)|u \in U_{\text{ad}}\} = \inf\{I(y(u),u)|u \in U_{\text{ad}}, y \in X(u)\}. \tag{8.2.29}$$

If $u_n \in U_{\text{ad}}$ is a minimizing sequence for $\mathbb{P}$, then as $n \to \infty$

$$E(u_n) \to q \tag{8.2.30}$$

and $\{u_n\}$ is bounded because of (8.1.21). Accordingly a subsequence can be determined such that

$$u_n \to u^* \text{ weakly in } U. \tag{8.2.31}$$

Since U_{ad} is convex and closed, $u^* \in U_{\text{ad}}$. We consider the corresponding solutions $\bar{y}(u_n)$ of $\mathbb{P}(u_n)$ which are also bounded. Hence a subsequence can be determined such that

$$\bar{y}(u_n) \to y^* \text{ weakly in } V. \tag{8.2.32}$$

Thus $y^* \in X(u^*)$. Moreover from (8.2.29) and (8.1.20) we obtain that

$$q = I(y^*, u^*),$$

q.e.d. Note that if U_{ad} is bounded then the assumption (8.1.20) can be omitted. Further we shall give some approximation results for the optimal control problem.

Let us denote by h the discretization parameter and let $V_h \subset V \cap L^\infty(\Omega)$, $U_h \subset U$ be finite dimensional subspaces of V and U respectively. Let also U_{ad}^h be a closed, convex subset of U_h not necessarily contained in U_{ad}. For any $u_h \in U_{\text{ad}}^h$ the following problem $\bar{P}_h(u_h)$ is formulated:
Find $y_h \in V_h$ such that

$$a_{u_h}(y_h, z_h) + \int_\Omega \chi_h z_h d\Omega = (f + Bu_h, z_h) \quad \forall z_h \in V_h \tag{8.2.33}$$

$$\chi_h \in L^1(\Omega) \cap V_h' \tag{8.2.34}$$

$$\chi_h(x) \in \tilde{\beta}(y_h(x)) \text{ a.e. on } \Omega. \tag{8.2.35}$$

We denote by $X_h(u_h)$ the set of all solutions of $\bar{P}_h(u_h)$. Using the same method as in the previous section we can prove the following results for the finite dimensional formulation.

Proposition 8.2.5 Suppose that (8.1.4) and (8.1.5) hold for any $u_h \in U_{\text{ad}}^h \cup U_{\text{ad}}$. Then $X_h(u_h)$ is nonempty.

Proposition 8.2.6 The following estimate holds

$$\sup_{y_h \in X_h(u_h)} \|y_h\| \leq \frac{1}{\alpha}(\|f\|_{V'} + \|B\|\,\|u_h\|_U) + (\frac{2}{\alpha}\bar{\rho}_1\bar{\rho}_2\text{mes}\Omega)^{1/2}. \tag{8.2.36}$$

8.2 Mathematical Studies

Proposition 8.2.7 $X_h(u_h)$ is compact for any $h > 0$ and any $u_h \in U_{ad}^h$.

Proposition 8.2.8 The mapping $u_h \to X_h(u_h)$ is upper semicontinuous.

Now let us define for any $u_h \in U_{ad}^h$ the problem $\mathbb{P}_h(u_h)$:
Find $\bar{y}_h \in X_h(u_h)$ such that

$$I(\bar{y}_h, u_h) \leq I(y_n, u_h) \quad \forall y_h \in X_h(u_h). \tag{8.2.37}$$

Then the following proposition holds (proof is obvious).

Proposition 8.2.9 For any $u_h \in U_{ad}^h$ there exists at least one solution of the problem $\mathbb{P}_h(u_h)$.

Now we can define the finite dimensional approximation $\mathbb{P}_h$ of the optimal control problem $\mathbb{P}$. We use the notation $E(u_h) = I(\bar{y}_h, u_h)$. Then $\mathbb{P}_h$ reads:
Find $u_h^\star \in U_{ad}^h$ such that

$$E(u_h^\star) \leq E(u_h) \quad \forall u_h \in U_{ad}^h. \tag{8.2.38}$$

Then if (8.1.21) holds for any $u \in U_{ad} \cup U_{ad}^h$ the following result can be obtained.

Proposition 8.2.10 $\mathbb{P}_h$ has at least one solution for any $h > 0$.

Note here that usually $U_{ad}^h \subset U_{ad}$ and thus Prop. 8.2.5 and 8.2.10 hold without any additional assumption. Next we shall investigate the relation between $\mathbb{P}$ and $\mathbb{P}_h$ when $h \to 0_+$.

Let $\{V_h\}, \{U_h\}$ be two families of finite dimensional subspaces of V and U respectively such that $V_h \subset V \cap L^\infty(\Omega)$ and let their dimensions tend to infinity if $h \to 0_+$. Let $\{U_{ad}^h\}$ be a family of convex closed subsets of U_h not necessarily contained in U_{ad}. We assume that

i) $\forall v \in V \cap L^\infty(\Omega)$ there exists $v_h \in V_h$ such that

$$v_h \to v \text{ strongly in } L^\infty(\Omega) \text{ and in } V; \tag{8.2.39}$$

ii) $\forall u \in U_{ad}$ there exists $u_h \in U_{ad}^h$ such that $u_h \to u$ weakly in U; (8.2.40)

iii) $u_h \in U_{ad}^h$ $u_h \to u$ weakly in $U \implies u \in U_{ad}$; (8.2.41)

iv) there exists $M > 0$ such that $|a_u(y,z)| \leq M\|y\|\,\|z\|$ $\forall y, z \in V$

$$\forall u \in U_{ad} \cup (\bigcup_h U_{ad}^h); \tag{8.2.42}$$

v) there exists $\alpha > 0$ such that $a_u(y,z) \geq \alpha\|y\|^2$ $\forall y, z \in V$

$$\forall u \in U_{ad} \cup (\bigcup_h U_{ad}^h); \tag{8.2.43}$$

vi) $\forall R > 0$ there exists $r > 0$ such that $\forall \|u\| \geq r$, $u \in U_{ad} \cup (\bigcup_h U_{ad}^h)$,

$$\forall y \in V \text{ it holds } I(u,y) \geq R; \tag{8.2.44}$$

vii) $V_h \ni y_h \to y$ weakly in V, $U_{ad}^h \ni u_h \to u$

weakly in $U \implies \lim_{h \to 0_+} I(y_h, u_h) = I(y, u).$ (8.2.45)

If $U_{ad}^h \subset U_{ad}$ then (8.2.41)÷(8.2.44) are superfluous. First the following proposition will be proved.

Proposition 8.2.11 Let $u_h \in U_{ad}^h$ be such that as $h \to 0_+$

$$u_h \to u \text{ weakly in } U \qquad (8.2.46)$$

and $y_h \in X_h(u_h)$. Then there exists a subsequence of $\{y_h\}$ and a $y \in X(u)$ such that as $h \to 0_+$

$$y_h \to y \text{ weakly in } V. \qquad (8.2.47)$$

Proof. From (8.2.46) and the estimate (8.2.36) it results that $\{y_h\}$ is bounded. Thus a subsequence again denoted by $\{y_h\}$ and $y \in V$ can be determined which satisfy (8.2.47). Let us show that $y \in X(u)$. If $\bar{z} \in V \cap L^\infty(\Omega)$ is an arbitrary element, then according to (8.2.39) a sequence $\{\bar{z}_h\}$ can be determined such that

$$\bar{z}_h \to \bar{z} \text{ strongly in } L^\infty(\Omega) \text{ and in } V. \qquad (8.2.48)$$

Since $y_h \in X_h(u_h)$, there exists $\chi_h \in L^1(\Omega) \cap V_h'$ such that

$$a_{u_h}(y_h, \bar{z}_h) + \int_\Omega \chi_h \bar{z}_h d\Omega = (f + Bu_h, \bar{z}_h) \qquad (8.2.49)$$

$$\chi_h(x) \in \tilde{\beta}(y_h(x)) \text{ a.e. in } \Omega. \qquad (8.2.50)$$

From (8.2.46), (8.2.47) and (8.2.48) we get that

$$a_{u_h}(y_h, \bar{z}_h) \to a_u(y, \bar{z}) \qquad (8.2.51)$$

$$(f + Bu_h, \bar{z}_h) \to (f + Bu, \bar{z}). \qquad (8.2.52)$$

Using the same method as in the previous section we can show that $\{\chi_h\}$ is weakly precompact in $L^1(\Omega)$ and thus there exists $\chi \in L^1(\Omega)$ such that

$$\int_\Omega \chi_h \bar{z}_h d\Omega \to \int_\Omega \chi \bar{z} d\Omega. \qquad (8.2.53)$$

From (8.2.49), (8.2.51)÷(8.2.53) we obtain that $\chi \in V'$ and that

$$a_u(y, \bar{z}) + \int_\Omega \chi \bar{z} d\Omega = (f + Bu, \bar{z}) \quad \forall \bar{z} \in V. \qquad (8.2.54)$$

It remains to show that $\chi(x) \in \tilde{\beta}(y(x))$ a.e. in Ω. This is proved by using the same procedure as in Prop. 8.2.2. q.e.d.

Let us denote further by $\tilde{X}(u)$ the set

$$\tilde{X}(u) = \{y \in V | \text{ there exists } \{u_h\}, \qquad (8.2.55)$$
$$u_h \in U_{ad}^h \text{ such that } u_h \to u \text{ weakly in } U$$
$$\text{and } y_h(u_h) \to y \text{ weakly in } V, \text{ where } y_h(u_h) \in X_h(u_h)\}.$$

Then the following proposition results immediately:

Proposition 8.2.12 It is true that $\tilde{X}(u)$ is nonempty and that

$$\tilde{X}(u) \subseteq X(u) \quad \forall u \in U_{ad}. \qquad (8.2.56)$$

Proposition 8.2.13 The set $\tilde{X}(u)$ is weakly compact in V.

Proof. By definition $\tilde{X}(u)$ is a weakly closed set in $X(u)$ q.e.d.

Let us now define the following problem $\tilde{\mathbb{P}}(u)$ for any $u \in U_{ad}$:
Find $\tilde{y} = \tilde{y}(u) \in \tilde{X}(u)$ such that

$$I(\tilde{y}, u) \leq I(y, u) \quad \forall y \in \tilde{X}(u). \qquad (8.2.57)$$

Note that a $\tilde{y}$ satisfying (8.2.57) exists as it results from Prop. 8.2.13. Let us introduce the notation $\tilde{E}(u) = I(\tilde{y}, u) \; \forall u \in U_{ad}$ and let us define the problem $\tilde{\mathbb{P}}$:
Find $\tilde{u}^\star \in U_{ad}$ such that

$$\tilde{E}(\tilde{u}^\star) \leq \tilde{E}(u) \quad \forall u \in U_{ad}. \qquad (8.2.58)$$

The following theorem concerns the relationship between the problems $\mathbb{P}_h$ and $\tilde{\mathbb{P}}$ as $h \to 0_+$.

Theorem 8.2.2: Let $u_h^\star$ be a solution of $\mathbb{P}_h$ and $y_h^\star \in V_h$ the corresponding solution of $\mathbb{P}_h(u_h^\star)$. Then there exist subsequence of $\{u_h^\star\}$, $\{y_h^\star\}$ such that as $h \to 0_+$

$$u_h^\star \to \tilde{u}^\star \text{ weakly in } U \qquad (8.2.59)$$
$$y_h^\star \to \tilde{y}^\star \text{ weakly in } V \qquad (8.2.60)$$

where $\tilde{u}^\star$ is a solution of $\tilde{\mathbb{P}}$ and $\tilde{y}^\star$ a solution of $\tilde{\mathbb{P}}(\tilde{u}^\star)$.

Proof. Both $\{u_h^\star\}$ and $\{y_h^\star\}$ are bounded. Thus there exist subsequences satisfying (8.2.59) and (8.2.60). From Prop. 8.2.11 and the definitions (8.2.55) of $\tilde{X}(u)$ we have that $\tilde{y}^\star \in \tilde{X}(\tilde{u}^\star)$. The definition of $\mathbb{P}_h$ implies that

$$I(y_h^\star, u_h^\star) \leq I(y_h, u_h) \quad \forall u_h \in U_{ad}^h, \; \forall y_h \in X_h(u_h). \qquad (8.2.61)$$

Let $\bar{u} \in U_{ad}$ and $\bar{y} \in \tilde{X}(u)$ be given. The definition of $\tilde{X}(\bar{u})$ implies that there exists a sequence $\{\bar{u}_h\}, \bar{u}_h \in U_{ad}^h$, such that as $h \to 0_+$

$$\bar{u}_h \to \bar{u} \text{ weakly in } U \qquad (8.2.62)$$

and for $\bar{y}_h \in X_h(\bar{u}_h)$ we have that

$$\bar{y}_h \to \bar{y} \text{ weakly in } V. \qquad (8.2.63)$$

Now we set $y_h = \bar{y}_h$ and $u_h = \bar{u}_h$ in the right-hand side of (8.2.61) we pass to the limit $h \to 0_+$ and use (8.2.45). It results that

$$I(\tilde{y}^*, \tilde{u}^*) \leq I(\bar{y}, \bar{u}) \quad \forall \bar{u} \in U_{ad}^h, \ \forall \bar{y} \in \tilde{X}(\bar{u}) \qquad (8.2.64)$$

i.e. that $\tilde{u}^*$ solves problem $\tilde{\mathbb{P}}$ and $\tilde{y}^*$ problem $\tilde{\mathbb{P}}(\tilde{u}^*)$. q.e.d.

Note that if $\tilde{X}(u) = X(u) \ \forall u \in U_{ad}$ then $\tilde{\mathbb{P}}(u)$ and $\mathbb{P}(u)$ are identical. The same holds for $\tilde{\mathbb{P}}$ and $\mathbb{P}$. In this case $\mathbb{P}_h$ is a real approximation of $\mathbb{P}$. In the general case, in which $\tilde{X}(u), \subseteq X(u)$ $\mathbb{P}_h$, is the approximation of a certain restricted problem as we have obtained also from Theorem 8.2.2.

8.3 Applications to Engineering Problems

The functional framework of the previous sections is quite general and allows the application of the propositions and theorems already proved to large classes of mechanical problems. Those are the optimal control (i.e. optimum weight or optimum deviation problems from a given state) and the parameter identification problems which can be formulated–with a cost functional having the properties (8.1.20) and (8.1.21)–with respect to the mechanical problems given in Sect. 6.5. (cf. also [Panag92]). Of course in some cases some minor modifications are necessary, as e.g. the L-operator of the hemivariational inequality (6.1.51).

As an example we shall study the optimal control or parameter identification problem for a quite general hemivariational inequality. It results if l elastic bodies are connected along k interfaces which introduce a nonmonotone multivalued stress-strain relation. The corresponding hemivariational inequality is given in (4.2.14).

With respect to this hemivariational inequality we formulate the optimal control problem. Let $V_{(m)}$ be a Hilbert space, $V'_{(m)}$ its dual space and $(.,.)$ the duality pairing between them $(m = 1, \ldots, l)$. We assume that $V_{(m)}$ is a set of functions defined on an open, bounded subset $\Omega^{(m)} \subset \mathbb{R}^2$. Let us assume that $V_{(m)} \subset L^2(\Omega^{(m)}) \subset V'_{(m)}$, where the injections are continuous and

$$V_{(m)} \subset L^2(\Gamma_S^{(m)}) \text{ is compact } m = 1, \ldots, l. \qquad (8.3.1)$$

Moreover let

$$\{y^{(m)} \in V_{(m)} | y^{(m)}_{|\Gamma_S^{(m)}} \in L^\infty(\Gamma_S^{(m)})\} \text{ be dense in } V_{(m)}, \qquad (8.3.2)$$

where $|_{\Gamma_S^{(m)}}$ denotes the restriction to $\Gamma_S^{(m)}$. Obviously for $V_{(m)} = H^1(\Omega^{(m)})$ these two last assumptions are satisfied. We consider the following hemivariational inequality:
Find $y = \{y^{(m)}\}, y^{(m)} \in V_{(m)}, m = 1, \ldots, l$, such that

8.3 Applications to Engineering Problems

$$\begin{cases} \sum_m a^{(m)}(y, y^\star - y) + \sum_{L=N,T} \int_{\Gamma^{(q)}} j^0_{L(q)}([y_L^{(q)}], [y_L^{(q)}]^\star - [y_L^{(q)}]) d\Gamma \geq (p, y^\star - y) \\ \forall y^\star = \{y^{\star(m)}\} \in V_{(1)} \times V_{(2)} \times \ldots \times V_{(l)}\} = V. \end{cases}$$

(8.3.3)

For every m, $a^{(m)}(\cdot, \cdot)$ is a bilinear form which is continuous, symmetric and coercive on $V_{(m)}$, $j_{N(q)}$ (resp. $j_{T(q)}$) is a nonconvex superpotential of the q-interface in the normal (resp. the tangential) direction to the interface, p represents the given loading and is defined by the expression

$$(p, y) = \sum_{m=1}^{l} (f^{(m)}, y^{(m)}) = f^{(m)} \in V'_{(m)} \quad m = 1, \ldots, l \tag{8.3.4}$$

and finally

$$\sum_{L=N,T} \int j^0_{L(q)}([y_L^{(q)}], [y_L^{(q)}]^\star - [y_L^{(q)}]) d\Gamma \tag{8.3.5}$$

$$= \sum_{q=1}^{k} \Big\{ \int_{\Gamma^{(q)}} j^0_{N(q)}([y_N^{(q)}], [y_N^{\star(q)}] - [y_N^{(q)}]) d\Gamma$$

$$+ \int_{\Gamma^{(q)}} j^0_{T(q)}([y_T^{(q)}], [y_T^{\star(q)}] - [y_T^{(q)}]) d\Gamma \Big\}.$$

Now we assume that $j_N : \mathbb{R} \to \mathbb{R}$ and $j_T : \mathbb{R} \to \mathbb{R}$ result from $b_N \in L^\infty_{\text{loc}}(\mathbb{R})$ and $b_T \in L^\infty_{\text{loc}}(\mathbb{R})$ as in (1.2.51), (1.2.52) and let $\tilde{b}_N$ and $\tilde{b}_T$ be the corresponding multivalued functions on $\mathbb{R}$ resulting from b_N and b_T by "filling in the jumps". Then (1.2.55) holds for $\tilde{b}_N, j_N$ and $\tilde{b}_T, j_T$ respectively. In order to define the optimal control and identification problem we introduce a control variable u. We assume further that the energy terms $a^{(m)}(\cdot, \cdot)$ depend on u and we denote them as $a_u^{(m)}(\cdot, \cdot)$.

Let us assume that $u \in U_{\text{ad}}$, where U_{ad} is not empty, closed convex subset of a Hilbert space U. For any $u \in U_{\text{ad}}$ we consider the following problem

$$(P(u)) \begin{cases} \text{Find } y = y(u) \in V \text{ such that} \\ \sum_m (A_m(d)y, y^\star) + \sum_{L=N,T} \int_{\Gamma^{(q)}} \chi_L^{(q)}[y_L^{\star(q)}] d\Gamma = (p + Bu, y^\star) \; \forall y^\star \in V. \\ \chi_L^{(q)} \in L^1(\Gamma^{(q)}) \\ \chi_L^{(q)} \in \tilde{b}_{L(q)}([y_L^{(q)}]) \text{ a.e. on } \Gamma^{(q)}, \quad q = 1, \ldots, k, \; L = N, T. \end{cases}$$

(8.3.6)

Here $B \in L(U, V')$ (i.e. a linear, continuous mapping). We assume that for any $u \in U_{\text{ad}}$, $A_m(u) \in L(V_{(m)}, V'_{(m)})$ is generated by the bilinear, symmetric continuous form $a_u^{(m)}(\cdot, \cdot)$, i.e. relations similar to (8.1.14) hold. Moreover, the bilinear forms $a_u^{(m)}(\cdot, \cdot)$ have the properties (8.1.4) (8.1.5) and (8.1.15). Finally we define the cost functional $J : V \times U \to \mathbb{R}$. Let $N : U \to U$ be a symmetric, linear continuous operator such that

$$(Nu, u)_U \geq \tilde{c}||u||^2_U, \quad \forall u \in U \quad \tilde{c} \text{ const } > 0. \tag{8.3.7}$$

8. Optimal Control and Identification Problems

Let H be the Hilbert space of observations and let $M \in L(V, H)$ be a given operator. Moreover let $h \in H$ be given. With every control $u \in U_{ad}$ we associate the cost functional

$$J(y(u), u) = (Nu, u)_U + ||My(u) - h||_H^2 \qquad (8.3.8)$$

which represents a very large class of optimal control and identification problems. The optimal control problem is defined as in (8.1.23).

For the problem which we have formulated all propositions and the two theorems proved in the previous section hold (cf. also [Panag90] for the corresponding necessary conditions).

Part IV
NUMERICAL APPLICATIONS

9. On the Numerical Treatment of Hemivariational Inequalities

This chapter is the first one of a series of chapters concerning the numerical solution of hemivariational inequalities. Since we are interested in engineering applications we deal with realistic problems, which have a large number of unknowns. After a short description of the first attempts to find the numerical solution of a hemivariational inequalities we deal mainly with four methods which seem today to be quite efficient for nonmonotone possibly multivalued engineering problems leading to hemivariational inequalities. The two first methods are, the "Microspring Approximation of the Decreasing Branches" and the "Decreasing Branch Approximation by Monotone Laws" which are described in this chapter. The second method is generalized in the next chapter for any hemivariational inequality. The two other methods are, the method of the "Substationarity Point Search" (Chapt. 11 and the "Decomposition into Two Convex Problems" Chapt. 12. It should be noted that the numerical methods presented in this part of the book do not treat all types of hemivariational inequalities. We prefer to sacrifice this in favour of the efficiency of the numerical treatment of more restricted classes of nevertheless, interesting problems. On the contrary the initial numerical efforts aimed to the maximum of generality; this was very soon abandoned. The chapters of the present part of the book do not treat the typical questions of convergence, stability etc. with respect to the developed numerical methods. This would be outside the scope of the present book even if it would be possible; indeed there are still many open questions concerning the rigorous numerical analysis of hemivariational inequalities. Accordingly we are compelled to confine ourselves to an engineering oriented numerical treatment of hemivariational inequalities enriched with representative numerical examples chosen to illustrate the efficiency and addressable span of problems of each approach. The convergence, stability etc. of the proposed numerical methods are checked on the basis of an abundance of existing numerical experience that has been aquired.

9.1 The First Numerical Attempts and the Questions of Stability and Uniqueness

The theory of hemivariational inequalities has been developed as an evolution of variational inequalities. Therefore the first attempts for the numerical treatment of hemivariational inequalities follow and generalize the methods applied for the numerical solution of the variational inequalities. Let us consider now a very simple problem leading to a variational inequality: this is the problem of a structure involving elastic cables in the framework of a small deformation theory. We recall [Nits67,71] that the cables may go slack and this fact is described by two inequalities (stress $s_i \geq 0$, nonnegative $v_i \geq 0$) and by the complementarity relation $s_i v_i = 0$ stating that a cable element either will be slack or will have a nonnegative, i.e. noncompressive, force (cf. e.g. [Pan85] p.350). The arising system of equalities and inequalities for a discretized structure includes the conditions of equilibrium, the compatibility relations, the constitutive laws of the elements of the structure and the two mentioned inequalities and the complementarity condition for each cable element, and is equivalent to a quadratic programming problem (Q.P.P.), i.e. to an inequality constrained minimization problem. The expression to be minimized is either the potential or the complementary energy of the structure.

The aforementioned equalities and inequalities are the Kuhn-Tucker conditions for the Q.P.P. which, if the matrix of the quadratic form is positive definite, are necessary and sufficient for the existence of the solution. The numerical treatment of such a cable-structure under a given loading seems to be difficult, because we do not a priori know which cables are slack and which are not, and this thought might lead someone to the use of a combinatorial-like method. However things are much more simple because the inequality constrained Q.P.P. when solved numerically by an appropriate quadratic programming (Q.P.) algorithm leads automatically to the solution, i.e. it "decides" which cables are slack and which are not and therefore transmit tensile forces. For a positive definite matrix the solution is unique.

Note also that any classical method, i.e. a method, which does not lead to a Q.P.P., for instance an incremental-iterative, or trial and error method, can be applied; its solution will be solution of the problem if it satisfies the complete set of Kuhn-Tucker conditions. Any attempts to obtain the solution through trial and error methods which do not make a final check for all the Kuhn-Tucker conditions, may lead to a wrong result (cf. [Pan85] Chapt. 10).

Now let us assume that the final value of the loading is attained through a sequence of load increments $\dot{p}$. Suppose then that by the use of some classical solution scheme we find a displacement, a strain and a stress field. Obviously we have to check whether this "solution" satisfies the Kuhn-Tucker conditions and then if it indeed is or not the solution of the problem due to the uniqueness of it. However the classical schemes of incremental methods may not be able to give a result due to the horizontal and vertical part of the graphs of the cable law e.g. if the law $s_i \geq 0$, $v_i \geq 0$, $s_i v_i = 0$ is considered (cf. Fig. 9.1.1).

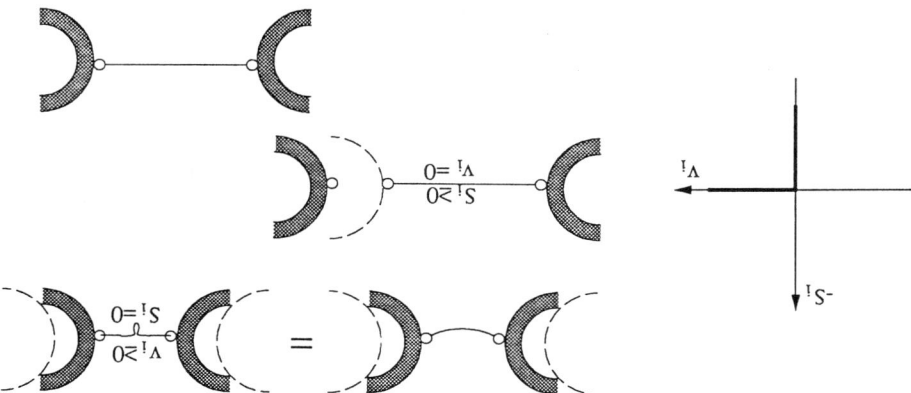

Fig. 9.1.1. The cable-element law.

Analogous questions arise in all problems expressed in terms of variational inequalities (monotone case), i.e. involving convex superpotentials. Again we may apply a minimization procedure, or any other classical incremental iterative method always with the restrictions cited before: the final result must satisfy all the static and kinematic relations of the problem and the incremental method may fail to work for horizontal or vertical parts of the assumed $\varepsilon - \sigma$ material laws. For these cases the use of Q.P. algorithms is advisable in order to overcome this difficulty.

Moreover, in the case of large deformations or lack of uniqueness of the solution, one must be careful in the application of an incremental method. The increments must be small enough because otherwise some equilibrium configurations can be lost during the incremental procedure. Indeed, because of stability reasons, the incremental algorithm may jump from one solution path to another.

For nonconvex superpotentials and hemivariational inequalities we have analogous but more difficult questions. Let us consider first the purely static problem: The total load p is given and we want to determine all possible positions of equilibrium. The problem is formulated as a substationarity problem and every such a point of the potential energy corresponds to an equilibrium configuration. There are several numerical difficulties concerning the determination of a local minimum of a nonconvex generally nondifferentiable function (cf. e.g. [Murty]). We mention here as a possible tool the bundle method of Lemarechal-Strodiot [Strod] which was the first we examined with numerical experiments concerning its applicability to this problem. This method generalizes the gradient method for optimization problem for the case in which, roughly speaking, the "gradient is multivalued". The algorithm it is appropriate for the numerical treatment of the substationarity problem $0 \in \bar{\partial} f(x)$, where $\bar{\partial}$ is the generalized gradient. For the description of this method we refer the reader to the [Strod]. Analogously to the case of cable structures, and, let us say, with respect to the diagram of Fig. 2.4.1f one could make some combinatorial attempts to find on which part of Fig. 2.4.1f the solution can appear. But the substationarity property of the solution gives the answer and no need for combinatorial thoughts exists. However we did not continue to apply the bundle method to the numerical solution of hemivariational inequalities, because, according to our experience one could not solve large problems (more than 100 unknowns) and there were additional difficulties in the case of generalized gradients involving complete vertical branches. Of course we cannot exclude the possibility that new, more efficient codes of the bundle method may exist which would be appropriate for large scale engineering problems.

Another method related to the previous one is the regularization method whose convergence has been already discussed in Chapters 6, 7 and 8 as part of the existence proofs. Let us describe the method with respect to the hemivariational inequality (4.2.14). We assume that superpotentials $j_{N(q)}$ and $j_{T(q)}$ are approximated by the smooth functions $j_{N\varepsilon}$ and $j_{T\varepsilon}$ depending on ε (we omit (q)), such that, as $\varepsilon \to 0$ $j_{N\varepsilon}$ (resp. $j_{T\varepsilon}$) tends to j_N (resp. j_T). Any heuristic method for the regularization can be applied: For instance, the graphs of the corresponding normal and tangential boundary reaction-displacement laws can be appropriately smoothened and any vertical line (multivaluedness) can be replaced by an inclined line. Due to the smoothness of $j_{N\varepsilon}$ and $j_{T\varepsilon}$, the directional differentials $j_N^0(\xi, z)$, $j_T^0(\xi, z)$ are replaced by $\dfrac{dj_N(\xi)}{d\xi}z$ and $\dfrac{dj_T(\xi)}{d\xi}z$ and thus we are led to the following problem:

Find $u_\varepsilon^{(m)}$, $m = 1, 2, \cdots, l$, such as to satisfy the variational equality

$$\sum_{m=1}^{l} \alpha(u_\varepsilon^{(m)}, v^{(m)}) + \sum_{q=1}^{k}\{\int_{\Gamma_q}\{\frac{dj_{N\varepsilon(q)}}{du_N}([u_{N\varepsilon}^{(q)}])[v_N^{(q)}] \quad (9.1.1)$$

$$+ \frac{dj_{T\varepsilon(q)}}{du_T}([u_{T\varepsilon}^{(q)}])[v_T^{(q)}]\}d\Gamma\}$$

$$- \sum_{m=1}^{l}[\int_{\Omega^{(m)}} f_i^{(m)}v_i^{(m)}d\Omega + \int_{\Gamma_F^{(m)}} F_i^{(m)}v_i^{(m)}d\Gamma] = 0$$

$$\forall v \in V_{ad}$$

By means of discretization, (9.1.1) is equivalent to a system of semilinear algebraic equations depending on ε. The solution of this system "tends" as $\varepsilon \to 0$ to the solution of the initial problem. The system of nonlinear algebraic equations has the general form

$$\boldsymbol{Ku} + \boldsymbol{B}_\varepsilon(\boldsymbol{Tu}) = \boldsymbol{p} \quad (9.1.2)$$

where $\boldsymbol{u}$ (resp. $\boldsymbol{p}$) is the total displacement (resp. load) vector and $\boldsymbol{K}$ is the stiffness matrix. Here $\boldsymbol{T}$ is a transformation matrix such that $\boldsymbol{Tu}$ gives the appropriate relative displacements and $\boldsymbol{B}_\varepsilon(\cdot)$ is the vector of the nonlinear terms

9.1 The First Numerical Attempts and the Questions of Stability and Uniqueness   243

depending on ε. Noteworthy is the sparsity of the nonlinear terms which in some cases only slightly influences the monotonicity of the problem. In this case the classical Newton-Raphson algorithm has given acceptable numerical results. However, it remains still an open question the testing of other types of algorithms like the fixed point algorithms and the homotopy algorithms. On the other hand the counterexample given by Demyanov for the regularization method in optimization problems (cf. [Demy86c],p.7), should be always taken seriously into account and one should verify if the solution found by solving the system (9.1.2) as $\varepsilon \to 0$, fulfills with an appropriate numerical error, the initial problem. All the aforementioned numerical methods are general and their main disadvantage is the large computation time even for small number of unknowns. In contrast to these general methods it very soon became apparent that one should develop numerical methods appropriate for the efficient treatment of restricted classes of hemivariational inequalities.

One such method, which is appropriate for onedimensional nonconvex superpotential laws involving strong nonmonotonicities but without complete vertical branches is the following method proposed by Koltsakis [Kol91]. It is called "the microspring or sawtooth approximation of the decreasing branches" and also holds for multidimensional nonconvex superpotential laws. This method is briefly described in Fig. 9.1.2: the dotted vertical parts of the graph denote

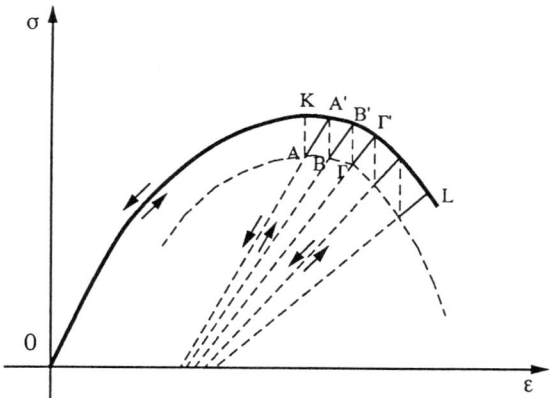

Fig. 9.1.2. On the microspring approximation of a decreasing branch.

local jumps and the decreasing branch is "attained" by small monotone increasing "stress recovery" parts AA', BB'. The physical explanation of this approximation is rather obvious: the replacement of the decreasing continuous branch KL by the discontinuous curve KAA'BB'$\Gamma\Gamma'\cdots$, where KA, A'B, B'Γ etc. are jumps, describes the gradual failure of the material on the decreasing branch. Recall at this point that the discontinuous graph KAA'BB'$\Gamma\Gamma'\cdots$ corresponds to a fan (cf. Sect. 3.5) thus giving rise to an F-hemivariational inequality. Thus the method of sawtooth approximation of a decreasing branch consists in the replacement of a hemivariational inequality by an F-hemivariational inequality, i.e. in the replacement of the sublinear directional differential of Clarke by the bisublinear energy form of the fans (cf. Sect. 1.4.) Small numerical applications solved by this method are given in [Pan87a,c]. Longer examples are given in [Kol91]. In the next section we shall describe the method and we shall give a numerical application of it.

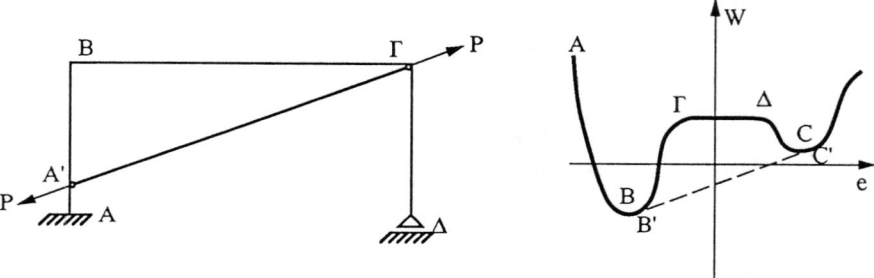

Fig. 9.1.3. On the questions of stability and uniqueness of the solution.

In all the methods we have until now mentioned there are considerable difficulties in obtaining numerically all the solutions of the hemivariational inequality holding at the position(s) of equilibrium of the structure under consideration. At this point we would like to remark that a structure may have a position of overall equilibrium in which one or some elements of the structure are in unstable equilibrium. For instance the frame $AB\varGamma\varDelta$ of Fig. 9.1.3a with strong linear elastic vertical and horizontal beams, is always in equilibrium under the loading P even if the rod $A'\varGamma$ has a stress-strain law derived by the potential function of Fig. 9.1.3b. Note that a simple rod-element having this strain energy would have as a position of stable equilibrium the point B and as other possible position of equilibrium the point C and all the points between $\varGamma$ and $\varDelta$ ($\varGamma\varDelta$ is an horizontal line segment). Indeed all these positions of equilibrium are substationarity points. Several authors, of both mathematical and engineering origin, apply a convexification procedure for the energy. It is also noteworthy that a convexification of the energy would eliminate several realizable positions of equilibrium (dotted line in Fig. 9.1.3b).

Let us now assume that the final value of the loading in a problem involving nonconvex energy functions is attained through a sequence of load increments. Let us apply a classical incremental-iterative procedure to find the solution of the arising hemivariational inequalities. Then the danger of obtaining a "wrong" result is greater than in the monotone case, due to the lack of monotonicity of the problem: it is well-known, for instance, that the classical incremental-iterative methods face serious numerical difficulties especially if one wants to take into account complete vertical jumps in a stress-strain diagram, for which the classical incremental-iterative schemes do not work. In this direction the following possibility should be taken into account analogously to the large displacement theory of cable structures, where an incremental-iterative procedure is combined with a Q.P.P. within each load increment, we may combine the incremental-iterative procedure with the solution of the hemivariational inequality within each load increment. Then the numerical solution of the hemivariational inequality within the considered load increment can be obtained using the methods developed in the forthcoming sections for the efficient numerical treatment of hemivariational inequalities.

9.2 The Microspring Approximation Method of the Decreasing Branch

Let us consider an onedimensional reaction displacement law which is linear up to the displacement u_0 and descending linearly until the displacement u_1 is reached. For $u > u_1$ we have zero reaction, i.e. the law is inactive.

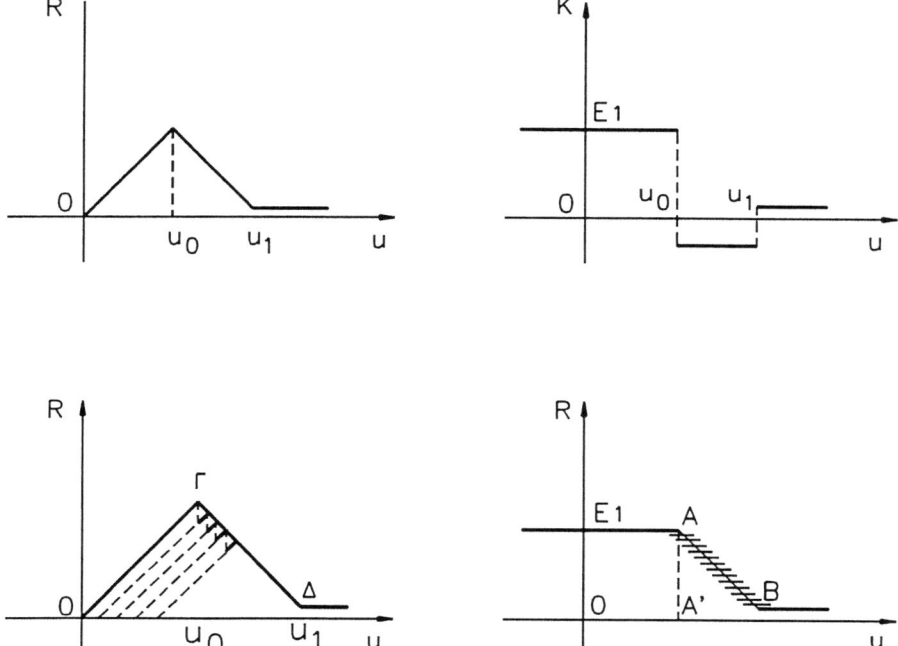

Fig. 9.2.1. About the microspring or sawtooth approximation.

The reaction is given by the formula (Fig. 9.2.1a)

$$R(u) = F_1(u) + F_2(u)H(u - u_0) + F_3(u)H(u - u_1) \qquad (9.2.1)$$

where $H(\cdot)$ is the Heaviside step function ($H(x) = 0$ for $x < 0$ and $H(x) = 1$ for $x > 0$) and F_1, F_2, F_3 are appropriate linear functions so as to produce the trilinear form of the diagram. The Heaviside functions represent the triggering of different physical mechanisms after characteristic values of deformation. In the diagram we have ascending slope equal to E_1, and descending slope $-E_2$. The exact forms of F_1, F_2, F_3, read: $F_1 = E_1 u$, $F_2 = (E_1 + E_2)u_0 - (E_1 + E_2)u$ and $F_3 = -E_2 u_1 + E_2 u (E_1 > 0, E_2 > 0)$ The stiffness of this law taken with a classical differentiation process (Fig. 9.2.1b) reads

$$K(u) = \frac{dR(U)}{du} = \begin{cases} E_1 & \text{for } u < u_0 \\ -E_2(<0) & \text{for } u_0 < u < u_1 \\ 0 & \text{for } u > u_1 \end{cases}. \qquad (9.2.2)$$

The resulting negative stiffness $-E_2$ of this model is the major drawback both mechanically and numerically. In order to remove the negative stiffness, which may lead to serious numerical difficulties, the present method replaces the decreasing branch by a sawtooth discontinuous diagram (Fig. 9.2.1c) and thus a stepwise decreasing stiffness results (Fig. 9.2.1d). Note that if the number of steps can be taken to tend to infinity the nonsmooth curve is approximated smooth "curve" (in Fig. 9.2.1d) (the segment AB). This introduces no numerical error as long as only values and not slopes of the approximated step-like function are required. This kind of approximation was influenced by the model of Dougill and Rida [Dugi76,80] concerning the progressively fracturing solids. Such a material: i) Begins to progressive loose stiffness with increasing deformation after some characteristic value of strain ii) During unloading its behavior is linear elastic, the modulus depending in a monotone decreasing way on the magnitude of maximum strain imposed. iii) Returns to a state of zero strain with total removal of stress. The microspring or sawtooth approximation model of the decreasing branch maintains these three, basic properties but this fact requires the existence of some new factor: this is the lack of smoothness of the softening part of the model. Although the macroscopic appearance of the reaction displacement curve is continuous, the existence of a nonsmooth nature (sawtooth softening) in the microscopic level is postulated. This means that the structural degradation process, although gradual, is the result of infinitely many instabilities in the microscale of the material. This is not of course a new idea but permits to avoid the problem of the negative stiffness of the smooth softening approach: in the negligible small interval between two microfractures the material remains stable and a positive modulus will locally exist. The overall appearance of the stiffness displacement graph will be that of a decreasing but always positive function and in the microscopic level there appear infinitely many horizontal steps. Formally, we can write the reaction as function of the displacement:

$$R(u) = \lim_{k \to \infty} \sum_{i=0}^{k} \frac{E}{k} u (1 - H(u - i\frac{u_1 - u_0}{k})) \quad \text{for } u > 0 \qquad (9.2.3)$$

9.2 The Microspring Approximation Method of the Decreasing Branch

and the stiffness as:

$$K(u) = \lim_{k \to \infty} \sum_{i=0}^{k} \frac{E}{k}(1 - H(u - i\frac{u_1 - u_0}{k})) \quad \text{or} \quad (9.2.4)$$
$$K(u) = E - \lambda(u - u_0)H(u - u_0) + \lambda(u - u_1)H(u - u_1)$$

where $\lambda = K_0/(u_1 - u_0)$, for the special case of "linear softening". Here $H(\cdot)$ is the Heaviside step function. A physical analogon for a linear softening behavior is a fiber bundle with a uniform distribution of the ultimate deformation of the fibers between two values u_0 and u_1. By u_0 we denote the strain of initialization of the decreasing branch (let us call it here "damage" initialization) by u_1 the strain where total destruction has occured and by K_0 the stiffness of the intact fiber. (cf. also [Hult]). Let us now study a discretized structure with a nonsmooth, linearly softening support Γ_S.

Keeping the usual structural analysis symbols we can write for a classical structure:

$$\boldsymbol{Gs = p}, \quad \boldsymbol{\sigma = K_0 \varepsilon}, \quad \boldsymbol{\varepsilon = G^T u}, \quad (9.2.5)$$

hence

$$\boldsymbol{G^T K G u = K u = p} \quad (9.2.6)$$

the equilibrium, elasticity, strain displacement and global equilibrium equations. $\boldsymbol{K}_0$ is the local stiffness or elasticity matrix of the whole structure. Let us now introduce a partitioning of the equations of the problem into two parts: the first one corresponds to the linear part of the structure and the second to the nonsmooth, softening terms of the support interface Γ_S. After splitting $\boldsymbol{u}$ into $\boldsymbol{u}_1$ and $\boldsymbol{u}_2$ and writing $\boldsymbol{K}$ as $\begin{bmatrix} \boldsymbol{K}_{11} & \boldsymbol{K}_{12} \\ \boldsymbol{K}_{21} & \boldsymbol{K}_{22} \end{bmatrix}$ we obtain the relation

$$(\boldsymbol{A + B})\boldsymbol{w} = \bar{\boldsymbol{p}} \quad (9.2.7)$$

where :

$$-\boldsymbol{K}_{12}\boldsymbol{K}_{11}^{-1}\boldsymbol{K}_{21} = \boldsymbol{A}, \; \boldsymbol{K}_{22} = \boldsymbol{B}, \; \boldsymbol{w} = \boldsymbol{u}_2, \bar{\boldsymbol{p}} = \boldsymbol{p}_2 - \boldsymbol{K}_{21}\boldsymbol{K}_{11}^{-1}\boldsymbol{p}_1 \quad (9.2.8)$$

Further some definitions that are usefull to characterize the "damage" status of the structural system are given ("damage" is understood here as previously explained)

i) Elastic regime E: The set of possible displacement vectors for which $u_i < u_{0i}$.

ii) Fully "damaged" regime Θ: The set of displacement vectors for which $u_i > u_{1i} \; \forall i \in \Gamma_S$.

iii) Progressing "damage" regime Φ: The set of displacement arrays for which $u_{0i} < u_i < u_{1i}$ for some (not necessarily all) $i \in \Gamma_S$.

Further one can define the sets $\tilde{E}, \tilde{\Phi}$ and $\tilde{\Theta}$ as:

i') $\tilde{E}$: the set of load vectors for which the structure remains in the E regime.

ii') $\tilde{\Theta}$: the set of load vectors that make the structure completely enter the Θ regime.

iii') $\tilde{\Phi}$: the set of load vectors causing displacements that belong to the set Φ.

iv') Φ_S: the strict Φ set, i.e. the set of displacement vectors u for which $u_{0i} < u_i < u_{1i} \, \forall i \in \Gamma_S$.

v') $\tilde{\Phi}_S$: the stict $\tilde{\Phi}$ set, i.e. the set of load vectors that when applied to the structure described by relations (9.2.5) produces a displacement vector belonging to Φ_S.

From eq. (9.2.6) we can see that if the structure "operates" in the E regime, the contribution of the nonlinear support is linear and is numerically effectuated by the matrix diag(K_{si}), where K_{si} stands for the contribution of the i-th nonlinear spring in its intact state. This contribution will be null when the system enters the Θ regime. Combining (9.2.5) with the explicit expressions of the linearly softening terms (cf. (9.2.4)) we obtain the nonlinear equilibrium equations of the softening support as:

$$(A + B_1 - \text{diag}(\lambda_i w_i H_{0i}))w = \bar{p} \qquad (9.2.9)$$

where

$$\begin{aligned} B_1 &= B_0 + \text{diag}(u_{0i} + \lambda_i u_{0i} H_{0i}), & (9.2.10) \\ H_{0i} &= H(u_i - u_{0i}), \\ B_0 &= B - \text{diag}(K_{si}), \\ H_{1i} &= H(u_i - u_{1i}). \end{aligned}$$

In (9.2.9) the components of w belong to the Φ_S regime only. We notice that if in (9.2.3) the number of microsprings is infinite, then the nonlinear equation (9.2.9) gives to us a numerical solution by means of the Newton-Raphson procedure. There are several other interesting results obtained from the study of (9.2.9), especially concerning the measure of the global "damage" of the system. We refer in this context to [Kol,Kolts].

As a numerical application we shall present the structure of Fig. 9.2.2a (from [Kol]. In this example we study the behaviour of an end plate, adhesively connected to a rigid substratum, subjected to forces normal to its support, ultimately causing the debonding. The plate is taken to be a simple composite plate with an isotropic NARMCO-core and orthotropic fiber glass facings (respectively E = 338000 t/m^2, ν = 0.35, for the core and E$_1$ = 556000 t/m^2, E$_2$ = 1854000 t/m^2, ν_{12} = 0.25, G$_{12}$ = 891000 t/m^2 for the facings made of epoxy glass F. Here E denotes the modulus of elasticity, ν denotes Poisson's number and G is the shear modulus. The dimensions are 0.4 m by 0.4 m and the total thickness is 0.03 m. For the adhesive material two cases

9.3 The Method of Decreasing Branch Approximation by Monotone Laws

are studied concerning the relation between intact region modulus (i.e. the segment OA' in Fig. 9.2.1d) and the "damaged" region (i.e. the segment A'B in Fig. 9.2.1d): the later region is taken to be equal and three times as long on the intact region. This makes two cases of different softening ductility. What can be drawn as a conclusion from Figs. 9.2.2b,c is the ability of the method to describe stress situations well within the decreasing branch regime Φ_S. Although the aim of the book is not to study the influence of the behaviour of different adhesives to the margins of safety of a structure in detail, one cannot fail to observe the gradual (ductile) transition into the "damaged" state that the mildly softening adhesives Fig. 9.2.2c,d exhibit over their more brittle counterparts Fig. 9.2.2a,b. The convergence rate proved [Kol] to be Newton-like. The example does not attempt to combine the unilateral friction and contact phenomena with the failing adhesion mechanism; relevant research is currently under way.

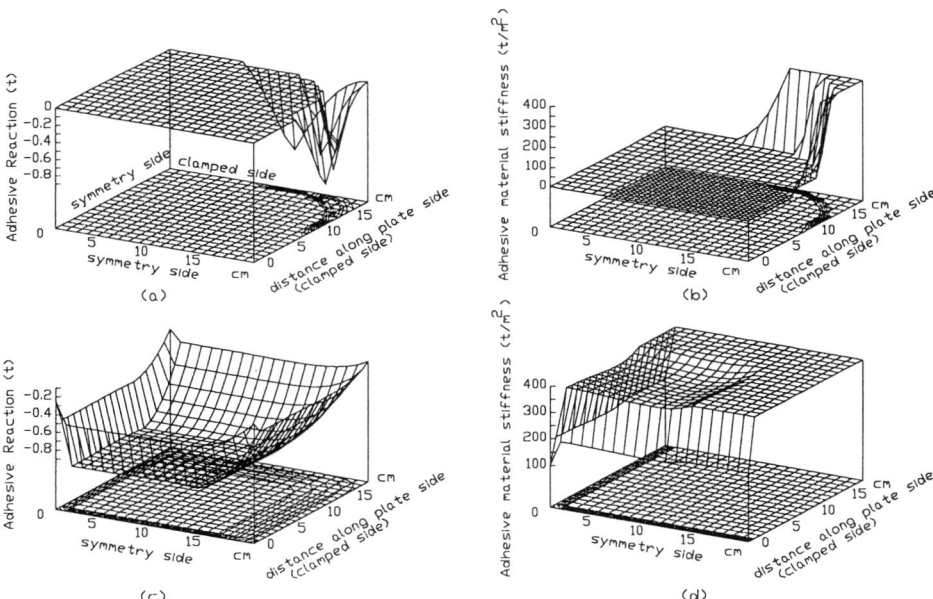

Fig. 9.2.2. Stresses (a,c) and adhesive layer stiffness distribution (b,d) of an all around clamped square plate (only the upper right quarter is depicted as symmetry allows). For a,b $u_0 = 2.5$ mm, $u_1 = 5.0$ mm; for c,d $u_0 = 2.5$ mm, $u_1 = 10.0$ mm.

9.3 The Method of Decreasing Branch Approximation by Monotone Laws

Zigzag stress-strain or reaction-displacement laws appear in several mechanical problems. They constitute a special case of the nonmonotone multivalued laws. We can mention here the reaction-displacement (or relative displacement) diagram of Fig. 9.3.1a which results in a sense tangential to the interface, if two bodies are in adhesive contact, i.e. if they are glued by an adhesive material. This material can sustain a small tension or compression and then undergoes either brittle fracture (dotted line) or semibrittle fracture. In the case of brittle fracture the law has complete vertical branches, i.e. it is a multivalued law.

The same effects may appear at the interface of sandwich beams and plates as well as in composite materials, e.g. in fiber reinforced materials between fibers and matrix and are called delamination effects. Also the nonmonotone variants of the well-known friction law of Coulomb (Fig. 9.3.1b) or the adhesiveless friction law of Fig. 9.3.1e or the friction law between reinforcement and concrete of Fig. 2.4.1c can also be mentioned. Similar is the situation with sawtooth stress-strain laws in reinforced concrete in tension (Scanlon's diagram) and in laminated and composite material structures (Fig. 3.2.2c, 9.3.1d). For this type of laws and for their physical meaning we refer to Sect. 2.4 and 3.2.

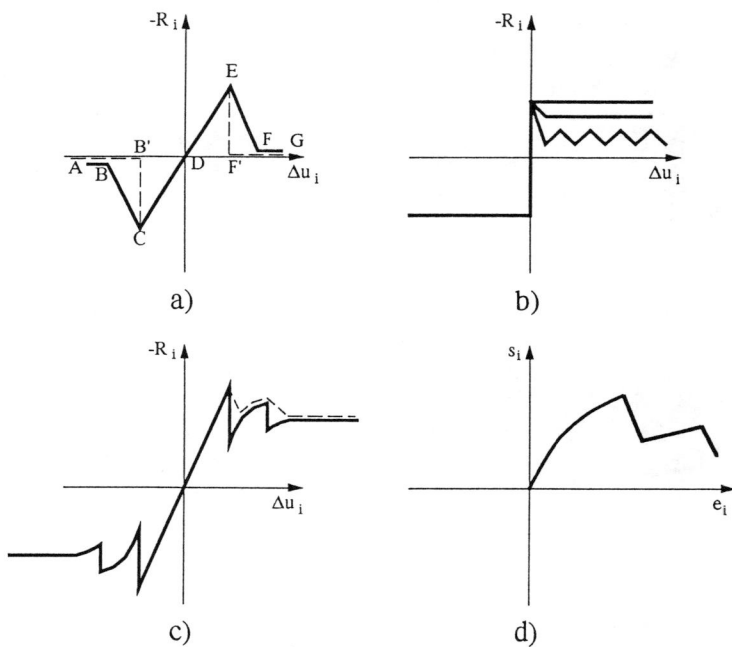

Fig. 9.3.1. Zigzag reaction-displacement and stress-strain laws.

As shown in the previous chapters nonmonotone possibly multivalued laws in structures and solids give rise to a new type of variational expresions in inequality form which the author has called hemivariational inequalities. They express the "principles" of virtual or complementary virtual work in inequality form.

The theory of hemivariational inequalities leads to the result that local minima of the potential or the complementary energy of the structure represent equi-

9.3 The Method of Decreasing Branch Approximation by Monotone Laws

librium positions of the problem. It is however possible, certain solutions of the problem not to be local minima but another more general type of points which make the potential or the complementary energy "substationary". As indicated those are solutions of the differential inclusion $0 \in \bar{\partial}\Pi(u)$ where Π is the potential energy, u is the displacement vector and $\bar{\partial}$ denotes the generalized gradient. Analogous results hold for the complementary energy expressed as function of the stresses.

Although the formulation of a nonmonotone problem as a hemivariational inequality has a lot of advantages concerning the theoretical mathematical study of this problem this formulation does not seem for the present to entail analogous advantages concerning the numerical treatment. Indeed the numerical determination of all local minima of a nonconvex function is still an open problem in numerical optimization [Pard],[Murty]. Moreover, only few results exist estimating the efficiency of the available nonconvex optimization algorithms for the determination of a local minimum. At this point we should mention that efficient numerical equilibrium path tracing methods for nonmonotone laws have been developed in [Cris82,86,88,91]. This last method is not based on a variational approach but on a direct enforcement of the fulfillment of nonmonotone stress-strain diagrams. For the sake of completeness we should also mention here the dissipation method (cf. e.g. [Frem83,88]), which has certain serious advantages from the standpoint of the physical interpretation of the arising nonmonotone laws.

However all the aforementioned methods do not treat the combinatorial aspect of the problem efficiently. Let us explain this fact here: Suppose that we have a structure containing certain elements obeying, the law of Fig. 9.3.1a. For a given loading we want to determine a corresponding equilibrium state. In an equilibrium state some elements may have a stress and strain state on the branch AB, other on BC, other on CDE etc., i.e. it is not a priori known which part of the stress-strain law will be realized. For a very small number of structural elements obeying this law one can determine the solution by considering all possible branch combinations of the stress strain law. For a larger number of elements one could apply an incremental back-and forward trial and error procedure for the loading in order to arrive to a solution of the problem (cf. e.g. the method of Crisfield [Cris91]).

Note that due to the lack of monotonicity no uniqueness is generally guaranteed. A difficult problem is the determination of the elements having stress and strains on AB or on BC etc. this is a problem of combinatorial nature. We recall at this point that in the case of monotone laws Fig. 9.3.2, e.g. in the case of a structure having only cables Fig. 9.3.2b, which may become slack, the solution of the inequality constrained problem of minimum potential or complementary energy gives those cables which are slack (on AB) or under tension (on BC) [Pan76]. This solution is obtained through a convex optimization algorithm [1]. The minimum is uniquely determined, due to the monotonicity of the stress-strain law. This approach works well for the more complicated law of Fig. 9.3.2c, as well as for its three-dimensional version, and determines, by the use of a convex programming algorithm, the solution of the problem. Note that for a continuous structure, e.g. if for instance a plate is glued with a support by an adhesive material having the law of Fig. 9.3.2c, we have to determine six regions on the plate according to the diagram; thus a convex programming (C.P.) algorithm determines all the free boundaries between the six a priori unknown regions . In the case of a monotone law of Fig. 9.3.2a,b the convex minimization problem reduces to a quadratic programming (Q.P) problem. Both C.P. and Q.P. problems may be treated by efficient algorithms yielding quickly a reliable solution of the problem. Concerning the numerical properties of Q.P. and C.P. algorithms see [Flet].

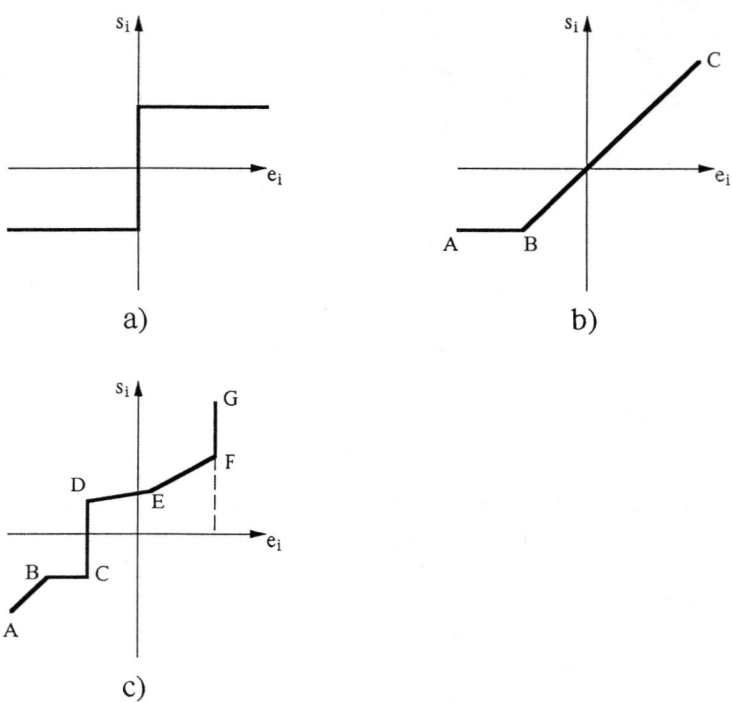

Fig. 9.3.2. Monotone laws leading to global minima.

The aim of the present section is to replace a nonmonotone law problem by a sequence of problems involving monotone laws, and in particular laws of the form of Fig. 9.3.2a leading to Q.P. problems. We thus succeed to extend the efficient treatment of the combinatorial character of the problem to nonmonotone laws and for large number of unknowns. This combinatorial task could be possibly fulfilled by the direct use of a nonconvex programming (N.C.P.) algorithm; but, in such a case no large scale problems could be treated, since the N.C.P. algorithms do not have the convergence rate and the robustness of the C.P. and the Q.P. algorithms.

9.3 The Method of Decreasing Branch Approximation by Monotone Laws 253

Comparisons of the present algorithm with the path-following numerical methods of [Cris91] have shown that the present method has considerable advantages concerning first the treatment of the complete vertical branches and second the combinatorial character of the problem. However, the method of [Cris82,86,88,91] can be more easily incorporated into general purpose finite element (F.E.) computer programs and gives more accurately the equilibrium path.

The case of unloading paths reduces to the repeated use of the algorithm presented here (cf. Sect. 3.4). In next chapter a generalization of the present algorithm to $3D$-nonmonotone laws is attempted. Note that this algorithm which first appeared in [Mis92a,b] is completely new in numerical analysis.

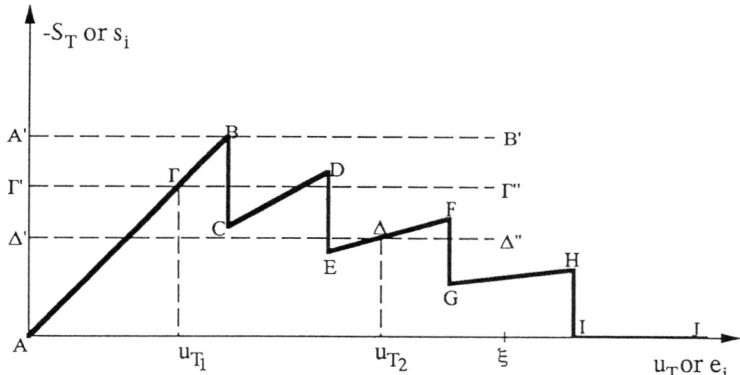

Fig. 9.3.3. On the description of the algorithm.

We shall first show the main idea of the present algorithm by means of Fig. 9.3.3. Let us assume that the diagram ABCD$\cdots$J of this figure represents the possible stress and strain law of certain elements of a structure, or a reaction displacement law, in the tangential direction with respect to the boundary of a body under consideration. We first assume that the fictitious law AA' BB' holds. If we deal with a reaction displacement zigzag law then AA'BB' holds on the boundary and it is a fictitious friction type law. If Fig. 9.3.3 describes a stress-strain law then AA'BB' is a fictitious rigid plastic law. Let us consider here the first case. Then the structure has a unique solution, as it is known from the solution of the variational inequality of the friction problem [Pan85]. Note that the arising problem is a friction type problem with a given fictitious normal force $S_N = C_N$, i.e. A' is prescribed (AA' $= \mu |C_N|$ where μ is the fictitious friction coefficient). It is worth noting that in the normal direction any other type of boundary condition may hold e.g $S_N = \tilde{C}_N$ given, where this $\tilde{C}_N$ may generally be different from the fictitious normal force C_N. Suppose that the friction type problem gives as a result certain u_{T_1}. Now we solve a new friction type problem, with given normal force, obeying the law A$\Gamma'\Gamma\Gamma''$. Suppose that the solution of this problem offers as a result a new u_{T_2} which gives rise to the new friction diagram A$\Delta' \Delta\Delta''$. This procedure is continued until in all elements the differences $|R_{i+1} - R_i|$ and $|u_{T_{i+1}} - u_{T_i}|$ become small enough. This covers the case of multivalued laws, i.e. of diagrams with complete vertical branches which are very common in composite materials and in adhesive contact problems. Note also that within each step the elements of the structure having the zigzag law may obey to different friction type laws, i.e. AA', AΓ'' and AΔ' is different from element to element.

Now we recall [Pan75,85] that in the case of an elastic body obeying a friction type law on a part Γ_S of its boundary (on Γ_S the normal forces are prescribed and equal to $\tilde{C}_N$), we have either to solve the minimum potential energy problem

$$\Pi(u) = \min\{\Pi(v) | v \in V_0\}, \qquad (9.3.1)$$

where $u = \{u_i\}$ and $v = \{v_i\}$, $i = 1, 2, 3$, represents the displacement of the body, or the minimum complementary energy problem

$$\Pi^c(\sigma) = \min\{\Pi^c(\tau) | \tau \in \Sigma_0\}, \qquad (9.3.2)$$

where $\tau = \{\tau_{ij}\}$, $i, j = 1, 2, 3$, represent the stress tensor of the body. Here V_0 is the kinematically admissible set, i.e. $V_0 = \{v | v_i = U_i \text{ on } \Gamma_U\}$, where $U_i, i = 1, 2, 3$, is the prescribed displacement on the rest part of the boundary denoted by Γ_U (Γ is made up of the three mutually disjoint parts $\Gamma_U, \Gamma_F, \Gamma_S$),

$$\Pi(v) = \frac{1}{2}a(v,v) - (l,v) + \int_{\Gamma_S} \mu |C_N| |v_T| d\Gamma \qquad (9.3.3)$$

with $a(\cdot, \cdot)$ the linear elastic strain energy, $(l, \cdot)$ the work of the external forces i.e of the volume forces f_i, the given normal forces $\tilde{C}_N$ on Γ_S and the given forces $S_i = F_i$ on the boundary part Γ_F. Moreover $\int_{\Gamma_S} \mu |C_N| |v_T| d\Gamma$ is the work of the frictional forces which extends over the boundary or the interface area Γ_S. In the diagram depicted in Fig. 9.3.3 u_T is the tangential boundary displacement. In the case of an interface problem, u_T is replaced by the tangential relative displacement of the two sides of the interface. Moreover μ may be different from point to point.

The complementary energy is given by

$$\Pi^c(\tau) = \frac{1}{2}A(\tau,\tau) - (\tau, U) \qquad (9.3.4)$$

where $A(\cdot, \cdot)$ is the elastic energy expressed in terms of the stress tensor $\tau = \{\tau_{ij}\}$, $i, j = 1, 2, 3$ and $(U, \cdot)$ is the work of the prescribed displacement $U = \{U_i\}$ of the body on the boundary part Γ_U. The statically admissible set is given by

$$\Sigma_0 = \{\tau | \tau = \{\tau_{ij}\},\ \tau_{ij} = \tau_{ji}, \tau_{ji,j} + f_i = 0,$$

$$T_N = \tilde{C}_N \text{ on } \Gamma_S, T_i = F_i \text{ on } \Gamma_F,\ |T_T| \le \mu |C_N| \text{ on } \Gamma_S\} \qquad (9.3.5)$$

9.3 The Method of Decreasing Branch Approximation by Monotone Laws

where $\{T_N, T_T\}$ are the normal and tangential components of the boundary traction $T = \{T_i\}$ with respect to the boundary. In the numerical implementations we prefer to solve the minimum complementary energy problem (9.3.2) (or dual problem) which leads to a Q.P. problem instead of the minimum potential energy problem (or primal problem) which contains the nondifferentiable term $\int_{\Gamma_S} \mu |C_N| |v_T| d\Gamma$. However, from the analytical point of view both the primal and the dual problems are equivalent because both of them yield the solution of the friction problem. We recall also that the primal friction problem with given normal force is equivalent to the following variational inequality expressing the principle of virtual work:

Find $u \in V_0$ such as to satisfy the inequality

$$a(u, v-u) + \int_{\Gamma_S} \mu|C_N|(|v_T| - |u_T|)d\Gamma \geq (l, v-u) \quad \forall v \in V_0. \tag{9.3.6}$$

Let us go now back to the nonmonotone multivalued zigzag law of Fig. 9.3.3. We consider the same linear elastic body Ω as in the previous friction problem under the action of $f = \{f_i\}$ in Ω, of $F = \{F_i\}$ on Γ_F, $U = \{U_i\}$ on Γ_U; we assume also that on Γ_S a normal force distribution $\tilde{C}_N$ is prescribed and that in the tangential direction with respect to the boundary part Γ_S the reaction-displacement law of Fig. 9.3.3 holds in the two directions of an intrinsic coordinate system on Γ (in the two-dimensional case $\Omega \subset \mathbb{R}^2$, this situation is more simplified). According to Sect. 4.1 an equilibrium position is characterized by the following hemivariational inequality problem:

Find $u \in V_0$ such as to satisfy the inequality

$$a(u, v-u) + \int_{\Gamma_S} j^0(u_T, v_T - u_T)d\Gamma \geq (l, v-u) \quad \forall v \in V_0. \tag{(9.3.7)}$$

Here $j(\cdot)$ is the nonconvex superpotential of the law of Fig. 9.3.3 and $j^0(\cdot, \cdot)$ is the corresponding directional differential in the sense of Clarke, i.e. in the case of the graph of the one-dimensional stress-strain law of Fig. 9.3.3 $\xi \to j(\xi)$ is the area between the horizontal axis and the graph until the point ξ of the axis.

The algorithm presented can be easily justified by comparing the variational inequality (9.3.6) with the hemivariational inequality (9.3.7). Let us write (9.3.7) in the form:

Find $u \in V_0$ such that

$$a(u, v-u) + \int_{\Gamma_S} \mu|C_N|(|v_T| - |u_T|)d\Gamma \geq \tag{9.3.8}$$
$$(l, v-u) + R(v_T, u_T, |v_T|, |u_T|) \quad \forall v \in V_0$$

where

$$R(v_T, u_T, |v_T|, |u_T|) = \tag{9.3.9}$$
$$\int_{\Gamma_S} \mu|C_N|(|v_T| - |u_T|)d\Gamma - \int_{\Gamma_S} j^0(u_T, v_T - u_T)d\Gamma.$$

Then the following iterative scheme is proposed:
Find $u^{(\varrho)} \in V_0$ such that

$$\alpha(u^{(\varrho)}, v - u^{(\varrho)}) + \int_{\Gamma_S} \mu |C_N|(|v_T| - |u_T^{(\varrho)}|) d\Gamma \qquad (9.3.10)$$
$$\geq (l, v - u^{(\varrho)}) + R(v_T, |v_T|, u_T^{(\varrho-1)}, |u_T^{(\varrho-1)}|) \quad \forall v \in V_0.$$

This iterative scheme is compatible with the Fig. 9.3.3 as it becomes obvious from the geometrical meaning of the quantity R. Indeed R expresses the difference of the area variation under the two graphs, i.e. it measures the difference of the reaction forces corresponding to the two laws, the nonmonotone one and the frictional one.

The mathematical convergence proof of the aforementioned iterative scheme is beyond the scope of the present section. The numerical experiments we have performed have always shown a good convergence, which from the physical point of view seems to be a reasonable result. However, we shall give a rough description of the convergence proof which assumes that the reader is familiar with the existence and approximation results for the friction variational inequality (9.3.6) (cf. e.g. [Pan85]) and for the general hemivariational inequality (9.3.7) (cf. e.g. Chapt. 6). Indeed on the assumptions needed to guarantee that (9.3.6) and (9.3.7) admit a solution, we may obtain that in (9.3.10) the sequence $u^{(\varrho)}$ is bounded in the norm of V_0, which is a subspace of the Sobolev space $[H^1(\Omega)]^3$. Thus we may extract a subsequence again denoted by $u^{(\varrho)}$ such that as $\varrho \to \infty, u^{(\varrho)} \to u$ weakly in V_0. Then we show by using the same estimates for the term $R(\cdot)$ as the ones needed for the existence proof that u is indeed a solution of the initial hemivariational inequality. We leave here the proof as an exercise for the reader. One can verify also that a growth assumption analogous to (6.1.44) may guarantee the strong convergence as $\varrho \to \infty$.

We close this section by pointing out that the proposed iterative scheme may be used to define and to treat numerically the threedimensional extensions of the nonmonotone law of Fig. 9.3.3 by approximating them by a sequence of friction problems with given normal force S_N. We recall here that threedimensional friction problems with given normal force are well-defined and give rise to a variational inequality of the type (9.3.6). Closing this section we would like to remark that if the law of Fig. 9.3.3 holds between any other reaction $-R_i$ and the corresponding displacement u_i, then the laws AA'BB' etc. do not give rise to friction type problems but to variational inequalities analogous to the ones of the friction type problem: in (9.3.6) $\int_{\Gamma_S} \mu |C_N|(|v_T| - |u_T|) d\Gamma$ is replaced by $\int_{\Gamma_S} |AA'|(|v_i| - |u_i|) d\Gamma$ etc. and in (9.3.5) $|T_T| \leq \mu |C_N|$ is replaced by $|R_i| \leq |AA'|$ where $|AA'|$ denotes the length of the segment (AA'). If the law of Fig. 9.3.3 holds between a stress s_i and a strain e_i of a structural element i then the law $AA'BB'$ etc. correspond to rigid plastic behaviour of the i-th element.

According to Maier [Ma68,69,78] the problem is formulated as a Linear Complementarity Problem (L.C.P.) which is equivalent to a minimum energy problem and to a variational inequality[Pan85].

9.4 Application I: Cleavage in Laminated Composites and the Nonmonotone Unilateral Contact Problem

The idea of ameliorating the mechanical response of a structure by combining several structural elements with different mechanical properties is old enough:we recall in Homer's Iliad the description of the manufacturing technique (Book Σ, verses 478-482) and of the Achille's shield (Book Y, verses 268-272) . Since the 50's, laminated composites have rapidly acquired-due to their broad range of applications-an explosively developing part of modern industrial technology. It is worth noting that laminated composites have significantly better properties than their constituents: Strength, stiffness, weight, fracture toughness, corrosion resistence, fatigue, crack growth, thermal conductivity and interlaminar cohesion are, for instance, properties which can be suitably predicted during manufacturing process, so that laminated composites with improved behaviour regarding one or some of the previously mentioned properties can be obtained. The better mechanical properties of laminates in comparison to those of conventional materials have as direct consequence weight and cost reduction. However their mechanical behaviour is much more complicated compared with the behaviour of homogeneous bodies. This is mainly due to the appearance of "limit" phenomena (as are e.g. cracking or crushing of interlaminar phases, as well as the delamination and or slip phenomena along the interfaces) under certain loading conditions. Among these cases, interlaminar fracture under cleavage loading (in the sequel we use delamination or cleavage as terms having the same meaning) received more attention than any other.

Throughout the present section the numerical method of Sect. 9.3 is applied to quantify the toughness of laminated products under cleavage loading, taking into account the delamination effects between the layers. Such a mechanical behaviour may be described by means of nonmonotone, possibly multivalued, stress-strain or reaction-displacement laws, i.e. laws including complete jumps or decreasing branches that correspond to the abrupt reduction of the strength of laminates after a critical value of stress or strain. We recall that these laws, defined along the whole length of the axis of strains or displacements are derived by uniaxial experimental tests, and are called complete stress-strain or force-displacement diagrams (see e.g. [Hult][Ban85]). Recently certain numerical approaches to the problem of structures having complete nonmonotone constitutive or boundary laws have been proposed by Bažant and Chang [Baža],Belytschko et al. [Bely], Clech et al. [Clec], Crisfield (1986), Crisfield and Willis [Cris88,91], Del Piero and Maceri [DelP], Schellekens and De Borst [Schel], Pagano [Pagan], and Williams et al. [Willi82,86] among others.All these numericall approaches accurately treat only some aspects of the delamination

258   9. On the Treatment of Hemivariational Inequalities

phenomenon (for instance, incomplete consideration of the complete vertical branches).

Note that stress-strain or force-displacement diagrams in composite laminates extended over the whole horizontal axis, have been recently obtained by using advanced technology testing machines that minimize the range of instability effects during experimental tests.We refer to Sect. 3.2 for examples. The nonconventional character of the previous material and/or boundary laws due mainly to the decreasing branches of the stress-strain or reaction-displacement diagrams and to the vertical complete jumps, does not permit the analysis of the cleavage problem in laminated composites to be performed elegantly and correctly by means of the classical structural analysis methods. The problem is formulated as a hemivariational inequality which is numerically solved by the method of Sect. 9.3.

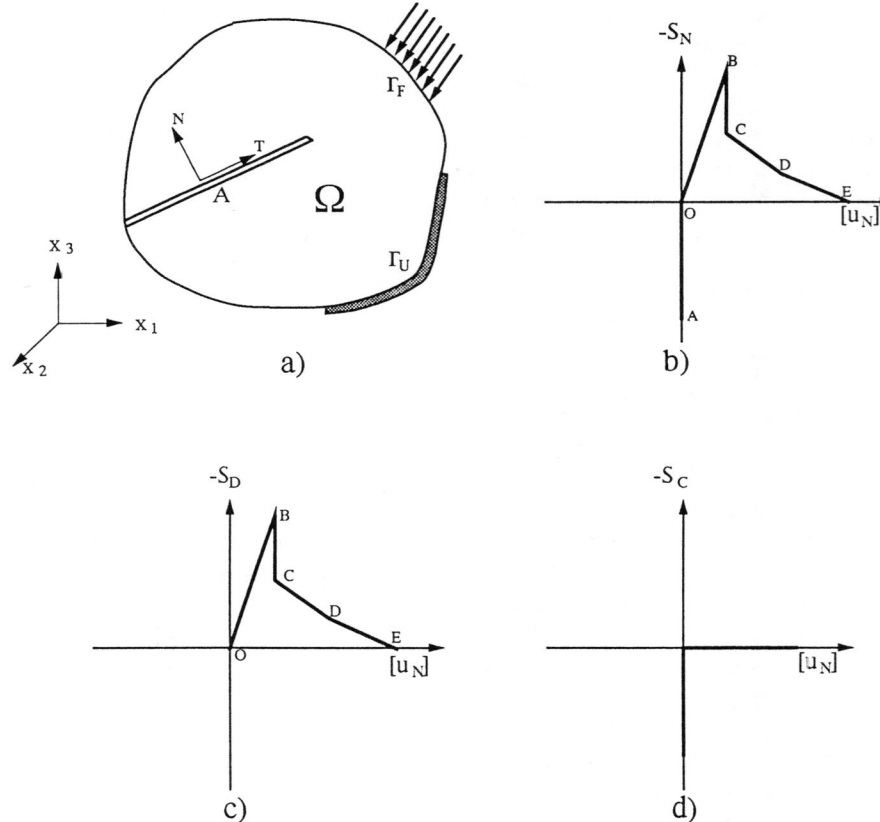

Fig. 9.4.1. Formulation of the cleavage problem.

9.4 Application I: Cleavage in Laminated Composites

Let us consider in an orthogonal Cartesian co-ordinate system $0x_1x_2x_3$ a linear elastic body Ω that occupies in its undeformed state a subset of the three-dimensional Euclidean space $\mathbb{R}^3$. The boundary Γ of Ω is piecewise smooth, and consists of two nonoverlapping parts Γ_U and Γ_F, where displacements and boundary forces are respectively prescribed. We assume further that Ω contains certain interfaces or surfaces denoted by A (Fig. 9.4.1a) where an adhesive material produces a behavior in the normal-to the interface-direction, according to the displacement-reaction law of Fig. 9.4.1b. The diagram contains a vertical branch OA and a nonmonotone branch OBCDE. The first is called contact branch and the second adhesive branch. The first (resp. the second) is realized for compressive (resp. noncompressive or cleavage) interface forces.

We shall split the problem into two subproblems by considering either the adhesive law of Fig. 9.4.1c, or the unilateral contact law of Fig. 9.4.1d. We denote by $[u_N]$ the relative normal displacement of the two fronts of the interface. Denoting by A_C (resp. A_D) the contact regions or regions in compression (resp. the regions where the adhesive forces appear with the simultaneous presence of detachment) on the interface A, the interface behaviour can be described by means of the following conditions:

$$\text{if} \quad [u_N] = 0 \quad \text{then} \quad S_N = S_C \geq 0 \quad \text{on} \quad A_C \quad (9.4.1)$$

$$\text{if} \quad [u_N] > 0 \quad \text{then} \quad S_N = S_D, S_D + g([u_N]) = 0 \quad \text{on} \quad A_D, \quad (9.4.2)$$

where S_C is the normal force on the contact region, S_D denotes the resistance of the adhesive in the normal to the interface direction and $g(\cdot)$ is a nonmonotone function that may also include complete jumps (as it is BC in Fig. 9.4.1b). Function $g(\cdot)$ may be different from point to point. As is obvious, condition (9.4.2) holds on the detached regions A_D of the interface A, whereas condition (9.4.1) on the contact regions A_C. We note here that A_C and A_D are not a priori known. Moreover $[u_N]$ is positive if it corresponds to an opening of the interface. Analogously two forces which tend to open the interface are considered as positive. In this section we make the hypothesis as it is dictated by the physical model (cf. e.g. [Davie]) that along the interface A the tangential stresses are given, i.e. that

$$S_{T_i} = C_{T_i} \quad (i = 1, 2). \quad (9.4.3)$$

Boundary conditions holding on Γ_U and Γ_F are written in the form

$$u_i = U_i \quad \text{on} \quad \Gamma_U \quad (9.4.4)$$

and

$$S_i = F_i \quad \text{on} \quad \Gamma_F, \quad (9.4.5)$$

where U_i, F_i are given functions defined respectively on Γ_U and Γ_F. On the assumption of small deformations, the equation of equilibrium, the strain-displacement relation and the elastic material law are written in the form

$$\sigma_{ij,j} + f_i = 0, \varepsilon_{ij} = \frac{1}{2}(u_{i,j} + u_{j,i}), \sigma_{ij} = C_{ijkl}\varepsilon_{kl} \quad \text{in} \quad \Omega \quad (9.4.6)$$

where the comma denotes partial differentiation, $f = \{f_i\}$ is the volume force vector and $C = \{C_{ijkl}\}$ the elasticity tensor having the well-known properties of symmetry and ellipticity properties (3.1.22a,b). According to Chapt. 2, the previous nonmonotone possibly multivalued law (9.4.2) of Fig. 9.4.1c, can be written in the form

$$-S_D \in \bar{\partial} j_N([u_N]) \quad \text{if} \quad [u_N] > 0, \tag{9.4.7}$$

where $j_N(\cdot)$ is a locally Lipschitz function resulting from g by integration (cf. (1.2.54),(1.2.55)) and $\bar{\partial}$ is the generalized gradient. The energy function $j_N(\cdot)$ is called delamination "superpotential" and is in general a nonconvex function. We introduce now the set X_{ad} of kinematically admissible displacements

$$X_{\text{ad}} = \{v | v = U \quad \text{on} \quad \Gamma_U, \quad [u_N] \geq 0 \quad \text{on} \quad A \} \tag{9.4.8}$$

and by applying the Green-Gauss theorem, we obtain as in Sect. 4.2 the following inequality constrained hemivariational inequality:
Find $u \in X_{\text{ad}}$ such as to satisfy the hemivariational inequality

$$a(u, v - u) + \int_A j_N^0 [u_N], ([v_N] - [u_N]) dA \geq \int_\Omega f_i(v_i - u_i) d\Omega + \tag{9.4.9}$$
$$+ \int_{\Gamma_F} F_i(v_i - u_i) d\Gamma + \int_A C_{T_i}([v_T]_i - [u_T]_i) dA \quad \forall v \in X_{\text{ad}}$$

Here $a(\cdot, \cdot)$ is the bilinear form of linear elasticity (see eq. (4.1.7)) and $[u_T]$ denotes the relative tangential displacement. Note that X_{ad} is a convex set and thus (9.4.9) is actually a variational hemivariational inequality.

For the numerical solution of (9.4.9) we approximate after discretization the nonmonotone law with a sequence of monotone ones as in Sect. 9.3. Thus we succeed to replace the hemivariational inequality by a sequence of variational inequality problems. From the numerical point of view, the latter kind of problems has the advantage to lead to Quadratic Programming (Q.P.) minimization problems, where the minimum is always uniquely determined and for which a lot of efficient algorithms have been developed [Flet].

In the following, the approximation procedure is briefly described. Let us assume that the nonmonotone diagram ABCDEF of Fig. 9.4.2 represents the possible stress and strain cleavage law of certain elements of the discrete structure.

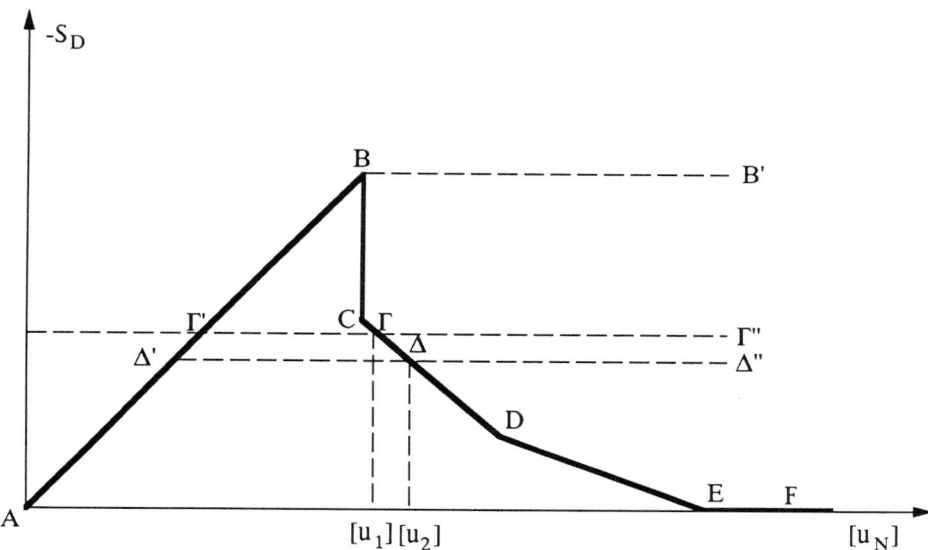

Fig. 9.4.2. Description of the algorithm.

At the first step we assume that all these elements obey the fictitious law ABB″. The solution of this monotone problem gives as a result certain $[u_1]$. Now we solve a new monotone problem, obeying the law $A\Gamma'\Gamma\Gamma''$. Suppose that the solution of this problem offers as a result a new $[u_2]$ which gives rise to the new diagram $A\Delta'\Delta\Delta''$. This procedure is continued until in all elements the differences $|S^{(i+1)} - S^{(i)}|$ and $|[u]^{(i+1)} - [u]^{(i)}|$ in two successive steps $i+1$ and i become small enough. This method can treat any kind of nonmonotone possibly multivalued interface laws which are common in composite materials and in problems involving adhesively connected interfaces. The above algorithm is combined with a fixed point type algorithm for the treatment of the more complicated law of Fig. 9.4.1b. The proposed algorithm is analogous to the one proposed in [Pan75] for the treatment of the unilateral contact problem with Coulomb friction (Kalker [Ka88b] named this algorithm PANA , and Nečas, Jarusek and Haslinger [Neč80] proved its convergence by applying a fixed point argument). In our case, first the pure unilateral contact problem is solved, i.e. the action of the adhesive is ignored and thus the problem obeys the diagram of Fig. 9.4.1d, (phase I algorithm) by assuming that $S_D = S_D^{(i)}$ is given on $A_D^{(i)}$ which is known from the step $(i-1)$. At the step $i=1$ we assume that $A_D^{(1)} = A$. Let now $S_C^{(i)}$ be the obtained normal reaction from the solution of the unilateral contact problem (Signorini problem) with additional loading $S_D = S_D^{(i)}$ on $A_D^{(i)}$. The reaction $S_C^{(i)}$ gives also the region $A_C^{(i)}$, which is the region where $[u_N^{(i)}] = 0$. Then, under the hypothesis that $S_C = S_C^{(i)}$ on $A_C^{(i)}$, the pure delamination problem, i.e. the problem obeying the diagram of Fig. 9.4.1c (phase II algorithm) is solved; let $S_D^{(i+1)}$ be the resulting cleavage force on A_D. Note that $S_D^{(i+1)}$ is assumed as nonpositive. If $S_D^{(i+1)} > 0$ then we pose $S_D^{(i+1)} = 0$ and we go to the next step. This additional correction is necessary because the cleavage law holds only for $[u_N] \geq 0$ The first subproblem is again solved with $S_D = S_D^{(i+1)}$ and the previously described procedure is repeated interactively until the differences $|S_D^{(i)} - S_D^{(i+1)}|$ and $|S_C^{(i)} - S_C^{(i+1)}|$ and the differences of the relative displacements $[u_N]$ resulting from the two phases of the algorithm become less or equal to an appropriately defined accuracy. The previously proposed algorithm, contains two phases :

1st step: Phase I: Calculate the laminated composite structure if on $A_D, S_D = S_D^{(0)}$ is given and the unilateral contact condition

$$S_C \geq 0, \quad [u_N] \geq 0, \quad S_C[u_N] = 0 \qquad (9.4.10)$$

holds at the interface. A Q.P. algorithm for the solution of this classic problem is applied(see e.g. [Pan75]). Usually it is assumed that $S_D^{(0)} = 0$. $S_C^{(1)}$ denotes the resulting normal force and $[u_N^{(1)}]$ the corresponding relative normal displacements on A.

2nd step: Phase II: This step deals with the behaviour of the adhesive and it contains the algorithm prescribed previously in Fig. 9.4.2. Note that each subproblem of this step, e.g. the subproblems corresponding to the law ABB' or to the law $A\Gamma'\Gamma\Gamma''$, are numerically treated by minimizing the corresponding potential or complementary energy. Indeed for any monotone possibly multivalued law of the type $A\Gamma'\Gamma\Gamma''$ the position of equilibrium is characterized by the minimum of the potential and of complementary energy as it has been shown first by Prager [Prag65] (see also for a modern approach [Pan76a,85]). After each such step of the phase II algorithm, we go back to the phase I contact algorithm to obtain a better estimation of the detachment zone A_D (i.e. immediately after the calculation of $[u_1]$ in Fig. 9.4.2 and not after the determination of the phase II final equilibrium solution). Indeed, since the unilateral contact region can considerably differ from the adhesive region, it is time consuming to perform all the steps of the phase II algorithm for a given normal force. So, we perform only one substep of the algorithm until the contact region is accurately estimated and then we apply all the substeps of the phase II algorithm. The phase I algorithm yields $[u_N^{(i)}]$. From $[u_N^{(i)}]$ and the $\{[u_N], S_D\}$ law, the corresponding $S_D^{(i)}$ is obtained; we set it as $S_D^{(i,1)}$, where the second superscript denotes the number of substeps performed in the phase II algorithm. Then we solve the corresponding monotone problem (path $A\Gamma'\Gamma\Gamma''$) for the given $S_C^{(i)}$ from the phase I algorithm. If the delamination area A_D has not been well established or

9.4 Application I: Cleavage in Laminated Composites

S_C varies considerably from step to step, then we go back to the phase I algorithm and so on. Otherwise we continue the substeps of the phase II algorithm: from the solution of the Q.P., a new $[u_N^{(i,2)}]$ is obtained. Then we go back to the $\{[u_N], S_D\}$ law and we obtain $S_D^{(i,2)}$. The corresponding monotone problem is once more solved (path $(A\Delta'\Delta\Delta''$ of Fig. 9.4.2) and a new $[u_N^{(i,3)}]$ is obtained. This procedure is continued until the convergence of the algorithm, say after k-substeps to $[u_N^{(i,k)}]$. In all these substeps, S_C is always taken equal to $S_C^{(i)}$.

3rd step: Return to phase I: The 1st step is repeated for $S_D = S_D^{(i,k)}$ obtained from the solution of the k-substep in the 2nd-step. The whole procedure is repeated until the differences between the values of S_C, S_D and u_N of two consecutive cycles become appropriately small. Suppose that $u_N^{(1)}$ belongs to a branch like the EF of Fig. 9.4.2. Then a small modification is proposed in the first execution of the phase II algorithm: $u_N^{(1)}$ is ignored and we take as $S_D^{(1,1)}$ the one defined by the upper and/or lower bound of the nonmonotone law. In the case of Fig. 9.4.2, $S_D^{(1,1)}$ is thus defined by the point B, i.e. the solution is calculated for the law ABB'. With this $S_D^{(1,1)}$, the monotone problem is solved with $S_C = S_C^{(1)}$ and a new $[u_N^{(1,2)}]$ is obtained. From the $\{[u_N], S_D\}$ law, the corresponding $S_D^{(1,2)}$ is then obtained and so on, as is previously described. The proof of the convergence of the proposed algorithm is still an open problem. However, the performed numerical experiments exhibit very good convergence properties.

The described algorithm has been applied on a HP-720 RISC workstation. For the solution of the Q.P. problem arising in phase II, the Hildreth and d'Esopo Q.P. algorithm has been applied, which, in comparison with other Q.P. algorithms is slower but yields some results needed in further steps, like inverse matrices, decompositions etc. This last fact reduces the total execution time. Note also that in order to calculate the structure obeying the monotone law $A\Gamma'\Gamma\Gamma'''$ we prefer to minimize the complementary energy (i.e. to apply the force method) with allows the direct treatment of the stress constraints of the type $S_D \leq S_D^0$ (cf. Fig. 9.4.2).

The proposed numerical scheme has been applied to the analysis of a laminated composite structure under cleavage loading of the Fig. 9.4.3a. The delamination process is described by a nonmonotone law holding at the interface. The discretization has been performed using constant stress triangular elements. The structure having an interface with 30 nodes, is loaded with forces in the normal-to the interface-direction and is completely determined kinematically. The eleven load cases of Fig. 9.4.3b have been numerically investigated for the two different nonmonotone laws of Fig. 9.4.4. The two laws have the same elastic part, but the second one is more "shallow" and this fact influences considerably the behaviour of the structure. In Figs 9.4.5a and 9.4.5b (resp. 9.4.6a and 9.4.6b) we present the distribution of the adhesive force (resp. normal relative displacements) along the interface. We notice the great influence of the nonmonotone part of the law to the behaviour of the structure.

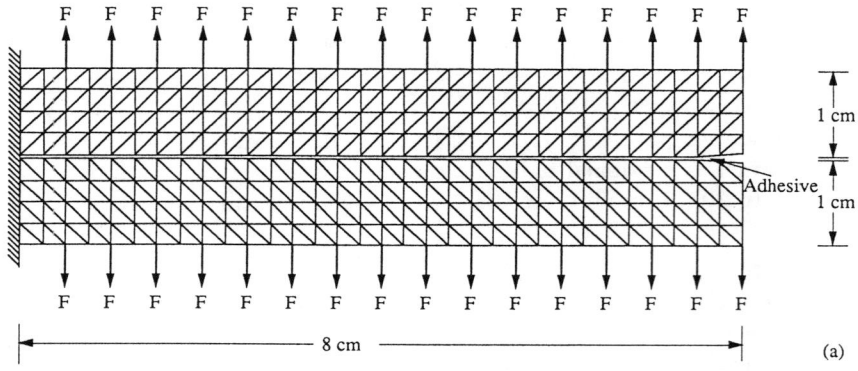

Fig. 9.4.3. The numerical example ($E=1.378 \times 10^8 \text{kN/m}^2$, $\nu = 0.3$, t (thickness) $= 0.005$m).

The solution for the first law shows a partial failure of the interface material between the 1st and the 2nd load case of Fig. 9.4.3b. Solving for the second law we note that the changes are more smooth and the failure. of the laminate happens between the 3rd and the 4th load cases. In Figs. 9.4.7a and 9.4.7b the progress of the damage of the adhesive for the eleven load cases of Fig. 9.4.3b is depicted. Darker shades represent higher resistance of the interface material. Here it is clear that while the load increases, the strength of the adhesive passes from branches to the left to branches to the right on the nonmonotone diagram of Fig. 9.4.2 loosing in this way the capability to undertake the cleavage loading. The convergence of the algorithm is very fast. All the previous problems have been solved in about 7-14 cycles of the phase I - phase II algorithm using a second order norm of the normal and delamination forces calculated, as a stopping criterion: the algorithm terminates when $\frac{\|S^{(i)} - S^{(i-1)}\|}{\|S^{(i)}\|}$ becomes smaller than 10^{-4}.

Analogous to the numerical treatment of the cleavage problem is the treatment of the nonmonotone unilateral contact problem. In this problem the normal forces to the boundary of the deformable body are related with the corresponding normal displacements by a nonmonotone possibly multivalued law of Winkler type in compression. Moreover the possibility of detachment of the body from the support must be taken into account. Thus the S_N and u_N at the boundary are related by a law of the form depicted in Fig. 2.4.1b which may be written as

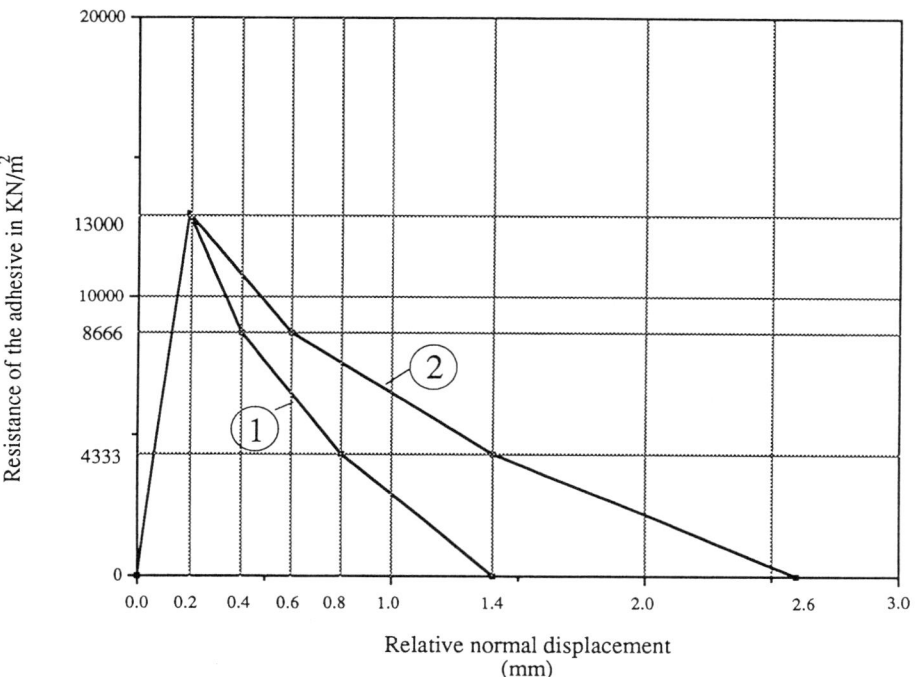

Fig. 9.4.4. The law of the adhesive behaviour.

$$\text{if} \quad u_N < 0 \quad \text{then} \quad S_N = 0 \tag{9.4.11}$$

$$\text{if} \quad u_N \geq 0 \quad \text{then} \quad -S_N \in \bar{\partial} j_N(u_N). \tag{9.4.12}$$

Due to (9.4.12) the compression force-displacement diagram can be multivalued. Thus we may take into account local crushing effects. The tangential forces are assumed as given. As it is obvious, (9.4.11) and (9.4.12) lead to a more easy hemivariational inequality than the cleavage problem; the solution does not need to belong to a convex set as in (9.4.9), because the kinematically admissible set is a linear subspace. In the next section we treat a much more complicated problem. It is the nonmonotone unilateral contact problem obeying a nonmonotone possibly multivalued friction law.

266   9. On the Treatment of Hemivariational Inequalities

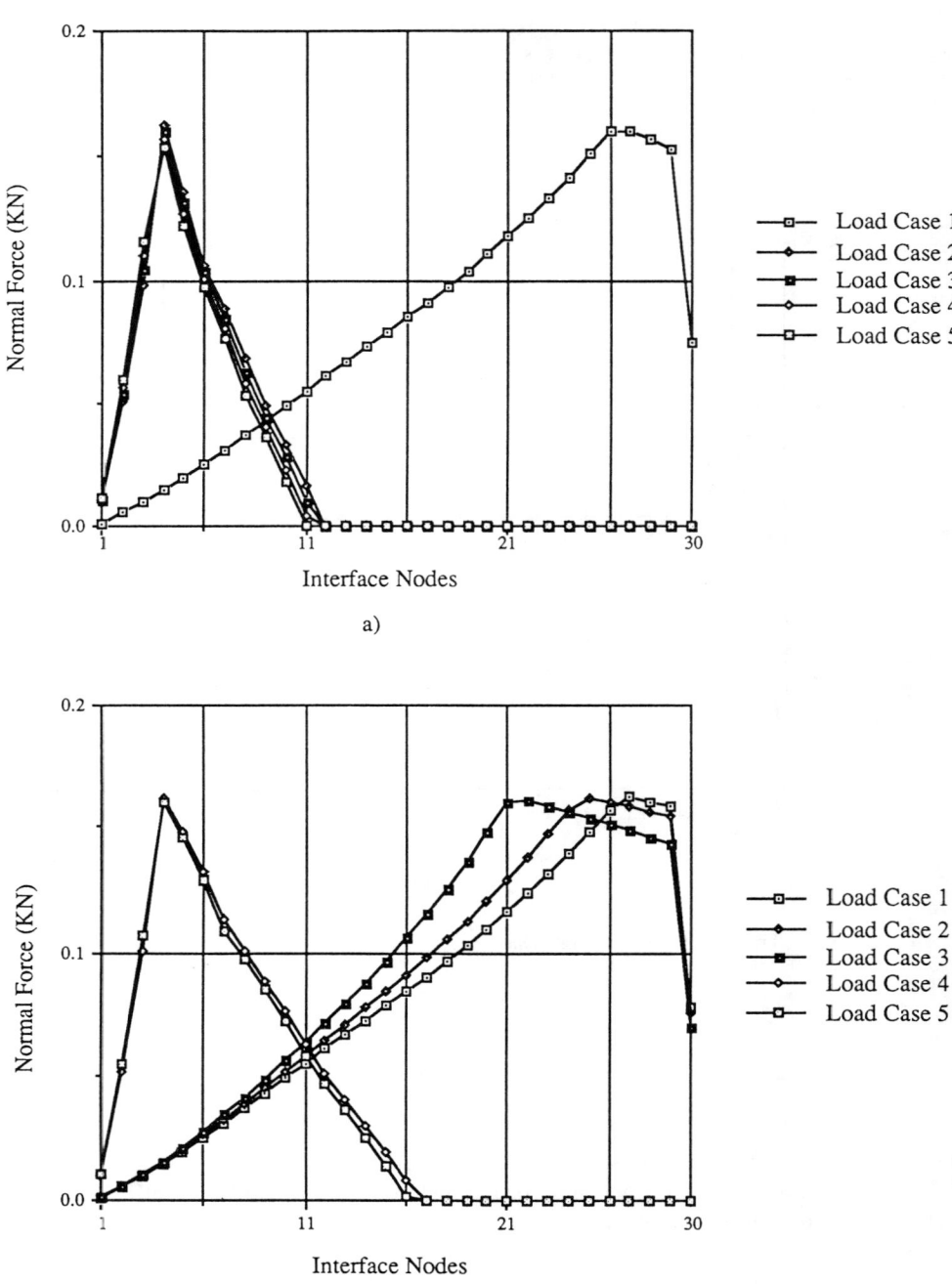

Fig. 9.4.5. The adhesive force distribution for the two laws of Fig. 9.4.4.

9.4 Application I: Cleavage in Laminated Composites

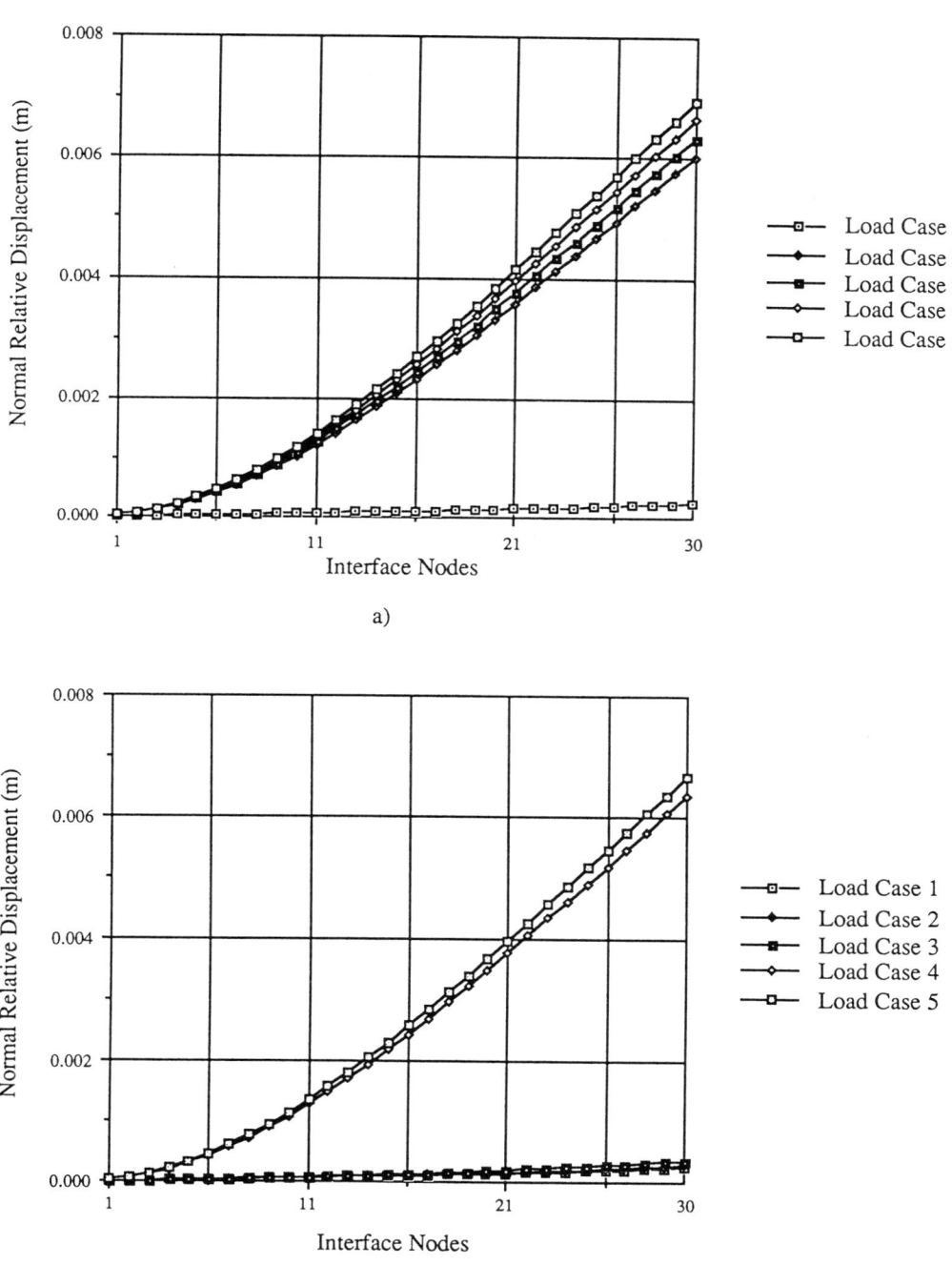

Fig. 9.4.6. The $[u_N]$ distribution for the two laws of Fig. 9.4.4.

268 9. On the Treatment of Hemivariational Inequalities

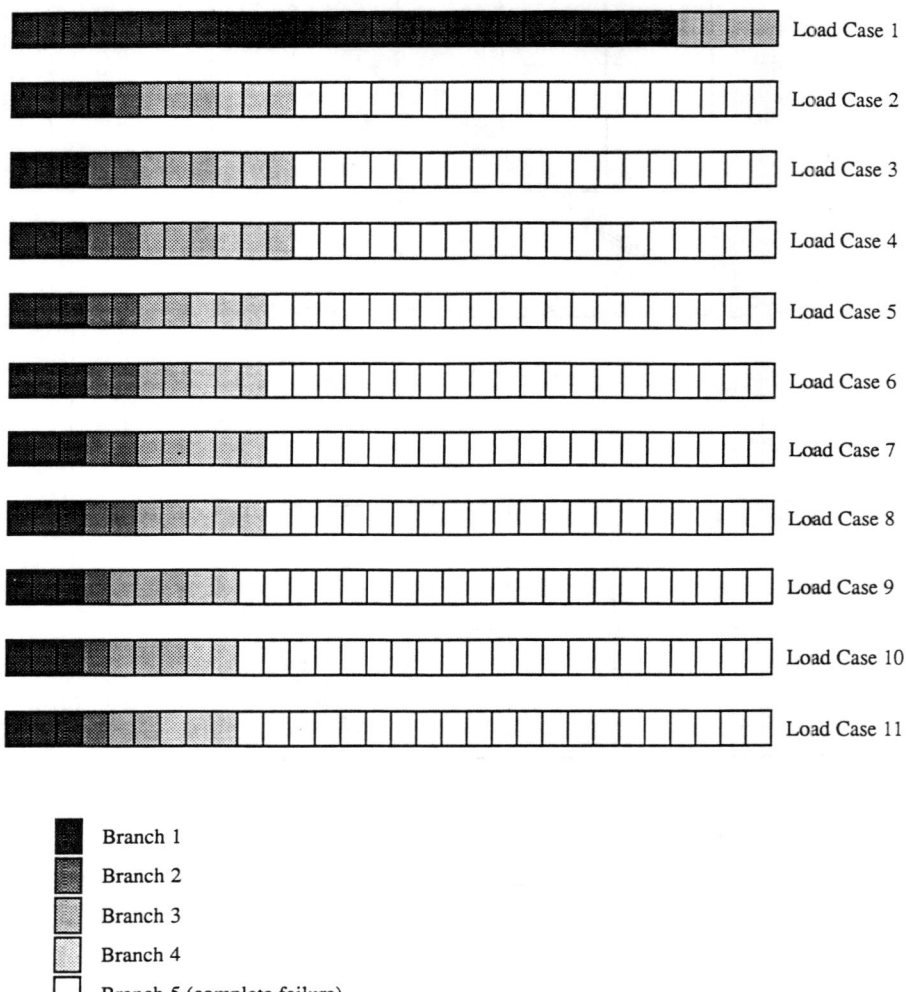

a)

Fig. 9.4.7. Progression of the adhesive failure (branch 1 corresponds to AB, branch 2 to BC etc. of Fig. 9.4.2).

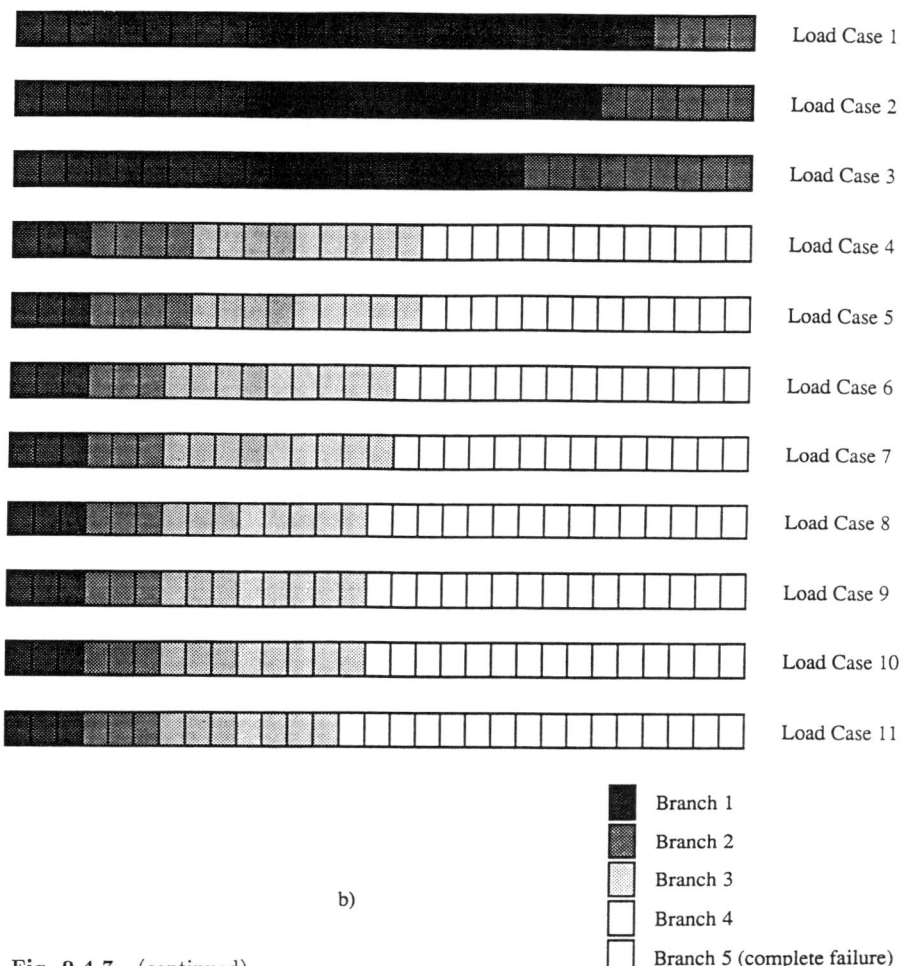

Fig. 9.4.7. (continued).

9.5 Application II: The Nonmonotone Friction Problem and the Combined Unilateral Contact Problem with Nonmonotone Friction

Let us consider here the following problem. Suppose that we have a linear elastic body, $\Omega \subset \mathbb{R}^2$ with the regular boundary Γ consisting of Γ_S, Γ_U and Γ_F. On Γ_S a nonmonotone friction law of the type of Fig. 9.5.1b holds at the points where the body remains in contact with a rigid or a deformable support (Fig. 9.5.1d). Thus we can write the following law:

$$\text{If } \quad u_N < 0 \quad \text{then } S_N = 0 \quad \text{and } S_T = 0, \tag{9.5.1}$$

if $u_N = 0$ then $S_N < 0$ (9.5.2)
and $\{S_T, u_T\}$ are according to Fig. 9.5.1b

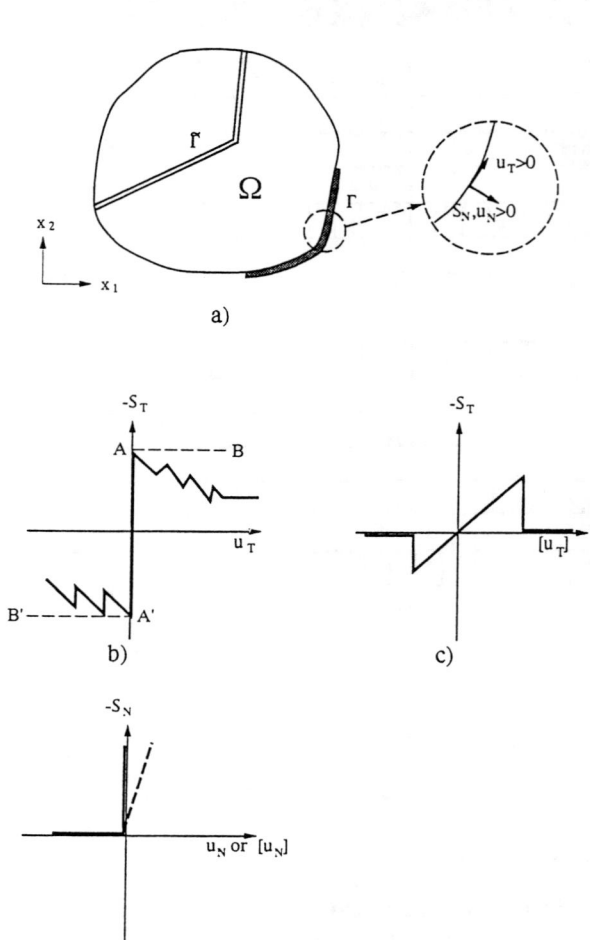

Fig. 9.5.1. The nonmonotone friction problem ($E = 2.1 \times 10^6 \text{t/m}^2, \nu = 0.16$, $t = 0.01$ m).

In the case of a linear deformable support with Winkler constant k,(9.5.2) has to be replaced by the relation

if $u_N \geq 0$ then $S_N + k u_N = 0$
and $\{S_T, u_T\}$ are according to Fig. 9.5.1b (9.5.3a)

9.5 Application II: The Nonmonotone Friction Problem

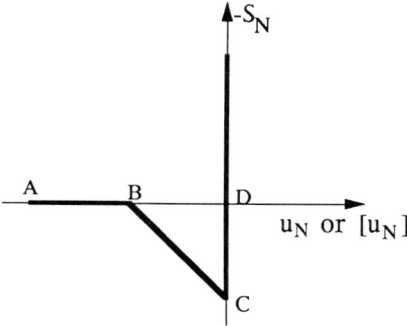

Fig. 9.5.2. The normal law.

Note that the possibility of a small resistance in traction in the direction normal to the interface is not taken here into account, i.e. the diagram ABCD in Fig. 9.5.2. In this section, in contrast to the previous, we consider $[u_N]$ as negative, if it corresponds to an opening of the interface. Analogously, if S_N tends to open the interface it is considered negative. We further assume that Ω contains certain interfaces denoted by $\tilde{\Gamma}$ in which an adhesive material introduces a shear behavior according to the law of Fig. 9.5.1c. Now $[u_T]$ denotes the relative tangential displacement of the two sides of the interface. In the normal direction the unilateral contact law of Fig. 9.5.1d holds. Thus we may write relations analogous to (9.5.1) and (9.5.2) with the difference that u_N and u_T are replaced by the relative displacements $[u_N]$ and $[u_T]$. We shall refer to this condition as (9.5.1a) and (9.5.2a). Of course Γ_S or $\tilde{\Gamma}$ may be empty. Finally on the boundary part Γ_U the displacements are prescribed and on Γ_F the boundary forces are given.

The boundary and interface condition defined by (9.5.1), (9.5.2) and (9.5.1a) (9.5.2a) do not directly lead to a hemivariational inequality. If, in place of the nonmonotone friction law a classical Coulomb's friction law without given normal force could be given (see the dotted line BAA'B' in fig 9.5.1b), then we could apply the following algorithm for the unilateral contact problem with Coulomb friction proposed in [Pan75]: First solve the pure (i.e without friction) unilateral contact problem (normal algorithm) with $S_T = S_T^{(i)}$ given on Γ_S and $\tilde{\Gamma}$ and let $S_N^{(i)}$ be the obtained normal reaction. Then the pure friction problem with prescribed normal force (tangential algorithm) is solved assuming that on Γ_S and $\tilde{\Gamma}$ $S_N = S_N^{(i)}$; let $S_T^{(i+1)}$ be the resulting tangential force on Γ_S and $\tilde{\Gamma}$. Again the first subproblem is solved with $S_T = S_T^{(i+1)}$ and soon until the differences $|S_T^{(i)} - S_T^{(i+1)}|$ and $|S_N^{(i)} - S_N^{(i+1)}|$ and the differences of the corresponding displacements and/or relative displacements become small enough. This algorithm has been proved to converge in [Neč80] by using a fixed point argument. This fixed point type algorithm used for the classical unilateral contact problem with Coulomb friction is extended here for the unilateral contact problem with nonmonotone (zigzag) possibly multivalued friction boundary condition and/or adhesive contact interface condition. We formulate the following algorithm:

1st step: Calculate the structure if on Γ_S and $\tilde{\Gamma}$, $S_T = S_T^{(0)}$ is given and the unilateral contact condition

$$S_N \leq 0, \quad u_N \leq 0, \quad S_N u_N = 0 \tag{9.5.3b}$$

holds. We use a Q.P. algorithm in the normal direction (normal algorithm). Usually we assume that $S_T^{(0)} = 0$. Let us denote by $S_N^{(1)}$ the resulting normal force and let $u_N^{(1)}$ and $u_T^{(1)}$ be the corresponding normal and tangential displacement on Γ_S (resp. relative displacement on $\tilde{\Gamma}$).

2nd step: This step uses the slight variation of the algorithm described in Sect. 9.4. After each step of the tangential algorithm, we go back to the normal contact algorithm to obtain a better estimation of the debonding region (i.e. after the calculation of u_{T_1} in Fig. 9.3.3 and not after the determination of the tangential subproblem equilibrium solution e.g. u_{T_5}). Indeed, because the contact and debonding regions can vary considerably from step to step it is time consuming to perform all the steps of the tangential algorithm for a given normal force. Also the numerical experience shows a high sensitivity of the whole algorithm in the case of the unstable contact regions. So we perform only one substep of the algorithm until the contact region is well estimated and apply all the substeps of the tangential algorithm afterwards. Suppose that the normal algorithm yields $u_T^{(i)}$. From $u_T^{(i)}$ and the $\{u_T, S_T\}$ law the corresponding $S_T^{(i)}$ is obtained; set it as $S_T^{(i,1)}$ where the second upper index denotes the number of substeps performed in the tangential algorithm. Then we solve the corresponding classical (or Coulomb) friction problem for $|S_T| \leq |S_T^{(i,1)}|$ and for the given $S_N^{(i)}$ from the normal algorithm. If the debonding area is not well established, or if S_N varies considerably from step to step then we go back to the normal algorithm and so on. Otherwise we continue the substeps of the tangential algorithm: from the solution of the friction Q.P. a new $u_T^{(i,2)}$ is obtained. Then we go back to the $\{S_T, u_T\}$ law and we obtain $S_T^{(i,2)}$. Again the corresponding classical (or Coulomb) friction problem is solved for $|S_T| \leq |S_T^{(i,2)}|$ and a new $u_T^{(i,3)}$ is obtained. This procedure is continued until the algorithm converges, say after k-substeps to $u_T^{(i,k)}$. In all these substeps S_N is always taken equal to $S_N^{(i)}$.

3rd step: The 1st step is repeated for $S_T = S_T^{(i,k)}$ obtained from the solution of the k- substep in the 2nd-step. The whole procedure is repeated until the differences between the S_N's, S_T's, u_N's and u_T's of two consecutive cycles become appropriately small. Suppose now that $u_T^{(1)}$ lands on a branch like the IJ of Fig. 9.3.3. Then $u_T^{(1)}$ is ignored and we take as $S_T^{(1,1)}$ the one defined by the upper and/or lower bound of the nonmonotone law. Thus in the case of Fig. 9.3.3 $S_T^{(1,1)}$ is defined by the point B i.e. it is taken as equal to AA'. With this $S_T^{(1,1)}$ the classical friction problem is solved with $S_N = S_N^{(1)}$ and $|S_T| \leq |S_T^{(1,1)}|$ and a new $u_T^{(1,2)}$ is obtained. Then from the $\{S_T, u_T\}$ law the corresponding $S_T^{(1,2)}$ results and so on as described before. In Fig. 9.5.3 a diagram is given describing this algorithm, which is also a fixed point type algorithm. The mathematical convergence proof of this last algorithm with debonding is still an open problem. However the numerical experiments we have performed have shown very good convergence properties.

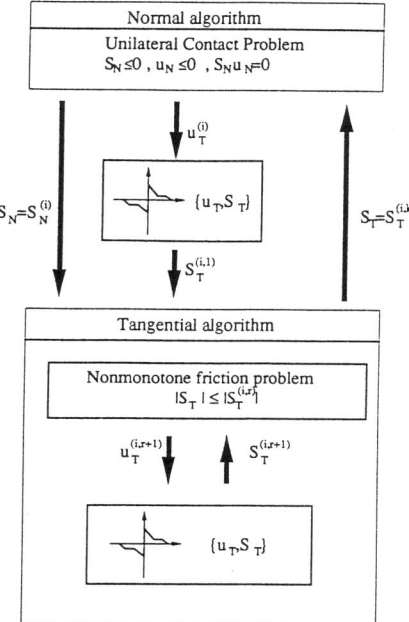

Fig. 9.5.3. The diagram of the algorithm.

According to the described algorithm, an appropriate code in FORTRAN was developed on a HP-720 workstation. The first step of the algorithm, i.e. the solution of the unilateral contact problem, was achieved using the direct stiffness method and a fast active set Q.P. technique (subroutine VE09 of the HARWELL Subroutine Library). The second step of the algorithm, i.e. the friction type problem was treated using the force method. Indeed this method allows the direct treatment of the stress constraints introduced. For the solution of this second Q.P. problem we have applied the Hildreth and d'Esopo Q.P. algorithm, as in Sect. 9.4. Several classical numerical techniques optimizing the performance of the finite element codes have been used. For instance a profile reduction algorithm is applied to reduce the bandwidth of the matrices, among others. The numerical scheme proposed, was used for the numerical solution of the structure depicted in Fig. 9.5.4 which represents a joint with nonmonotone friction law and debonding holding at the interfaces. The six load cases of Fig. 9.5.4 were considered, for values of friction coefficient μ equal to 0.1 and 0.2. The discretization was performed using constant stress triangular elements. The structure has two interfaces with 31 couples of nods, it is loaded with forces in the normal and the tangential to the interfaces directions and is well determined kinematically. The nonmonotone friction law depicted in Fig. 9.5.5 is assumed to hold at the interfaces. In order to obtain the two nonmonotone friction laws

274   9. On the Treatment of Hemivariational Inequalities

considered we have to multiply at each point of the boundary or the interface the vertical coordinates of the diagram of this figure by $\mu|S_N|$ where S_N is the normal force at the point considered. Thus we have here a genuine nonmonotone friction law depending on the unknown normal forces S_N.

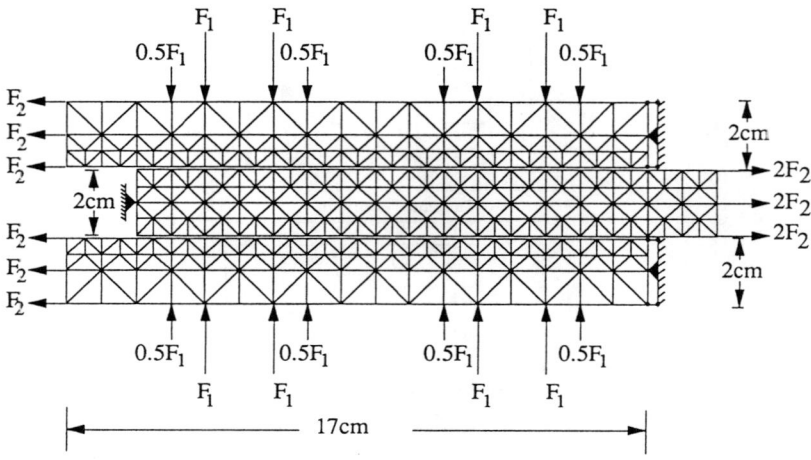

Load Case	1	2	3	4	5	6
F_1	1.00	1.00	1.00	1.00	1.00	1.00
F_2	0.20	0.40	0.60	0.80	1.00	1.20

Fig. 9.5.4. The numerical application.

Here with the notation of Sect. 9.3, $C_N = \tilde{C}_N$, is not given. The distribution of the contact forces S_N with respect to the 31 distinct positions on the one of the interfaces, is given in Fig. 9.5.6a,b. The results are almost identical. However the influence of the nonmonotone diagram on the results of the friction forces given in Fig. 9.5.7 is very considerable. When the tangential forces F_2 are small, the behavior of the joint is linear and all the results lie on branch AB of Fig. 9.5.5. As the tangential forces increase, the equilibrium points on the diagram move to the branches at the right (BCDE...) and finally the joint fails (e.g. Fig. 9.5.7b, load case 5, points 22-31). The points of abrupt changes of the curves in Fig. 9.5.7a,b correspond to the points where the strength of the joint passes from the one branch of the zigzag diagram to the other, thus leading to a progressive failure as the tangential displacement increases. It is important to note here the great influence of the friction coefficient on the results. This

9.5 Application II: The Nonmonotone Friction Problem   275

is clearly seen in Fig. 9.5.8, where the relative tangential displacements at each point are depicted for the various load cases.

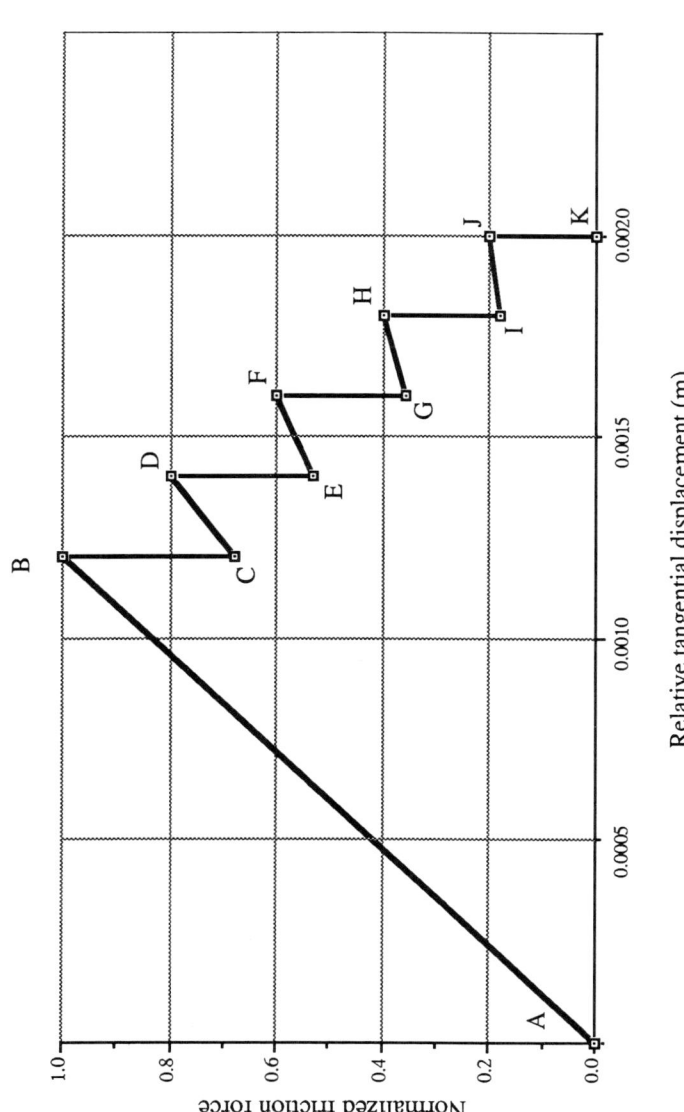

Fig. 9.5.5. The nonmonotone friction law.

276 9. On the Treatment of Hemivariational Inequalities

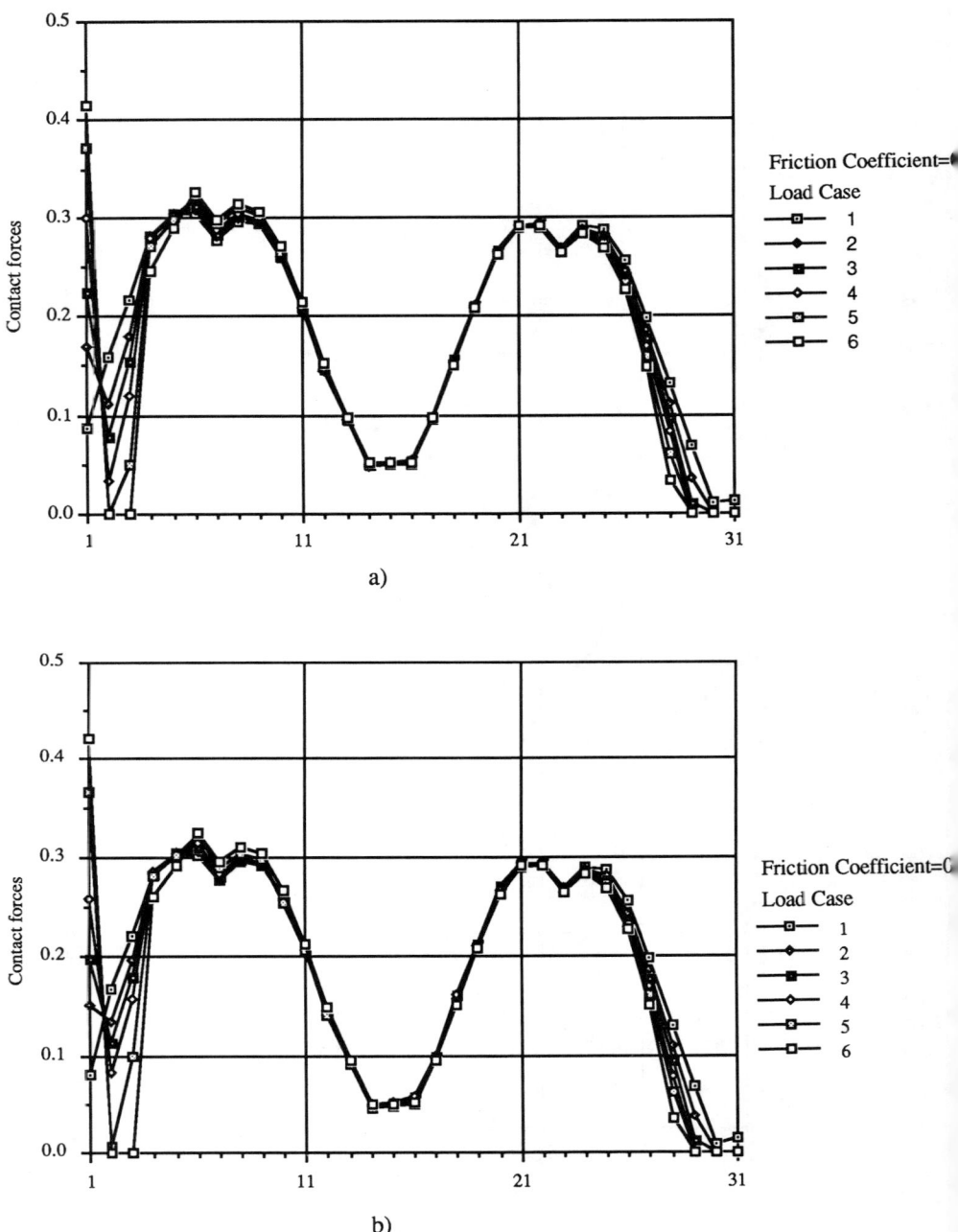

Fig. 9.5.6. Contact force (S_N) distribution.

9.5 Application II: The Nonmonotone Friction Problem

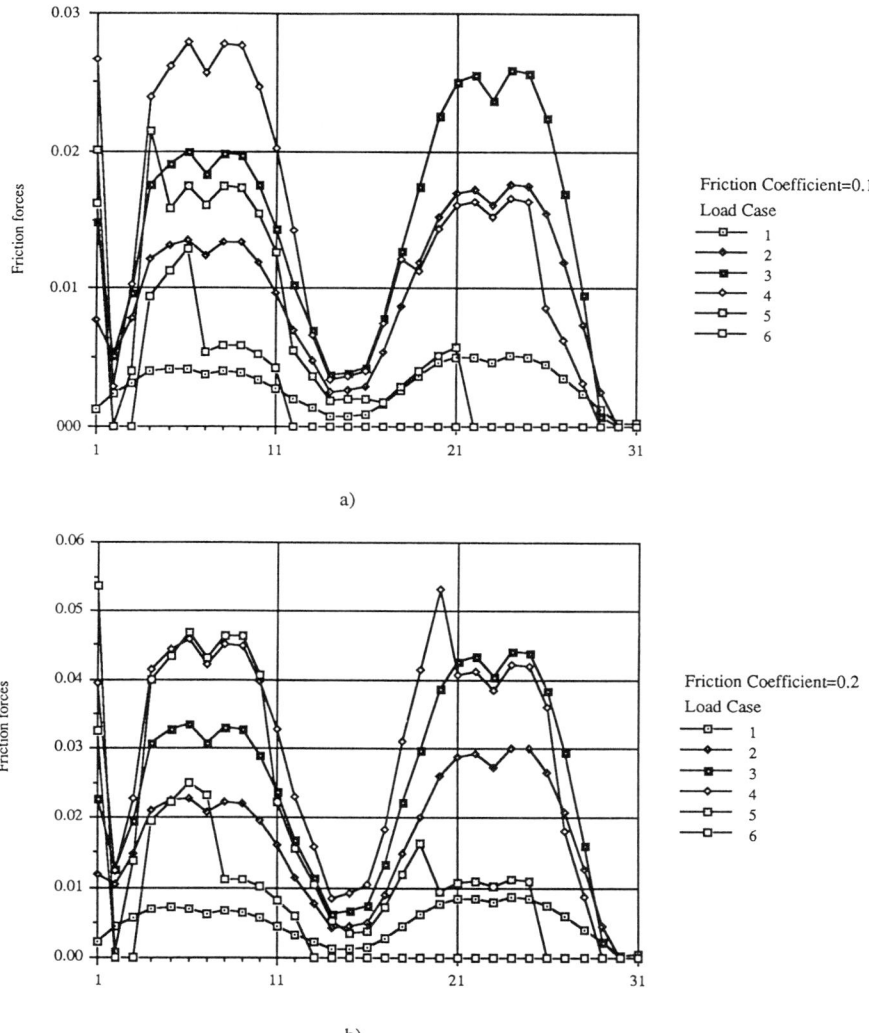

Fig. 9.5.7. Friction force (S_T) distribution.

The convergence of the algorithm is very fast. All the previous problems have been solved in about 7-14 cycles of the normal-tangential algorithm using a second order norm of the normal and tangential forces calculated as a stopping criterion: the algorithm terminates when $\frac{||S^{(i)} - S^{(i-1)}||}{||S^{(i)}||}$ becomes smaller than 10^{-4}. In Fig. 9.5.9 we give a diagram representing the convergence rate of the algorithm.

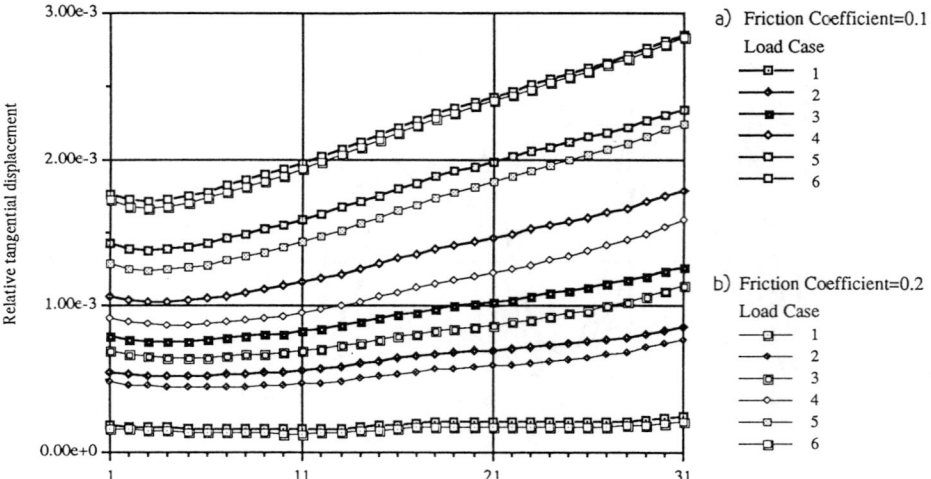

Fig. 9.5.8. Influence of the friction coefficient on the results.

The differences of the relative tangential displacements between two consecutive steps with respect to the step are depicted and the convergence rate seems to be linear. Numerical experiments have shown that the present form of the algorithm is 30-70% faster than the initial form, in which, after each estimation of the normal force, say $S_N^{(i)}$, we leave the whole tangential algorithm to converge.

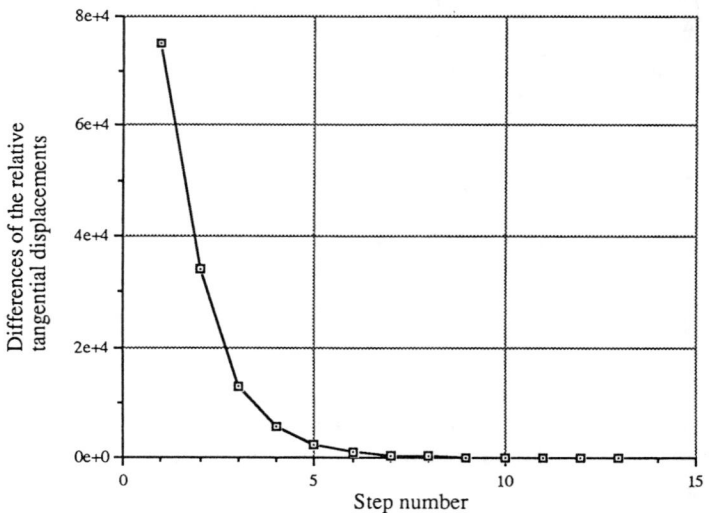

Fig. 9.5.9. About the covergence rate of the algorithm.

9.5 Application II: The Nonmonotone Friction Problem

Load Case		Proposed form			Initial form		
		number of cycles	number of substeps	execution time (min)	number of cycles	number of substeps	execution time (min)
$\mu=0.1$	1	9	1	35	6	27	48
	2	9	1	33	6	30	51
	3	8	1	32	6	29	53
	4	7	1	30	6	22	45
	5	8	1	32	7	24	50
	6	7	1	31	5	14	37
$\mu=0.2$	1	14	1	40	7	87	104
	2	14	1	38	8	110	105
	3	14	1	40	9	110	132
	4	11	1	35	8	84	103
	5	9	1	34	8	39	61
	6	8	1	33	8	36	58

Fig. 9.5.10. Comparison of the proposed form of the algorithm with its initial form.

This seemingly more accurate initial form gives the same results in a less number of cycles but each cycle needs much more time to terminate (see Fig. 9.5.10). Note that the presented algorithm here can be applied more easily to the adhesive contact problem, where the diagram of Fig. 9.5.5 does not depend on μ and S_N, and is fixed. Then considerably more fast convergence was obtained, a fact which was obviously expected.

10. On the Approximation of Hemivariational Inequalities by Variational Inequalities

The present chapter deals with a generalization of the method presented in Sect. 9.3, i.e. the approximation of a decreasing branch by monotone laws. After a general description of the method based on [Mis92a,b,93] we give some numerical applications. Moreover a comparison with the path following method [Cris91] is attempted. The material of the present chapter permits us to consider zigzag nonmonotone multivalued material laws and boundary conditions for twodimensional and threedimensional bodies.

10.1 General Formulation of the Method

Let $\Omega \subset \mathbb{R}^3$ be an open bounded connected subset with a regular boundary Γ, which is occupied by a deformable body Ω in its undeformed state. The body is assumed to have a boundary Γ divided into the three nonoverlapping parts Γ_U, Γ_F and Γ_S. On Γ_U (resp. on Γ_F) the displacements u_i (resp. the tractions S_i), $i = 1, 2, 3$, are prescribed and are equal to U_i (resp. to F_i) and on Γ_S a nonmonotone possibly multivalued boundary condition holds. Moreover let us assume that the body is nonlinear elastic obeying the subdifferential material law

$$\sigma \in \partial w(\varepsilon) \quad \text{in} \quad \Omega, \qquad (10.1.1)$$

where $w : \mathbb{R}^6 \to (-\infty, +\infty]$, $w(\varepsilon) \not\equiv \infty$, is a convex l.s.c proper functional. We assume that on Γ_S the normal traction is given, i.e. that

$$S_N = C_N, \quad C_N = C(x) \quad \text{on} \quad \Gamma_S \qquad (10.1.2)$$

and that in the tangential direction

$$-S_T \in \bar{\partial} j(u_T), \qquad (10.1.3)$$

where $j : \mathbb{R}^2 \to (-\infty, +\infty)$ is a locally Lipschitz superpotential. The twodimensional reaction-displacement law (10.1.3) is assumed to be the extension in the two dimensions of the onedimensional nonmonotone possibly multivalued laws of Fig. 10.1.1a,b. The extension in the two dimensions is achieved by the methods of Sect. 3.2. Here we assume that the superpotential j is given. Let us introduce the kinematically admissible set $V_0 = \{v | v = \{v_i\} \quad v_i = U_i \text{ on } \Gamma_U\}$ and let W be defined by

$$W(\varepsilon) = \begin{cases} \int_\Omega w(\varepsilon(u)) d\Omega & \text{if } w(\varepsilon(u)) \in L^1(\Omega) \\ \infty & \text{otherwise.} \end{cases} \quad (10.1.4)$$

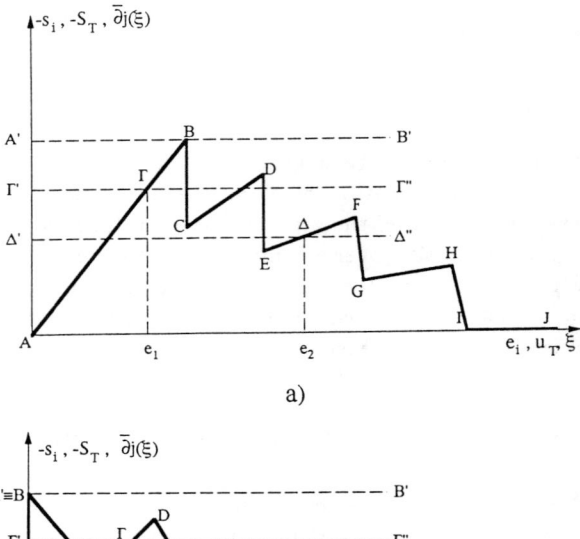

Fig. 10.1.1. The nonmonotone law.

Then at the position of equilibrium (assumption of small deformations, the first two equations of (9.4.6) hold) the following variational hemivariational inequality holds (cf. Sect. 4.2):
Find $u \in V_0$ such as to satisfy

$$W(\varepsilon(v)) - W(\varepsilon(u)) + \int_{\Gamma_S} j_T^0(u, v_T - u_T) d\Gamma \geq (l, v - u) \quad \forall v \in V_0. \quad (10.1.5)$$

Here (l, v) denotes the work of the volume forces and the forces F_i, $i = 1, 2, 3$, and C_N on Γ_F and Γ_S respectively. Let us define the corresponding problem with friction boundary conditions with given normal force:
Find $u \in V_0$ such that

$$W(\varepsilon(v)) - W(\varepsilon(u)) + \int_{\Gamma_S} \mu|C_N|(|v_T| - |u_T|)d\Gamma \geq (l, v - u) \quad \forall v \in V_0. \quad (10.1.6)$$

Here $\mu|C_N|$ is equal to AA' in Fig. 10.1.1a or b and this fact yields the value of μ if C_N is given. In the twodimensional law μ results from $\bar{\partial}j(0)$ analogously. On Γ_S one may consider more general nonmonotone laws by assuming that between the boundary stress vector $S = \{S_i\}$ and the displacement $u = \{u_i\}$, a law of the form

$$-S \in \bar{\partial}j(u) \text{ on } \Gamma \quad (10.1.7)$$

holds. Then the position of equilibrium is characterized by a hemivariational inequality similar to (10.1.5) with the difference that the term $\int_{\Gamma_S} j^0(u_T, v_T - u_T)d\Gamma$ is replaced by the term $\int_{\Gamma_S} j^0(u, v - u)d\Gamma$. Note also that if such a law holds at an interface Γ_1 then in $j(u)$ and in $j^0(u, v - u)$, u and v denote the relative interface displacements.

The proposed approximation method can be obtained by comparing the variational inequality (10.1.6) with the hemivariational inequality (10.1.5). We write (10.1.6) in the form:

$$W(\varepsilon(v)) - W(\varepsilon(u)) + \int_{\Gamma_S} \mu|C_N|(|v_T| - |u_T|)d\Gamma \geq \quad (10.1.8)$$
$$(l, v - u) + R(v_T, u_T, |v_T|, |u_T|) \quad \forall v \in V_0,$$

where

$$R(v_T, u_T, |v_T|, |u_T|) = \quad (10.1.9)$$
$$\int_{\Gamma_S} \mu|C_N|(|v_T| - |u_T|)d\Gamma - \int_{\Gamma_S} j^0(u_T, v_T - u_T)d\Gamma.$$

Then the algorithm we have proposed consists of the following iterative scheme: Find $u^{(\rho)} \in V_0$ such that

$$W(\varepsilon(v)) - W(\varepsilon(u^{(\rho)})) + \int_{\Gamma_S} \mu|C_N|(|v_T| - |u_T^{(\rho)}|)d\Gamma \geq \quad (10.1.10)$$
$$(l, v - u^{(\rho)}) + R(v_T, |v_T|, u_T^{(\rho-1)}, |u_T^{(\rho-1)}|) \quad \forall v \in V_0.$$

This iterative scheme is compatible with the graphic description of the approximation steps in Fig. 10.1.1 a,b, since R expresses the difference of the area variations under the two graphs, and measures the difference of the reaction forces corresponding to the initial nonmonotone law and the frictional one.

The foregoing approximation scheme can be extended by considering any monotone possibly multivalued law and not only the friction law. Let us assume that we have a problem for which the hemivariational inequality (10.1.5) holds. Moreover we chose to approximate the nonmonotone multivalued law (10.1.3) by the monotone law (the convex superpotential law)

$$-S_T \in \partial\Phi(u_T) \text{ on } \Gamma \quad (10.1.11)$$

where Φ is a convex, lower semicontinuous energy functional taking values in the interval $(-\infty, +\infty]$, $\Phi \not\equiv \infty$, and ∂ is the subdifferential. For this problem the variational inequality

$$u \in V_0, \quad W(\varepsilon(v)) - W(\varepsilon(u)) + \Phi(v_T) - \Phi(u_T) \geq (l, v - u) \; \forall v \in V_0 \quad (10.1.12)$$

holds. Thus we are led to consider the following iterative scheme:
Find $u^{(\rho)} \in V_0$ such that

$$W(\varepsilon(v)) - W(\varepsilon(u^{(\rho)})) + \Phi(v_T) - \Phi(u_T^{(\rho)}) \geq (l, v - u^{(\rho)}) + R(v_T, u_T^{(\rho-1)}) \quad (10.1.13)$$

where

$$R(v_T, u_T^{(\rho-1)}) = \Phi(v_T) - \Phi(u_T^{(\rho-1)}) - \int_{\Gamma_S} j^0(u_T^{(\rho-1)}, v_T - u_T^{(\rho-1)})d\Gamma. \quad (10.1.14)$$

The proof of convergence of the aforementioned iterative scheme is outside the scope of the present Section. The numerical examples we have solved have always shown a fast convergence, something which from the physical point of view seems quite reasonable.

10.2 Application III: Nonmonotone Friction Interface Conditions with Debonding

This application is similar to the problem of Sect. 9.5. The structure is depicted in Fig. 10.2.1a. In Fig. 10.2.1b the nonmonotone friction law is given and in Fig. 10.2.1c the loading cases. The structure has E=2.1×10^{8}kN/m^2, $\nu = 0.16$ and thickness t=0.01 m. The friction law can be put in the form

$$S_T = g([u_T], \mu, S_N) \quad (10.2.1)$$

where g is a nonmonotone function, $[u_T]$ is the relative displacements of the two parts of the interface, S_N and S_T are the normal and tangential to the interface forces respectively and μ is the friction coefficient. Note that this diagram may be different from point to point, i.e.

$$S_T = g([u_T], \mu, S_N, x) \quad (10.2.2)$$

where x is the position on the interface. Thus, at every point, the friction force is calculated by multiplying the vertical coordinate of the diagram by $\mu|S_N|$ where $\mu = \mu(x)$ and $|S_N| = |S_N(x)|$ are the friction coefficient and the normal contact force respectively, at the considered point. Obviously these conditions hold at the parts of the interface which come in contact. Thus, the interface conditions read (the sign convention is obvious)

$$\text{If } [u_N] > 0 \text{ then } S_N = 0 \text{ and } S_T = 0 \quad (10.2.3)$$

10.2 Application III: Nonmonotone Friction Interface Conditions with Debonding

If $[u_N] = 0$ then $S_N < 0$ and $S_T = g([u_T], \mu, S_N, x)$. (10.2.4)

where $[u_N]$ are the relative normal displacements of the interface. The algorithm described in Sect. 10.1 must be appropriately completed for the treatment of this more complicated problem. The coupling of the unilateral contact conditions with the nonmonotone friction conditions follows the same steps as in Sect. 9.5. Moreover for the friction step the more general algorithm described by (10.1.10) will be applied.

Step 1: Subproblem (a). This subproblem consists of a pure unilateral contact problem. Calculate the structure with the tangential forces $S_T = S_T^{(0)}$ given, (an upper index denotes the iteration number in which the corresponding quantity occurs). Usually it is assumed that $S_T^{(0)} = 0$. This problem is treated by a Q.P. algorithm which minimizes the potential energy of the structure. The solution of this problem yields the normal forces $S_N^{(1)}$ and the normal and tangential to the interface displacements $[u_N^{(1)}]$ and $[u_T^{(1)}]$.

Step 2: Subproblem (b). The problem corresponding to the tangential interface conditions is solved with the assumption of constant normal forces $S_N = S_N^{(1)}$. The iterative scheme described in Sect. 10.1 (relation 10.1.10) is applied in order to approximate the given nonmonotone law. Thus the relative tangential displacements $[u_T^{(1)}]$ and the respective tangential forces $S_T^{(1,1)}$ are calculated. Here the second upper index denotes the number of the approximation of the nonmonotone law. Within each step a classical monotone problem is solved. This problem is again treated by a Q.P. algorithm. The solution gives as a result the relative tangential displacements $[u_T^{(1,2)}]$ from which new values for the tangential forces $S_T^{(1,2)}$ are obtained. A new monotone problem is then solved and so on until the differences $|S_T^{(1,k)} - S_T^{(1,k+1)}|$ and $|[u_T^{(1,k)}] - [u_T^{(1,k+1)}]|$ between two consecutive steps k and $k+1$ become sufficiently small. The tangential forces $S_T^{(1,k+1)}$ and the respective displacements $[u_T^{(1,k+1)}]$ are the solution of the considered subproblem. In the following these quantities are set as $S_T^{(1)}$ and $[u_T^{(1)}]$.

Step 3: Subproblem (a). The pure unilateral contact subproblem is again solved with constant tangential forces $S_T = S_T^{(1)}$, which gives as a result the normal forces $S_N^{(2)}$ and the relative displacements $[u_N^{(2)}]$ and $[u_T^{(2)}]$. The procedure is continued until the differences $|S^{(i)} - S^{(i+1)}|$, where $S = \begin{bmatrix} S_N \\ S_T \end{bmatrix}$, between two sequential solutions i and $i+1$ become smaller than a predefined accuracy. Numerical experience shows a high sensitivity of the whole algorithm in the case of unstable contact regions.

286   10. On the Approximation of Hemivariational Inequalities

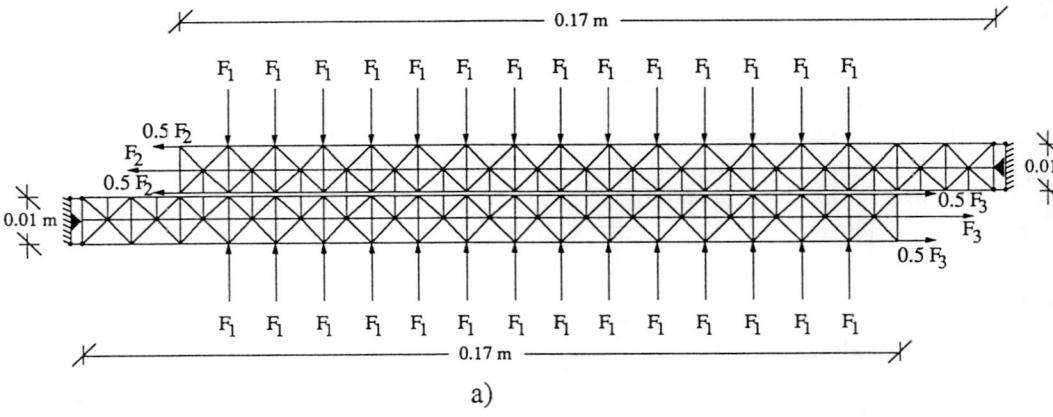

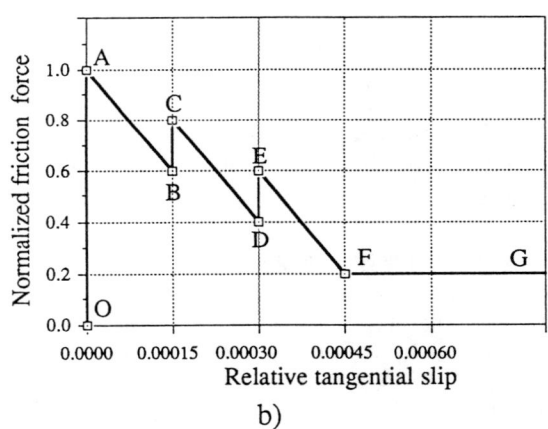

Load Case	F_1	F_2	F_3
1	5.00	2.00	10.00
2	5.00	4.00	20.00
3	5.00	6.00	30.00
4	5.00	8.00	40.00
5	5.00	10.00	50.00

c)

Fig. 10.2.1. Application III.

10.2 Application III: Nonmonotone Friction Interface Conditions with Debonding

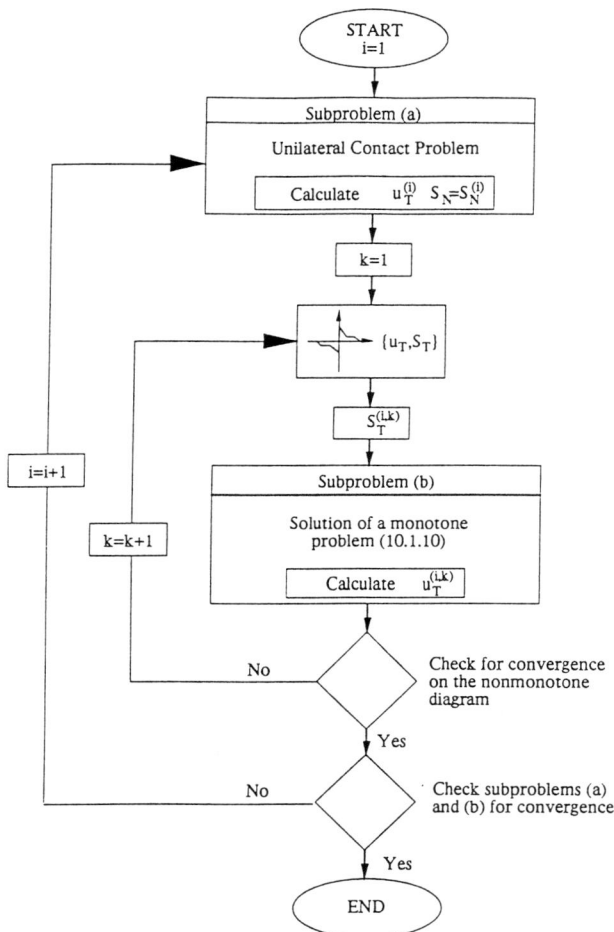

Fig. 10.2.2. Description of the algorithm.

The algorithm is described in Fig. 10.2.2. A modified form of the algorithm which proved to be much faster is the following: we perform only one substep inside the second step of the algorithm, until the contact region is well established, i.e. after the calculation of $[u_T^{(k,2)}]$ we continue to step 3. This modified form of the algorithm is presented in Fig. 10.2.3. The algorithm is now applied to the analysis of the structure which is discretized through constant stress triangular elements. Moreover, the friction coefficient μ equals 0.3. Loads F_1 remain constant, while loads F_2 and F_3 increase in order to obtain a parametric solution. The five load cases of Fig. 10.2.1c are analyzed. The distribution of the normal forces along the interface is given in Fig. 10.2.4a. It is noticed only a slight influence of the tangential forces F_2 and F_3 and of the nonmonotone diagram on the results. On the other hand the influence of the nonmonotone diagram on the distribution of the friction forces (Fig. 10.2.4b) is strong. For the first load case, where the tangential loads are small, the friction forces lie on the vertical branch OA of Fig. 10.2.1b and in a region between the nodes 1 to 14. This is the adhesive contact region, where the respective displacements are equal to zero (see Fig. 10.2.5, load case 1). For the same load case, the friction forces lie on the branch AB and the respective relative displacements increase from point to point at the remaining part of the interface. For the rest of the load cases, we note that zigzag shapes appear on the distribution of the shear forces this being an obvious consequence of the zigzag form of the law of Fig. 10.2.1b. As the tangential loading increases, the total friction force diminishes and this is a result of the softening form of the law. In the curve corresponding to the fifth load case we note that the friction forces are equal from point 13 to point 7 and lie on the branch FG of the nonmonotone diagram. It is observed that the existence of the complete vertical branches BC, DE does not lead to numerical instabilities. The numerical experimentation has shown that by means of a fine discretization it is indeed possible to obtain solution points on each vertical branch of the diagram of Fig. 10.2.1b.

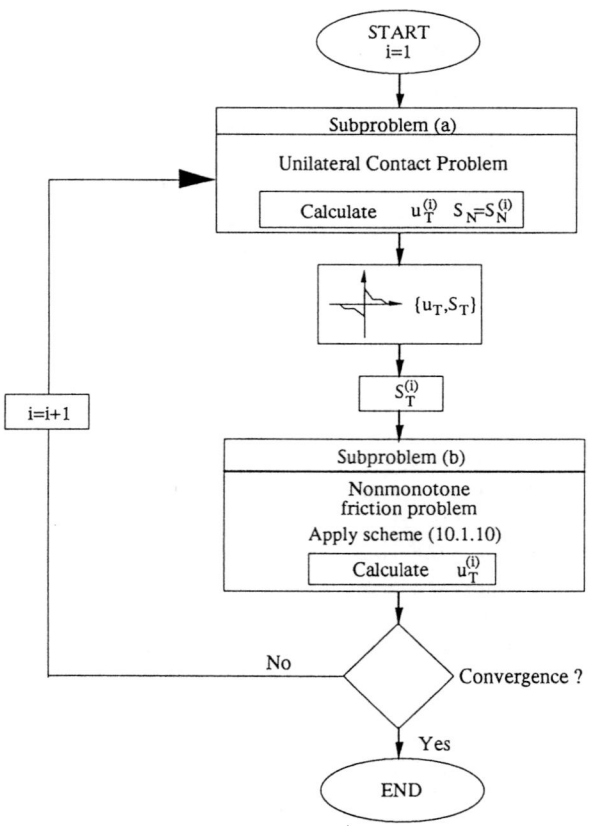

Fig. 10.2.3. Description of the modified algorithm.

10.2 Application III: Nonmonotone Friction Interface Conditions with Debonding

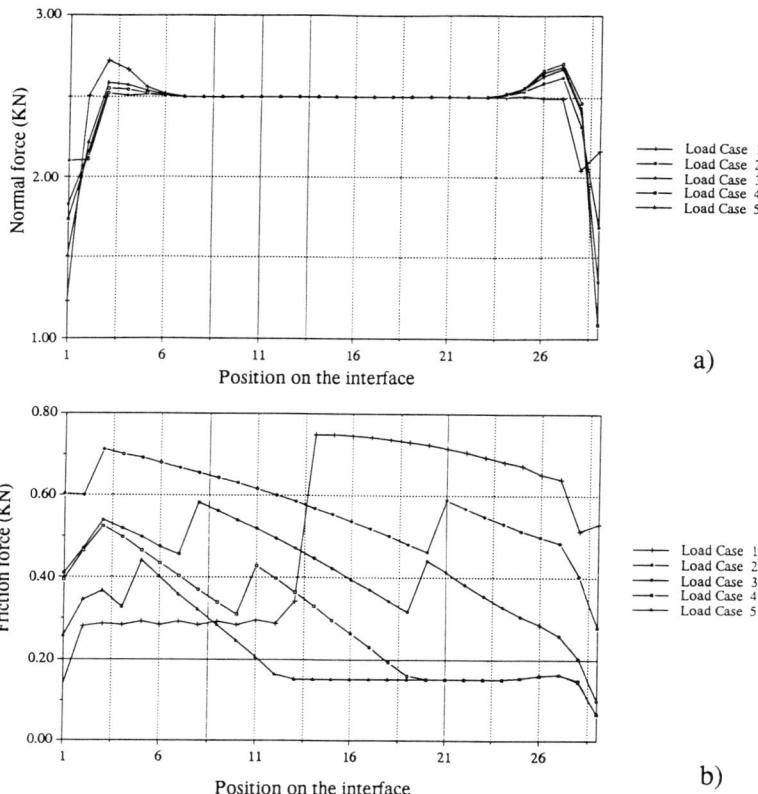

Fig. 10.2.4. Distributions of S_N and S_T on the interface.

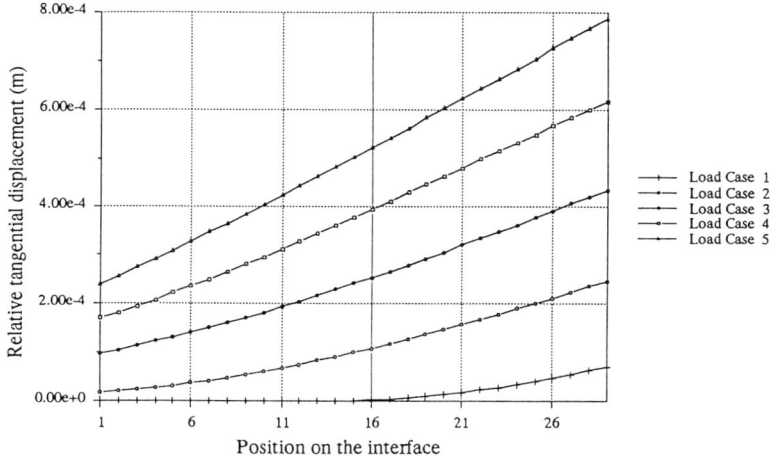

Fig. 10.2.5. Relative tangential displacements.

10.3 Application IV : Adhesive Joints in Structural Mechanics

Here we will study a problem much more complicated than the one of Sect. 9.4. It concerns the calculation, under arbitrary loading, of laminate structures which are glued with some adhesive material obeying to a nonmonotone law. Such stress-strain or reaction-displacement laws have been recently obtained as consequence of the use of advanced technology testing machines. We mention here the force-displacement diagrams of Fig. 10.3.1 given by Green and Boyer [Gree] during slow pull of tests of fiber reinforced specimens of laminated

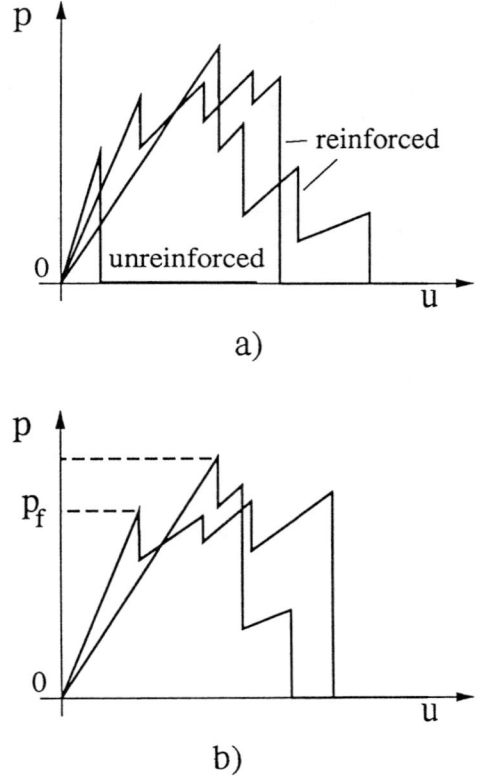

Fig. 10.3.1. Certain nonmonotone multivalued laws in laminated composites.

10.3 Application IV : Adhesive Joints in Structural Mechanics

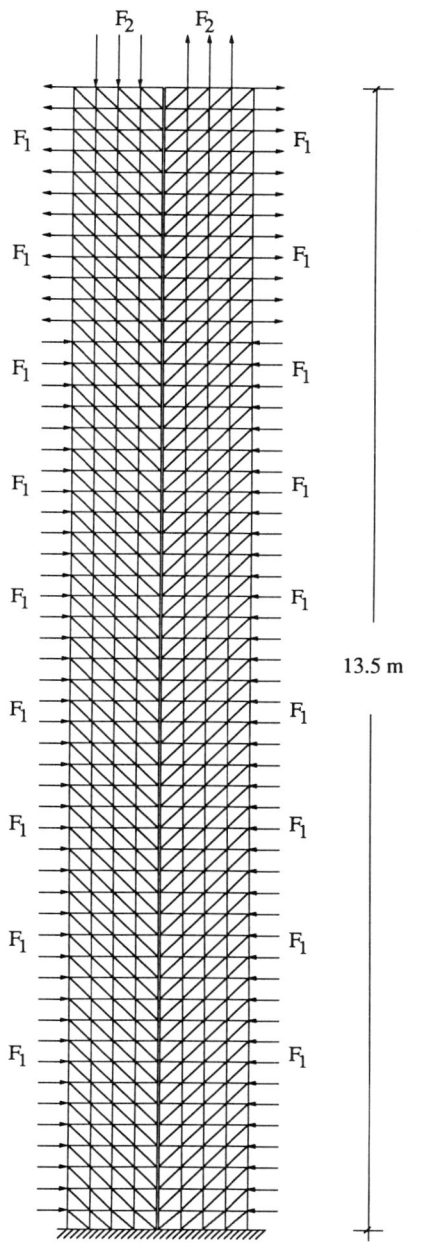

Load Case	F_1	F_2
1	0.10	1.0
2	0.08	0.8
3	0.06	0.6
4	0.04	0.4
5	0.02	0.2

Fig. 10.3.2. Structures connected with an adhesive material. F.E. discretization and load cases. (E= $2.1 \times 10^6 \text{kN/m}^2$, $\nu = 0.30$, t=0.01 m).

292   10. On the Approximation of Hemivariational Inequalities

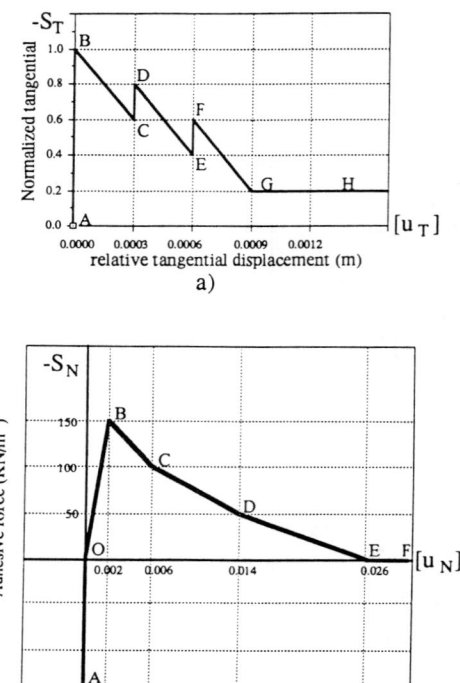

Fig. 10.3.3. The nonmonotone multivalued behaviour of the adhesive tangentially and normally to the interface.

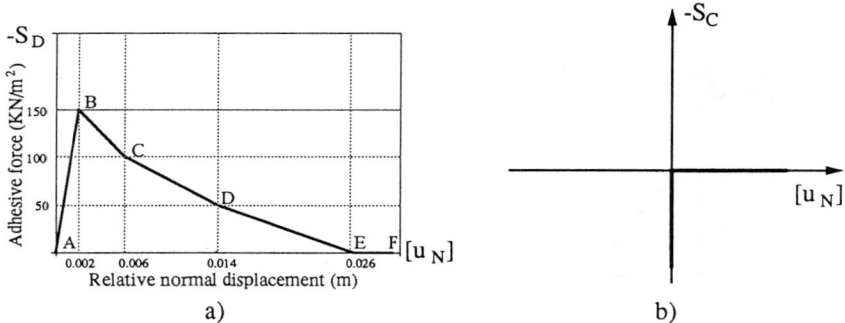

Fig. 10.3.4. The splitting of the $S_N - [u_N]$ law.

10.3 Application IV : Adhesive Joints in Structural Mechanics

products and the diagrams obtained by Roman et. al [Roman] (Fig. 10.3.1b) during several types of loading tests of specimens of glass fiber reinforced epoxy laminates, among others (see also Sect. 3.2). In the sequel the method presented in Sect. 10.1 will be applied. As an example let us consider the structure of Fig. 10.3.2. The two parts of the interface Γ are glued with an adhesive material. Furthermore, let us suppose that this material exhibits a nonmonotone multivalued response both in the normal and in the tangential direction with respect to the interface. Tangentially we assume that a law of the form of Fig. 10.3.3a holds, whereas the law of Fig. 10.3.3b holds normally to the interface. This diagram contains the vertical part OA and the nonmonotone branch OBCDEF. The first is called contact branch and the second adhesive branch. The first (resp. the second) is realized for compressive (resp. noncompressive) interface forces. We split the normal to the interface problem into two subproblems by considering either the adhesive law of Fig. 10.3.4a or the unilateral contact law of Fig. 10.3.4b. We denote by $[u_N]$ the relative normal displacements of the two fronts of the interface and by Γ_C (resp. Γ_D) the contact region (resp. the region where adhesive forces appear with the simultaneous beginning of detachment) of the interface Γ. We note here that the contact and non-contact areas are not a priori known. Then the interface behaviour can be described by the following conditions (sign convention is as in Sect. 9.4):

$$\text{if } [u_N] = 0 \text{ then } S_C \geq 0 \text{ and } S_T = g([u_T], \mu, S_N, x) \text{ on } \Gamma_C \quad (10.3.1)$$

$$\text{if } [u_N] > 0 \text{ then } S_D = h([u_N], x) \text{ on } \Gamma_D \quad (10.3.2)$$

where S_C is the normal force to the contact region (contact force), S_D is the resistance of the adhesive in the normal to the interface direction and g, h are nonmonotone functions that may also include jumps (e.g branch CD in Fig. 10.3.3a). Forces S_C are always positive while forces S_D are always negative. Here we have made the assumption that the adhesive material has zero ability to transmit tangential forces.

The solution of this problem is obtained by approximating both the nonmonotone laws by two sequences of monotone ones. Thus we apply twice and in an appropriate order the procedure of Fig. 9.4.2 or we apply twice the method of Sect. 10.1. In the following we propose an algorithm for the treatment of the problem. We split the problem into two subproblems. Subproblem (a) concerns the solution of the problem which arises in the normal to the interface direction. The solution of the problem arising in the tangential direction corresponds to subproblem (b). Furthermore, subproblem (a) is splitted into two more subproblems, (a1) and (a2). Subproblem (a1) corresponds to the unilateral contact problem and subproblem (a2) corresponds to the detachment problem. The algorithm is depicted in Fig. 10.3.5. Starting the algorithm, set $i = 1$ and $k = 1$.

Step 1: Subproblem (a1). Calculate the structure with given tangential forces $S_T = S_T^{(i)}$ and given detachment forces $S_D = S_D^{(i,k)}$ and with the assumption that the unilateral contact conditions hold on the interface. Usually we assume that $S_T^{(1)} = 0$ and $S_D^{(1,1)} = 0$. Let $S_N^{(i,k)}$ and $[u_N^{(i,k)}]$ be the resulting forces and relative displacements respectively normal to the interface. They are obtained through a Q.P. algorithm which minimizes the potential energy of the structure.

Step 2: Subproblem (a2). Suppose that the contact forces $S_C = S_C^{(i,k)}$ and the tangential forces $S_T = S_T^{(i)}$ are given. From the nonmonotone diagram of Fig. 10.3.4a we find the detachment forces that correspond to $[u_N^{(i,k)}]$. We solve the arising monotone problem using a Q.P. algorithm which minimizes the complementary energy of the structure, with the restrictions $|S_D| \leq |S_D^{(i,k)}|$. Let $[u_N^{(i,k+1)}]$ be the new values of the normal displacements.

Step 3 : We check the convergence with respect to the subproblems (a1) and (a2) using some predefined criteria. If convergence has been achieved we proceed to step 4, otherwise we return to step 1 with $k = k + 1$.

Step 4 : Subproblem (b). We suppose that the normal forces to the interface $S_C = S_C^{(i,k)}$ and $S_D = S_D^{(i,k)}$ are given. From the calculated tangential relative displacements $[u_T^{(i)}] = [u_T^{(i,k)}]$ and the nonmonotone diagram of Fig. 10.3.3a we find the respective tangential forces $S_T^{(i)}$. We solve the arising monotone problem using again a Q.P. algorithm which minimizes the complementary energy of the structure, with the restrictions $|S_T| \leq |S_T^{(i)}|$. Let $[u_T^{(i+1)}]$ be the new values for the relative tangential displacements.

Step 5 : We check the convergence with respect to the subproblems (a) and (b). In the case of convergence we terminate the algorithm, otherwise we return to step 1 with $i = i + 1$.

In this algorithm we have directly used for each nonmonotone subproblem the modified version of the previous section (cf. Fig. 10.2.3). The numerical results are depicted in Fig. 10.3.6a,b. Fig. 10.3.6a gives the distribution of the normal forces $-S_N$ along the interface; positive values on the diagram, correspond to adhesive forces, while the negative ones correspond to the contact forces. The contact and non-contact regions are accurately established for each load case. When the value of F_1 is small the interface is completely "closed". As the value of F_1 increases, the distribution of the normal forces appears positive values (load cases 4,3). Further increase of the forces has as a result a sudden opening of the interface, and the adhesive forces appear now at the middle part of it (load cases 3,2,1).

10.3 Application IV : Adhesive Joints in Structural Mechanics

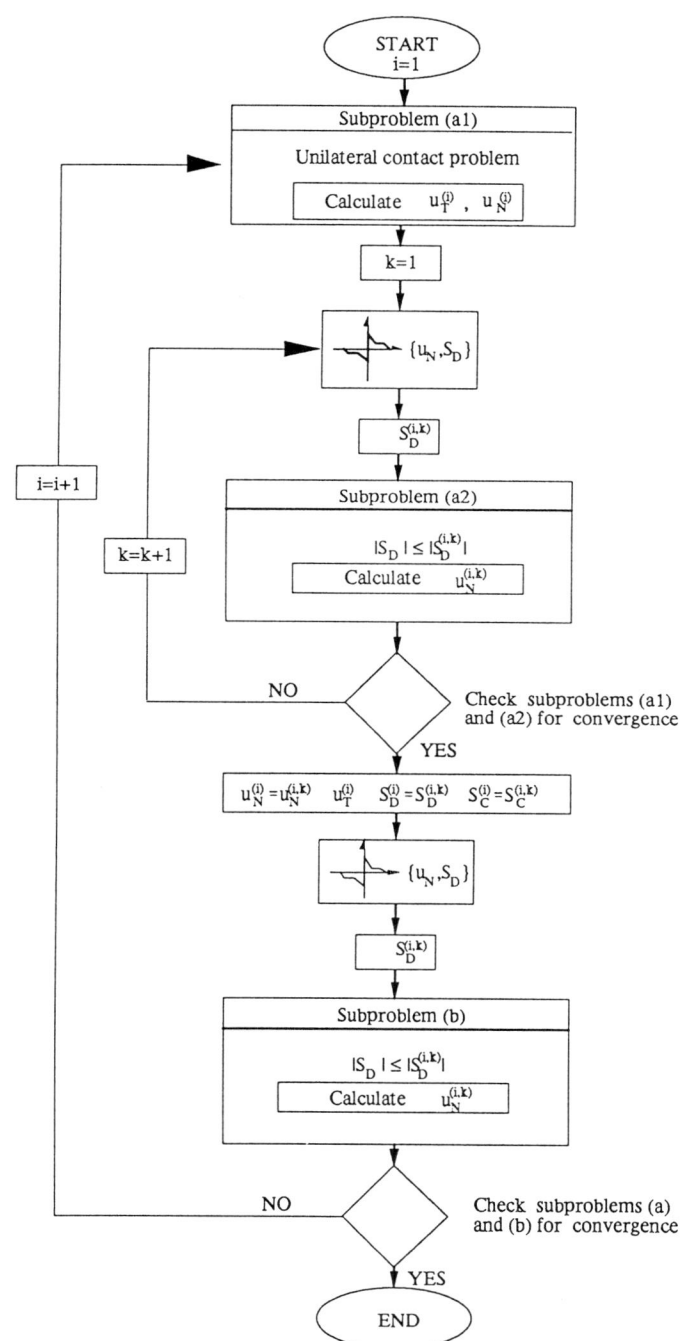

Fig. 10.3.5. Description of the algorithm.

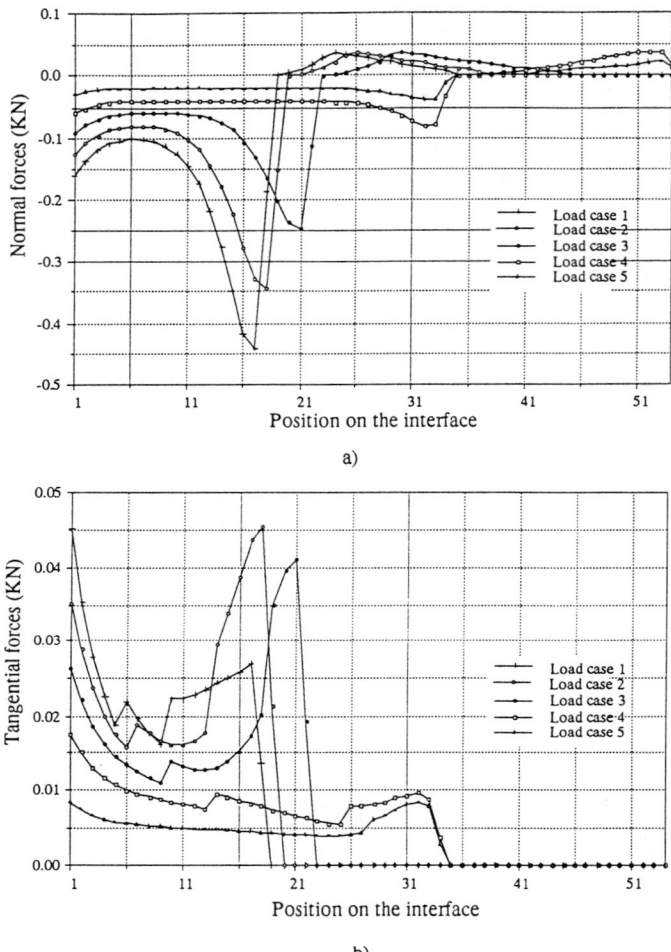

Fig. 10.3.6. Distribution of normal and tangential forces $-S_N$ and $-S_T$ on the interface.

The distribution of the contact forces has a maximum just after the point from which the interface is completely closed. This fact is quite reasonable from the engineering point of view.

The distribution of the tangential forces along the interface if given in Fig. 10.3.6b. Zero tangential forces correspond to the parts of the interface which do not come in contact. In general the distribution of the friction forces follows the shape of the distribution of the contact forces. But the zigzag shape of the tangential law of Fig. 10.3.3a influences the shape of the distribution. The abrupt changes of the values have their explanation on this fact. Indeed, at these points, the adhesive force passes on the nonmonotone diagram of Fig. 10.3.3a from one branch to another. Characteristic is the first load case: points 1-5 of this curve lie on branch BC, points 6-9 lie on branch DE and points 10-18 lie on branch FG of the nonmonotone diagram of Fig. 10.3.3a.

10.4 Application V : Comparison with the Path Following Method

In this section we will compare the results of the decreasing branch approximation method developed here with those of Crisfield's path following method (see e.g. [Cris91]). The proposed algorithm will be applied to the analysis of a model problem described in [Cris88]p. 271. The structure (Fig. 10.4.1a) consists of bar elements, with elements 1 and 2 obeying to the nonmonotone material law OAB of Fig. 10.4.1b.

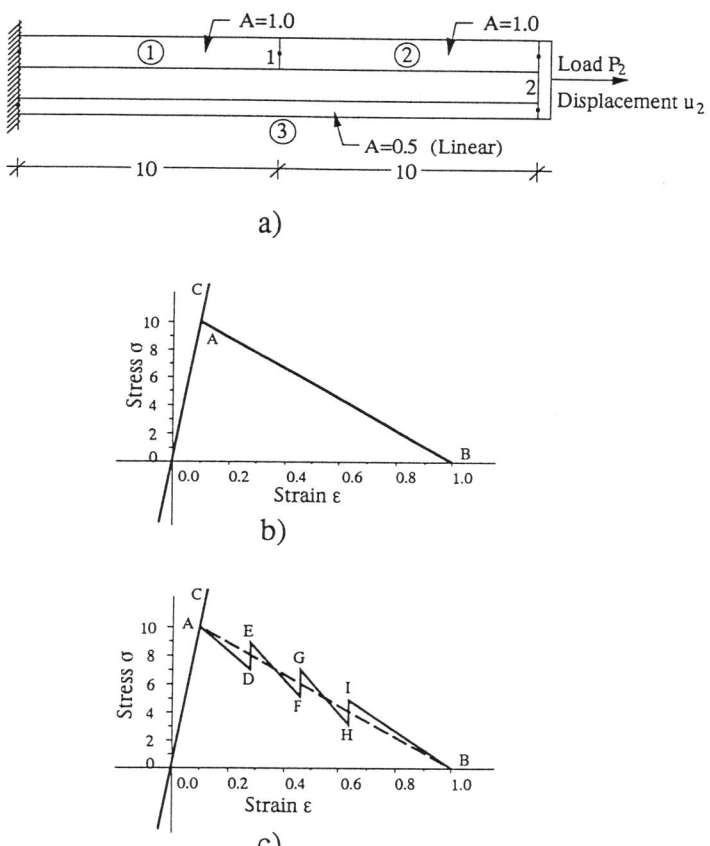

Fig. 10.4.1. Three-bar model problem : a) The structure (A = Area of Cross-section, E=100-use of compatible units) b) The nonmonotone law c) A nonmonotone possibly multivalued law.

Element 3 is assumed to be linearly elastic with modulus of elasticity corresponding to the branch OAC. One could consider the more realistic material law of Fig. 10.4.1c. However the path following methods cannot handle the complete vertical branches DE, FG, HI. Thus for the sake of compasiron of the results, we will adopt the simple law of Fig. 10.4.1b. We note that the proposed here method can easily treat the exact form of the law, with no increase of the complexity of the calculations.

The structure is loaded with the force P_2. Using an appropriate modification of the iterative scheme described in Sect. 10.1, the displacement u_2 is calculated for increasing values of the load. For small values of the load P_2 (less than 15), the behaviour of the structure is linear and the plots of the stress-strain values of elements 1 and 2 on the diagram of Fig. 10.4.1b lie on the branch OA. For greater values, softening occurs in the elements 1 and 2 and the response is nonlinear. The obtained results are identical with those of [Cris88]. The load-displacement curve of the structure is given in Fig. 10.4.2. It is interesting to calculate the convergence rate for this simple example. For a given load level ($P_2 = 20$), the differences of the displacements u_2 between two consecutive steps are calculated.

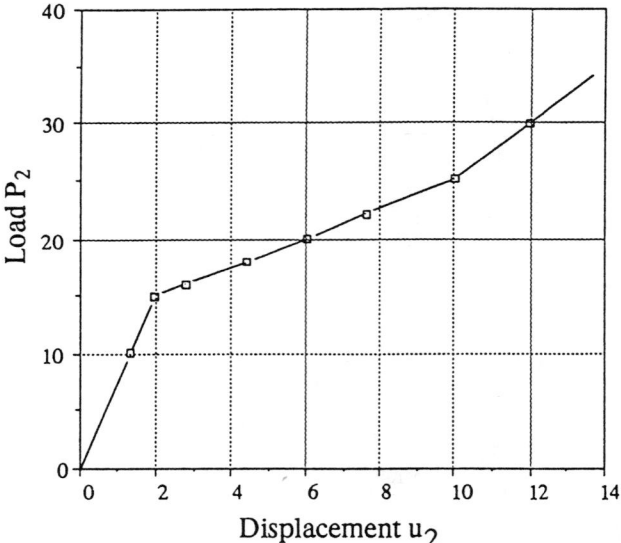

Fig. 10.4.2. Load-displacement curve.

10.5 Application VI : The Sawtooth Behaviour of Composites

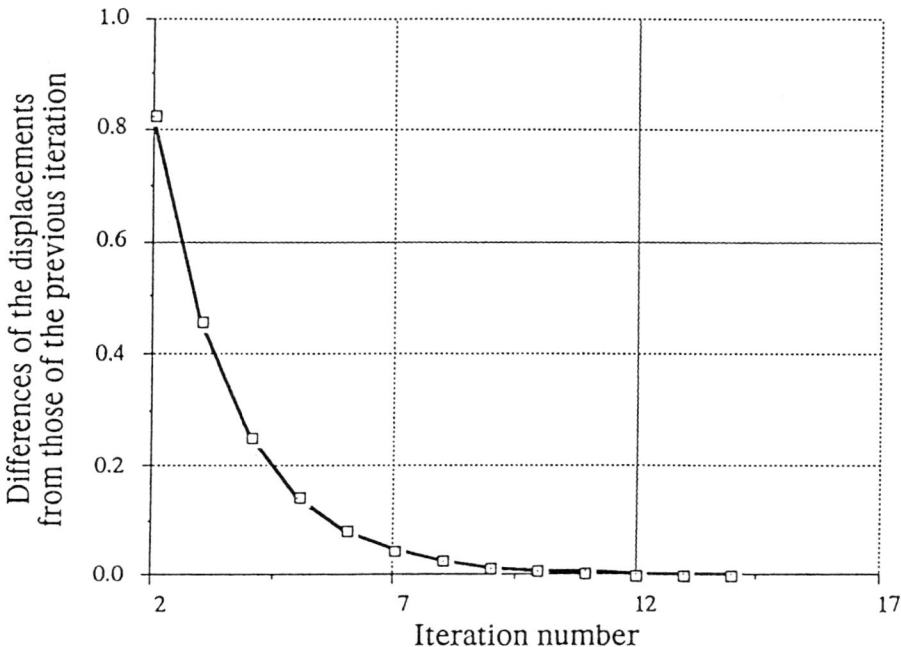

Fig. 10.4.3. The convergence rate of the algorithm.

The results are depicted in Fig. 10.4.3. The convergence rate calculated for several numerical experiments was found to be linear. We recall here that if $h^{(k)}$ is the difference between the solution of step k and the real solution, i.e. $h^{(k)} = u^{(k)} - u^*$, then the convergence rate is defined to be linear when the following relation holds [Flet]

$$\frac{||h^{(k+1)}||}{||h^{(k)}||} \leq \alpha, \tag{10.4.1}$$

where $\alpha > 0$. The iterative scheme we have proposed has a very satisfactory convergence rate for the present problem.

10.5 Application VI : Nonmonotone Stress-Strain Laws. The Sawtooth Behaviour of Composites

This application concerns a structure whose upper half part is fiber reinforced with unidirectional horizontal fibers. The lower half part is homogenous linear elastic. The interaction of the fibers with the matrix and the arising delamination phenomena are described by considering in the Ox direction the nonmonotone $\varepsilon_x - \sigma_x$ law of Fig. 10.5.1. For each fiber a law of the form of Fig. 10.5.2a holds but ε_m and ε_r are different from fiber to fiber. The analysis of a continuum consisting of a large number of parallel fibers yieds a diagram of the form of Fig. 10.5.2b according to Hult and Travniček [Hult]. This diagram can be put in the form

$$\sigma \in \bar{\partial} w(\varepsilon). \tag{10.5.1}$$

According to the notation introduced in Sect. 4.1 we have to find the solution of the following hemivariational inequality:
Find $u \in V_0$ such that

$$\int_\Omega w^0(\varepsilon(v), \varepsilon(v-u))d\Omega \geq (l, v-u) \quad \forall v \in V_0. \tag{10.5.2}$$

Here V_0 is the kinematically admissible set and (l, u) denotes the work of all applied loads. The application to the above problem of the main idea of the numerical method of Sect. 10.1 i.e the iterative scheme of (10.1.10), yields the following algorithm which is analogous to the one used also for the solution of the problem of the previous Sections (see Fig. 10.5.3). Note that the same iterative scheme applies to threedimensional superpotentials (e.g. to (3.2.19).

Step 1: Calculate the structure with the assumption that the law ABB' (see Fig. 10.5.2b) holds. Then we have to find the solution of the following variational inequality:
Find $u \in V_0$ such that

$$\int_\Omega [w(\varepsilon(v)) - w(\varepsilon(u))]d\Omega \geq (l, v-u) \quad \forall v \in V_0. \tag{10.5.3}$$

This variational inequality corresponds to the minimization of the potential energy of the structure with the constraints $\sigma \leq |AB''|$, which can be treated with a Q.P. algorithm. Let us denote by $\varepsilon^{(1)}$ the solution of this problem. Set $i = 1$.

10.5 Application VI : The Sawtooth Behaviour of Composites

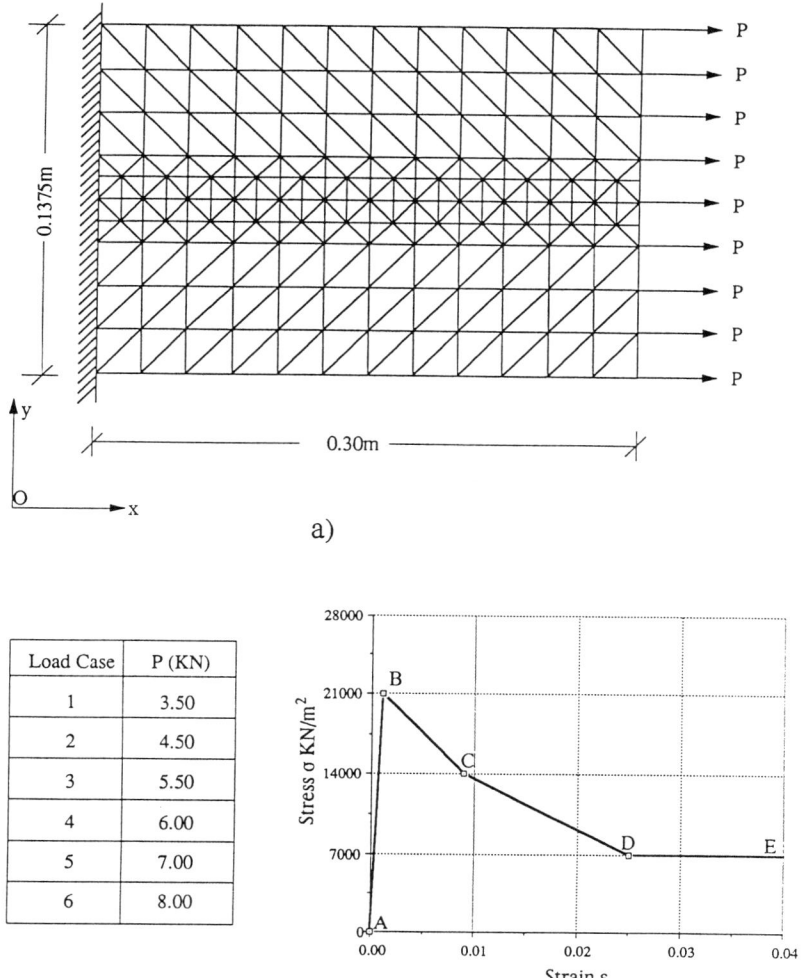

Fig. 10.5.1. a) F.E. discretization of the structure (E=2.1 ×10⁶kN/m²,
$\nu = 0$, t=0.01 m) b) The loading cases c) The nonmonotone $\sigma_x - \varepsilon_x$ stress-strain law.

Step 2: From the diagram of Fig. 10.5.2b we calculate the stress $\sigma^{(i)}$ that corresponds to the calculated value of the strain $\varepsilon^{(i)}$.

Step 3 : We solve the arising new monotone problem (with the constraint $\sigma \leq |\sigma^{(i)}|$) and let $\varepsilon^{(i+1)}$ be its solution.

302   10. On the Approximation of Hemivariational Inequalities

Step 4 : If convergence on the nonmonotone diagram has been achieved, the algorithm is terminated, if not set $i = i + 1$ and return to step 2.

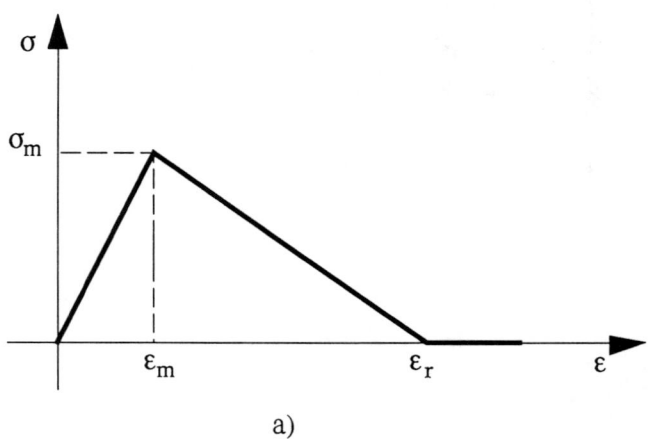

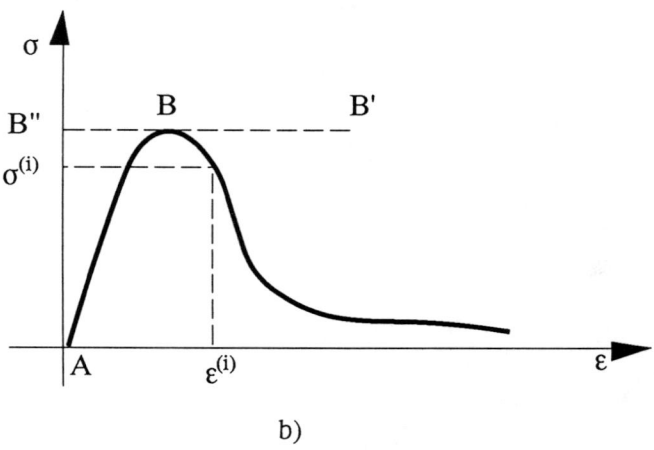

Fig. 10.5.2. a) Stress-strain law of a single fiber b) Stress-strain law for the fiber reinforced body.

10.5 Application VI : The Sawtooth Behaviour of Composites

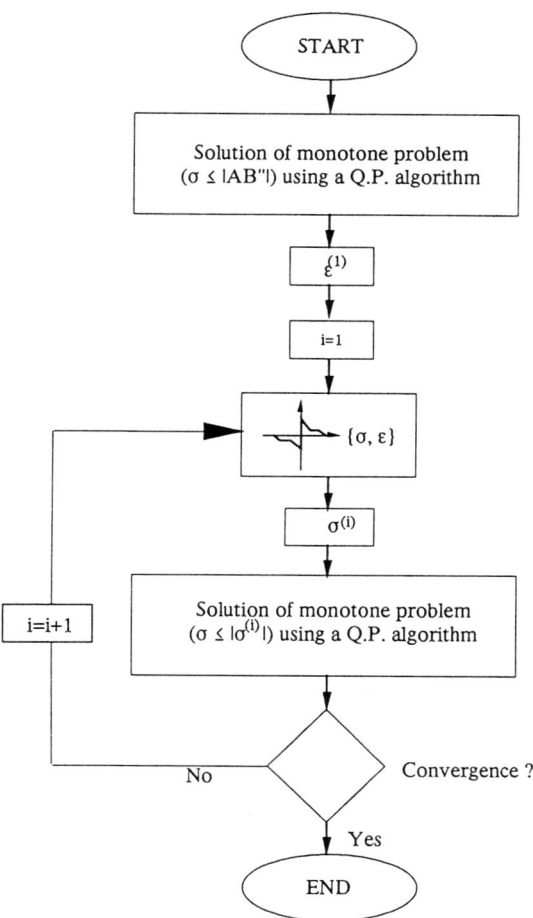

Fig. 10.5.3. Description of the algorithm.

The application of the above algorithm to the problem of Fig. 10.5.1a gives the results depicted in Fig. 10.5.4. For small values of the load (P=3.50 KN) the behaviour of the structure is linear and the stress field is symmetric with respect to the axis Ox. As the loading increases the stress fields are influenced by the nonmonotone law and the symmetry is destroyed. We note that the area between the boundary and the stress contour "F" (the one that corresponds to the branch DE of the nonmonotone diagram of Fig. 10.5.1c) increases with the load. At the last load case (P=8.00 KN) almost all the fibers are at softening state (branch BCDE of Fig. 10.5.1c).

304   10. On the Approximation of Hemivariational Inequalities

The method presented in this Section can be directly applied to all sawtooth laws of composite materials. In this case the 3D-generalization of the sawtooth laws according to Sect. 3.2 (see e.g. eq.(3.2.19)) replaces (10.5.1).

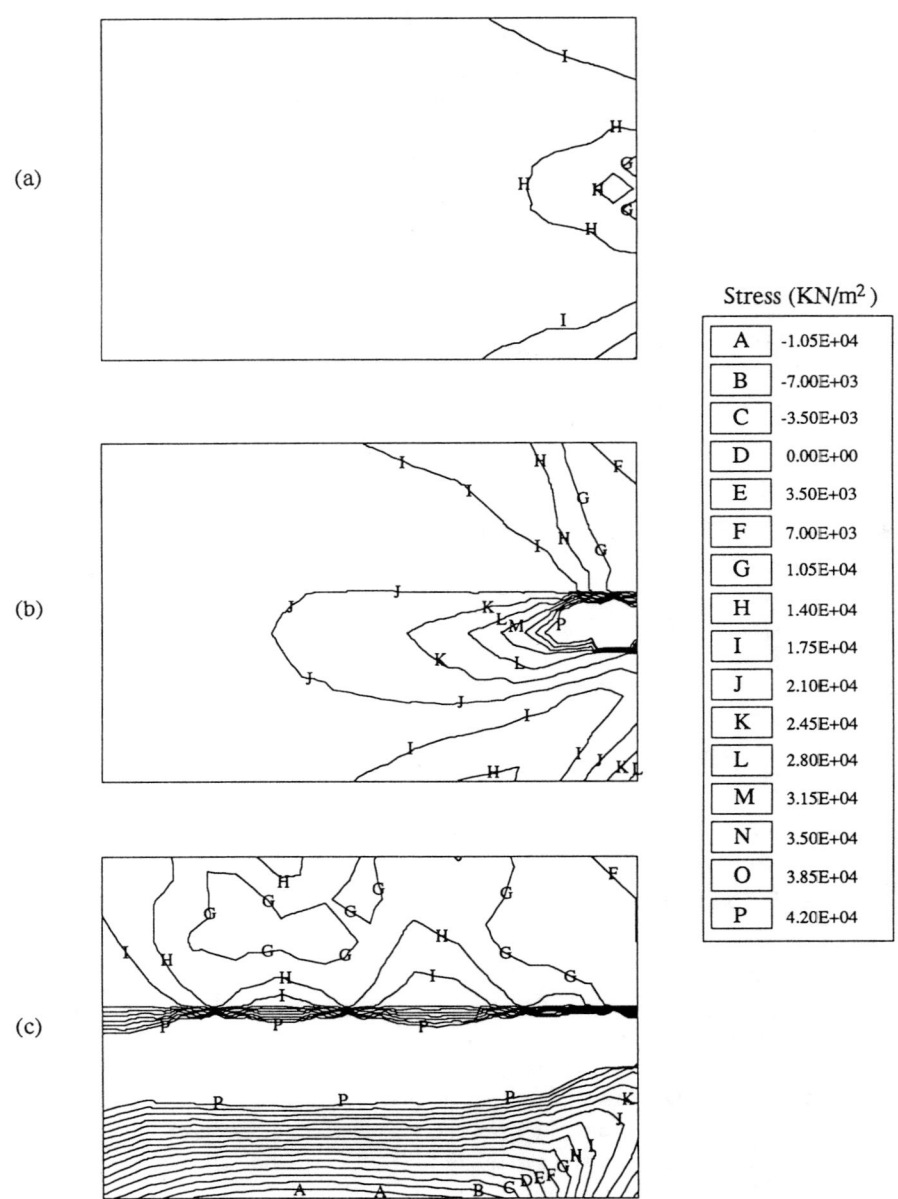

Fig. 10.5.4. σ_x-Stress fields for the six load cases.

10.5 Application VI : The Sawtooth Behaviour of Composites   305

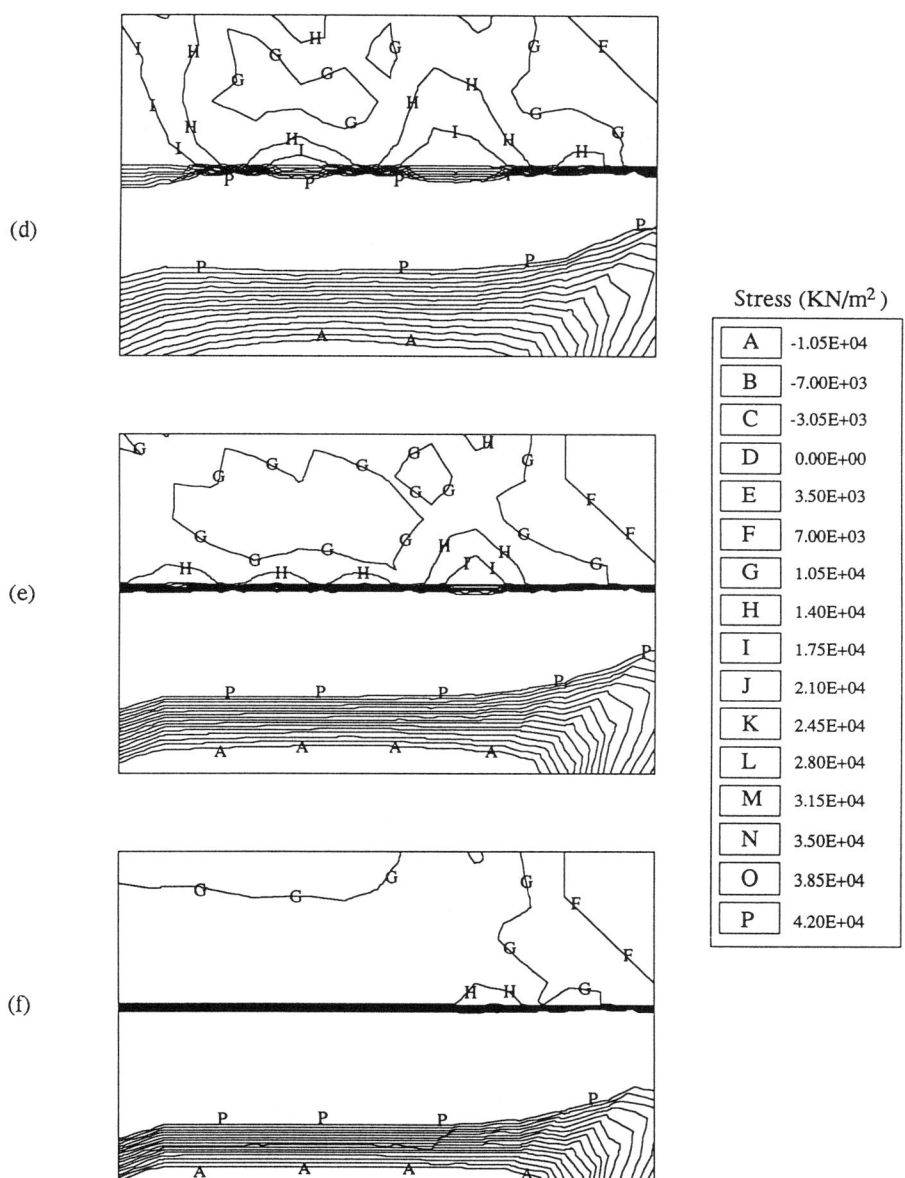

	Stress (KN/m^2)
A	-1.05E+04
B	-7.00E+03
C	-3.05E+03
D	0.00E+00
E	3.50E+03
F	7.00E+03
G	1.05E+04
H	1.40E+04
I	1.75E+04
J	2.10E+04
K	2.45E+04
L	2.80E+04
M	3.15E+04
N	3.50E+04
O	3.85E+04
P	4.20E+04

Fig. 10.5.4. (continued)

10.6 Application VII : Shear Connectors in Composite Beams

A very important part in the design of composite beams is the design of shear connectors. The connection between the steel beam and the concrete slab is, in most cases, complete, in the sense that any debonding, slip or uplift, at the interface of the two elements is neglible. Shear connection in composite beams is defined as complete when [ECC] the beam has a bending strength that does not increase by the addition of further connectors. However, in many cases, the full bending strength need not be used. This is the case for example, when the dimensions of the beam are dictated by serviceability rather than strength criteria, as e.g. the reduction of deformations or vibrations. In such cases we can use a smaller number of connectors than the required for the complete shear connection. Then the connection between steel beam and concrete is partial. Then nonneglible slip at the steel beam-concrete slab interface occurs, which has a considerable influence on the strength and the deformations of the composite beam.

Design rules for the determination of the deflection and the strength of the composite beams with partial connection were given in [Johns], [Mof]. The ultimate strength of composite beams with partial connection taking into account the strain hardening of the steel, the softening branch of the strain-deformation curve of concrete and the uplift possibility of concrete slab was studied in [Arib]. Recently, the behaviour of composite beams with partial connection spaning over 20.0m, which is the limit given by the recommendations [Euroc], was studied in [Gatt].

The aim of the present section based on [Mist93a] is to contribute to the analysis of composite beams spaning more than 20.0 m taking into account the exact force-deformation diagram of the shear connectors used. The existence of the softening branch is verified by numerous experimental data [Dehl],[Kuhn]. Composite beams which are used for long spans are slender; therefore we will not apply plastic analysis for the determination of their ultimate strength, because their ultimate strength is usually dictated by the local buckling of the steel beam plates.

The accurate analysis of the behaviour of composite beams with partial connection is very important as the mutual slip between steel and concrete may be big enough to cause fracture of some connectors at serviceability state. Appropriate ductility of the shear connectors is the only way to sustain the big slip deformations without their total failure. However we should note that the most commonly used shear connectors, the headed studs, do not exhibit big ductility. Accordingly the accurate study of the composite beams with partial connection is important for the developement of safe design rules.

In the solution model of composite beams that will be used in this Section, we consider for the shear connectors a force-deformation law containing a softening branch. The corresponding curves are given in Fig. 10.6.1 (diagrams (i),(ii)). The simplified law having the shape of diagram (iii) will also be used for

10.6 Application VII : Shear Connectors in Composite Beams

the sake of comparison. Since the beam under consideration is slender, a linear stress-strain law for the steel and the concrete is accepted. The ultimate moment of slender cross sections is reached when a single fiber of the steel member, subjected to the total loading (i.e. dead and live loads), starts yielding.

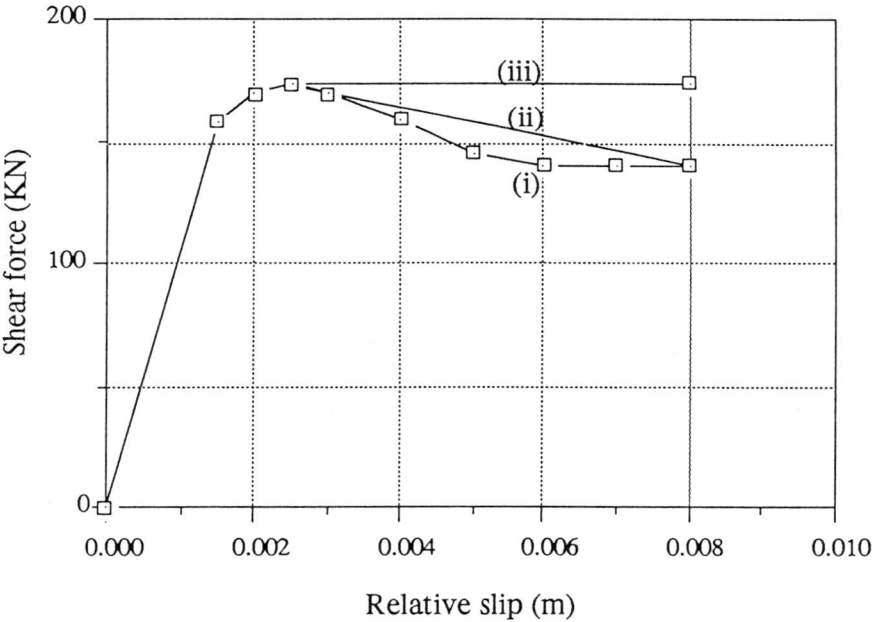

Fig. 10.6.1. Force-deformation diagram of the shear connectors.

308   10. On the Approximation of Hemivariational Inequalities

The aim of this Section is the calculation of the maximum slip between steel and concrete at serviceability limit state. The number of shear connectors (N_f) needed to obtain complete connection between steel and concrete is calculated in such a way that the full shear flow at the ultimate limit state can be sustained. For this number (N_f) of shear connectors the maximum slip at the serviceability limit state is calculated. Then, the maximum slips for various numbers (N) of shear connectors are calculated, until the first fracture of a shear connector at serviceability limit state occurs. For the calculation of the model, the finite element method has been used. Thus, the beam is simulated by finite elements as shown in Fig. 10.6.2a. The elements of type A and B are bending elements with axial deformation possibility, which correspond to the concrete slab and the steel beam respectively. The element types C_1 and C_2 are completely rigid having length $\frac{h_1}{2}$ and $\frac{h_2}{2}$ respectively; this rigidity is imposed by the principle of plane sections. The elements C_3 are "spring" elements of zero length which can transmit only shear forces and obey to the force-deformation law of the shear connectors. The positions of the elements C_1, C_2 and C_3 coincide with the positions of the shear connectors.

The calculation of the structure would be easy if the force-deformation law of the shear connector had no softening branch. Now due to the softening branch we shall use the algorithm of Sect. 9.3 (see also Fig. 10.6.3). We recall that this algorithm has a clear mathematical justification as it approximates the hemivariational inequality of the problem with a sequence of variational inequalities.

The beam of Fig. 10.6.2a is considered as non-propped, carrying its own weight and the weight of the concrete slab during casting. The dead and live loads carried by the beam are given in Fig. 10.6.2c. The maximum strength of the shear connectors was calculated according to the rules of Eurocode 4 [Euroc]. Then we assume that the "fracture" of the connectors occurs. The shear connectors along the beam were distributed either according to the longitudinal distribution of the shear flow ("triangular" distribution), or uniformly leading to equal spaces between the connectors ("uniform" distribution), so that a comparison between these two cases can be made. The load of the beam corresponds to the serviceability limit state.

In Figs. 10.6.4a,b,c the change of the slip value along the beam is given for triangular distribution of the shear connectors, (i.e. according to the longitudinal distribution of the shear flow) while Fig. 10.6.5a,b,c give the change of the slip value for uniform distribution of the shear connectors. In both the figures the cases a,b,c correspond to the force-displacement curves (i), (ii), (iii) respectively. Significantly bigger slip values occur in the case of uniform distribution of the shear connectors. This fact is independent of the force-deformation curve used. Moreover in the aforementioned case, the increase of slip values is larger as the number of the shear connectors decreases. For ratio $\frac{N}{N_f} > 0.69$ the slip values are actually independent of the force-deformation curve. On the contrary, for ratio $\frac{N}{N_f} < 0.69$, the slip values are greater in the case of (ii) compared to the slip values corresponding to (iii). Finally the slip values corresponding to (i)

10.6 Application VII : Shear Connectors in Composite Beams

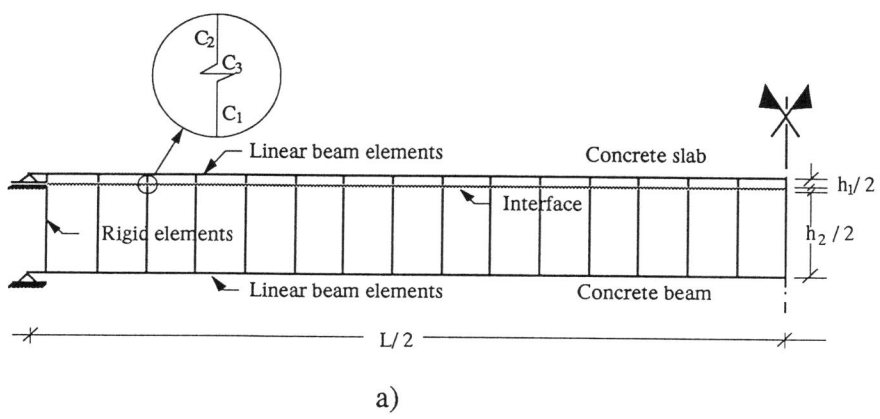

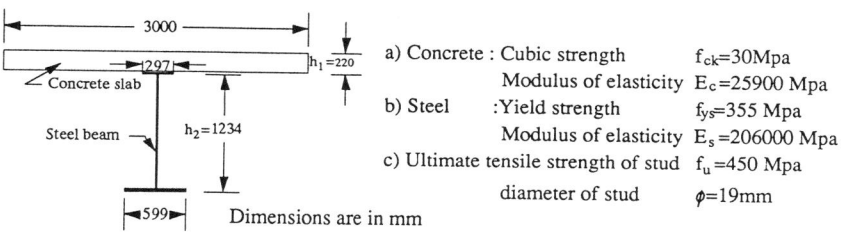

a) Concrete : Cubic strength $f_{ck}=30$ Mpa
Modulus of elasticity $E_c =25900$ Mpa
b) Steel : Yield strength $f_{ys}=355$ Mpa
Modulus of elasticity $E_s =206000$ Mpa
c) Ultimate tensile strength of stud $f_u =450$ Mpa
diameter of stud $\varphi=19$ mm

Weight of structural steel (KN/m)	3.6
Weight of concrete slab (KN/m)	16.5
Dead loads (KN/m)	20.6
Superimposed dead loads (KN/m)	7.0
Live loads (KN/m)	30.0

The partial safety factors for ultimate loads : 1.35 for dead loads
1.50 for live load
For service loads the corresponding factors are taken equal to 1.0

c)

Fig. 10.6.2. Dimensions, discretization and loading.

are the biggest of all. On the other hand in the case of triangular distribution of shear connectors and for $\frac{N}{N_f} > 0.61$ it was noticed that the slip values are independent of the force-deformation curves used for the solution. For ratio $\frac{N}{N_f} < 0.61$, the slips increase with the consideration of the curves (iii), (ii) and (i) respectively. The forces of the shear connectors for the "triangular" and the "uniform" distribution are given in kN in Fig. 10.6.6 and 10.6.7 respectively. Note that these figures give the forces of the connectors and not the shear flow at the steel-concrete interface.

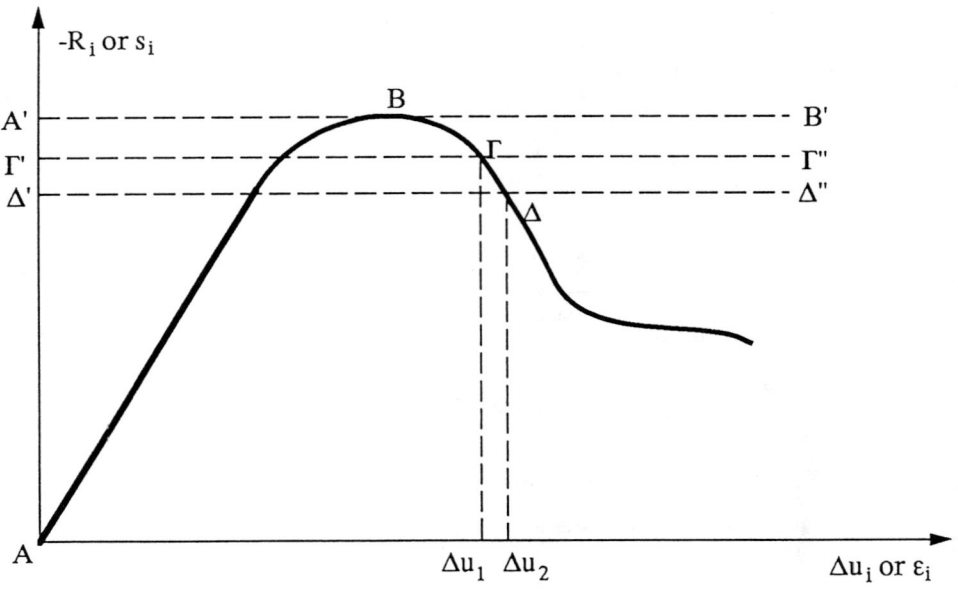

Fig. 10.6.3. Approximation of the nonmonotone law by monotone ones (Δu = relative slip).

10.6 Application VII : Shear Connectors in Composite Beams

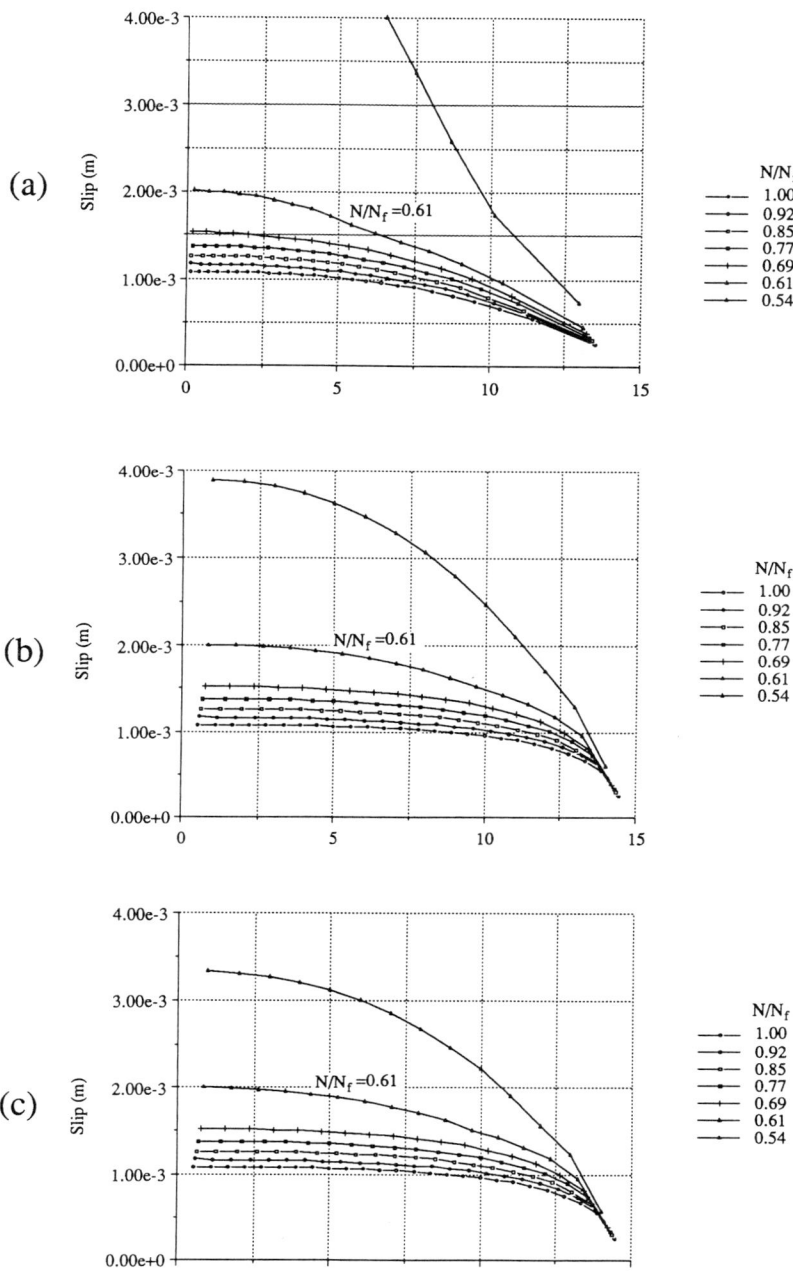

Fig. 10.6.4. Slip value variation along the beam ("triangular" distribution).

312   10. On the Approximation of Hemivariational Inequalities

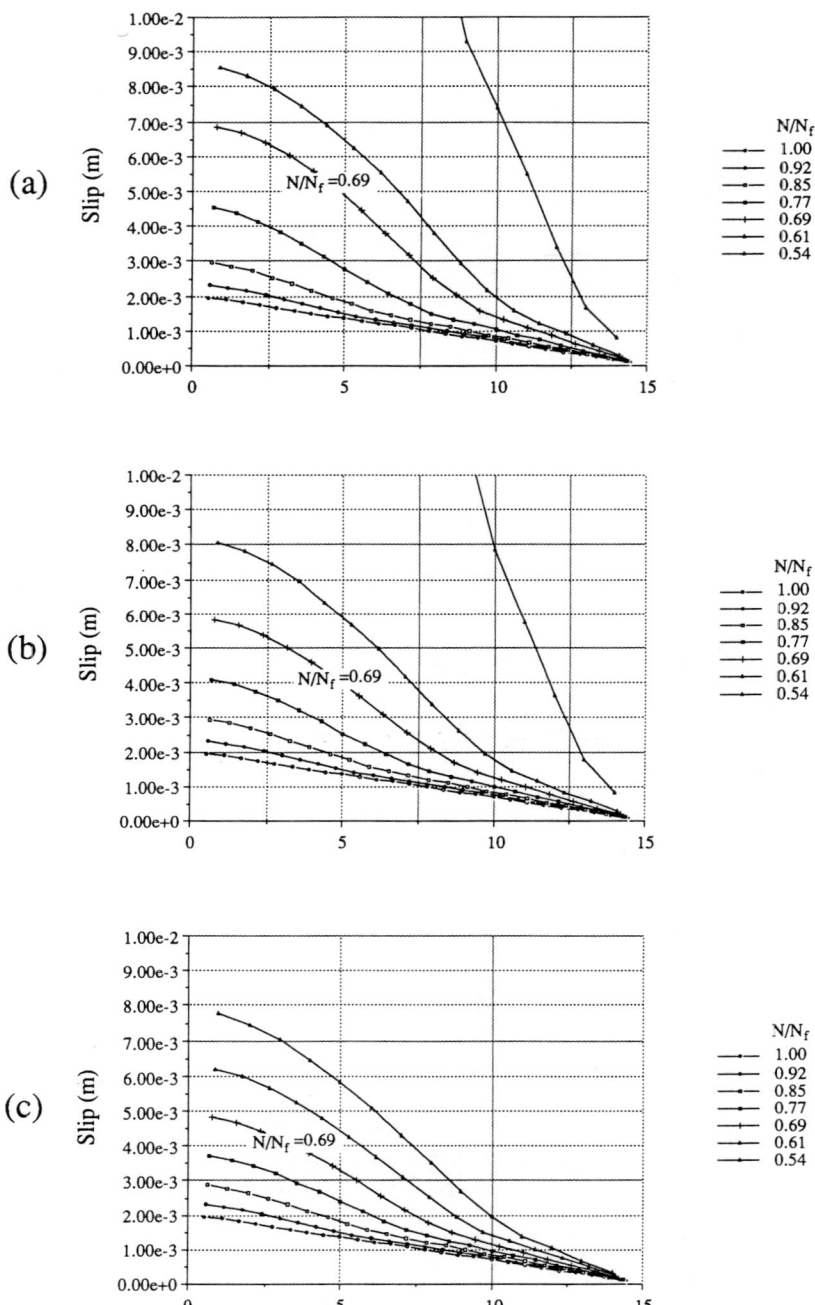

Fig. 10.6.5. Slip value variation along the beam ("uniform" distribution).

10.6 Application VII : Shear Connectors in Composite Beams

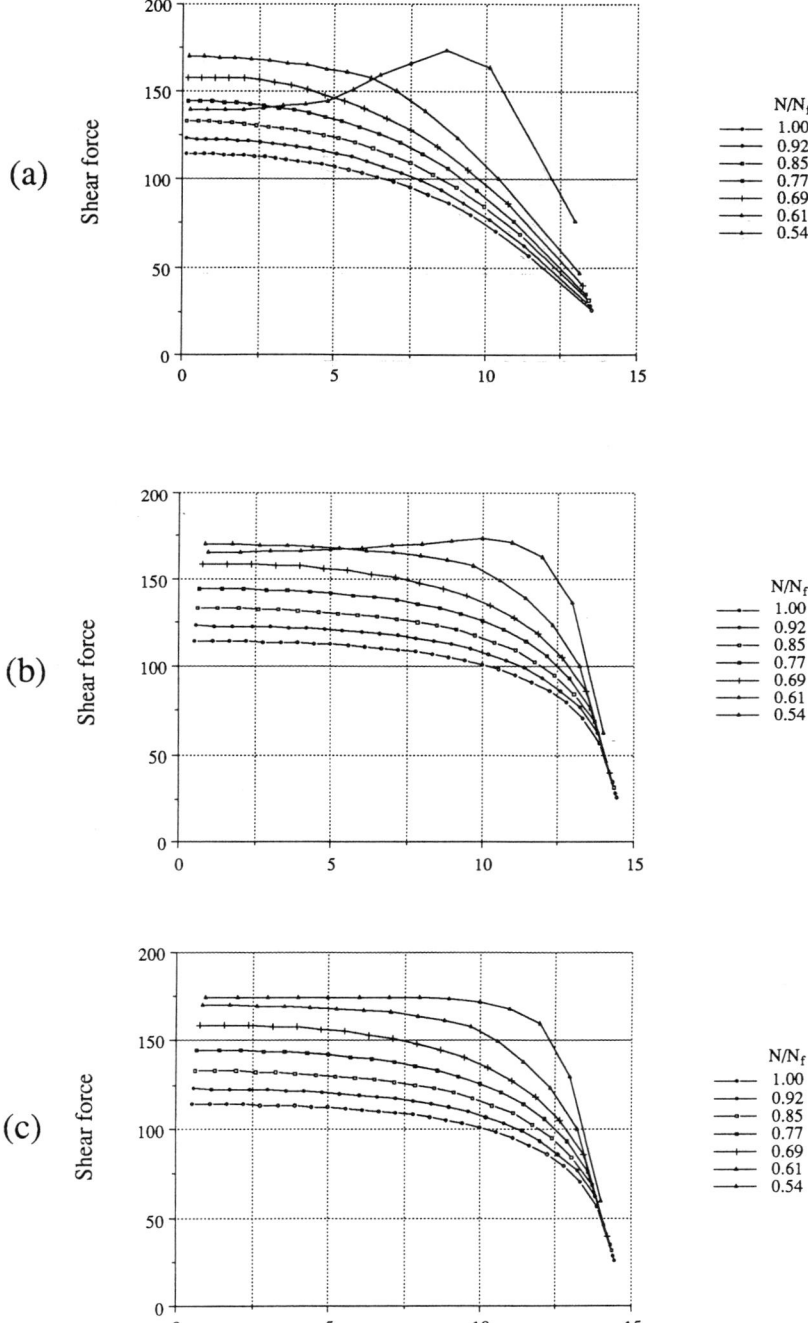

Fig. 10.6.6. Forces on the shear connectors ("triangular" distribution).

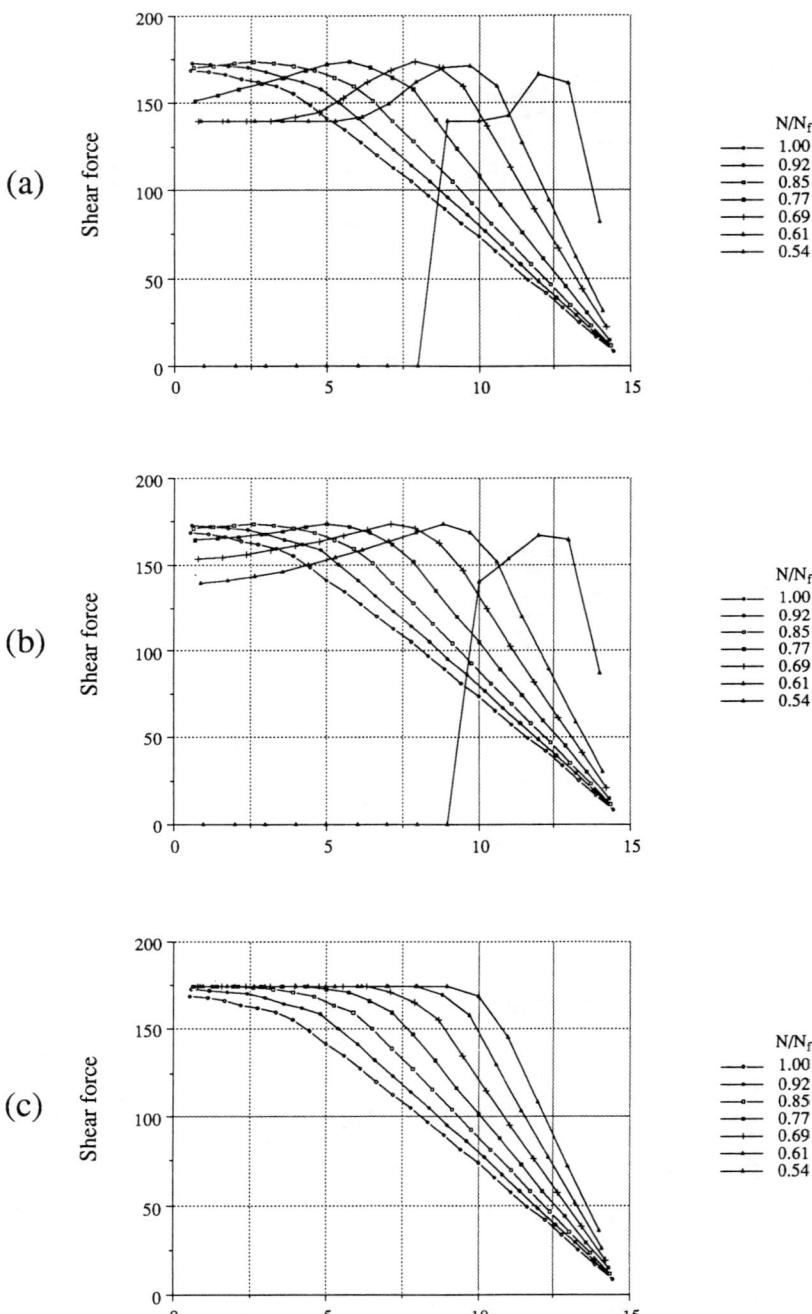

Fig. 10.6.7. Forces on the shear connectors ("uniform" distribution)

10.6 Application VII : Shear Connectors in Composite Beams

It may be observed that the "fracture" of the connectors begins for a smaller ratio $\frac{N}{N_f}$ in the case of the "triangular" distribution compared with the case of "uniform" distribution.

The study of the ratio $\frac{N}{N_f}$ for which the first fracture of connectors is observed with respect to the curves (i), (ii), (iii), presents a special interest. It can be easily seen from Fig. 10.6.6 and Fig. 10.6.7 that the first "fracture" is observed for a higher ratio $\frac{N}{N_f}$ in the cases of force-deformation curves (i) or (ii) which have softening branch, compared to the curve (iii) without softening branch. Recall that all the curves have the same maximum value corresponding to the ultimate strength of the shear connector. Thus, the "fracture" does not depend only on the ultimate strength of the shear connector but also on the form of the force-deformation curve and more particularly on the existence of the softening branch. An important result that arises is that if one does not take into account the softening branch of the force-deformation curve, the calculations may lead to less conservative conclusions concerning the real safety of the beam at the serviceability state.

11. The Method of Substationary Point Search

The present chapter is devoted to the presentation of another numerical method for the treatment of hemivariational inequalities. This method deals with the search for the substationarity points of the potential or the complementary energy of the structure. We recall (cf. Sect. 4.3) that besides all local minima, substationarity points are also the classical stationary points and some local maxima (cf. Sect. 1.2). Note that a hemivariational inequality is generally not equivalent to the corresponding substationarity problem. However, Prop. 6.3.1 holds for most practical applications and we have a complete equivalence between the hemivariational inequality and the substationarity problem. The method presented here makes use of efficient algorithms of numerical optimization and is an extension of a method presented in [Tzaf91,91a,93]. The hemivariational inequality is decomposed into a finite number of variational inequalities (monotone problems). This is achieved if the epigraph(s) of the superpotential(s) involved can be split into convex parts. This is possible, e.g. in the onedimensional case, if the superpotential is the minimum of a finite number of convex functionals. After a description of the algorithm and a discussion of its numerical properties we give a number of numerical applications.

11.1 General Formulation of the Method

The classical methods of linear or linearized analysis encounter serious and rather forbidding difficulties in the numerical approximation of problems involving nonmonotone, possibly multivalued stress-strain or reaction-displacement laws. This is caused by the fact that the laws are multivalued, i.e. in the onedimensional case they contain complete vertical branches on which a solution point may be determined. Moreover the energy functionals involved are nonconvex and nonsmooth, thus leading to hemivariational inequalities and to corresponding substationarity problems.

The determination of all solutions of a substationarity problem, even when only smooth functionals are involved, yet remains in the general case, an open problem and is an area of active research in the optimization theory. Indeed, this also holds for a global optimization problem, which is a particular case of the general substationarity problem (cf. [Flet],[Pard],[Gill],[Strod]). In addition,

318   11. The Method of Substationary Point Search

the problems encountered in Mechanics usually have very large dimension, compared to those presented in the books of mathematical programming, where the algorithms are usually tested for small size problems. Moreover, in contrast to global optimization methods all the stable and unstable solutions on the loading path are sought and not only the global minimizer. Thus the existing nonconvex optimization algorithms lead only to a partial solution for engineering problems expressed in terms of hemivariational inequalities and even then only for a small number of unknowns. The methods applied until now for the numerical solution of problems involving nonmonotone laws are the following:

i. Penalization of the energy expression through additional energy terms [Frém87],

ii. Application of bundle-type methods of Nonsmooth Optimization [Strod], [Stav91],

iii. Use of mollifiers (regularization) to eliminate the nonsmooth parts, followed by the solution of the resulting sequence of variational equalities [Pan87a],

iv. Path following methods assisted by a step-length control procedure to recognize limit points [Cris85,88],[Riks],[Schel],

v. Nonconvex minimization methods [Tzaf91],

vi. Use of the quasidifferentiability notion to replace the nonconvex problem by an appropriate combination of convex problems (cf. next Chapter). We call this method "Multilevel Decomposition into two Convex Problems",

vii. The method of "Microspring Approximation of the Decreasing Branches", (cf. Sect. 9.2)

viii. The method of "Decreasing Branch Approximation by Monotone Laws" (cf. Sect. 9.3 and Sect. 10.1).

Methods (i), (ii), (iii) are descendants of convex minimization iterative processes that today find only very limited use in practical applications, due to numerical accuracy and stability problems (cf. [Flet] on the earlier convex minimization processes) and the fact that they are able to treat only small numbers of unknowns.

The last five methods lead to the solution of subproblems whose solution is obtained by existing algorithms of verified reliability and high convergence rates. An additional advantage is that the combination of these algorithms with the methods developed in modern nonlinear F.E. analysis [Argy][Bath][Stein] and with efficient iterative minimization procedures (quasi-Newton, conjugate gradient etc.), specialized preconditioning schemes and versatile step control procedures (in order to determine the position and nature of limit points), may lead to the construction of numerical schemes that are able to operate even after the onset of instability even in the large scale systems, that arise in engineering.

In the present Section we propose a method which generalizes the method of (v), and is called "the Substationarity Point Search". The method decomposes certain types of hemivariational inequalities into a finite number of variational inequalities related to convex functionals. This is achieved if the epigraph of the nonconvex superpotentials involved may be decomposed into convex parts (see for preliminary information Fig. 11.1.1a). Numerically the method works even if this decomposition is approximative. Thus the problem reduces to a finite number of variational inequalities or convex optimization problems defined on different subsets of the space of the displacements (e.g. if the potential energy is considered).

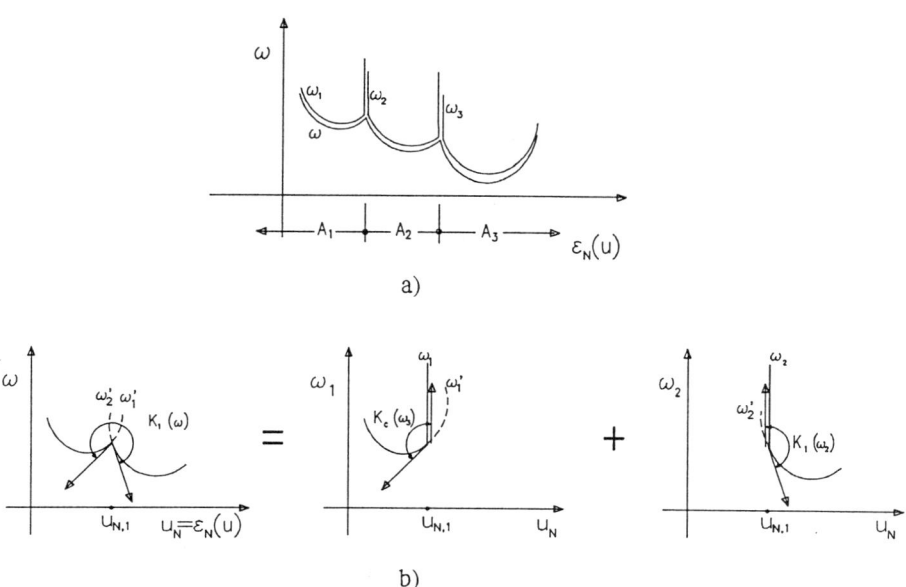

Fig. 11.1.1. The contingent cone decomposition and the resulting convex energy functions.

Each arising variational inequality can be treated effectively by specialized convex minimization algorithms (mostly based on Nonsmooth Sequential Quadratic Programming [Flet],[Womer]).

The proposed decomposition method and the resulting algorithm do not hold for all types of hemivariational inequalities but only for those for which the epigraphs of the superpotentials involved may be decomposed into convex subsets. This is the case in several important engineering problems, as e.g. in the delamination problem of multilayer structures glued by a softening adhesive material etc.

Let us consider an open, bounded, connected subset Ω of $\mathbb{R}^3$ (structure) and let Γ_k be open, mutually nonoverlapping parts of the boundary of Ω. The following problem is considered (Problem P_1):
Find $u \in U_{\text{ad}}$ satisfying the inequality :

$$\int_\Omega w^0[e(u), e(v-u)]d\Omega + \sum_k \int_{\Gamma_k} j_N^0[\varepsilon_N(u), \varepsilon_N(v) - \varepsilon_N(u)]d\Gamma \quad (11.1.1)$$

$$+ \sum_k \int_{\Gamma_k} j_T^0[\varepsilon_T(u), \varepsilon_T(v) - \varepsilon_T(u)]d\Gamma \geq (l, v-u) \quad \forall v \in U_{\text{ad}}$$

Here $U_{\text{ad}} \subset U$ is the kinematically admissible set, where U denotes the vector space of the displacements, e denotes the strain tensor, ε_N (resp. ε_T) maps the displacement vector u or v to u_N (resp. u_T), which, as usual, denote the normal (resp. the tangential) component of u on the boundary, and (l, v) denotes the work of the volume force vector and of the given boundary tractions. Moreover w, j_N and j_T are nonconvex superpotentials.

Let us introduce now the "potential energy" functional

$$\Pi(u) = W(u) + J(u) - (l, u), \quad (11.1.2)$$

where (provided that the integrals make sence)

$$W(u) = \int_\Omega w(e)d\Omega \quad (11.1.3)$$

$$J(u) = \sum_k \int_{\Gamma_k} j_N(\varepsilon_N)d\Gamma + \sum_k \int_{\Gamma_k} j_T(\varepsilon_T)d\Gamma \quad (11.1.4)$$

and the following "substationarity" problem (Problem P_2):
Find $u \in U_{\text{ad}}$ such that $0 \in \bar{\partial}\Pi(u)$. \hfill (11.1.5)

The numerical solution of either P_1 or P_2, even when the functionals involved are smooth, is yet an open problem. Thus, we focus our attention on a wide subclass of nonsmooth and nonconvex functionals which are often encountered in many engineering applications. More specifically, we will assume that w, j_N, j_T belong to the class of functionals whose epigraphs can be decomposed into a finite number of convex subsets (cf. the diagrams of Fig. 11.1.1b). Further we will denote by $K(\omega(u(x), x))$ the contingent cones related to a functional $\omega(u, x)$. The set of all contingent cones (denoted by $K(\omega)$) which is related to a nonconvex, nonsmooth superpotential, is a characteristic measure for the nonsmoothness of ω. For each nonconvex superpotential, ω (i.e. w, j_N, j_T) involved in Π, we define at every point $x \in \Omega$ a set of auxiliary, nonsmooth but convex superpotentials ω_i, $i \in L(\omega) = \{1, 2, \ldots, l\}$ (l depends on the specific form of ω) assuming that the following properties hold:

$$\omega(u, x) = \min_i \{\omega_1(u, x), \omega_2(u, x), \ldots, \omega_l(u, x)\} \quad (11.1.6)$$

$$\alpha(\omega) = \{\, i \in L(\omega) : \omega_i(u, x) = \omega(u, x) \,\}. \quad (11.1.7)$$

In the onedimensional case (also in other special cases) we will have (Fig. 11.1.1b)

$$K[\omega(u, x)] = \bigcup_{i \in \alpha(\omega)} K[\omega_i(u, x)] \quad (11.1.8)$$

11.1 General Formulation of the Method

$$K[\omega_i(u,x)] \cap K[\omega_j(u,x)] = 0 \quad \forall \ i \neq j, \quad i,j \in \alpha(\omega). \tag{11.1.9}$$

For every point u in the admissible set of the displacements space U_{ad} the set $\alpha(\omega)$ exists, containing the indices of the auxiliary functionals ω_i that are "active" (in the sense of pointwise equality with ω) at u. The mutually nonoverlapping (convex) contingent cones cover, in the onedimensional case, exactly the contingent cone of ω at u. More than one auxiliary functional is needed for the decomposition of $K(\omega(\cdot))$ at the nonconvexity points u of ω where $K(\omega(\cdot))$ is nonconvex too, while at all remaining points $u(x)$ of U_{ad}, where ω is convex, $K(\omega(\cdot))$ is the contingent cone of a single convex functional $\omega_1 \equiv \omega$ and $l = 1$, and is identified with the tangent cone (Sect. 1.2). Relations (11.1.6),(11.1.7) define a partitioning of the domain of definition of ω into subsets characterized by common active indices. Within every such subset $\bigcup_{i \in \alpha(\omega)} \partial \omega_i(u,x)$ gives the same information as $\bar{\partial}\omega(u,x)$. In special cases we have that

$$\bar{\partial}\omega(u,x) = \bigcup_{i \in \alpha(\omega)} \partial \omega_i(u,x) \tag{11.1.10}$$

$$\bar{\partial}\omega_i(u,x) \cap \bar{\partial}\omega_j(u,x) = 0 \quad \forall \ i \neq j, \quad i,j \in \alpha(\omega). \tag{11.1.11}$$

A characteristic partitioning case is depicted in Fig. 11.1.1 for u_N onedimensional.

The proposed method includes two interrelated iterations: the first with respect to the decomposition of the energy functions and the second with respect to a partitioning of the structure Ω into parts corresponding to common indices. To explain this let us consider Fig. 11.1.1b. The energy function is decomposed into two convex parts ω_1 and ω_2. The first (resp. the second) part introduces the inequality $u_N \leq u_{N,1}$ (resp. $u_N \geq u_{N,1}$). Thus we may denote the set of points of Ω, where the first (resp. the second) inequality if fulfilled by Ω_1 (resp. by Ω_2). The idea of the proposed algorithm is the following: iterations with respect to the problems $0 \in \partial \omega_1(u), 0 \in \partial \omega_2(u)$ ($\partial \equiv \bar{\partial}$ due to the convexity) and also with respect to the partitioning of Ω into Ω_1 and Ω_2 (i.e. by changing Ω_1 and Ω_2 we try to find a better local minimum) lead to the solution set. Note that, besides the kinematic conditions we must check the static conditions (Lagrange multipliers) at any solution point. Moreover if at some points of Ω, $u_N = u_{N_1}$, the solution of the minimum problem of ω_1 or of ω_2 will accurately supply the value of the corresponding stress or reaction on the filled-in vertical branch (it corresponds to the kink between A_1 and A_2 in Fig. 11.1.1a) of the multivalued nonmonotone law under consideration.

Suppose now that the displacement field u is given. Then the set of indices $\alpha[\omega(u,x)]$ that are active at every point x within the "structure" Ω for the given u, are assumed to create a partitioning of Ω in m ($m \geq l$) subsets with common indices. For the special case depicted in Fig. 11.1.2 we have that

$$\Omega = \Omega_1 \cup \Omega_2 \cup \ldots \cup \Omega_m \tag{11.1.12}$$

$$\Omega_j = \{x | x \in \Omega, \ j \in \alpha[\omega(u,x)]\}, \quad j = 1, \ldots, m \tag{11.1.13}$$

$$\Psi_i = \Omega_i \cap \Omega_{i+1} \quad i = 1, \ldots, m-1. \tag{11.1.14}$$

Further we denote by $\Phi_1, \Phi_2, \ldots, \Phi_m$ the corresponding open sets and let us denote by $\Psi_\rho, \rho = 1, \ldots, \mu$, the arising boundaries between the Φ_i's, $i = 1, \ldots, m$,

$$\Phi_i = \Omega_i - \Psi_{i-1} - \Psi_i \quad i = 2, \ldots, m. \tag{11.1.15}$$

Moreover it holds that

$$\Omega = \Phi_1 \cup \Psi_1 \cup \Phi_2 \cup \Psi_2 \cup \ldots \cup \Psi_{m-1} \cup \Phi_m, \tag{11.1.16}$$

and $\quad \Phi_i \cap \Phi_j = \Psi_i \cap \Psi_j = \Phi_i \cap \Psi_j = 0 \quad i, j = 1, \ldots, m, \tag{11.1.17}$

where Φ_i the interior of a partition with "borders" Ψ_i and Ψ_{i+1}.

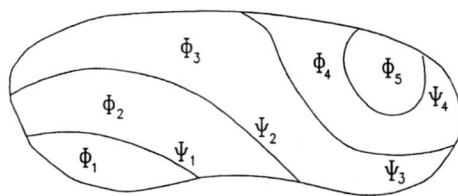

Fig. 11.1.2. Partitioning of Ω in parts with common indices. A special case of regular topology.

We would like here to point out that the determination of the properties of the sets Φ_i and Ψ_ρ is a very difficult, still open, mathematical problem. Although the theory does not exclude the possibility of very complicated partitioning topologies in most numerical applications a "certain continuity" of mechanical behaviour leads to rather simple topological forms concerning the partitioning of Ω. We note that the maximum number $l(\Pi)$ of the auxiliary convex functionals is assumed to be finite. Applying the decomposition (11.1.6),(11.1.7) and (11.1.12),(11.1.13) to the functional W we may write at any step of the calculation the functional W in (11.1.13) as:

$$\begin{aligned}\tilde{W}(u) &= \int_{\Phi_1} w_1(e(u))d\Omega + \int_{\Phi_2} w_2(e(u))d\Omega + \ldots + \int_{\Phi_m} w_m(e(u))d\Omega + \\ &+ \sum_{\rho=1}^{\mu} \int_{\Psi_\rho} w|_{\Psi_\rho}(e(u))d\Psi \end{aligned} \tag{11.1.18}$$

where $w|_{\Psi_\rho} = w_i|_{\Psi_\rho}$ or $w|_{\Psi_\rho} = w_{i+1}|_{\Psi_\rho}$. Here Ψ_ρ is, e.g., the common boundary between Φ_i and Φ_{i+1}. We assume that all the integrals in (11.1.18) make sense. Then the obvious relation

$$0 \in \bar{\partial}\tilde{W}(u) \quad \Leftrightarrow \quad 0 \in \bar{\partial}W(u) \tag{11.1.19}$$

holds. However the generalized gradient of $\tilde{W}$ can be "easily" computed in comparison with the calculation of $\bar{\partial}W(u)$. Especially, if during an iteration step one varies u only on Φ_1, for instance, the substationarity problem on the left hand side of (11.1.19) reduces to a global minimum problem over Φ_1. Analogously it is more easier to compute the feasible descent directions for $\tilde{W}$ than for W. The thoughts that led us to (11.1.19) can be easily extended to include also the functional J of (11.1.4) and finally the potential energy expression Π of (11.1.2). We may write that

$$0 \in \bar{\partial}\tilde{\Pi}(u) \quad \Leftrightarrow \quad 0 \in \bar{\partial}\Pi(u). \tag{11.1.20}$$

Roughly speaking, (11.1.20) results from the decomposition of the nonmonotone stress-strain curves into monotone parts, which correspond to the decomposition of the epigraph into convex parts. Relation (11.1.20) assures that all solutions of the substationarity problem P_2 may be obtained through an iterative procedure from the (unique) minima of the convex auxiliary functionals included in $\tilde{\Pi}(u)$. This fact is used to construct an algorithm resolving the nonconvex substationarity problem P_2 through a sequence of convex minimization subproblems. Let us make things more concrete. To simplify the presentation of the method we assume that J in (11.1.4) is zero and that the problem is governed simply by the following hemivariational inequality:
Find $u \in U_{\text{ad}}$ such that

$$\int_\Omega w^0(e(u), e(v) - e(u))d\Omega \geq (l, v - u) \quad \forall v \in U_{\text{ad}}. \tag{11.1.21}$$

Let us assume that the δ-iteration of the solution $u^{(\delta)}$ is known and we want to determine the $(\delta + 1)$-iteration $u^{(\delta+1)}$. The iteration $u^{(\delta)}$ also supplies the sets $\Phi_1^{(\delta)}, \Phi_2^{(\delta)}, \ldots, \Psi_1^{(\delta)}, \Psi_2^{(\delta)}, \ldots$. Let us split Ω into $\Phi_1^{(\delta)}, \Phi_2^{(\delta)}, \ldots$ etc. and let us consider the hemivariational inequality

$$u^{(\delta+1)} \in \tilde{U}_{\text{ad}} \quad \int_{\Phi_1^{(\delta)}+\Delta_1} w_1^0(e(u^{(\delta+1)}), e(v - u^{(\delta+1)}))d\Phi_1 \tag{11.1.22}$$

$$+ \int_{\Phi_2^{(\delta)}+\Delta_2} w_2^0(e(u^{(\delta+1)}), e(v - u^{(\delta+1)}))d\Phi_2 + \ldots$$

$$\geq (l, v - u^{(\delta+1)}) \quad \forall v \in \tilde{U}_{\text{ad}}$$

Here Δ_1, Δ_2 must be determined and

$$\tilde{U}_{\text{ad}} = \{u | u = \{u_i\} \quad u \in U_{\text{ad}} \, u|_{\Phi_j^{(\delta+1)}}(x) = u|_{\Phi_{j+1}^{(\delta+1)}}(x) \tag{11.1.23}$$
$$\text{for } x \in \Psi_\rho^{(\delta+1)}, \, \rho = 1, \ldots, \mu\},$$

where it is assumed that $\Psi_\rho^{(\delta+1)}$ is the common boundary of $\Phi_j^{(\delta+1)} = \Phi_j^{(\delta)} + \Delta_j$ and $\Phi_{j+1}^{(\delta+1)} = \Phi_{j+1}^{(\delta)} + \Delta_{j+1}$. Setting in (11.1.22) $v = u^{(\delta+1)}$ on $\Phi_1^{(\delta+1)}, \Phi_2^{(\delta+1)}, \ldots$,

$\Phi_{j-1}^{(\delta+1)}, \Phi_{j+1}^{(\delta+1)}, \ldots, \Phi_{m}^{(\delta+1)}$ we obtain due to the convexity of w_i the variational inequality

$$u^{(\delta+1)} \in \tilde{U}_{\text{ad}} \quad \int_{\Phi_i^{(\delta)}+\Delta_i} \left[w_i(\varepsilon(v)) - w_i(\varepsilon(u^{(\delta+1)}))\right] d\Phi_i \quad (11.1.24)$$
$$\geq (l, v - u^{(\delta+1)})|_{\Phi_i^{(\delta)}+\Delta_i} \quad \forall v \in \tilde{U}_{\text{ad}}$$

for $i = 1, 2, \ldots, m$. Let us consider the corresponding minimum problems for the subsets $\Phi_j^{(\delta+1)}$ and $\Phi_{j+1}^{(\delta+1)}$ which have, for instance, the common boundary $\Psi_\rho^{(\delta+1)}$. They read, if one introduces the kinematical compatibility condition on $\Psi_\rho^{(\delta+1)}$ as a subsidiary condition,

$$\Pi_j(u_j^{(\delta+1)}) = \min\{\int_{\Phi_j^{(\delta)}+\Delta_j} w_j(\varepsilon(v))d\Phi_j - (l,v)|_{\Phi_j^{(\delta)}} \Big| v \in U_{\text{ad}}\} \quad (11.1.25)$$

$$\Pi_{j+1}(u_{j+1}^{(\delta+1)}) = \min\{\int_{\Phi_{j+1}^{(\delta)}+\Delta_{j+1}} w_{j+1}(\varepsilon(v))d\Phi_{j+1} - (l,v)|_{\Phi_{j+1}^{(\delta)}} \Big| v \in U_{\text{ad}}\} \quad (11.1.26)$$

$$u_{j+1}^{(\delta+1)}(x) = u_j^{(\delta+1)}(x) \quad \text{on} \quad \Psi_\rho^{(\delta+1)}. \quad (11.1.27)$$

Now we may consider the following problem:
Find $u_j^{(\delta+1)}$ and $u_{j+1}^{(\delta+1)}$ solution of

$$\min\{\Pi_j(u_j^{(\delta+1)}) + \Pi_{j+1}(u_{j+1}^{(\delta+1)}) \quad (11.1.28)$$
$$- \int_{\Psi_\rho^{(\delta+1)}} \lambda_\rho(u_{j+1}^{(\delta+1)}(x) - u_j^{(\delta+1)}(x))d\Psi_\rho \mid u_j^{(\delta+1)}, u_{j+1}^{(\delta+1)} \in U_{\text{ad}}\}.$$

For the solution of (11.1.28) we follow a modification of the "nonfeasible gradient controller" method of Lasdon and Schoeffler ([Baum],[Pan85] p.356). Suppose that λ_ρ has a given value, say $\lambda_{\rho,\nu}$ for $\nu = 1, 2, \ldots$ and that $\Psi_\rho^{(\delta+1)}$ (or equivalently Δ_j or Δ_{j+1}) have a given geometry say $\Psi_{\rho,\nu}^{(\delta)}$. We consider the first level problems

$$\min\{\Pi_j(u_{j,\nu}^{(\delta+1)}) + \int_{\Psi_{\rho,\nu}^{(\delta)}} \lambda_{\rho,\nu} u_{j,\nu}^{(\delta+1)} d\Psi_\rho \,\Big|\, u_{j,\nu}^{(\delta+1)} \in U_{\text{ad}}\} \quad (11.1.29)$$

$$\min\{\Pi_{j+1}(u_{j+1,\nu}^{(\delta+1)}) - \int_{\Psi_{\rho,\nu}^{(\delta)}} \lambda_{\rho,\nu} u_{j+1,\nu}^{(\delta+1)} d\Psi_\rho \,\Big|\, u_{j+1,\nu}^{(\delta+1)} \in U_{\text{ad}}\} \quad (11.1.30)$$

and then the second level corrections

$$\lambda_{\rho,\nu+1} \text{ (resp. } \Psi_{\rho,\nu+1}^{(\delta)}) = \lambda_{\rho,\nu} \text{ (resp. } \Psi_{\rho,\nu}^{(\delta)}) \quad (11.1.31)$$
$$+ k \text{ (resp. } \bar{k})\left(u_{j+1,\nu}^{(\delta+1)} - u_{j,\nu}^{(\delta+1)}\right), \quad k > 0, \ \bar{k} > 0$$

where k (resp. $\bar{k}$) is an appropriately [Baum] chosen constant. First (11.1.29), (11.1.30) are solved, then $\lambda_{\rho,\nu}$ (resp. $\Psi_{\rho,\nu}^{(\delta)}$) is corrected through (11.1.31), the new value $\lambda_{\rho,\nu+1}$ (resp. the new geometry of the interface $\Psi_{\rho,\nu+1}^{(\delta)}$) is passed to (11.1.29)(11.1.30) and so on until the differences $|u_{j,\nu+1}^{(\delta+1)} - u_{j,\nu}^{(\delta+1)}|$, $|\lambda_{\rho,\nu+1} - \lambda_{\rho,\nu}|$ become appropriately small for every x. This procedure is continued until $|u_j^{(\delta+k)} - u_j^{(\delta+k-1)}|$ becomes appropriately small and the sets $\Phi_i^{(\delta+k)}$ $i = 1,\ldots,m$ do not vary considerably from step to step.

Note that $u^{(1)}$ may be obtained by assuming that the whole body Ω has the energy density function w_1. Moreover the solution of the arising convex minimization problems is performed by a sequential quadratic programming algorithm [Flet], [Womer], which makes use of nonsmooth quadratic approximations of the convex functionals. Finally we must check the points at the kink between e.g. w_1 and w_2. At these points the compatibility condition (11.1.27) is satisfied. Moreover the values of the Lagrange multipliers of the minimization at these points give the corresponding constraint forces. In the stress-strain or reaction-displacement diagram the stresses or reactions lie on the vertical filled-in branch of the considered law.

In connection with the described algorithm, the following scheme may serve to correct the solution. It can also be used as an independent algorithm, for the determination of solution of a hemivariational inequality. Note that the algorithm we proposed before can supply several solutions of the hemivariational inequality through repeated trials (e.g. in the first step we minimize w_2 over Ω, instead of w_1), whereas the scheme we will present here permits the quick amelioration of a solution obtained by the previous algorithm. The iterative scheme we will present holds in the cases where the hemivariational inequality has the following form (a special case of (11.1.21)):
Find $u \in U_{\text{ad}}$ such that

$$\alpha(u, v - u) + \int_\Omega w_*^0(e(u), e(v - u))d\Omega \geq (l, v - u) \quad \forall v \in U_{\text{ad}}. \quad (11.1.32)$$

Here $\alpha(u,v)$ is the linear elasticity bilinear form. Obviously (11.1.32) holds if w in (11.1.21) contains a quadratic part corresponding to a basic linear elastic behaviour on which the inelastic behaviour is superimposed, i.e. if

$$\int_\Omega w(e(u))d\Omega = \int_\Omega w_*(e(u))d\Omega + \frac{1}{2}\alpha(u,u). \quad (11.1.33)$$

Moreover a hemivariational inequality analogous to (11.1.32) holds for a linear elastic body subjected to nonconvex superpotential laws on the boundary Γ. Suppose that $u^{(p)}$ is an approximation to the solution which must be improved. This is achieved by the scheme:
Find $u^{(p+1)} \in U_{\text{ad}}$ such that

$$\alpha(u^{(p+1)}, v - u^{(p+1)}) + \int_\Omega w_*^0(e(u^{(p)}), e(v - u^{(p)}))d\Omega \quad (11.1.34)$$
$$\geq (l, v - u^{(p+1)}) \quad \forall v \in U_{\text{ad}}.$$

Then (11.1.34) leads to a Q.P. Problem, as we will see further. Note that in the case of the hemivariational inequality:
Find $u \in U_{\text{ad}}$ such that

$$a(u, v - u) + \int_{\Gamma} j^0(u, v - u) d\Gamma \geq (l, v - u) \quad \forall v \in U_{\text{ad}} \qquad (11.1.35)$$

the iterations as indicated in (11.1.35) have a precise mechanical meaning. Instead of the variational inequality

$$a(u^{(p+1)}, v - u^{(p+1)}) + \int_{\Gamma} j^0(u^{(p)}, v - u^{(p)}) d\Gamma \qquad (11.1.36)$$

$$\geq (l, v - u^{(p+1)}) \quad \forall v \in U_{\text{ad}}$$

leading to a Q.P. problem, we may consider the classical linear problem

$$a(u^{(p+1)}, v) = (l, v) + \int_{\Gamma} S_i^{(p)} v_i d\Gamma \quad \forall v \in U_{\text{ad}} \qquad (11.1.37)$$

where $S_i^{(p)}$ is the reaction of the nonconvex superpotential law calculated in the previous p-step. Analogously one may proceed in the case of (9.3.10), (10.1.10) or (10.1.13); this is the justification in terms of mechanics of the arising minimum problems within each iteration. The convergence of the scheme (11.1.36) called "superpotential updating algorithm" can be proved if the multivalued law is the one considered in Prop. 6.3.1 (onedimensional law, satisfying a growth assumption). In the next Section some aspects of the practical applications of the algorithm are illustrated.

11.2 On the Numerical Implementation of the Algorithm.

In the previous chapters we explained how nonmonotone multivalued laws lead to hemivariational inequalities. Many nonmonotone multivalued laws arising in engineering problems (e.g. the delamination or the adhesive contact laws) can be expressed in the form (11.1.6), are onedimensional and therefore permit a decomposition of the contingent cones into convex cones according to Fig. 11.1.1b. Moreover they contain only a few number of convex parts $w_1, w_2, \ldots$ and this fact is of importance for the algorithm for the substationarity point search to converge rapidly and correctly. Let us first give some practical aspects concerning the algorithm.

We deal with discretized problems, e.g. a truss or a beam structure or a plate discretized by finite elements. If the number of elements, as well as the number of energy functions w_i defining the behaviour of each structural element, is relatively small, one can simply consider all possible combinations of energy functions with respect to the structural elements. Note that, e.g. in a truss,

11.2 On the Numerical Implementation of the Algorithm.

the stress and strain of each rod will correspond to only one energy function ω_i, $i = 1, \ldots, l$, at the position of equilibrium. Thus the nonconvex substationarity problem is reduced to a finite number of convex minimization subproblems by the following algorithm:

CSPS : Complete Substationarity Problem Solver

1. Create a complete partitioning of the discretized structure.

2. Select a combination of the ω_i's , $i = 1, \ldots, l$.
   Stop, if combinations are exhausted,

3. If there exists u satisfying the left hand side of (11.1.20), then accept it as a solution of the substationarity problem for the discretized structure.

4. Repeat step 2.

Note that, in the case e.g. of a truss or a beam structure, the left hand side of (11.1.20) reduces to a convex minimization problem. This is also the case in some special F.E. discretization schemes.

Although finite, the number of the convex subproblems (combinations at step 2) rises exponentially with the dimension of the original problem, and for usual structures becomes very high. Fortunately, the computation for the complete set of combinations of energy functions is rarely necessary, because in most cases the initial description of the physical problem is setting restrictions on the range and the properties of the solutions. If, for instance a nondecreasing loading assumption is made, the initial nonconvex problem reduces to fairly a small number of problems. If for an element the strain obtained lies on the kink between two energy functions, then we can determine the stress by simply taking the condition of equilibrium locally with the neighboring elements. Thus we avoid the calculation of Lagrange multipliers in the arising numerical optimization problems.

Allowing for a Sequential Quadratic Programming type of algorithm [Flet] [Womer] for the treatment of the convex minimization subproblems, and incorporating the previous modifications in *CSPS*, leads to the following algorithm:

PSPS : Partial Substationarity Problem Solver

1.a. Create a complete partitioning of the discretized structure.

  b. Select the initial combination of the ω_i's , $i = 1, 2, \ldots, l$.

2.a. Compute a nonsmooth quadratic approximation $L(u)$ (convex) of $\tilde{\Pi}(u)$ in (11.1.20).

b. Solve the quadratic subproblem :

$$0 \in \partial L(u) \qquad (11.2.1)$$

to compute the direction of minimum d from the current point v; $d = u - v$.

c. Compute the Lagrange multipliers λ of the convex minimization problems.

d. Perform a univariate search along d, in order to reduce the value of $\tilde{\Pi}$.

e. If the termination criteria are satisfied, move to step 3.

f. Repeat step 2.a.

3. If termination criteria are satisfied (depending on λ, d, the ω_i-combination and the loading) accept current point as a solution of the substationarity problem for the discretized structure and stop.

4.a. Select new combination of the ω_i's.

b. Repeat step 2.

Although these quadratic algorithms are robust and versatile, their main disadvantage is the requirement for the solution of a number of quadratic subproblems at step 2.b. Indeed the solution of a sparse (banded) Q.P. problem is many times more expensive than that of comparable system of linear equations ("unconstrained" quadratic problem).

We observe that the quadratic approximation of step 2.a is nonsmooth either in order to model nonsmoothness of the convex energy function ω_i, or to enforce the partition boundary inequalities (e.g. $u_N \leq u_{N,1}$ in Fig. 11.1.1b in F.E. discretized continuous structures).

Due to the specific form of the considered interlaminar laws, each auxiliary potential is smooth within the respective domain, a property that does not hold in the general case of nonmonotone laws. Thus the primary reason for imposing constraints in the quadratic approximation does not hold and thus a simpler model may be employed. Moreover the enforcement of the current partition boundary inequalities may take place during the univariate search (step 2.d) and not in the step 2a. This fact permits the replacement of the nonsmooth local approximation $L(u)$, by a simple smooth quadratic function and consequently step 2.b is equivalent to the solution of a system of linear equations. Thus the following algorithm results:

SSPS : Simplified Substationarity Problem Solver

1.a. Create a complete partitioning of the discretized problem.

b. Select the initial combination of the ω_i's, $i = 1, \ldots, l$.

2.a. Compute a local smooth quadratic approximation $L(u)$ of $\tilde{\Pi}(u)$ in (11.1.20).

b. Solve the linear equations system

$$\text{grad } L(u) = 0 \qquad (11.2.2)$$

to find the direction of minimum d from the current point v. Note that $d = u - v$.

c. Compute the maximum step length α along d, combatible with the partitioning of the problem.

d. Perform a univariate search along d up to αd, in order to reduce the value of $\tilde{\Pi}$.

e. If "termination criteria" (see further in this Section) are satisfied (depending on d and α), move to step 3.

f. Repeat step 2.a.

3. If the solution belongs to the interior of the current partition, accept current point as solution of the discretized substationarity problem and stop.

4.a. Select new combination of the ω_i's.

b. Repeat step 2.

The function $L(u)$ in (11.2.2) is a quadratic function whose quadratic term has a matrix A having the structure of the stiffness matrix of elasticity. Thus, the efficient element stiffness computation and global stiffness factorization update methods already in use in F.E. codes, can be directly employed for the construction of A and its successive updates, during the quadratic approximation steps. It can also be seen that A is banded, symmetric, positive definite and hence the existing methods for treatment of linear equations systems of this type can be used in step 2.b. Note that *SSPS* can be considered as a generalization of the well known Newton-Raphson method of the nonlinear smooth elasticity [Bath], extended to allow for the lack of convexity of the potential energy functional.

As an iterative method, *SSPS* shares all the respective practical benefits, such as liberty to be interrupted at any moment (for revision of the tolerances or modifications of the model or the loading) and afterwards to be resumed without significant loss of convergence speed, if a minimal volume of information is preserved (current contents of active sets only). When nonmonotone filled-in segments are absent from the functions included in the constitutive relations, then the constraints of step 2.c disappear and the method degenerates to the classical Newton - Raphson approximation (simultaneously the static problem degenerates to a smooth nonlinear elasticity problem).

Numerical experience with iterative methods (especially those related to nonconvex energy functions), indicates that even if a procedure is proved to converge, there exists a strong possibility for it to diverge in actual applications or to exhibit convergence rates significantly lower from those theoretically predicted (cf. e.g. [Flet], [Gill], [Pow]). For that reason greater importance is attributed to the verification of the present algorithm through numerical experiments. The appropriate termination criteria at step 2.e of *SSPS* have a critical influence on the accuracy and efficiency of the algorithm.

The algorithm exits from the inner loop (2.a till 2.f) when either the norm of αd is less than a predefined tolerance e_d, or the decrease of the auxiliary functional $\tilde{\Pi}$ attained during the line search is lower than another limit e_δ. During the local approximation of *SSPS*, it may happen that the computed search direction intersects in the vicinity of nonconvexity kinks the nearest constraint at a very small distance, thus giving a small value of step length α during the line search and causing a premature termination of the inner loop. This deficiency is corrected [Tzaf91] by performing a projection of d on the constraint computed at the step 2.c and by recomputing α. The projection can be easily computed using either the procedures employed in linearly constrained optimization (see e.g. [Gill][Berts]) or the affine transformation introduced in [Bis86] that leads to a simple minimum distance problem. The latter is based on the *LU* decomposition of the matrix A which will be anyway applied in order to treat the linear equations system (11.2.2). In applications, a repetition of the projection (due to the subsequent activation of another kink) was rarely needed, since only a few elements (usually only one) operate near a softening branch at every moment. We note that computation of the projection is simple if the small deformations assumption holds for discretized structures, because in that case the kinks are polyhedral surfaces. If a large deformations model is employed, the projection will be performed on a linear approximation of the kink at the point of intersection of d. Then a second order correction step [Tzaf91] in analogy to [Flet82] might be necessary, if the curvature of the kink is considerable. A computed or performed step value lower than the accuracy limit e_d and a low expected drop in the value of $L(d)$, also signals the termination of the outer loop of *SSPS* (step 3). Concerning the line search termination criteria employed in step 2.d of the algorithm, the usual Wolfe-Powell conditions were found sufficient [Flet] [Gill]. These criteria must be completed with the step length restriction arising in step 2.c. After incorporating the amendments discussed in this paragraph, the following algorithm, called also *SSPS* was formulated.

SSPS : Simplified Substationarity Problem Solver

    1.a. Create a complete partitioning of the discretized problem.

       b. Select an initial displacement vector and define the initial combination of the ω_i's, $i = 1, \ldots, l$.

2.a. Compute the matrix A of the quadratic form and the vector b of the linear form of $L(u)$ of L.

b. Solve the linear equations system (11.2.2) and find d.

c. Compute the number α such that αd is the distance from the nearest border induced by the current combination of the w_i's. If $|d|_1 > e_d$, $|\alpha d|_1 \leq e_d$ project d on the bound and recompute α.

d. Compute expected decrease ΔL.

e. Terminate if $|\alpha d|_1 \leq e_d$ or $\Delta L \leq e_\delta$ and move to step 3.

f. Perform a univariate search along d up to αd, in order to reduce the value of $\tilde{\Pi}$.

g. Repeat the procedure from step 2.a

3. If $\Delta L \leq e_\delta$ accept current point as a solution of the discretized substationarity problem and stop.

4.a. Select next combination of the w_i's.

b. Repeat step 2.a.

11.3 Application VIII: Delamination and Adhesive Joints in Structural Mechanics

In the applications to be described next, the nodes on the opposing surfaces of each pair of laminae are arranged in couples (after adjusting accordingly the structure's mesh coarseness near the edges) and fictitious nonmonotone elements are introduced connecting the node couples. These elements describe the delamination effect or the action of an adhesive material between the laminae. The numerical scheme described in Sect. 11.2 has been applied for the analysis of a simple laminated structure under cleavage loading. The geometry of the structure and the loading are depicted in Fig. 11.3.1. For the discretization of the lamina we used 144 constant stress quadrilateral elements with linear elastic material properties. The binding layer was modeled by 18 couples of nodes.

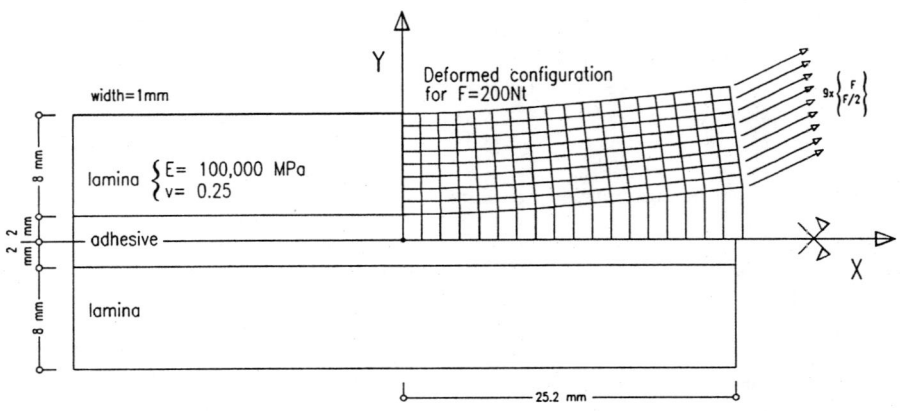

Fig. 11.3.1. Geometry and discretization of the structure.

The loading is applied at the nine nodes of the right side in the form of parallel forces $f_i = \{F, F/2\}, i = 1, \ldots, 9$ with the load intensity parameter F varying from 0 to 350 N. The two layer constant stress model was preferred instead of more realistic multilayer representation (through anisotropic shell or plate elements, or threedimensional bricks), with the purpose to show the influence of the softening behaviour of the interface. Furthermore, the particular structure-loading configuration was selected because of the high sensitivity it exhibits at the initial loading stages. This sensitivity is expected to magnify any numerical instabilities. Indeed for a load slightly lower than the one for which the outmost point of the interlaminar material reaches the softening branch, more than half of the remaining joints operate in adjacent reaction levels. When the first connection looses a fraction of its resistance, part of its load is transferred (magnified) to the next one, thus causing an abrupt chain delamination (cf. [Carls]). This effect is called "inherent instability" of the composite structure.

The nonmonotone laws of Fig. 11.3.2.a,b were used in order to describe the softening behavior of the binding material in the normal direction. In the tangential direction, the interaction between the laminae was considered negligible, i.e. the tangential forces are zero.

11.3 Application VIII: Delamination and Adhesive Joints in Structural Mechanics

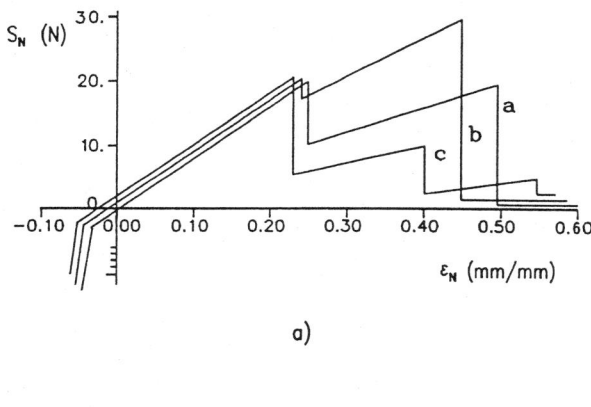

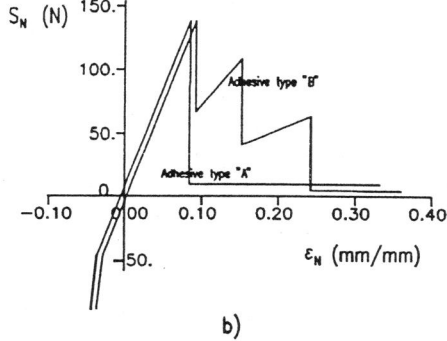

Fig. 11.3.2. Nonmonotone interface laws.

The predicted stability and sensitivity properties of the *SSPS* were verified by comparing the computed response of the structure for a wide range of loads (up to complete delamination) with solutions provided by the *PSPS*. The two sets of solutions were almost identical independently of the way the load was applied (either in varying increments or as a whole). We note that the diagram b in Fig. 11.3.2a proved particularly demanding from the viewpoint of the precision of the algorithm. It was also observed that the efficiency of the method did not depend on the initial displacements. In order to display the numerical stability and the physical correctness of the model, the structure was analyzed using the laws of Fig. 11.3.2.b for a varying cleavage load. The response at the uppermost point of the edge is depicted in Fig. 11.3.3 for a linear elastic interlaminar material and for two types of adhesives. In the last two cases, the delamination begins at a load F=149.6 N. We observe that the displacements increase suddenly for the critical load in the case of the "brittle" binding material "A", while the material law "B" is inducing a smoother softening behavior. In Fig. 11.3.4 the normal tractions along the boundary at an intermediate delamination stage (F=200 N) are depicted. The three decreasing traction peaks in the graph corresponding to material B are caused by the multiple softening limits of the respective law, and indicate the existence of progressive delamination "zone" of decreasing resistance instead of a simple front. For the gradual delamination at the crack tip of DCB (Double Cantilever Beam, Mode I) we refer to [Davie]. *SSPS* was substantially faster than *PSPS* in all tests performed, since the approximation step is much simpler. Computing time ratios of 15 to 1 were usual during the numerical tests. Even higher gains are expected in problems involving more complicated interfaces. The speed of *SSPS* becomes apparent from Fig. 11.3.5 concerning calculations with the adhesive material "B". In this figure we have depicted the distance from the solution (square norm) as a function of the computational time that each procedure needed to reach the solution "B" of Fig. 11.3.4. For comparison we note that the time refers to a workstation HP750, where the computations were performed "in core". The same computations are about $9 \div 10$ times faster on a parallel computer KSR 1-16 (16 parallel processors)[1]. Note that the proposed method is well-suited for implementation in parallel systems.

For the solution of the quadratic programming subproblems, arising in each step of *PSPS*, the FORTRAN code VE09 (sparse linearly constrained quadratic programming) of the A.E.R.E. Harwell subroutine library has been employed. The same subroutine (free of linear constraints) has also been used to solve the linear equations system of the *SSPS* approximation, in order to provide a common precision and speed reference. Considering that VE09 is a general purpose code, we conclude that better results should be expected if it is replaced by an appropriately optimized band matrix linear equations solver. If the geometry of a multilayer composite member is simple as in the previous example and the loading uniform, then the delamination along the free edges constitutes a major cause of bearing capacity loss, together with the matrix or fiber fracture. However, delamination effects away from free edges (called internal delamination) may also cause a strength degradation especially in structural members of complex shapes subjected to complicated combinations of boundary conditions and forces.

In the following application, we have used the proposed method to simulate the initiation and propagation of an internal delamination in a curved member. The geometry of the composite structure in the vicinity of a bend is depicted in Fig. 11.3.6 together with the boundary conditions and the loading. This structure-load combination is commonly encountered in practice. For example, internal delaminations around sharp bends are usual when levers are used to remove a laminated composite element from the cast. The lower part of member is assumed rigidly attached to the mould. Furthermore, the action of the lever on the element is idealized as a set of vertical parallel forces to the left

---

[1]These computations have been performed on KSR 1-16 computer of the Institute of Steel Structures at the Aristotle University, Thessaloniki in the framework of the STRIDE program on "Massively Parallel, Neurocomputing Environment with Applications in Engineering and Medicine". The financial support from the E.E.C. and the Greek General Secretariat of Research and Technology is acknowledged.

11.3 Application VIII: Delamination and Adhesive Joints in Structural Mechanics   335

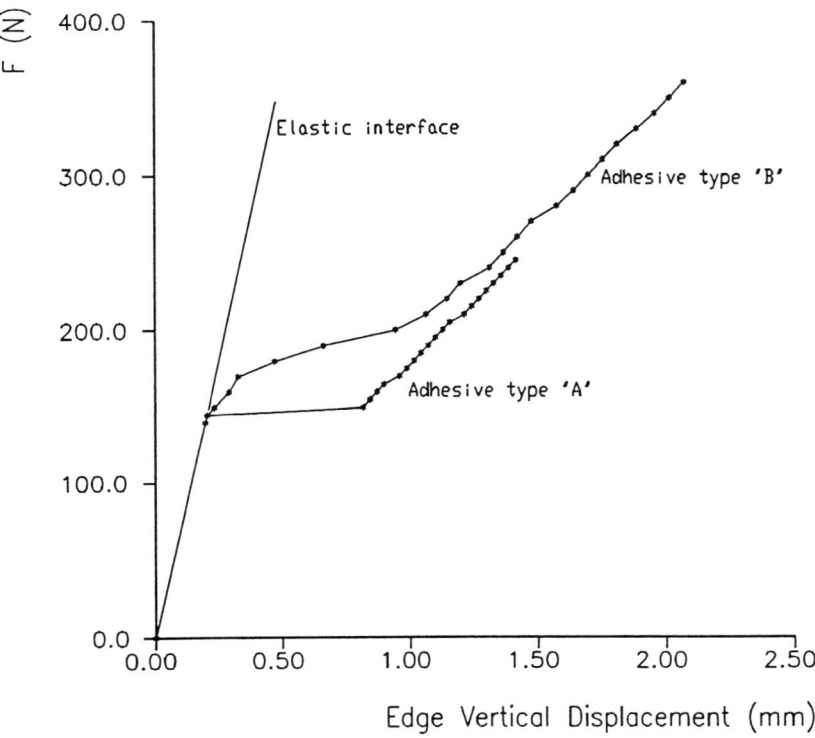

Fig. 11.3.3. Response of the structure under varying loads.

edge. The shear forces between the two laminae was assumed to be negligible. Considering the inherent instability of the load controlled delamination [Carls] we also restricted the vertical displacement of the nodes on the free edge to 1.5 mm. We used 174 constant stress isoparametric quadrilaterals for the structure modeling and 30 node pairs for the binding layer. The material characteristics and the nonmonotone law of the interface are given in Fig. 11.3.6.

336   11. The Method of Substationary Point Search

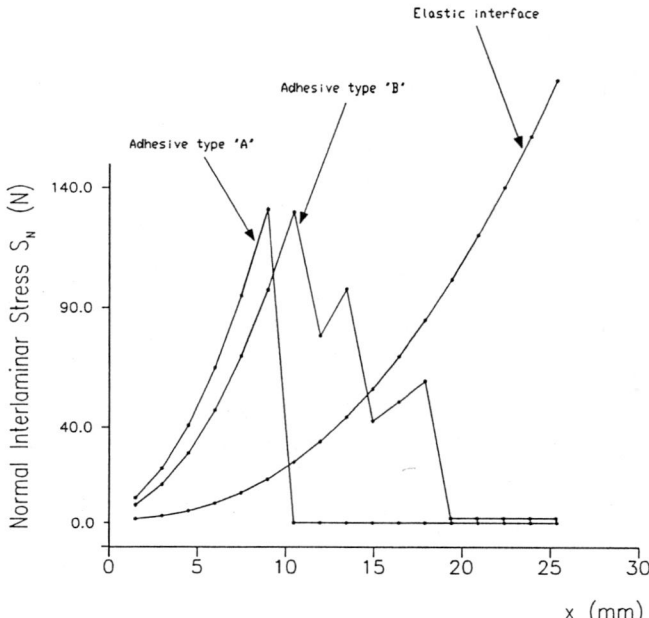

Fig. 11.3.4. The distribution of normal tractions along the interface.

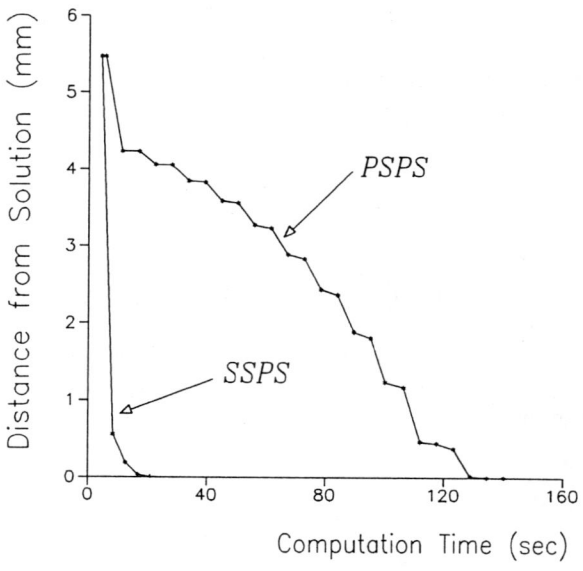

Fig. 11.3.5. About the computation time for each method.

11.3 Application VIII: Delamination and Adhesive Joints in Structural Mechanics 337

The deformed configuration of the structure is shown in Fig. 11.3.6b. The radial deformations ε_N and the normal tractions S_N along the interface are depicted in Fig. 11.3.7. Letters a, b, c, d denote the corresponding segments of the binding material law (see Fig. 11.3.6a).

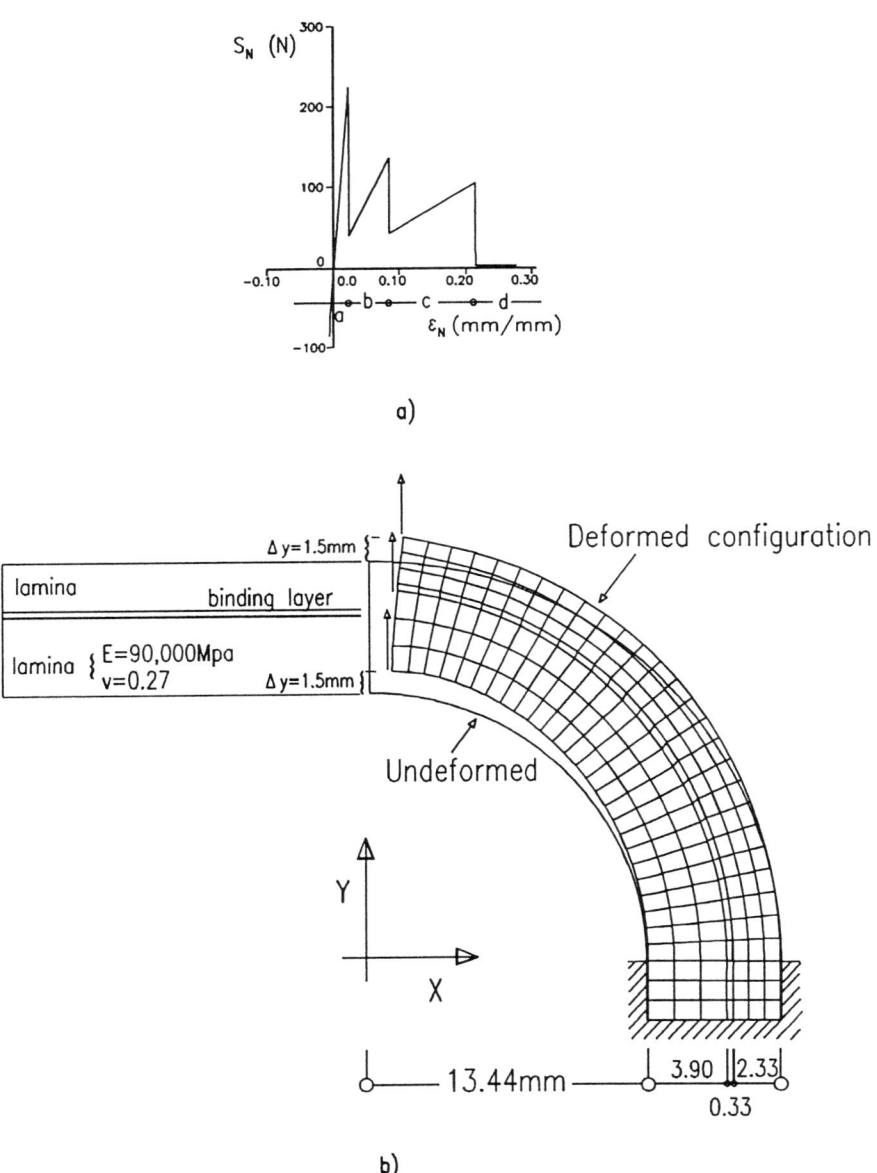

Fig. 11.3.6. A curved laminated structure.

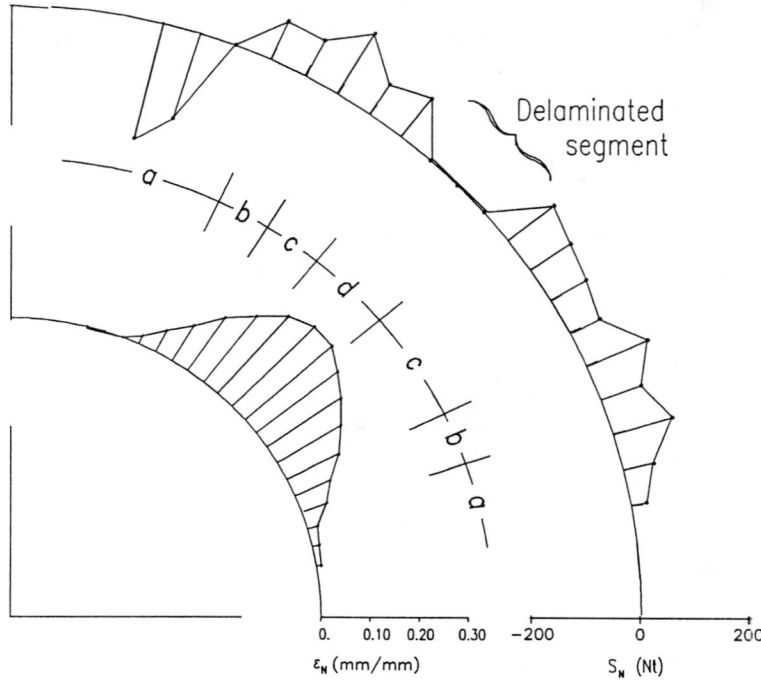

Fig. 11.3.7. Deformation and normal tractions on the interface.

11.4 Application IX: Semirigid Connections in Steel Structures

In conventional steel frame analysis and design methods it is assumed that steel frame joints are fully rigid or ideally pinned. It has long ago been recognized that most types of steel connections currently in use satisfy neither of the assumptions, but exhibit a finite degree of rotational stiffness, depending on the loading (Fig. 11.4.1, Fig. 11.4.2). Although the assumptions mentioned drastically simplify the analysis, lead to a wrong estimation of the structure flexibility and an erroneous internal forces distribution [Gerst]. Consequently, to be able to calculate the actual behaviour of a frame, it is necessary to incorporate the effect of connection flexibility to the structure model. An intensive effort has been devoted to the experimental investigation of joint characteristics, their modeling and verification in recent years. Fig. 11.4.2 shows moment-rotation $(M - \phi)$ curves, typical for a variety of commonly used bolted or welded semirigid joints, that were obtained from experiments [Davis].

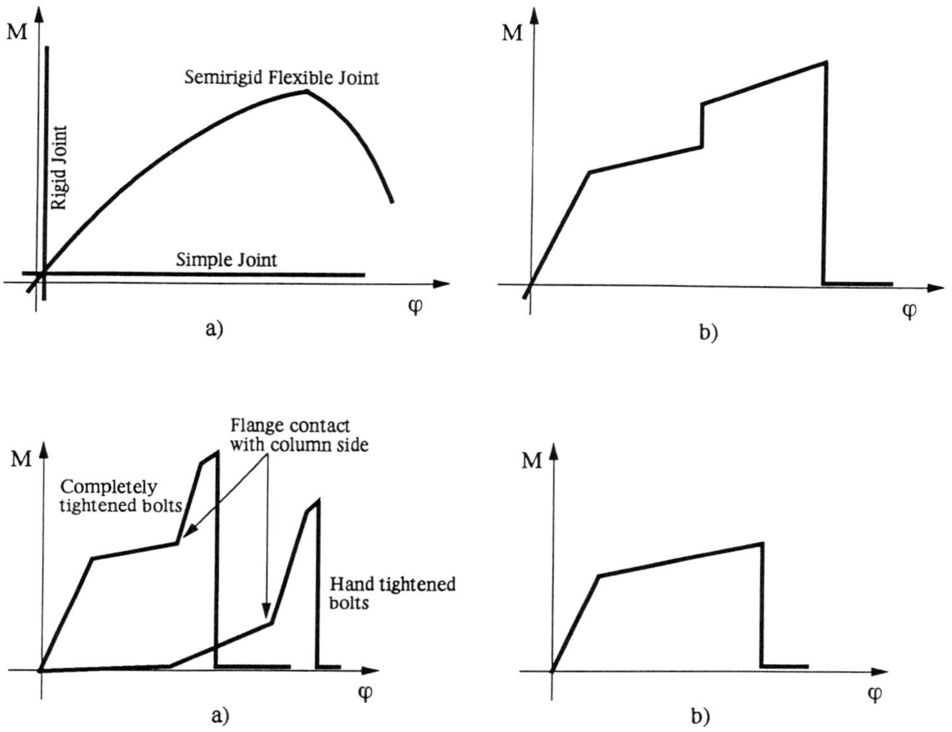

11.4 Application IX: Semirigid Connections in Steel Structures

Fig. 11.4.1. Moment-rotation diagrams. General cases.

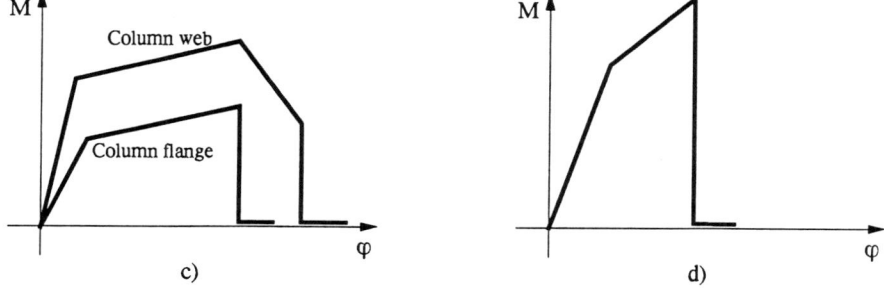

Fig. 11.4.2. Moment-rotation diagrams:
a) web cleats b) flange cleats c) flush end plate d) extended end plate.

It can be easily observed that the connection behaviour in all cases is highly nonlinear and is characterized by sharp slope changes. There exist also decreasing and vertical falling branches, especially in the segment preceding failure. Note that the jumps in the curves are complete.

Various models have been proposed until now in recent years for the description of the $M - \phi$ curve. They are based either on curve fitting of the experimental data through polynomials, B-splines, exponential sums [Lui] and linear or bilinear approximations [Gerst],[Goto]. The procedure of curve fitting an extended data set, or the simplifications of the linearized models, have constituted a source of computational problems such as reduced accuracy and range of application, numerical instability, lengthy and/or specialized computations e.t.c. [Lui],[Nether].

All existing models assume that the connection resistance curve is smooth and monotone. But these assumptions are not valid for real joints, where the contact of initially separated parts of the connection generates sharp slope changes and decreasing vertical segments. Furthermore, existing models cannot describe inclined or vertical decreasing branches of the diagram, observed before the loss of moment transferring capability. Here we introduce general model for the semirigid connections that permits the consideration of $M - \phi$ curves of the general form of Fig. 11.4.1b and does not suffer from any of the mentioned drawbacks. Moreover the model may take into account nonmonotonicities and any multivaluedness of the $M - \phi$ curve, i.e. vertical filled-in segments. We consider a $M - \phi$ law of the form

$$M(\phi) \in \bar{\partial} R(\phi), \qquad (11.4.1)$$

where R is a nonconvex superpotential describing the moment-rotation law and $\bar{\partial}$ is the generalized gradient. Further we will derive the hemivariational inequality governing the problem. Let us consider a plane frame with n nodes, discretized in m linear elastic elements (beams and columns). Some of the elements are connected at their end nodes through flexible joints (k in total) introducing the stress-strain law (11.4.1). The configuration of the structure in reference to a global cartesian orthogonal coordinate system is described by the nodal displacement vector u. The stress and strain fields in each element are described by a natural [Argy65,66,69] generalized stress vector and the corresponding strain vector. Thus the strains are not affected by a rigid body motion whereas the stresses within each element are selfequilibrated [Nits78] [Prze]. The natural generalized strain and stress vectors are denoted by s_i, e_i, $i = 1, \ldots, m$ for a structural member i, and by $s_j \in \mathbb{R}$, $e_j \in \mathbb{R}$, $j = 1, \ldots, k$ for the flexible joint ($s_j \equiv M_j$ and $e_j \equiv \phi_j$). In an attempt to simplify the mathematical expressions without loss of generality and in order to show the effect of the connection nonlinearities on the frame behavior in the numerical examples, we will develop a geometrically linear theory. The static equilibrium problem for the structure consists of the following relations : The equation of equilibrium

$$\boldsymbol{G}\,\boldsymbol{s} = \boldsymbol{p} \qquad (11.4.2)$$

11.4 Application IX: Semirigid Connections in Steel Structures

the strain-displacement relations

$$e = e_0 + \varepsilon = G^T \cdot u \tag{11.4.3}$$

and the material laws which for the beam and columns have the form

$$\varepsilon_i = F_{0i} \cdot s_i, \quad i = 1, \ldots, m \tag{11.4.4}$$

and for the flexible connections the form

$$s_j \in \bar{\partial} R_j \cdot (\varepsilon_j), \quad j = 1, \ldots, k. \tag{11.4.5}$$

Here G and F_{0i} are the equilibrium matrix and the natural flexibility matrix for the i-th element respectively, e_0 denotes the initial strain vector, p is the loading of the structure and R_j is a nonconvex onedimensional superpotential. From the "principle" of virtual work we have that for $u \in V$

$$s^T(e^* - e) = p^T(u^* - u) \quad \forall u^* \in V. \tag{11.4.6}$$

Here V denotes the kinematically admissible set (e.g. fixed nodes have $u_i = 0$) and s is the stress vector of the whole structure. Now the left hand side in (11.4.6) is written as

$$s_1^T(e_1^* - e_1) + s_2^T(e_2^* - e_2) = p^T(u^* - u) \quad \forall u^* \in V \quad \text{and for } u \in V \tag{11.4.7}$$

where s_1 contains the stress vectors s_i $i = 1, \ldots, m$ of the beams and columns and s_2 the stress vectors of the flexible connections i.e $s_2^T = [s_1, \ldots, s_j, \ldots, s_k]^T$. From (11.4.3) we obtain that $e_1 = G_1^T u$, where $G = [G_1 : G_2]$ and G_1 (resp. G_2) corresponds to s_1 (resp. to s_2). Moreover due to (11.4.5) we have that for $e_2 \in \mathbb{R}^k$

$$\sum_{j=1}^{k} R_j^0(\varepsilon_j, \varepsilon_j^* - \varepsilon_j) \geq s_2^T(e_2^* - e_2) = s_2^T(e_2^* - e_2) \quad \forall \varepsilon_2^* \in \mathbb{R}^k.$$

From (11.4.7), (11.4.4) and (11.4.8) we obtain the following hemivariational inequality:
Find $u \in V$ such that

$$u^T G_1 K_0 G_1^T (u^* - u) + \sum_{j=1}^{k} R_j^0(e_j - e_{0j}, e_j^* - e_j) \tag{11.4.9}$$
$$\geq p^T(u^* - u) \quad \forall u^* \in V.$$

here $K_0 = \text{diag}\{F_{01}^{-1}, \ldots, F_{0m}^{-1}\}$ and e_j or e_j^* are connected with u or u^* respectively through (11.4.3). The corresponding substationarity problem reads: Find $u \in V$ such as to satisfy $(K = G_1 K_0 G_1^T)$

$$0 \in \bar{\partial}\Pi(u), \quad \Pi(u) = \frac{1}{2} u^T K u + \sum_{j=1}^{k} R_j(e_j - e_{0j}) - p^T u. \tag{11.4.10}$$

Here Π denotes the potential energy of the whole structure. Due to the nonconvexity of Π there may exist more than one point in the admissible displacement set V that satisfy the substationarity condition (11.4.10) and thus are acceptable solutions. If the functional Π is convex in a neighborhood of a solution then this is a local minimum and represents a stable configuration of the structure. The remaining solutions (substationarity points) correspond to locally unstable but realizable configurations. For the numerical calculations we have applied the method of substationarity point search explained in Sect. 11.1 and Sect. 11.2. Due to the small number of branches in the nonmonotone multivalued laws describing the nonrigid behaviour of the steel frame nodes, we have only a comparatively small number of convex minimization subproblems.

In Fig. 11.4.3a a simple portal frame, subjected to a uniformly distributed load p and a lateral load H is depicted. To model the beam-column connection we have used a rigid joint, a linear flexible stiff joint (nearly rigid joint) and a semirigid joint with resistance limit. The moment-rotation curves for these connections are shown in Fig. 11.4.3b.

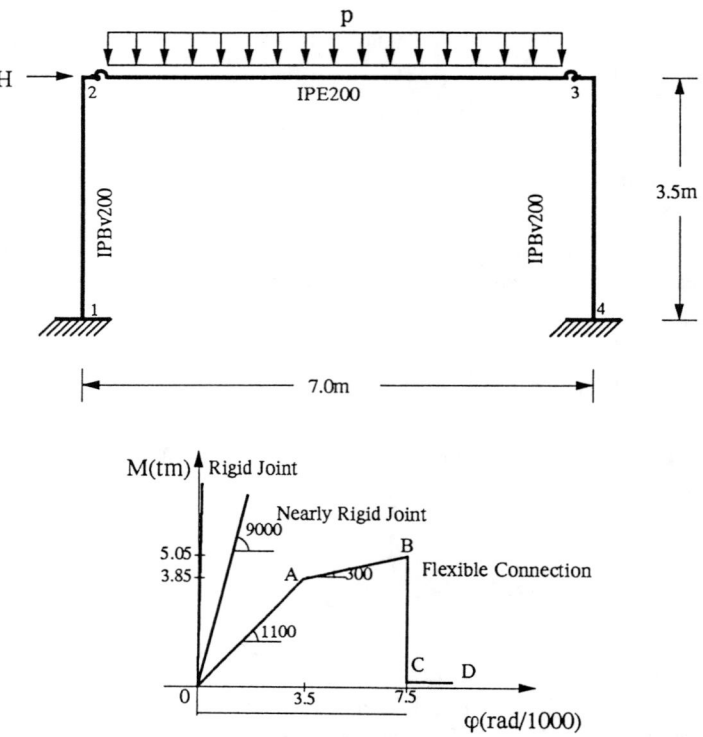

Fig. 11.4.3. Frame geometry and the $M - \phi$ curve of the joints 2 and 3.

The results of the analysis for the vertical load p varying from 0 to 4 t/m (without lateral load) are presented in Fig. 11.4.4.

11.4 Application IX: Semirigid Connections in Steel Structures 343

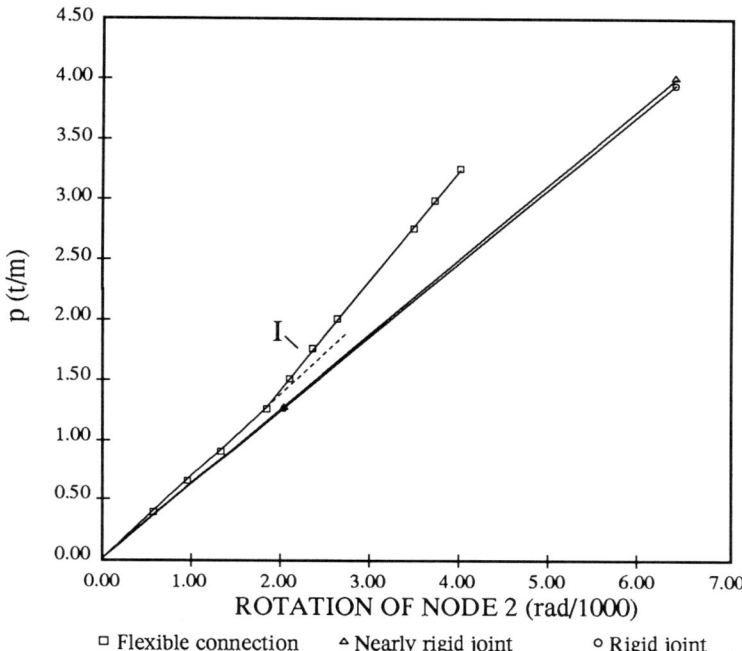

□ Flexible connection   ▲ Nearly rigid joint   ○ Rigid joint

Fig. 11.4.4. The frame response for a variation of the vertical load p.

The same frame has also been analyzed for constant vertical load p=1 t/m and for the lateral load varying from 1 to 9t. In Fig. 11.4.5a,b the horizontal deflection and the end moment of the left column as functions of the load H, for the three types of joints examined, are shown. We have marked with 'I' the points on the curves corresponding to the situation in which, some joints start operating in the less stiff part AB of the $M - \phi$ curve, and with 'II' the points where one or more joints lose their moment carrying capacity (falling branch BC of the $M - \phi$ curve) and start behaving as ideal pinned connections.

The change in the flexibility of the joints leads to a change of the response of the frame. For a rotation of the joints greater than 3.5 mrad or for a moment greater than 3.85 tm we observe a slope change in the curves of Fig. 11.4.5. When one or more joints surpass their resistance limit (at 7.5 mrad rotation or 5.05 tm bending moment (cf. Fig. 11.4.3b)), an almost vertical branch is observed in the load-deflection diagram due to the abrupt increase of the frame's flexibility. At the same time the moment transferred from the beam to the column vanishes.

344   11. The Method of Substationary Point Search

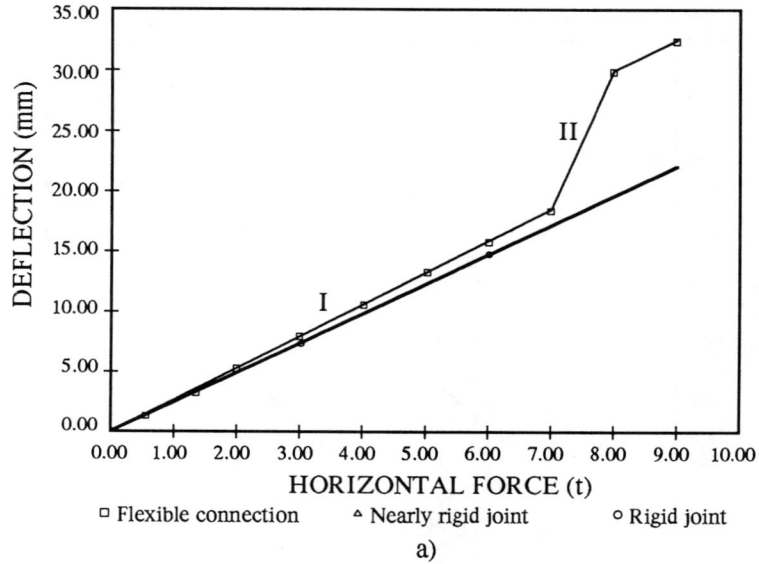

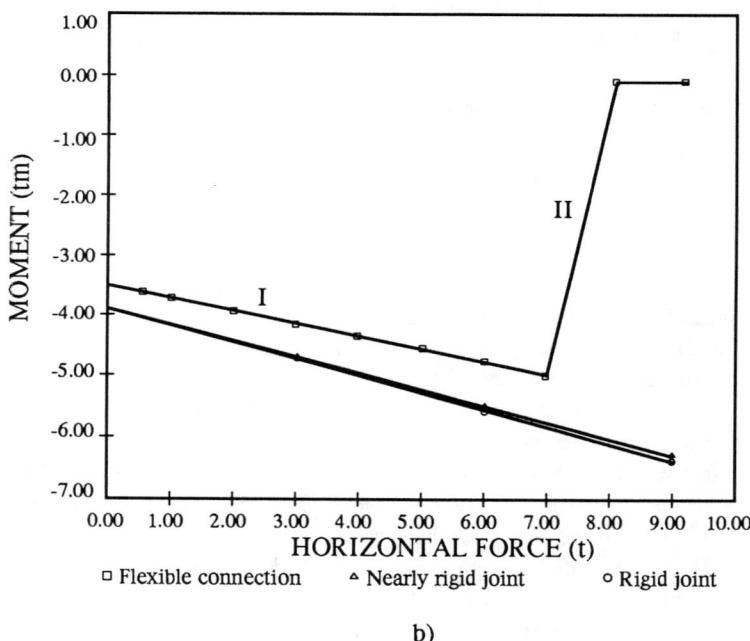

Fig. 11.4.5. The framed response for a variation of the lateral load H

12. On a Decomposition Method into Two Convex Problems

This chapter presents a numerical method related to the quasidifferentiability concept which has been studied in Sect. 1.4, 3.5 and 4.5. We deal especially with problems for which the nonconvex nonsmooth superpotential can be expressed as the difference of two convex functions. In this case the problem can be decomposed into two variational inequalities or equivalently into two convex minimization problems. The method presented here leads to an efficient algorithm for large classes of hemivariational inequalities arising in engineering problems. We base our presentation mainly on [Stav91,93a,b] and on some results of Auchmuty [Auch83,89].

12.1 General Formulation of the Method

In the present section we assume that the nonsmooth and nonconvex superpotential $\bar{\varphi}([u])$ can be expressed as the difference of two convex functions (a d.c. functions), i.e. we have that

$$\bar{\varphi}([u]) = \varphi_1([u]) - \varphi_2([u]), \tag{12.1.1}$$

where φ_1 and φ_2 are convex, possibly nonsmooth energy functions, and $[u]$ denotes a relative displacement. This type of functions has been studied in Sect. 3.5 and 4.5. For this type of functions an appropriate extension of the convex subdifferential operator to the nonconvex case has been introduced by Auchmuty [Auch89]. We may write that (cf. also (1.4.27),(3.5.9),(4.5.27))

$$\Theta\bar{\varphi}([u]) = \{-S = w_1 - w_2 | w_1 \in \partial\varphi_1([u]) \text{ and } w_2 \in \partial\varphi_2([u])\}. \tag{12.1.2}$$

Accordingly we can consider interface laws of the form

$$-S \in \Theta\bar{\varphi}([u]). \tag{12.1.3}$$

Note that $\Theta\bar{\varphi}(\cdot)$ in (12.1.2) denotes a generally nonconvex set which describes exactly the set of directional derivatives of the function $\bar{\varphi}$ at $[u]$. Thus a general nonmonotone possibly multivalued interface law is expressed through (12.1.2),(12.1.3) as the "difference" of more simple monotone possibly multivalued laws derived from the convex potentials φ_1, φ_2, through subdifferentiation.

Thus the interface law (12.1.3) can be written in the following two equivalent forms

$$-S \in \partial\varphi_1([u]) - w_2 \quad \text{and} \quad w_2 \in \partial\varphi_2([u]) \tag{12.1.4}$$

or

$$-S \in \partial\varphi_1([u]) - w_2 \quad \text{and} \quad [u] \in \partial\varphi_2^c(w_2). \tag{12.1.5}$$

Here w_2 is understood as an artificial interface variable (which has the mechanical meaning of a correcting interface traction) and φ_2^c denotes the conjugate function (cf. Sect. 1.3) of the convex function φ_2. Thus, the definition (12.1.2), (12.1.4) (or (12.1.5)) of the law (12.1.3) gives rise to the following system of variational inequalities:

$$\varphi_1([v]) - \varphi_1([u]) \geq -S^T([v]-[u]) + w_2^T([v]-[u]), \quad \forall [v] \in \mathbb{R}^3 \tag{12.1.6a}$$

$$\varphi_2([v]) - \varphi_2([u]) \geq w_2^T([v]-[u]), \quad \forall [v] \in \mathbb{R}^3. \tag{12.1.6b}$$

Let us consider further an interface law of the form

$$[u] \in \Theta\, \bar{\bar{\varphi}}\,(-S), \tag{12.1.7}$$

where $\bar{\bar{\varphi}}$ is a d.c. function. The same techniques as for (12.1.3) can be applied in this case. Let us assume that

$$\bar{\bar{\varphi}}\,(-S) = \bar{\varphi}_1(-S) - \bar{\varphi}_2(-S), \tag{12.1.8}$$

where $\bar{\varphi}_1$ and $\bar{\varphi}_2$ are convex functions. Thus for (12.1.7) we have that

$$\theta\, \bar{\bar{\varphi}}\,(-S) = \{[u] = \bar{w}_1 - \bar{w}_2 | \bar{w}_1 \in \partial\bar{\varphi}_1(-S),\; \bar{w}_2 \in \partial\bar{\varphi}_2(-S)\}. \tag{12.1.9}$$

The artificial variable $\bar{w}_2$ introduced in the decomposition of the law has in this case the physical meaning of a correcting interface relative displacement. All the above laws can be formulated as well between $-S_N$ and $[u]_N$, and $-S_T$ and $[u]_T$ respectively, on the usual assumption that the interface action is splitted into normal and tangential action.

Let the structure be discretized by means of an appropriate finite element scheme and let the stress and deformation vectors of the finite element assemblage be denoted by $s_i, e_i, i = 1, \ldots, n$. Let u be the nodal displacement vector and p be the loading corresponding to the nodal force vector of dimension m. Discrete stress and relative displacement interface quantities are assembled in the vectors S_N, S_T and $[u]_N, [u]_T$ respectively. They have the dimensions g and $3g$ in the normal and in the tangential direction respectively, where g is the number of couples of nodes that model the interface of the structure. The case "node to element-side contact" is reduced to the "node to node" contact by the well-known techniques of interpolation [Kik], [Bis90]. For the whole structure (including the interfaces) the stress vector $\bar{s}$ and the deformation vector $\bar{e}$ are written in the form

12.1 General Formulation of the Method

$$\bar{s} = \begin{bmatrix} s \\ -S_N \\ -S_T \end{bmatrix} \quad \bar{e} = \begin{bmatrix} e \\ [u]_N \\ [u]_T \end{bmatrix}. \tag{12.1.10}$$

Then the static analysis problem of the structure with interfaces takes the following form. We write first the equilibrium equation as

$$\bar{G}\bar{s} = \begin{bmatrix} G \vdots G_1 \end{bmatrix} \begin{bmatrix} s \\ -S_N \\ -S_T \end{bmatrix} = p. \tag{12.1.11}$$

Here G is the equilibrium matrix of the discretized structure and $\bar{G}$ is the enlarged equilibrium matrix such as to take into account the interface tractions S_N and S_T. Then the strain-displacement equations read

$$\bar{e} = \begin{bmatrix} e \\ [u]_N \\ [u]_T \end{bmatrix} = \bar{G}^T u = \begin{bmatrix} G^T \\ G_1^T \end{bmatrix} u \tag{12.1.12}$$

and the linearly elastic law for the material of the structure (not of the interfaces) is

$$e = e_0 + F_0 s \tag{12.1.13}$$

$$\text{or equivalently} \quad s = K_0(e - e_0). \tag{12.1.14}$$

Here F_0 (resp. $K_0 = F_0^{-1}$) denote the natural flexibility (resp. stiffness) matrix of the unassembled structure and e_0 is the initial strain vector. We write now the interface laws (decomposed normally and tangentially to the interface) using the decomposition (12.1.2). They take the following form

$$-S_a = w_{1,a} - w_{2,a}, \quad a = N, T, \tag{12.1.15}$$

with

$$w_{1,a} \in A_{1,a}([u]_a) = \begin{bmatrix} \partial\varphi_{1,a}([u]_a^{(1)}) \\ \vdots \\ \partial\varphi_{1,a}([u]_a^{(\lambda)}) \end{bmatrix}, \quad a = N, T, \tag{12.1.16}$$

and

$$w_{2,a} \in A_{2,a}([u]_a) = \begin{bmatrix} \partial\varphi_{2,a}([u]_a^{(1)}) \\ \vdots \\ \partial\varphi_{2,a}([u]_a^{(\lambda)}) \end{bmatrix}, \quad a = N, T, \tag{12.1.17}$$

where $\lambda = g$ if $a = N$ and $\lambda = 3g$ if $a = T$. Analogously the interface law (12.1.7) is written as

$$[u]_a = \bar{w}_{1,a} - \bar{w}_{2,a}, \quad a = N, T, \tag{12.1.18}$$

with

$$\bar{w}_{1,a} \in B_{1,a}(-S_a) = \begin{bmatrix} \partial\bar{\varphi}_{1,a}(-S_a^{(1)}) \\ \vdots \\ \partial\bar{\varphi}_{1,a}(-S_a^{(\lambda)}) \end{bmatrix}, \quad a = N, T, \tag{12.1.19}$$

and

$$\bar{w}_{2,a} \in B_{2,a}(-S_a) = \begin{bmatrix} \partial \bar{\varphi}_{2,a}(-S_a^{(1)}) \\ \vdots \\ \partial \bar{\varphi}_{2,a}(-S_a^{(\lambda)}) \end{bmatrix}, \quad a = N, T, \qquad (12.1.20)$$

where $\lambda = g$ if $a = N$ and $\lambda = 3g$ if $a = T$. The relations describing the problem are completed by the classical boundary conditions of the structure. They are written in the form

$$E u = u_0 \qquad (12.1.21)$$

$$Z s = F. \qquad (12.1.22)$$

where E and Z are appropriate transformation matrices and u_0, F denote the given boundary displacement (e.g. of a support) and the boundary loading of the structure.

First we shall formulate the problem in terms of the displacements (displacement method). For the structure with interfaces the discrete virtual work equation is written in the form

$$s^T(e^* - e) = p^T(u^* - u) + s_N^T([u]_N^* - [u]_N) + s_T^T([u]_T^* - [u]_T) \quad (12.1.23)$$
$$\forall e^*, u^*, [u]_N^*, [u]_T^* \in V_{\text{ad}}$$

where $V_{\text{ad}} = \{v \in \mathbb{R}^n \,|\, E v = u_0\}$ is the kinematically admissible displacement field. The linear elasticity law (12.1.14) with (12.1.23) and (12.1.12) give rise to

$$u^T G K_0^T G^T(u^* - u) - (p + G K_0 e_0)^T(u^* - u) \qquad (12.1.24)$$
$$= s_N^T(\tilde{N} u^* - \tilde{N} u) + s_T^T(\tilde{T} u^* - \tilde{T} u), \quad \forall u^* \in V_{\text{ad}}$$

where $\tilde{N}$ and $\tilde{T}$ are appropriate transformation matrices. As usual $K = G^T K_0 G$ denotes the stiffness matrix of the structure and $\bar{p} = p + G^T K_0 e_0$ denotes the nodal equivalent loading vector (including initial strains). The following functionals are now introduced:

$$\Phi_{1,N}(\tilde{N} u) = \sum_{i=1}^{g} \varphi_{1,N}([u]_N^{(i)}), \quad \Phi_{2,N}(\tilde{N} u) = \sum_{i=1}^{g} \varphi_{2,N}([u]_N^{(i)}), \qquad (12.1.25)$$

$$\Phi_{1,T}(\tilde{T} u) = \sum_{i=1}^{3g} \varphi_{1,T}([u]_T^{(i)}), \quad \Phi_{2,T}(\tilde{T} u) = \sum_{i=1}^{3g} \varphi_{2,T}([u]_T^{(i)}). \qquad (12.1.26)$$

From the relations (12.1.25) and the inequalities introduced in (12.1.6a,b) we obtain the following inequalities holding at the interfaces

$$s_N^T(\tilde{N} u^* - \tilde{N} u) \geq \Phi_{1,N}(\tilde{N} u) - \Phi_{1,N}(\tilde{N} u^*)$$
$$+ w_{2,N}^T(\tilde{N} u^* - \tilde{N} u), \quad \forall u^* \in V_{\text{ad}} \qquad (12.1.27a)$$

12.1 General Formulation of the Method

$$\Phi_{2,N}(\tilde{N}\,u^\star) - \Phi_{2,N}(\tilde{N}\,u) \geq w_{2,N}^T(\tilde{N}\,u^\star - \tilde{N}\,u), \quad \forall u^\star \in V_{\text{ad}}. \quad (12.1.27b)$$

Analogous expressions are obtained in the tangential direction combining (12.1.26) with the inequalities (12.1.6a,b). Thus the following variational problem results P_1):
Find the kinematically admissible displacement $u \in V_{\text{ad}}$ and the interface correcting tractions $w_2 = \{w_{2,N}, w_{2,T}\} \in \mathbb{R}^{g+3g}$ such as to satisfy the following system of variational inequalities:

$$u^T K\,(u^\star - u) - \bar{p}^T(u^\star - u) + \Phi_{1,N}(\tilde{N}\,u^\star) \quad (12.1.28)$$
$$- \Phi_{1,N}(\tilde{N}\,u) + \Phi_{1,T}(\tilde{T}\,u^\star) - \Phi_{1,T}(\tilde{T}\,u)$$
$$- w_{2,N}^T(\tilde{N}\,u^\star - \tilde{N}\,u) - w_{2,T}^T(\tilde{T}\,u^\star - \tilde{T}\,u) \geq 0, \quad \forall u^\star \in V_{\text{ad}},$$

and

$$\Phi_{2,N}(\tilde{N}\,u^\star) - \Phi_{2,N}(\tilde{N}\,u) + \Phi_{2,T}(\tilde{T}\,u^\star) - \Phi_{2,T}(\tilde{T}\,u) \quad (12.1.29)$$
$$\geq w_{2,N}^T(\tilde{N}\,u^\star - \tilde{N}\,u) + w_{2,T}^T(\tilde{T}\,u^\star - \tilde{T}\,u), \quad \forall u^\star \in V_{\text{ad}}.$$

The variational problem in terms of stresses (force method) is derived analogously on the assumption that the interface law (12.1.9) holds. In this case superpotentials like the $\bar{\Phi}_{1,N}(-S_N)$ of the interface laws are introduced resulting from $\bar{\varphi}_1$ and $\bar{\varphi}_2$. Then the following variational problem P_2 is formulated where the set $\Sigma_{\text{ad}} = \{s^\star | s^\star \text{ fullfils (12.1.11) and (12.1.22)}\}$ is the statically admissible stress field:
Find the stress $s \in \Sigma_{\text{ad}}$ and the vector of correcting interface relative displacements $\bar{w}_2 = \{\bar{w}_{2,N}, \bar{w}_{2,T}\} \in \mathbb{R}^{g+3g}$ such as to satisfy the following system of variational inequalities:

$$e_0^T(s^\star - s) + s^T F_0^T(s^\star - s) + \bar{\Phi}_{1,N}(-s_N^\star) - \bar{\Phi}_{1,N}(-s_N) \quad (12.1.30)$$
$$+ \bar{\Phi}_{1,T}(-S_T^\star) - \bar{\Phi}_{1,T}(-S_T) - \bar{w}_{2,N}^T(-S_N^\star + s_N)$$
$$- \bar{w}_{2,T}^T(-S_T^\star + S_T) \geq 0, \quad \forall s^\star \in \Sigma_{\text{ad}},$$

and

$$\bar{\Phi}_{2,N}(-S_N^\star) - \bar{\Phi}_{2,N}(-S_N) + \bar{\Phi}_{2,T}(-S_T^\star) - \bar{\Phi}_{2,T}(-S_T) \quad (12.1.31)$$
$$\geq \bar{w}_{2,N}^T(-S_N^\star + S_N) + \bar{w}_{2,T}^T(-S_T^\star + S_T), \quad \forall s^\star \in \Sigma_{\text{ad}}.$$

Note that P_1 and P_2 are not connected through the classical theory of duality of convex analysis (see in this context also [Tol]). If we assume monotone interface superpotential laws, P_1 and P_2 reduce to convex variational inequality problems like the ones studied in [Duv72], [Pan85], [Kik], [Glow]. In this case P_2 is the dual of P_1.

The potential energy for the considered structure is written in the form:

$$\Pi(u) = \frac{1}{2}u^T K\,u - \bar{p}^T u + \Phi_1(u) - \Phi_2(u), \quad (12.1.32)$$

where $\Phi_1 = \Phi_{1,N} + \Phi_{1,T}$ and $\Phi_2 = \Phi_{2,N} + \Phi_{2,T}$ are the convex and the concave parts respectively of the total interface potential, which is a d.c. function. The functions Π in (12.1.32) is a difference of convex functions due to the convexity of the terms $\Phi_i, i = 1, 2$. We recall here that K is generally positive semidefinite and thus the corresponding bilinear form is convex. The first two terms in the right hand side of (12.1.32) constitute the well-known, from the linear structural analysis theory, expression of the potential energy. Let us consider now the following minimum problem (P_3):

Find kinematically admissible displacement field $u \in V_{ad}$ which constitutes a (possibly local) minimum of the (d.c.) potential energy function (12.1.32), i.e.:
Find
$$u \in V_{ad} \quad \text{such that} \quad \Pi(u) = \inf\{\Pi(v)|v \in V_{ad}\}. \tag{12.1.32}$$

According to [Dem85], [Poly] a necessary condition for (12.1.32) is (cf. also Sect. 1.4)
$$-\bar{\partial}'\Pi(u) \subset \underline{\partial}'\Pi(u), \tag{12.1.33}$$
or, due to the d.c. decomposition of (12.1.32)
$$\partial \Phi_2(u) \subset K u - p + \partial \Phi_1(u), \tag{12.1.34}$$

if one applies the formula (1.4.27). But (12.1.34) implies the validity of (12.1.30) (12.1.31). Moreover we are led to the following problem (P_4):
Find a solution of the following system of multivalued differential equations (or differential inclusions). Calculate $u \in V_{ad}, w \in \mathbb{R}^{g+3g}$ such that

$$w \in \partial \Phi_2(u) \subset K u - p + \partial \Phi_1(u) \tag{12.1.35}$$

or equivalently such that

$$0 \in K u - p + \partial \Phi_1(u) - w, \quad \forall w \in \partial \Phi_2(u). \tag{12.1.36}$$

From P_3 and by using partial convex conjugacy in the sense of Auchmuty [Auch], the following inf-inf problem is formulated for an appropriately defined Lagrangian function L (of type II according to Auchmuty [Auch83,89]): Let us denote this problem as P_5. It reads:
Find $u \in V_{ad}$ and $w \in \mathbb{R}^{g+3g}$ such that

$$L_{II}(u, w) \leq L_{II}(u, w), \quad \forall v \in V_{ad}. \tag{12.1.37}$$

$$L_{II}(u, w) \leq L_{II}(v, \tilde{w}), \quad \forall \tilde{w} \in \mathbb{R}^{g+3g}, \tag{12.1.38}$$

Here the Lagrangian function $L_{II}(u, w) : V_{ad} \times \mathbb{R}^{g+3g} \to (-\infty, +\infty]$ is defined by

$$L_{II}(u, w) = \frac{1}{2} u^T K u - p^T u + \Phi_1(u) + \Phi_2^c(w) - u^T w, \tag{12.1.39}$$

and is obtained from the potential energy function $\Pi(\cdot)$ by using the convex conjugate function $\Phi_2^c(\cdot)$ of $\Phi_2(\cdot)$. Indeed from the (12.1.32) and (1.3.9) we obtain:

12.2 Application X: The Stamp Problem and the Interfacial Debonding in Composites   351

$$
\begin{aligned}
\Pi(u) &= \inf_{v \in V_{ad}} \{\frac{1}{2}v^T K v - p^T v + \Phi_2(v) - \Phi_2(u)\} \quad (12.1.40)\\
&= \inf_{v \in V_{ad}} \{\frac{1}{2}v^T K v - p^T v + \Phi_1(v)\} - \sup_{w \in \mathbf{R}^{g+3g}} \{w^T v - \Phi_2^c(w)\}\}\\
&= \inf_{v \in V_{ad}} \{\frac{1}{2}v^T K v - p^T v + \Phi_1(v)\} + \inf_{w \in \mathbf{R}^{g+3g}} \{-w^T v - \Phi_2^c(w)\}\}\\
&= \inf_{v \in V_{ad}} \inf_{w \in \mathbf{R}^{g+3g}} \{L_{II}(v, w)\}.
\end{aligned}
$$

The Lagrangian $L_{II}(.,.)$, which is in general a nonconvex function, is convex in each one of its arguments when the other is considered as constant, i.e. the functions $L_{II}(\cdot, w)$ and $L_{II}(v, \cdot)$ are convex. Analogous problems can be formulated for the complementary energy function. The last problem P_5 will be used for the numerical solution of the problem, as it will be explained in the next section.

12.2 Application X: The Stamp Problem and the Interfacial Debonding in Composites

The main idea of this chapter is to decompose the nonconvex problem into two convex problems. This is achieved by minimizing the Lagrangian given in (12.1.39) first with respect to u keeping w constant and then with respect to w keeping u constant. Each problem is a convex minimization problem. General methods for the solution of minimization problems concerning d.c. functions or generally quasidifferentiable functions (e.g. the problem P_3) have already been proposed in the mathematical programming literature (see [Auch83,89], [Dem86a], [Poly], [Tuy86,87] among others). Worthnoting is also the following formulation which would lead to a new problem denoted as P_6 which is not presented here: a d.c. minimization problem can equivalently be written in the form of a convex minimization problem with inequality constraints, where some of the constraints are concave (see e.g. [Tuy86,87]). All these approaches can be used for the solution of the engineering problems formulated in this chapter. The common idea of all the aforementioned approaches is the exploitation of every possible convexity information in the analysis of the noncovex problem studied. (cf. also [Hiri]).

In this section we apply an algorithm proposed by Auchmuty [Auch89] for the solution of the Lagrangian optimization P_5. The main advantage of this algorithm is that it is of a two level type and that within each level "classical" unilateral structural analysis problems arise, i.e. problems involving convex strain energy functions or monotone stress-strain relations. Thus we may use in a two level scheme certain algorithms already developed for the solution of variational inequality problems or equivalently of convex minimization problems and we finally obtain the solution of the studied nonclassical problem with a nonconvex, nondifferentiable potential energy function. The schematic two level optimization algorithm has the following steps:

Step 1: Set the iteration counter $k = 0$
Choose initial values $u^{(0)}, w_2^{(0)}$

Step 2: Iteration $k = k + 1$

Step 3: Solve the convex energy subproblem

$$\min_{v}\{L_{II}(v, w_2^{(k)}) | v \in V_{ad}\}. \qquad (12.2.1)$$

This problem corresponds to the potential energy of a linear elastic structure having prescribed nodal forces $w_2^{(k)}$ and subjected to an interface constraint related to the convex superpotential Φ_1. Let the solution of problem (12.2.1) be denoted by $u^{(k+1)}$.

Step 4: Solve the subproblem

$$\min_{w_2}\{L_{II}(u^{(k+1)}, w_2) | w_2 \in \mathbb{R}^{g+3g}\}. \qquad (12.2.2)$$

Let the solution of the problem (12.2.2) be denoted by $w_2^{(k+1)}$.

Step 5: Convergence check. If

$$||w_2^{(k+1)} - w_2^{(k)}|| \leq \varepsilon_1, \quad \text{and} \quad ||u^{(k+1)} - u^{(k)}|| \leq \varepsilon_2 \qquad (12.2.3)$$

then the solution of the problem has been found, otherwise continue with *Step 2*.

In (12.2.3) $||\cdot||$ denotes the L_∞-norm and $\varepsilon_1, \varepsilon_2$ denote appropriately choosen numerical accuracies. The two level algorithm is a generalization of the one proposed by the author in [Pan78] for large quadratic programming problems. The convex subproblems in the present method have a unique solution if K is positive definite and Φ_2^c strictly convex and the whole algorithm presents according to the numerical experiments a quadratic rate of convergence. This fact seems to be reasonable if one takes into account the mathematical investigations concerning convex programming problems (cf. [Flet], [Womer], [Tzaf]. As already noticed in Chapt. 11 the numerical experience with calculations related to nonconvex minimization problems indicates that, if for an algorithm one has proved its mathematical convergence, there exists the possibility that the same algorithm may diverge in certain applications or may exhibit lower convergence rates than those theoretically estimated [Pow]. This possibility here is minimal due to the decomposition into two convex subproblems; the behaviour of the algorithm in all numerical calculations was very satisfactory and its convergence was achieved after a few iteration cycles.

The method presented here has nothing in common with the path-following method with step-length control introduced by Crisfield [Cris91] but is equally effective. In comparison, however, with the path-following method the present method cannot so easily be incorporated into a general existing F.E. framework.

12.2 Application X: The Stamp Problem and the Interfacial Debonding in Composites

However the present method can treat problems with multivalued stress-strain or reaction-displacement laws. This is a quite important feature, which does not have the path-following method.

The method of this chapter can treat multidimensional nonmonotone problems. It is sufficient that the potential or the complementary energy is given as the difference of two convex functions which is indeed the case in most practical engineering problems (nonmonotone zig-zag friction, delamination problems, plasticity with softening etc). Now concerning the computational efficiency of the algorithm, we make here the following additional comments. The algorithm requires the repeating solution of two subproblems: The first subproblem (12.2.1) is usually a quadratic programming problem (e.g. a unilateral contact problem, a friction problem or a classical convex plasticity problem). There exist a lot of numerical experience for solving large scale problems of this kind. The second subproblem (12.2.2) is a simple convex unconstrained minimization problem. Note that the proposed algorithm presents good convergence and stability properties and combined with a load incremental procedure may lead also to more than one (stable or unstable) positions of equilibrium. For a fixed loading the algorithm converges to one of all possible positions of equilibrium. The problem of obtaining all the positions of equilibrium does not find a satisfactory solution by means of the present algorithm especially for a prescribed loading.

Let us now give some numerical applications. First we deal with the nonmonotone Stamp problem and then with the debonding in composite materials.

i) The Stamp Problem.

Let us consider the discretized elastic stamp problem which is shown in Fig. 12.2.1a. For the elastic structures "A" and "B" plane stress conditions are assumed with elasticity modulus (all quantities are in compatible units) $E = 9.32 \times 10^7$, Poisson's ratio $\nu = 0.30$ and thickness of the plate $t = 1.0$. For the interface between the two parts of the structure (parts A and B, Fig. 12.2.1a), unilateral contact law and Coulomb-type static friction laws are assumed to hold (see Figs. 12.2.1b,c, respectively). We study two cases, the monotone and nonmonotone friction law.

Under the assumption of classical, monotone friction law (see Fig. 12.2.1c, law I) the interface normal (compressive) and tangential stresses are given in Fig. 12.2.2a,b. Here the following coefficients of friction are used (in parentheses we give the interface nodes subjected to sticking, see also the node numbering in Fig. 12.2.1a): $\mu = 0.0$ (stick in nodes 7,8,9), $\mu = 0.1$ and $\mu = 0.3$ (stick in node 8), $\mu = 0.5$ (stick in nodes 7,8,9). Further we solve the rigid-elastic stamp problem of Fig. 12.2.1a (i.e. part "A" of the structure is assumed now as rigid), under the assumption of nonmonotone friction law (see Fig. 12.2.1c, law II). We consider the following friction laws. First the classical monotone friction law I in Fig. 12.2.1c with $\mu = 0.4$ (case A in Fig. 12.2.3), and then the nonmonotone friction law II with $\mu = 0.4$, $\mu_1 = 0.2$ and for the case B: $\alpha = 2 \times 10^{-6}$, for the case C: $\alpha = 1 \times 10^{-7}$ and for the case D: $\alpha = 2 \times 10^{-7}$. The numerical results are depicted in Fig. 12.2.3a,b. Interface nodes 6 to 10 remain in stick condition

354   12. On a Decomposition Method into Two Convex Problems

for all the considered cases. It is important to note that the normal interface forces are not considerably influenced by the friction law. This is not the case for the frictional forces.

ii) The Material Inclusion Problem. Debonding in Composite Materials.

Let us consider the rigid inclusion problem of Fig. 12.2.4a. For the elastic material of the matrix (region Ω) the following constants are used: Elasticity modulus E = 2000.0, Poisson's ratio ν = 0.25 and thickness t = 0.025 all in compatible units.

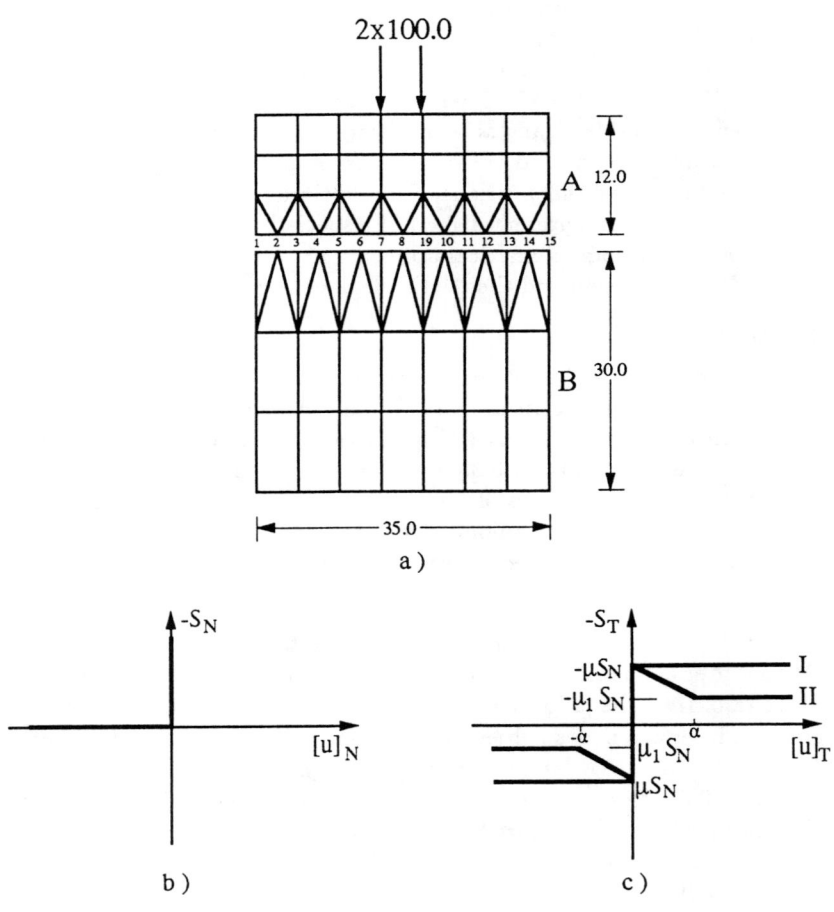

Fig. 12.2.1. The stamp problem a) discretized structure b) c) interface laws.

12.2 Application X: The Stamp Problem and the Interfacial Debonding in Composites

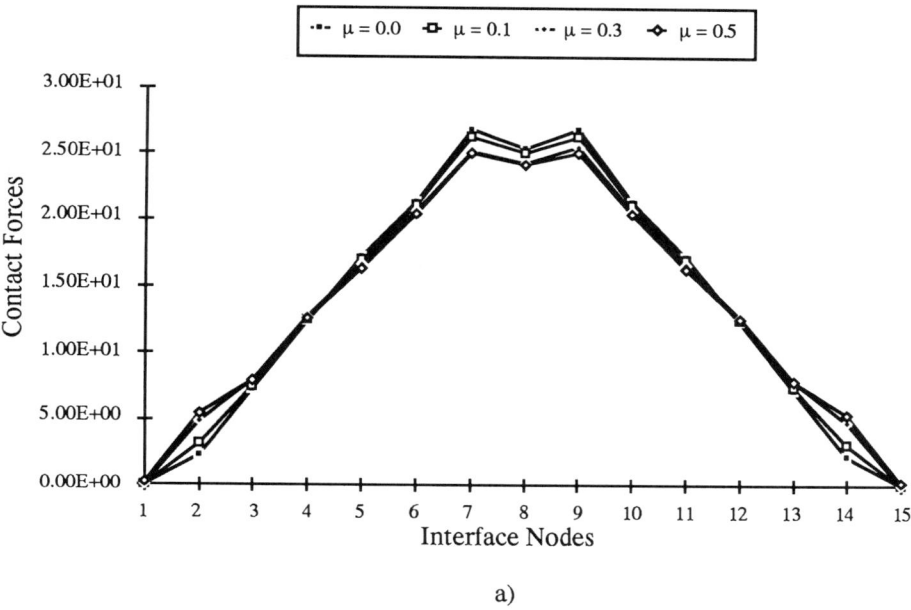

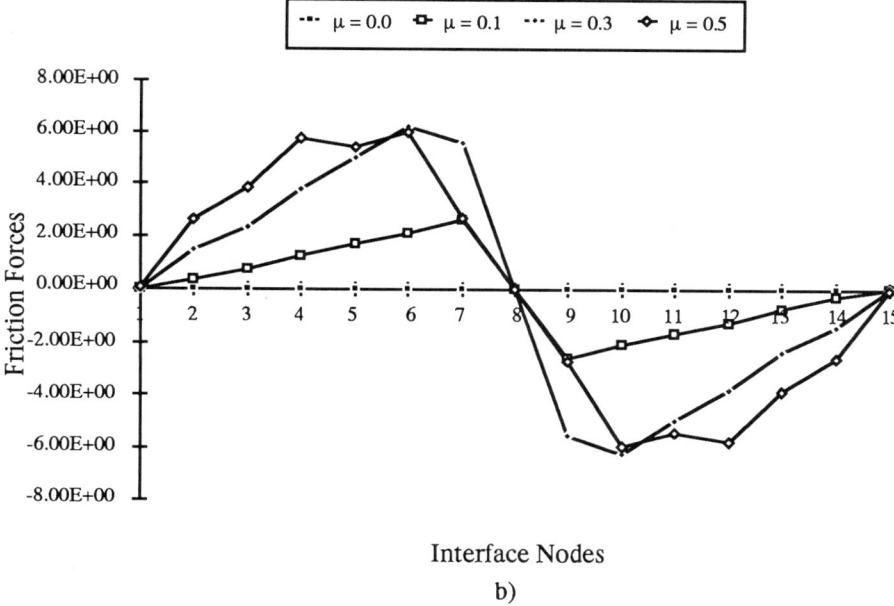

Fig. 12.2.2. The monotone friction problem for the stamp a) normal interface forces b) frictional interface forces.

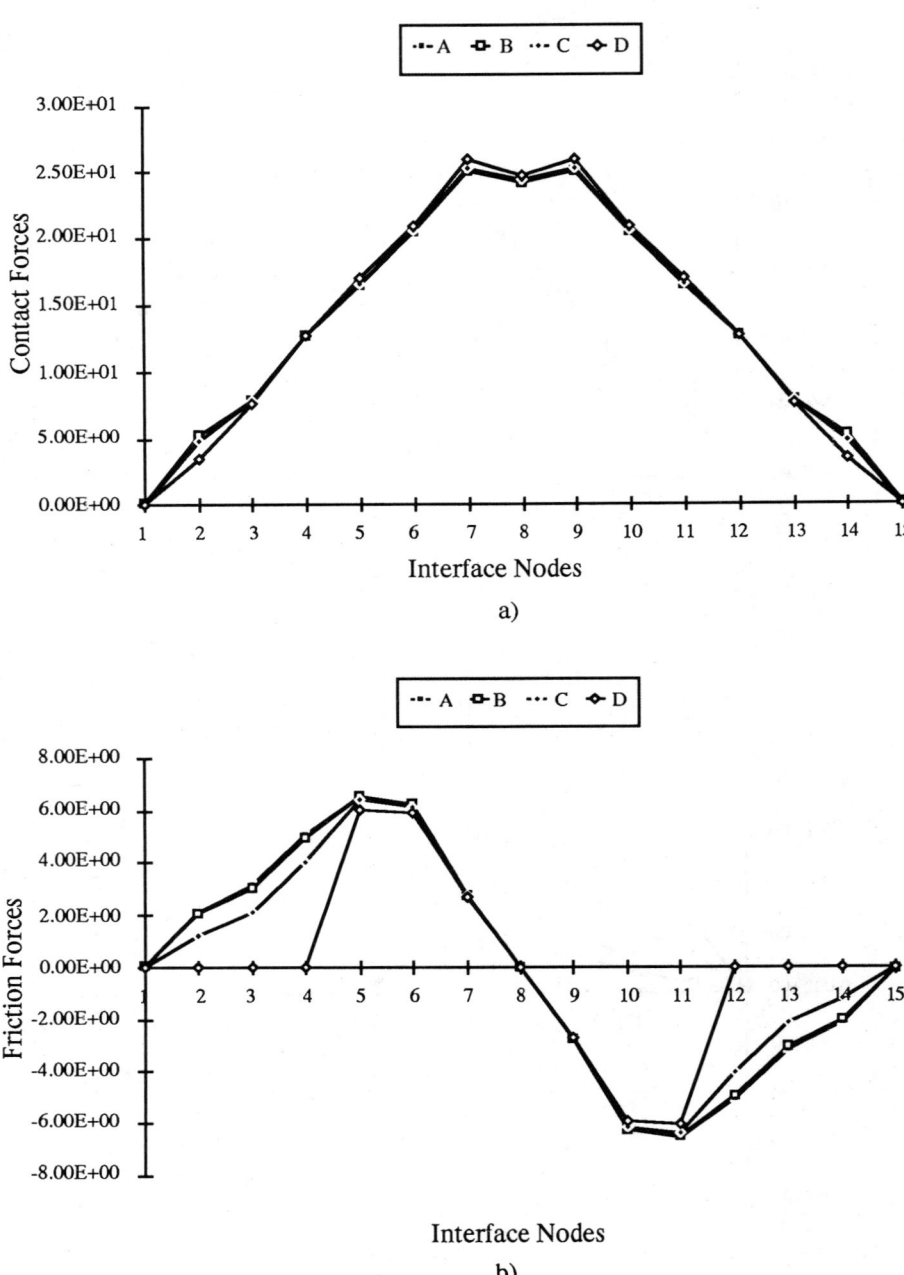

Fig. 12.2.3. The nonmonotone friction problem for the stamp a) normal interface forces b) frictional interface forces.

12.2 Application X: The Stamp Problem and the Interfacial Debonding in Composites

The cyclic inclusion is considered as rigid and constitutes the idealization for a fiber in the matrix of a composite material. An external displacement $\Delta w_{AB} = -0.025$ along the upper and the lower boundary of the structure is considered (see Fig. 12.2.4a). Due to the double symmetry the one quarter of the initial structure is solved; each quarter has 19 interface nodes. The finite element discretization of the matrix is shown in Fig. 12.2.4b. Under the assumption of unilateral contact condition and Coulomb law (Fig. 12.2.1c, law I) at the interface, the normal and the tangential interface forces are depicted in Fig. 12.2.5a,b.

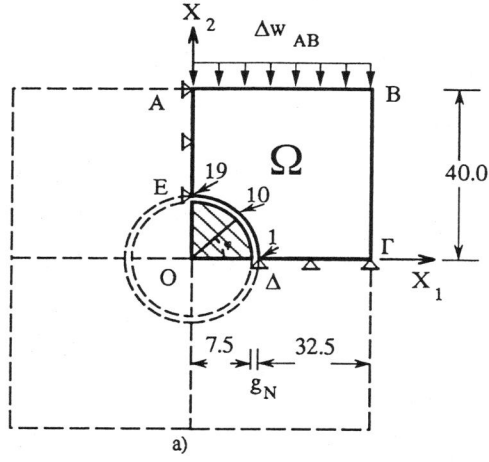

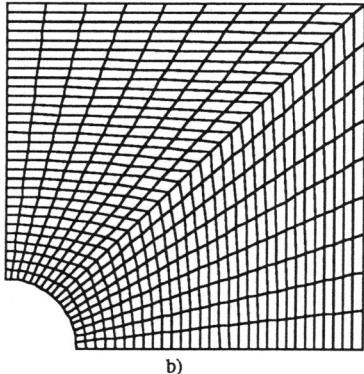

Fig. 12.2.4. The rigid inclusion problem and the discretization of the matrix.

The following friction coefficients are assumed (in parentheses are given the stick interface nodes, see Fig. 12.2.4a for the numbering of the interface nodes) For the case A: $\mu = 0.0$ (stick nodes: 19), for the case B: $\mu = 0.1$ (stick nodes: 19), for the case C: $\mu = 0.2$ (stick nodes: 19), for the case D: $\mu = 0.3$ (stick nodes: 19,18,17), for the case E: $\mu = 0.4$ (stick nodes: 19 till 15), for the case F: $\mu = 0.5$ (stick nodes: 19 till 10), for the case G: $\mu = 0.6$ (slick nodes 19 till 8).

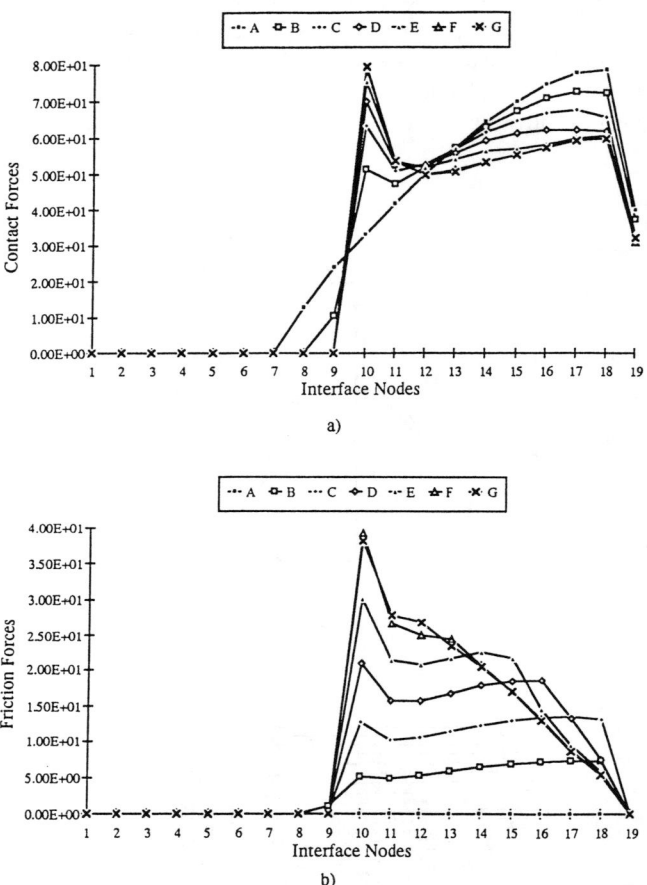

Fig. 12.2.5. The debonding problem for a rigid inclusion a) normal interface forces b) frictional interface forces for monotone friction.

From the numerical results of Fig. 12.2.5a,b we see that for a relatively high friction coefficient ($\mu = 0.6$) the stick and contact regions are approximately the same. Then as the structure reaches this condition, a peak is observed in both the normal and in the tangential tractions at the beginning of the noncontact region of the interface, i.e. a stress concentration results, which is analogous to the one observed in cracks.

Under the assumption of unilateral contact and nonmonotone friction law (see Fig. 12.2.1c, law II) the normal and the tangential interface stresses are schematically given in Fig. 12.2.6a,b. In Fig. 12.2.6a,b the following constants are used for the nonmonotone friction law: $\mu = 0.30$, $\alpha = 1 \times 10^{-2}$ and for the

12.2 Application X: The Stamp Problem and the Interfacial Debonding in Composites

case A: $\mu_1 = 0.20$, for the case B: $\mu_1 = 0.10$ and for the case C: $\mu_1 = 0.0$. In all the above problems the proposed algorithm needed 3-5 iteration cycles to converge. For the arising Q.P. subproblems the Hildreth and d'Esopo algorithm has been applied.

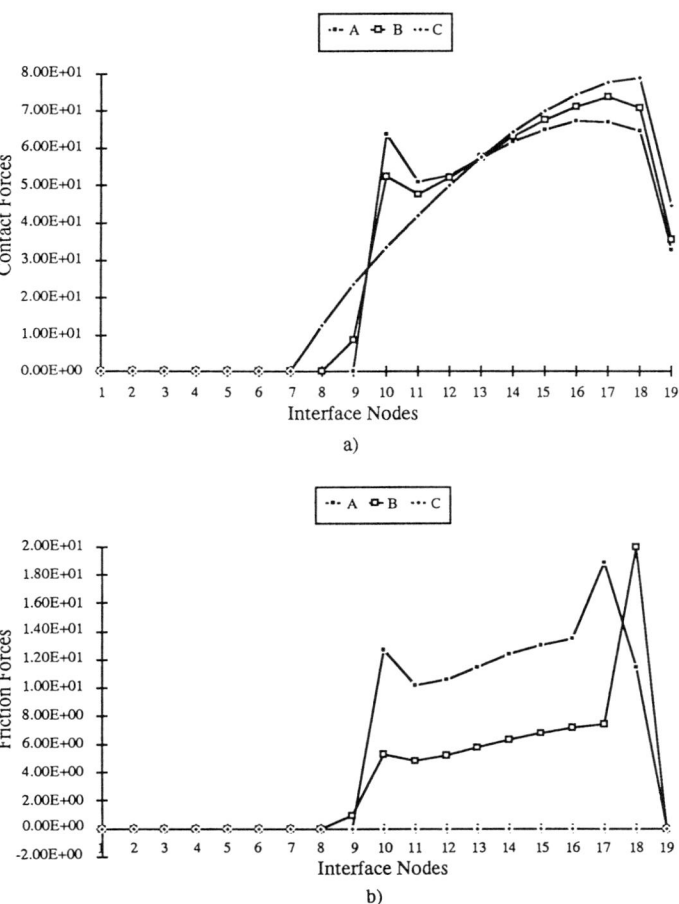

Fig. 12.2.6. The debonding problem for a rigid inclusion a) normal interface forces b) frictional interface forces for nonmonotone friction.

The consideration of both the unilateral contact and the friction laws is achieved by the fixed point type algorithm illustrated in Sect. 9.5 and proposed for the monotone case in [Pan75].

13. Dynamic Hemivariational Inequalities and Crack Problems

In this chapter we present certain numerical applications concerning the numerical treatment of dynamic hemivariational inequalites and of hemivariational inequalities arising in crack problems. In this area the theory of hemivariational inequalities permits the formulation and study of new and very interesting engineering problems, as is the analysis of cracks which are repaired by an adhesive material. The dynamic hemivariational inequalities are "transformed" through time discretization into static hemivariational inequalities which can be treated by one of the methods developed in the previous chapters.

13.1 Application XI: Numerical Treatment of Dynamic Hemivariational Inequalities

The dynamic hemivariational inequalities give rise to a sequence of static hemivariational inequalities only after an appropriate time discretization. In this way, a substationarity problem results within each time step. Moreover in the dynamic hemivariational inequalities arising in contact problems of two bodies the question of impact has to be carefully taken into account in a numerical method. In realistic problems we have to determine all the free boundaries as they vary with time and, in the case of contact, the time and the geometry of all the configurations in which the body is subjected to impact forces. Due to the heavy numerical calculations it is important from the numerical point of view to formulate the problem only with respect to the "ambiguous" degrees of freedom. Taking as an example the interface problem obeying the nonmonotone law of Fig. 9.5.2 (unilateral adhesive contact), the term "ambiguous" means that it is not a priori known which parts of the body are in contact with the support or with the other parts of body in the case of an internal interface; moreover, it is not a priori known which parts of the areas being in contact, stay in contact due to compressive forces or due to the action of the adhesive. In other classes of problems, e.g. in the plastic softening hinge problems in plates, other physical parameters like the extent of the plastic hinge along the boundary is not a priori known.

From the numerical point of view, an algorithm for hemivariational inequalities is considered as efficient if it can treat a large number of ambiguous points. Therefore the presented in Chapt. 5 B.I.E. approach gains importance, because it refers to the ambiguous degrees of freedom of a problem leading to a hemivariational inequality. Indeed the technique of Sect. 5.1 can be extended to dynamic hemivariational inequalities. Within each time step the application of Betti's theorem leads to a static hemivariational inequality holding, e.g. on the boundary, which is equivalent to a multivalued boundary integral equation within each time step, i.e. we get a time-difference multivalued B.I.E. Including of partial or total velocity reversal into our model permits the rational consideration of impact shocks. We consider a threedimensional linear elastic body; the method presented is general and also holds for plates, beams, etc., i.e. for all structures for which Betti's theorem holds in the equilibrium problem.

Let Ω be an open bounded subset of the threedimensional Euclidean space $\mathbb{R}^3$ with a boundary Γ. Ω is occupied by a linear elastic body in its undeformed case which is referred to an orthogonal Cartesian coordinate system $Ox_1x_2x_3$. Γ is decomposed as usual into three mutually disjoint parts Γ_U, Γ_F and Γ_S. On Γ_U (resp. Γ_F) the displacements (resp. the tractions) are given and on Γ_S boundary conditions giving rise to hemivariational inequalities hold. We assume that the time t takes values in the time interval $[0,T]$.

On Γ_U
$$u_i = U_i, \quad U_i = U_i(x,t), \qquad (13.1.1)$$

and on Γ_F
$$S_i = F_i, \quad F_i = F_i(x,t). \qquad (13.1.2)$$

The boundary condition on Γ_S reads
$$-\bar{S} \in \bar{\partial} j(u,t) \text{ on } \Gamma_S \times [0,T]. \qquad (13.1.3)$$

Here j is a locally Lipschitzian (with respect to u) nonconvex superpotential depending on t. Note that (13.1.3) might be replaced without affecting the method of the present chapter by the two conditions

$$-S_N \in \bar{\partial} j_N(u_N,t) \text{ and } -S_T \in \bar{\partial} j_T(u_T,t) \text{ on } \Gamma_S \times [0,T]. \qquad (13.1.4)$$

The generalized gradient is taken with respect to the displacement variable. The following equations of motion hold on the assumption of small displacements and small strains

$$\sigma_{ij,j} + f_i(t) = \rho \ddot{u}_i + c\dot{u}_i \qquad \text{in } \Omega \times (0,T) \qquad (13.1.5)$$

$$\epsilon_{ij} = \epsilon_{ij}(u) = \frac{1}{2}(u_{i,j} + u_{j,i}) \qquad \text{in } \Omega \times (0,T) \qquad (13.1.6)$$

$$\sigma_{ij} = C_{ijhk}\epsilon_{hk} \qquad \text{in } \Omega \times (0,T) \qquad (13.1.7)$$

$$u_i = u_{i0}(x) \text{ at } t=0 \qquad (13.1.8)$$

13.1 Application XI: Dynamic Hemivariational Inequalities

$$\dot{u}_i = u_{i1}(x) \text{ at } t = 0 \qquad (13.1.9)$$

where u_{i0} (resp. u_{i1}) denotes the initial displacements (resp. velocities), $f = \{f_i\}$ represents the volume force vector, the comma denotes the partial derivation, $\ddot{u}_i$ is the acceleration vector and ρ is the mass density. Let us assume that the damping term is proportional to the velocity $\dot{u}_i$ and let us denote the damping coefficient by $c > 0$. We apply the method of time discretization in order to reduce the problem into a static problem. The method of m-step linear difference operators is applied for the time discretization of the problem with respect to time. At time instant $t^{(p)}$ we obtain (cf. also [Fel], [Mit83[, [Pan85])

$$\sum_{r=0}^{q} \alpha^{(r)} u^{(p-r)} = \Delta t \sum_{r=0}^{q} \beta^{(r)} \dot{u}^{(p-r)} \qquad (13.1.10)$$

and

$$\sum_{r=0}^{q} \gamma^{(r)} u^{(p-r)} = \mu \Delta t^2 \sum_{r=0}^{q} \beta^{(r)} \ddot{u}^{(p-r)}, p \geq q, p > 1 \;. \qquad (13.1.11)$$

The coefficients $\alpha^{(r)}, \beta^{(r)}, \gamma^{(r)}$, and μ depend on the chosen finite-difference scheme. We assume that the time step size Δt remains constant and in order to have an implicit integration scheme that $\beta^{(0)}$ is nonzero. Thus the relations

$$\alpha^{(0)} u^{(p)} - \Delta t \beta^{(0)} \dot{u}^{(p)} = -U\,\alpha + \Delta t \dot{U}\,\beta \qquad (13.1.12)$$

$$\mu(\Delta t)^2 \beta^{(0)} \ddot{u}^{(p)} = \gamma^{(0)} u^{(p)} + U\,\gamma - \mu(\Delta t)^2 \ddot{U}\,\beta = \gamma^{(0)} u^{(p)} + \tilde{U}\;, \qquad (13.1.13)$$

where

$$\alpha = [\alpha^{(1)}, \cdots, \alpha^{(q)}]^T, \qquad \beta = [\beta^{(1)}, \cdots, \beta^{(q)}]^T$$
$$\gamma = [\gamma^{(1)}, \cdots, \gamma^{(q)}]^T, \text{ and } \qquad U = [u^{(p-1)}, \cdots, u^{(p-q)}]$$

are obtained. Accordingly within the p-time interval $(t^{(p)}, t^{(p)} + \Delta t)$ we may write after applying the above time discretization to the equations of motion, that (we omit index p)

$$\sigma_{ij,j} + f_i(t) = g_i(t) + Au_i \qquad A > 0 \text{ in } \Omega \times (t, t + \Delta t). \qquad (13.1.14)$$

Here A is a constant equal to $\gamma^{(0)} \rho/\mu(\Delta t)^2 \beta^{(0)} + \alpha^{(0)} c/\Delta t \beta^{(0)} > 0$. The term g_i contains the terms of the previous time steps resulting from (13.1.12), (13.1.13). Note that thermal terms and terms due to dislocations may be included in the model provided we know their evolution with time. We denote further by $\bar{f}_i$ the term

$$\bar{f}_i = f_i - g_i$$

and we consider from now on only the behaviour of the structure in the time interval $(t, t + \Delta t)$. The above discretization process "replaces", roughly speaking, the dynamic hemivariational inequality formulation of the problem by a static hemivariational formulation. In the realistic case of shocks, treated in the present chapter, the mathematical problem of convergence is still open.

Let V_{ad} be the set

$$V_{ad} = \{v | v = \{v_i\},\ v \in \tilde{V},\ v_i = \bar{U}_i \quad i = 1,2,3 \text{ on } \Gamma_U\} \tag{13.1.15}$$

of the kinematically admissible displacements within the time interval $(t, t+\Delta t)$, where $\tilde{V}$ is the basic vector space for the displacements. We denote by $(\bar{f}, v)$ and by $[F, v]_{\Gamma_F}$ etc. the corresponding work expressions as in Ch.5 and let

$$(u, v) = A \int_\Omega u_i v_i d\Omega \qquad A > 0 \tag{13.1.16}$$

and

$$a(u, v) = (C\epsilon(u), \epsilon(v)) = \int_\Omega C_{ijhk}\epsilon_{ij}(u)\epsilon_{hk}(v)d\Omega \tag{13.1.17}$$

be the bilinear form of elasticity. Now let Π be the potential energy within the interval $(t, t + \Delta t)$

$$\Pi(v) = 1/2 a(v,v) + 1/2(v,v) - (\bar{f}, v) + \Phi(v) - [F, v]_{\Gamma_F} \tag{13.1.18}$$

where $\Phi(v) = \int_{\Gamma_S} j(v)d\Gamma$ on the assumption that the integral makes sense. We can show by the method explained in Sect. 4.1 for the derivation of hemivariational inequalities that within time interval $(t, t+\Delta t)$ the position of equilibrium is given by the following hemivariational inequality: Find $u \in V_{ad}$ such as to satisfy the inequality.

$$a(u, v-u) + (u, v-u) + \int_{\Gamma_S} j^0(u, v-u)d\Gamma \geq \tag{13.1.19}$$

$$(\bar{f}, v-u) + [F, v-u]_{\Gamma_F} \quad \forall v \in V_{ad}.$$

The procedure of Chapt. 5 is repeated within each time interval for the bilinear form $a(u,v) + (u,v) = \bar{a}(u,v)$ and for the volume forces $\bar{f}$ and for the boundary forces F. Obviously these changes do not affect the validity of all transformations applied in Chapt. 5 for the derivation of the hemivariational inequality (5.1.58) holding on Γ_S. If instead of (13.1.3) a boundary condition of the type (5.1.3) holds on Γ_S then within each time step a hemivariational inequality analogous to (5.1.33) results. To give the forms of these hemivariational inequalities let us denote by $\tilde{\beta}$, $\tilde{\gamma}$, $\tilde{\delta}$ and $\tilde{\zeta}$ the bilinear and linear forms within the time interval which correspond to the bilinear and linear forms β, γ, δ and ζ of Chapt. 5.

Then we obtain the following substationarity problem with respect to the unknown boundary tractions μ (cf. (5.1.34))

$$\mu \in L, \quad 0 \in \bar{\partial}\Pi_1(\mu), \quad \Pi_1(\mu) = \frac{1}{2}\tilde{\beta}(\mu,\mu) + \int_{\Gamma_S} \tilde{j}(-\mu)d\Gamma - \tilde{\gamma}(\mu). \tag{13.1.20}$$

Problem (13.1.20) corresponds to the following hemivariational inequality: Find $\lambda \in L$ such as to satisfy

13.1 Application XI: Dynamic Hemivariational Inequalities

$$\tilde{\beta}(\lambda, \mu - \lambda) + \int_{\Gamma_S} \tilde{j}^0(-\lambda, -\mu + \lambda) d\Gamma - \tilde{\gamma}(\mu - \lambda) \geq 0 \quad \forall \mu \in L. \tag{13.1.21}$$

We denote by λ the solution of this problem. Using the definition of the generalized gradient, (13.1.20) corresponds to a "multivalued" integral equation similar, to (5.1.35). It reads

$$\tilde{\gamma} - \frac{1}{2} \operatorname{grad} \tilde{\beta}(\lambda, \lambda) \in \bar{\partial} \int_{\Gamma_S} \tilde{j}(-\lambda) d\Gamma \quad \text{on } \Gamma_S. \tag{13.1.22}$$

The substationarity problem with respect to the unknown displacements u on Γ_S reads in the time interval $(t, t + \Delta t)$ (cf. 5.1.58)

$$v \in N, \quad 0 \in \bar{\partial} \Pi_2(v), \quad \Pi_2(v) = \frac{1}{2} \tilde{\delta}(v, v) + \int_{\Gamma_S} j(v) d\Gamma - \tilde{\zeta}(v). \tag{13.1.23}$$

To (13.1.23) corresponds the following hemivariational inequality holding in $(t, t + \Delta t)$:
Find $u \in N$ such as to satisfy the inequality

$$\tilde{\delta}(u, v - u) + \int_{\Gamma_S} j^0(u, v - u) d\Gamma - \tilde{\zeta}(v - u) \geq 0 \quad \forall v \in N. \tag{13.1.24}$$

Then the corresponding multivalued B.I.E. becomes

$$\tilde{\zeta} - \frac{1}{2} \operatorname{grad} \tilde{\delta}(u, u) \in \bar{\partial} \int_{\Gamma_S} j(u) d\Gamma. \tag{13.1.25}$$

Now we shall give a numerical application. The solution of the dynamic hemivariational inequality problems considered, is obtained after an appropriate time discretization. Here we apply for the numerical solution, the weighted residual time discretization algorithm proposed by Zienckiewicz, Wood and Taylor [Zie], which is a special case of the algorithm defined by eq. (13.1.10), (13.1.11). The algorithm is implicit and unconditionally stable. The algorithm interpolates independently the displacement and velocity vectors, and therefore computation of acceleration terms is avoided. This is an advantage for the present problems, because a calculation of "initial" accelerations in the case of impact is not necessary.

As we have shown previously two hemivariational inequalities hold, within each time interval, on the boundary Γ_S of the system. The two inequalities and the corresponding substationarity problems are dual in the sense that the first has as unknowns the boundary forces on Γ_S, whereas the second has as unknowns the boundary displacements on Γ_S, with the remark that the work of the fictitious springs with constant A must also be considered, as well as the fictitious volume forces $\bar{f}_i$, in order to take into account the influence of the time discretization scheme.

To calculate the discrete forms of Π_1 (resp. Π_2) we apply the same method proposed in [Ant92] which is based on unit force (resp. unit displacement) loadings of a bilateral structure and calculation of it by the classical B.E.M. or the F.E.M. Let us calculate the discrete form of Π_1. We discretize first the boundary of the elastic body under consideration by a B.E. or F.E. scheme. For this discretized system, the nonmonotone, multivalued, constraint defined by (5.1.3) on the boundary Γ_S is assumed to be realized, e.g. through fictitious springs attached at the m nodes of this boundary. We consider then the system Ω_0 obtained from the discretized one by assuming only the kinematical constraints on Γ_U. The resulting structural system is also appropriately modified by the fictitious springs of constant A introduced by the time discretization scheme. This discrete system is "solved" for a unit force corresponding to a constrained degree of freedom, on the first node of Γ_S, and for zero forces corresponding to the remaining degrees of freedom of the same node and of all the other nodes of Γ_S. In the case of interfaces we have pairs of nodes and pairs of unit forces corresponding to these nodes. The solution of the resulting underconstrained structure Ω_0 supplies the corresponding displacements in the directions of the constrained degrees of freedom of the m nodes of Γ_S. They constitute the first column of a matrix $\tilde{B}$. This procedure is repeated for all the m nodes and thus the whole matrix $\tilde{B}$ of the influence coefficients is calculated.

Within each time interval the displacements in the directions of the constrained degrees of freedom of the nodes of Γ_S, or in the case of interfaces of the node pairs, due to the external actions constitute a vector $\tilde{g}$. Then the discrete form of Π_1 reads (symmetric or symmetrized problem)

$$\Pi_1^d(\boldsymbol{\mu}) = \frac{1}{2}\boldsymbol{\mu}^T \tilde{B} \boldsymbol{\mu} + \sum_{i=1}^m \tilde{j}_i(-\boldsymbol{\mu}_i) - \tilde{g}^T \boldsymbol{\mu} \qquad (13.1.26)$$

$$\text{for} \quad \boldsymbol{\mu} = \{\boldsymbol{\mu}_i\} \in \mathbb{R}^{3m}, \quad i=1,\ldots,m, \quad \boldsymbol{\mu}_i \in \mathbb{R}^3.$$

Here $\boldsymbol{\mu}$ is the vector of the unknown reactions on Γ_S, which are constrained, i.e. they are subjected to the boundary conditions (5.1.3) resulting from a nonconvex superpotential. The discrete form of Π_2 is written analogously for the symmetrized problem as

$$\Pi_2^d(\boldsymbol{v}) = \frac{1}{2}\boldsymbol{v}^T \tilde{D} \boldsymbol{v} + \sum_{i=1}^m \tilde{j}_i(\boldsymbol{v}_i) - \tilde{z}^T \boldsymbol{v} \qquad (13.1.27)$$

$$\text{for} \quad \boldsymbol{v} = \{\boldsymbol{v}_i\} \in \mathbb{R}^{3m}, \quad i=1,\ldots,m, \quad \boldsymbol{v}_i \in \mathbb{R}^3.$$

Here $\boldsymbol{v}$ is the vector of the unknown displacements on Γ_S, which are constrained. The matrix $\tilde{D}$ and the known vector $\tilde{z}$ are obtained as follows: The discrete structural system Ω is solved by the classical F.E.M. or the B.E.M. by imposing a unit displacement corresponding to a constrained degree of freedom on the first node of Γ_S and by zeroing the displacements of the other degrees of freedom of the same node and of the degrees of freedom of all the other nodes on Γ_S. The solution of the resulting overconstrained structure Ω_0', by the classical B.E.M. or the F.E.M., supplies the corresponding reactions corresponding to all the

13.1 Application XI: Dynamic Hemivariational Inequalities

degrees of freedom of the m-nodes of Γ_S. They constitute the first column of a matrix $\tilde{D}$. This procedure is repeated for all the nodes of Γ_S and thus we obtain the whole matrix $\tilde{D}$ of influence coefficients. Within each time interval we fix all the degrees of freedom of the nodes of Γ_S. Then the corresponding reactions due to the external actions constitute the vector $\tilde{z}$. Matrices $\tilde{B}$ and $\tilde{D}$ should be symmetric due to Betti's theorem. However this symmetry may be lost because of the numerical method used for the unit load or the unit displacement calculations. For instance, if a displacement F.E. scheme is used for the calculation of $\tilde{B}$ or $\tilde{D}$ and the nodes of the F.E. scheme on Γ_S coincide with the m nodes of Γ_S, then $\tilde{B}$ and $\tilde{D}$ are symmetric. If on the contrary a B.E.M. code is used, then, generally, $\tilde{B}$ and $\tilde{D}$ are not symmetric. Note that in the numerical calculations of large scale problems the problem may be symmetrized by considering instead of $\tilde{B}$ and $\tilde{D}$ the matrices $(\tilde{B} + \tilde{B}^T)/2$ and $(\tilde{D} + \tilde{D}^T)/2$. We assume further that $\tilde{B}$ and $\tilde{D}$ are full symmetric or symmetrized positive definite matrices.

In order to solve the arising hemivariational inequality within each time interval we have applied the method developed in Sect. 9.3 which consists in the approximation of the decreasing branch by monotone laws. If the approximating monotone laws are of the type of friction law, as e.g. in Fig. 9.3.3 or in Fig. 9.4.2, then within each time step the solution of the hemivariational inequality (13.1.21) is obtained by solving an inequality constrained Q.P.P. for the function $\Pi_1^d(\boldsymbol{\mu})$ given in (13.1.26). Note that if one formulates the hemivariational inequality (13.1.24) with respect to the displacements then its solution will be approximated by a sequence of solutions of friction type variational inequalities. These last variational inequalities contain terms with the absolute value of the displacements and for this reason it is preferable to use again the dual problem which is a Q.P.P. for the function $\Pi_1^d(\boldsymbol{\mu})$ with simple inequality constraints.

Next a dynamic hemivariational inequality is numerically studied. In the examples within each time interval $(t, t + \Delta t)$ the Hildreth and d' Esopo's Q.P. algorithm is applied for the solution of the quadratic programming problems whose solutions are approximating the solution of the arising static hemivariational inequalities.

As an application we examine the problem of Fig. 13.1.1. It is a channel burried in a linear elastic homogeneous soil supported by a rigid bedrock on which a seismic excitation is given. A sinusoidal acceleration acts on the bedrock. The adhesive contact conditions of Fig. 10.3.3b with unprevented sliding are assumed to hold between the channel and the soil.

The vibration is considered to be undamped in the sense that c in (13.1.5) is zero. When evaluating the dynamic response of the structure it is necessary to consider the collisions at the interface between channel and soil. The velocity towards the support of a point i before impact is $\dot{u}_i^-$. When the point has a contact with the support a part of the kinetic energy is lost due to impact. It is reasonable to accept a perfectly inelastic collision which dissipates the whole kinetic energy. Thus the velocity of the point i just after impact $(\dot{u}_i^+)$ is assumed to be zero. We assume that immediately after the impact the adhesive acts, and in the next step again the law of Fig. 10.3.3b holds. Thus although c was assumed to be zero, an amount of damping is taken into account due to perfectly plastic collision and the action of the adhesive.

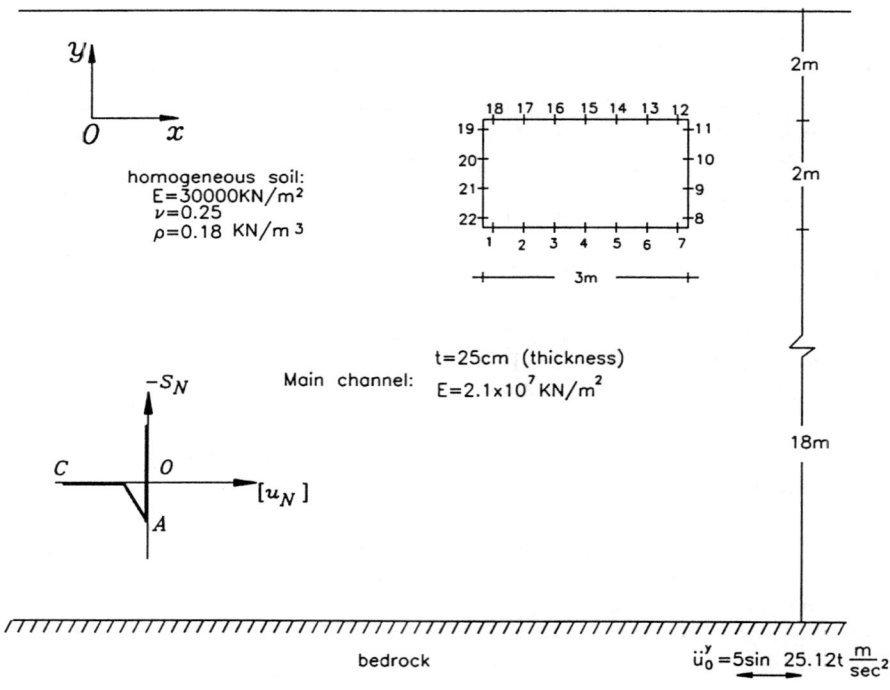

Fig. 13.1.1. Application. For the adhesive contact the law of Fig. 10.3.3b holds.

For the time discretization of the equations of motion the algorithm of Zienkiewicz, Wood and Taylor with $a = \theta = 0.5$ (cf. [Zie]) is applied. At each time step the convergence of the algorithm of Sect. 9.3 is very rapid. In most cases two or three Q.P. problems are solved within each time step for the treatment of the nonmonotonicity OAC (Fig. 13.1.1). When a node i has a contact with the support the condition $\dot{u}_i^+ = 0$ is imposed as "initial" condition for the next time step. Any other type of velocity changes due to impact (e.g. velocity reversal in the case of elastic impact etc.) can be taken into account for the solution of the arising hemivariational inequality. We have eliminated all the degrees of freedom with the exception of the interface degrees of freedom and we have obtained a hemivariational inequality of the type (13.1.24) holding on the interface between the two bodies. For the determination of the matrices $\tilde{B}$ and $\tilde{g}$ (the two last are necessary because we are working with the dual problem) of the discretized problem, which are necessary for the formulation of

13.1 Application XI: Dynamic Hemivariational Inequalities

the resulting Q.P. problems, a classical (i.e. bilateral) direct B.E.M. code using boundary elements with linear interpolation was used. In Fig. 13.1.2 the time history of the displacements of certain points at the channel-soil interface is depicted. In Fig. 13.1.3 the displacements in the y-y direction of the two corners of the channel are plotted. Finally in Fig. 13.1.4 the displacements of the channel obtained by assuming everywhere nondebonding (i.e. bilateral contact) are compared with the same results for unilateral contact (diagram COB in Fig. 13.1.1) and for adhesive contact (diagram CAOB in Fig. 13.1.1). The present application has been presented in [Mit91] for the case of the unilateral contact diagram COB (Fig. 13.1.3), i.e. without adhesive contact. (see also [Zer] for applications in a seismic design). The influence of the adhesive action is clear in the numerical results.

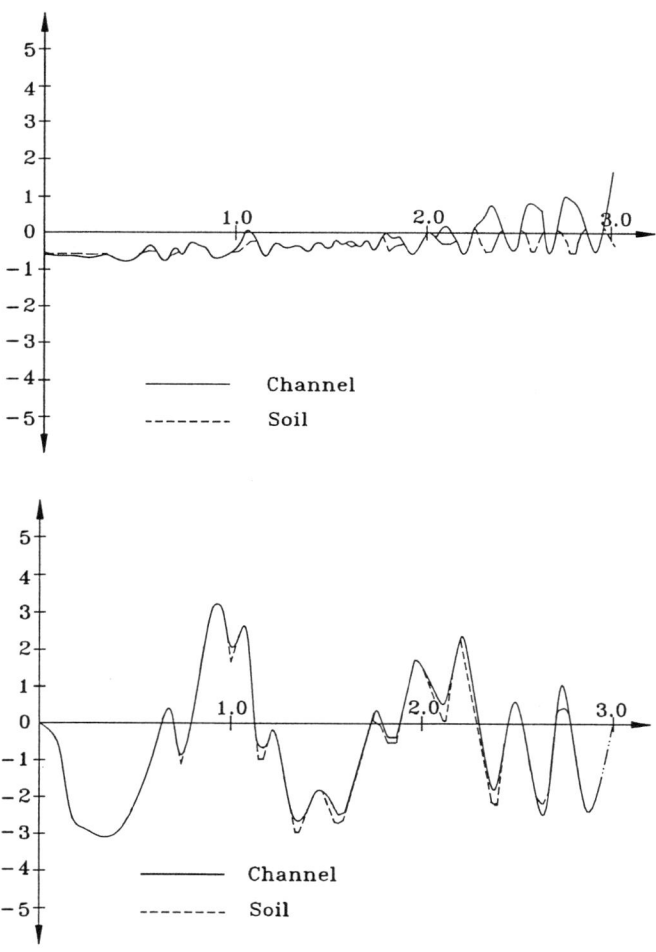

Fig. 13.1.2. The displacement evolution of points 1 and 19.

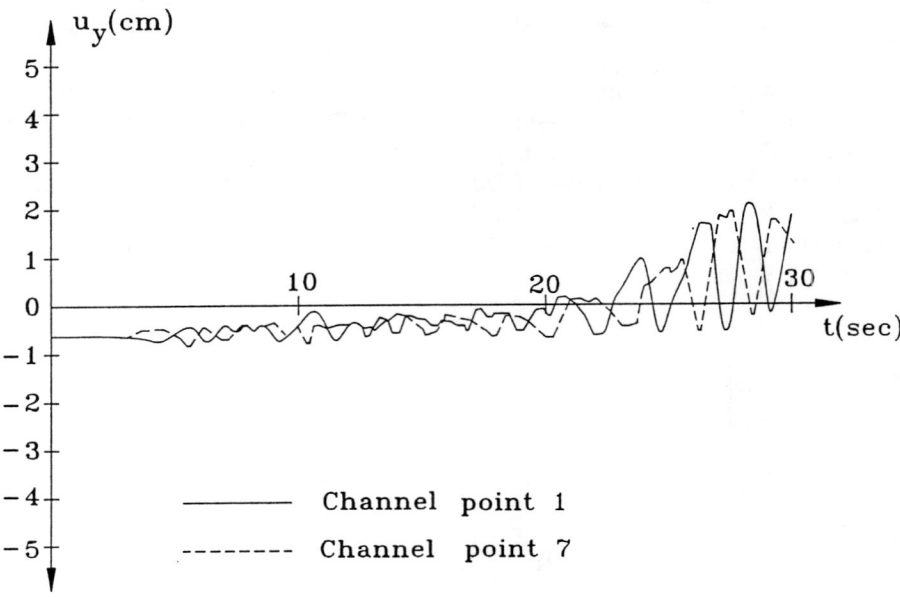

Fig. 13.1.3. The displacement evolution of points 1 and 7.

In this application we deal with the impact problem simply by following the evolution of the dynamic phenomenon and by making appropriate changes in the initial velocities, as already explained. For a more profound study of the impact phenomenon we refer to [Mon] and to the contribution of J.J. Moreau in [Mor88b,c]. In this context we have presented in [Panag88] a method with which one can determine the time moment of the impact by appropriately varying the time in a Hamilton-like variational expression. In this context we also refer to the theory of the "billiard problem" developed in [Kozl].

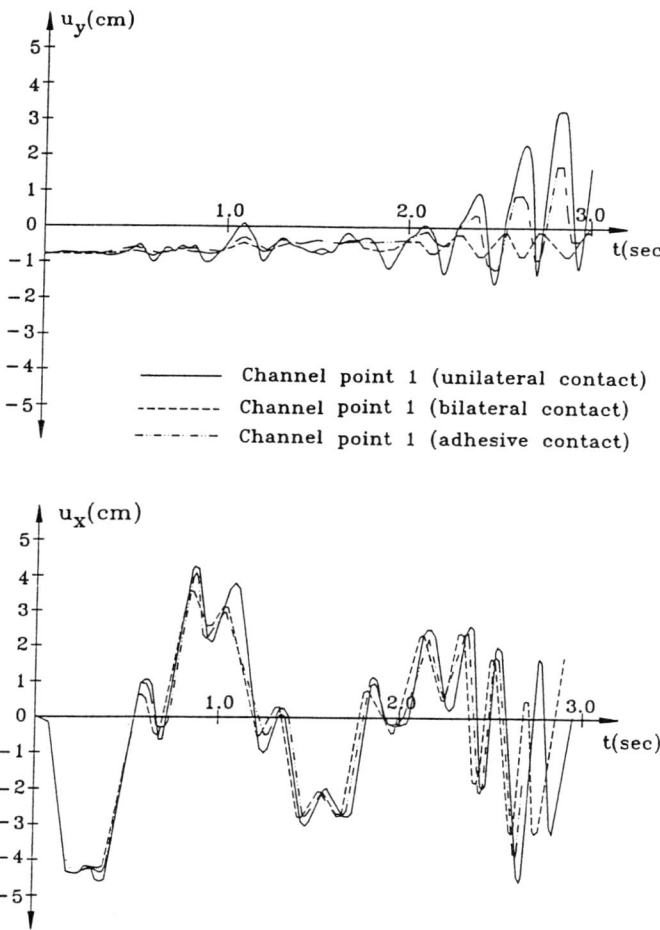

Fig. 13.1.4. Comparison of the results of the assumptions of unilateral, bilateral and adhesive contact.

13.2 Application XII: The Unilateral Contact and Nonmonotone Friction Problem in Cracks

Cracks in solids form interfaces where we have coupling of both the unilateral contact effect in the direction normal to the interface and the frictional effect in the tangential sense. This nonclassical and highly nonlinear behaviour of the cracks has not yet been extensively studied. Concerning the frictional contact in cracks, the works by Comninou and Dundurs [Com, Dun] are of importance; however they make an assumption concerning the contact zone, something not necessary in the framework of an inequality theory (cf. also [Dub89b] p.79). Note also in this context that, as shown in [Pan80;85], the trial and error methods, e.g. assumptions on contact zones and/or sliding zones, in large scale problems containing inequalities like the unilateral contact and the classical (monotone, i.e. Coulomb) friction problem, may lead to erroneous results. Unilateral contact problems for cracks including monotone friction have been studied by Dubourg et al [Dub88;89a]. In these papers the classical indirect B.I.E.M. (cf. [Ant92] Ch. 2) based on dislocations has been combined with the algorithm of Kalker for the treatment of the unilateral contact and the monotone friction problem [Ka88a,b;90].

In [The92] the method of [Pan83b] has been adapted to crack problems with unilateral contact and Coulomb friction at the interface. We have applied the indirect B.I.E.M. for the modelling of the crack, but for the numerical treatment we combine this method with the inequality contact algorithm of [Pan75], [Bis90]. In [The93] we have developed the direct B.I.E.M. for the same problem.

Here we deal with the nonmonotone friction problem. The friction diagram of Fig. 10.3.3a is taken into account in the numerical application. We should note that for the numerical calculations the method of Sect. 9.3, which approximates the nonmonotone friction law by monotone classical friction laws, is applied. The algorithm for the whole problem, i.e the treatment of both the unilateral contact and the nonmonotone friction is exactly the same as the algorithm of Sect. 9.5. The only difference in crack problems is that we prefer to eliminate the internal degrees of freedom and to deal with a hemivariational inequality of the type (5.2.19) or (5.2.43). Following then the method of Sect. 9.3 we approximate the solution of the hemivariational inequality by the solution of inequality constrained Q.P. problems solved by an optimization algorithm.

Let us note here that there also exist other types of algorithms which may treat the inequality constraints, as e.g. the classical and augmented Lagrangian method [Bat, Wri, For], the perturbed Lagrangian method [Sim, Ju, Kik] and the penalty method (cf. e.g. [Kik]). Most of the above methods cannot treat effectively the inequality constraints and the multivaluedness in the classical friction (Coulomb) law and this fact causes certain inaccuracies or additional numerical effort. The method proposed here is more appropriate for a crack, i.e., responds to the increased accuracy requirements concerning the determination of the stress intensity factors. In the case of cracks with unilateral contact and friction, the numerical implementation makes necessary the use of special singular elements for the consideration of the arising crack singularities.

We have treated numerically an orthogonal metallic plate sized 0.67 by 0.8015 m with a thichness of 0.1 m, a Poisson ratio equal to 0.3 and an elasticity modulus equal to $2.1 \times 10^7 t/m^2$. The plate contains a central crack as in Fig. 13.2.1. The crack is assumed to have an 1 m clearance. The plate was discretized in 325 plane stress elements and 33 special interface elements that simulate the unilateral contact and friction situation at the interface. As a friction law we have assumed the nonmontone multivalued law of Fig. 10.3.3a. It is a normalized friction law (multiply the ordinates by $\mu|S_N|$). The number of degrees of freedom taking part to the unilateral contact and nonmonotone friction phenomenon is 124. In front of the crack tip, six singularity elements (twelve node collapsed cubic isoparametric elements) take into account the singularity.

The method developed by [Bisb90] was used for the unilateral contact subproblem with Coulomb friction and the program was run on a Hewlett-Packard 9000/750 computer. We do not intend to present exact data about computation times but, we can say that the solution of this part of the problem took about 5.2 sec on the 750 and about 8.7 sec on a smaller model of the same series, namely the 720. These numbers correspond to the CPU-execution time of the algorithm after the stiffness matrix has been assembled. These times were obtained without having used profile optimization of the stiffness matrix. This procedure has been repeated $7 \div 8$ times maximum for each friction coefficient in order to get solution for the nonmonotone friction problem with an acceptable accuracy.

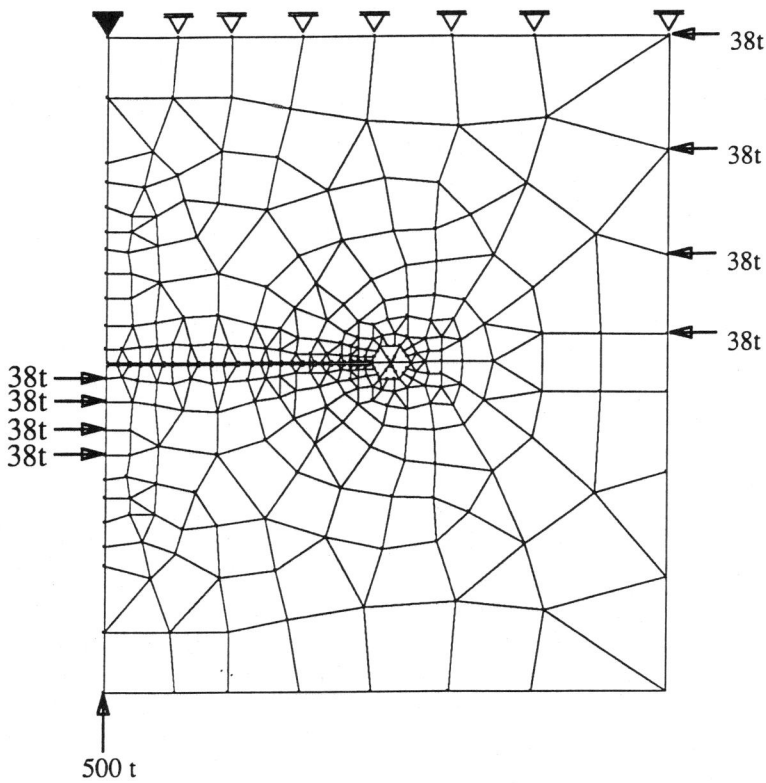

Fig. 13.2.1. A crack interface problem.

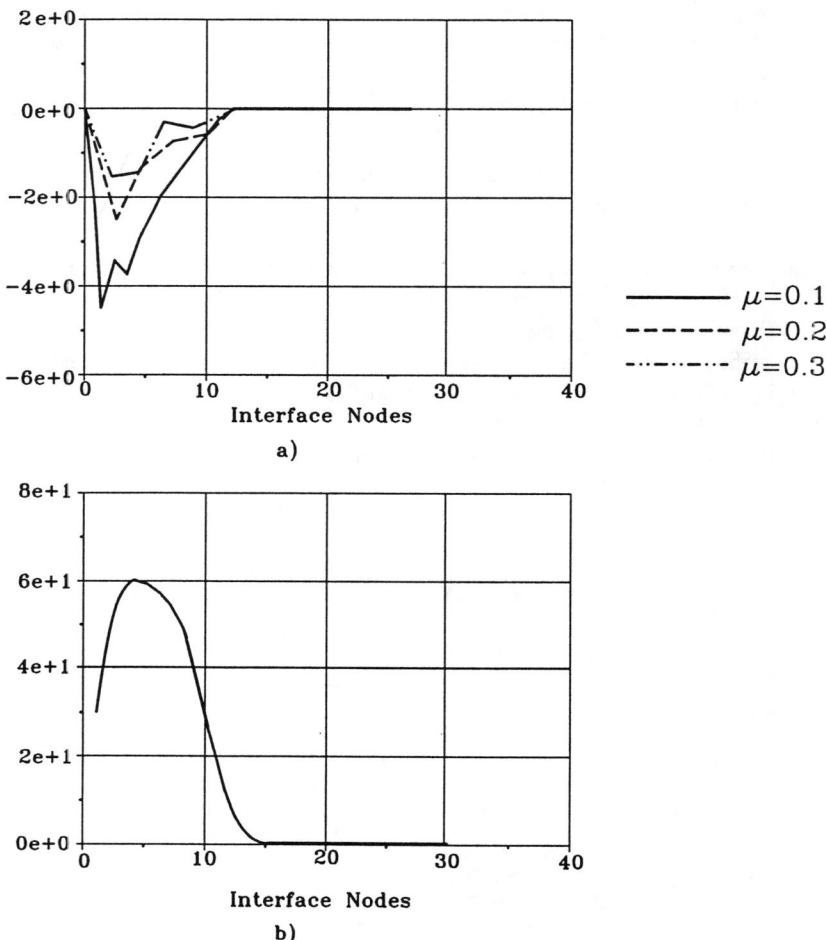

Fig. 13.2.2. Friction forces and normal force in t.

To disperse any impressions that seem to be overwhelming in the literature about unilateral and frictional behaviour of cracks being numerically tedious, we can cite here the fact that the machine time dedicated to the editing of the graphical output for the figures, was an order of magnitude more than the time needed for all the computations for one monotone subproblem, i.e for a unilateral contact problem with Coulomb friction. In fact, the example given here is considered rather small for the capabilities of the unilateral contact and (Coulomb) friction algorithm as those are only hardware limited. To give some rough idea of the respective requirements, this algorithm requires for n contact node pairs $4n^2 + 60n$ words (double presision). This fact leads for 4000 nodes (i.e. 2000 contact pairs) to a memory requirement of 130Mbytes.

This example actually is a numerical study of the effect the friction coefficient μ and the friction law can have on the stress intensity factors. These

13.2 Application XII: Nonmonotone Friction in Cracks

factors were computed by an energy method because other popular methods require the crack surface to be stress free. As one can see in Fig. 13.2.1a the friction forces rather rapidly converge to a stable pattern along the interface. We have given the corresponding curves for $\mu= 0.1, 0.2, 0.3$. The normal forces are not affected very much by the friction coefficient and what one sees in Fig. 13.2.2b is an overlay of three almost coinciding curves. In Fig. 13.2.3 the variation of K_I and K_II against μ is depicted and one sees a rather fast stabilisation of the stress intensity factors with increasing friction coefficient values.

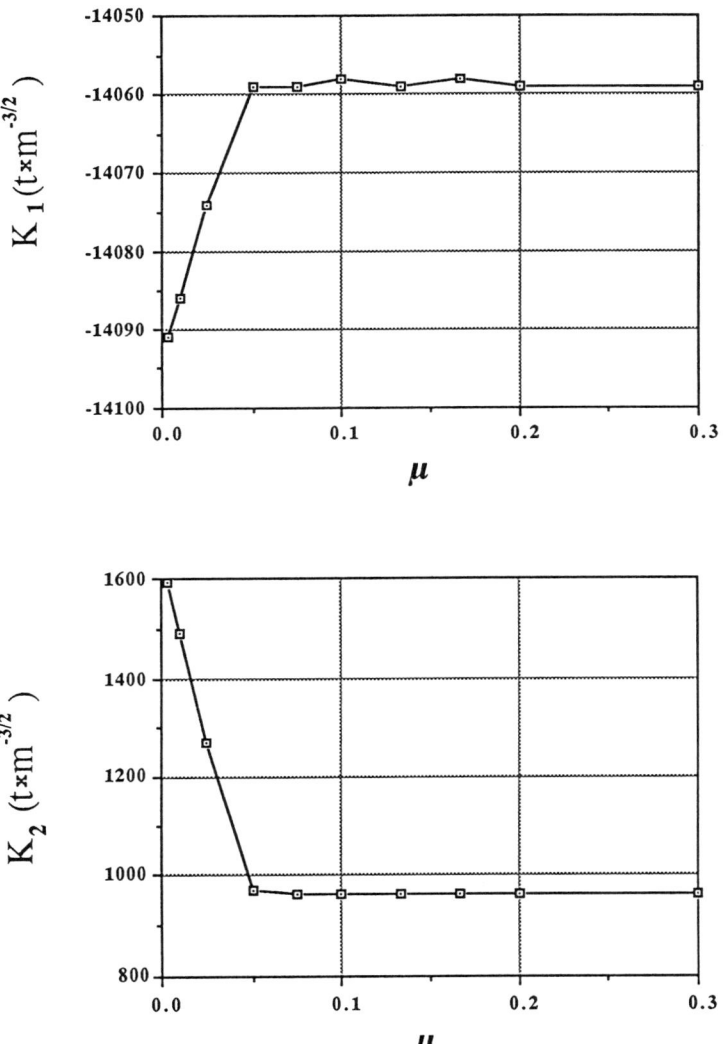

Fig. 13.2.3. Variation of K_I and K_II with μ.

13.3 Application XIII: Fracture of Cracks Repaired by an Adhesive Material

The theory of this interesting problem is given in Sect. 5.2, where the hemivariational inequalities along the crack (5.2.19) and (5.2.43) have been derived. In [The91] we have given some numerical results by the bundle method.

Due to the necessity of large calculations we have assumed that the action of the adhesive normal to the crack is independent of the action in the tangential sense. Here, using the algorithm of Sect. 10.3, we repeated the calculations for the same example, additionally taking into account the action of the adhesive as described by the diagrams of Fig. 10.3.3a in the normal direction and of Fig. 10.3.3b in the tangential direction. The crack singularity has been taken into account by means of the same singular elements as in Sect. 13.2. The coefficients K_I and K_{II} have been calculated by the formulas of [Bland] expressing these coefficients as functions of the net opening and relative sliding of the two sides of the crack. For a more accurate estimation of the stress intensity factors that takes into account the mutual influence of the kinked crack singularities we refer to [The86,87]. In Fig. 13.3.1a the cracked plate we have calculated, is depicted.

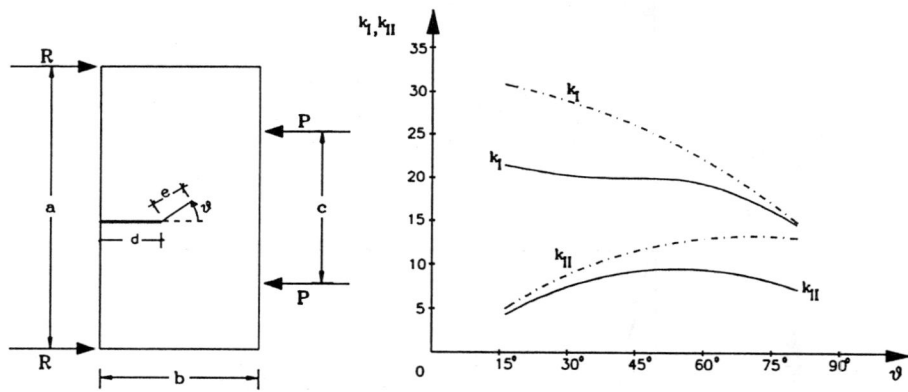

Fig. 13.3.1. The numerical application.

We assume that d=1 mm, b=2 mm, a=9.5 mm, c=6 mm, e=0.04 mm, the modulus of elasticity E=2.1 ×10^{4}kp/mm^2, Poisson's ratio ν =0.3 and R =P=5 kp. In Fig. 13.3.1b we give the variation of the stress intensity factors K_I and K_{II} (in kpmm$^{-3/2}$) with the angle θ. For the same example and without the adhesive material the stress intensity factors have been obtained by a numerical method in [Dat] and by an analytical method in [Cott]. In Fig. 13.3.1b their variation is shown by dotted lines.

14. Applications of the Theory of Hemivariational Inequalities in Robotics

This chapter deals with certain applications of the theory of hemivariational inequalities to the grasping problem of multifingered grippers of robots. We assume that the action of the support on each finger is described by a nonmonotone possibly multivalued relationship, or more generally by a nonconvex superpotential law. A special case of the general problem studied is the grasping of a robot gripper with adhesive action, or the grasping of a robot gripper on the assumption of nonmonotone frictional forces. The final aim is the solution of the corresponding optimal control problem. We base our method on the formulation of the classical grasping problem of a robot gripper as a nonsymmetric linear complementarity problem (L.C.P.). Under the term "classical" we mean that the grasping action is subjected to the classical Signorini condition (i.e nonpenetration inequality with the object assumed to be rigid) and to Coulomb frictional forces depending on the normal reaction between fingertip and object. The formulation of this problem as a L.C.P. was achieved in [Al-Fah91a,b,92a,b,93] where we have tried to introduce into the area of robotics certain methods of the Inequality Problems of Mechanics. Here we derive the corresponding hemivariational inequality by "superimposing" to the Signorini law and to the Coulomb friction law the necessary nonmonotonicities. The numerical solution proposed for both the arising hemivariatonal inequalities and the corresponding optimal control problem is achieved by a repeated solution of the L.C.P. according to a scheme analogous to the algorithm of Sect. 10.3.

14.1 Application XIV: Adhesive Grasping Problem in Robotics

In a robotic hand the effective study of the optimal control of the hand-object system is important for the tuning of the robot hand action. However, this problem is rather a difficult one since the criterion to be optimized, and which has a standard form, is subjected to the state relations describing the grasping effect. This effect does not give rise to classical state equations, but in the case of adhesive contact to state hemivariational inequalities. This is a nonclassical optimal control problem of the type studied in Ch. 8. Here we shall "superimpose" the nonmonotonicities of the adhesive grasping action to the Signorini contact condition and to Coulomb friction. Therefore first we shall study the problem of fingertips subjected to the two aforementioned interrelated contact and friction conditions. But even in this more simple case the state of the system is governed by variational inequalities thus defining a nonclassical optimal control problem. Indeed, if the fingertips are in unilateral frictional contact with the object, this problem is reduced to the solution of an appropriately defined variational inequality coupled with a fixed point search [Pan75,85], or to a quasivariational inequality [Tel], or to a game problem [Bis90,92]. However, in a discrete problem with the friction cone replaced by a pyramid, the unilateral contact problem with Coulomb friction gives rise [Kla88] to a nonsymmetric Linear Complementarity Problem(L.C.P.) which will constitute here the state relation for the basic optimal control problem. The iterative solution of the last problem will supply the solution of the optimal control problem for an adhesive gripper.

Firm grasping of an object by a multifingered gripper has been studied by many authors (see e.g. among others [Hana], [Mason], [Kerr], [Kuma], [Nguy], [Marken], [Al-Fah91a,b,92a,b,93], [Stav91a], [Panag93a]) in the area of robotic research. To give some more information we refer to Mason and Salisbury [Mason] in 1985, Kerr and Roth [Kerr] in 1986, where for rigid fingers only the equilibrium equations are satisfied constrained by the inequalities resulting from the unilateral contact conditions, to Kerr and Roth (1986), where this procedure is more systematized by considering the friction through the use of Linear Programming techniques (see also [Kuma]), and to Markenscoff et al [Marken] in 1990, where the special features of the object geometry have been taken into account with a clever geometrical technique. In [Al-Fah91b,92a,93], [Panag93a] the frictional unilateral contact problem has been treated for a first time using the methods of Inequality Mechanics, which fully exploit the nature of the problem.

The main purpose of the first two sections of this chapter is to study, for the robot gripper with adhesive contact, the equilibrium problem and the corresponding optimal control and identification problem. We consider here only the static problem. Indeed this problem is the basic subroutine in the solution of the corresponding dynamic problem for which, due to the nonclassical nature of the state relations, we are obliged to apply a time discretization procedure (cf. e.g. [Tabak]) and to solve within each time increment a static optimal control problem. Of course, the impact phenomena and the corresponding total or partial velocity reversing have to be considered separately during this incremental procedure.

Coming now to the nature of the grasping problem we distinguish two types of fingers, the hard (Fig. 14.1.1b) and the soft (Fig. 14.1.1c). The Kerr-Roth model is only a rough approximation to the soft finger contact for which one has to consider the finite contact area and the sliding force, normal force and torsional moment interaction (see e.g. [Goyal89,91], [Howe88]). In the hard finger the contact area reduces to a point and we have only unilateral contact with friction.

We would like to recall here two definitions from the robot grasping theory: the force and form closure. Force closure arises when the contact between the object and the fingers is maintained by the action of a given set of forces applied externally on the object. When the contact is maintained for any set of external forces by the reactions of the fingers, then we have form closure. Finally we point out that the developed theory holds not only for the problem treated here, but for all kinds of structural analysis problems which give rise to a nonsymmetric, positive semidefinite Linear Complementarity Problem (L.C.P.) as it is the case in the theory of plasticity (cf. [Ma71]), when the rigid body degrees of freedom are not excluded. This last result concerning the plasticity is crucial for the extension of the theory developed here to elastoplastic fingers with or without hardening.

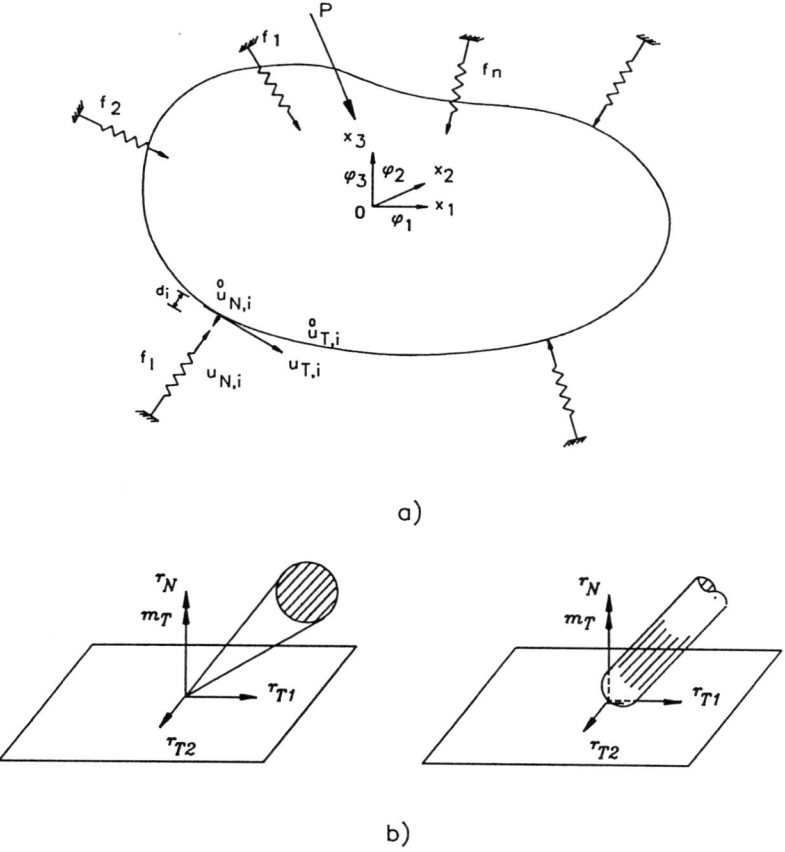

Fig. 14.1.1. The grasping problem a) rigid object and an elastic frictional multifingered gripper b) hard finger contact c) soft finger contact.

In Fig. 14.1.1a we show a rigid object which is grasped by a multifingered gripper with n frictional elastic fingers. The system is referred to an orthogonal cartesian coordinate system $Ox_1x_2x_3$. The two cases of "hard finger" contact (Fig. 14.1.1b), and "soft finger" contact (Fig. 14.1.1c) according to [Kerr] are considered. The hard-finger contact is simply a unilateral point contact with Coulomb friction. The soft finger contact introduces in addition constraints on the object for the rotations about the normal axis to the boundary which passes through the point of contact, if the torque remains smaller than a certain number which is equal to the normal force multiplied by a rotational coefficient of friction. This coefficient of friction depends actually on the tangential contact force, a fact ignored by the Kerr-Roth theory. Thus the relations which govern the object-gripper system are the following:

i) The global equilibrium equations

$$G_N r_N + G_T r_T = P. \tag{14.1.1}$$

Here G_N is $m \times n$ equilibrium matrix which corresponds to the normal contact reactions, G_T is $m \times 2n$ (resp. $m \times 3n$ for soft finger case) equilibrium matrix which corresponds to the friction forces (resp. plus the frictional torques in the soft finger case), $r_N = \{r_{N_1}, r_{N_2}, \ldots, r_{N_n}\}^T$ are the normal forces, or reactions, which exert the fingertips on the object, and r_T are the friction forces which are tangential to the boundary. For hard finger contact $r_T = \{r_{T_{11}}, r_{T_{12}}, r_{T_{21}}, r_{T_{22}}, \ldots, r_{T_{n1}}, r_{T_{n2}}\}^T$, whereas for soft finger contact $r_T = \{r_{T_{11}}, r_{T_{12}}, m_{T_1}, r_{T_{21}}, r_{T_{22}}, m_{T_2}, \ldots, r_{T_{n1}}, r_{T_{n2}}, m_{T_n}\}^T$ Here m_T is the torque about the axis normal to the object surface, exerted by the finger onto the body, if one assumes soft finger contact.

ii) The compatibility conditions: With respect to the origin of the cartesian orthogonal coordinate system the rigid body displacement and rotations which form the vector u^0, are measured. For threedimensional problems u^0 reads:

$$\overset{0}{u} = \{\overset{0}{u}_1, \overset{0}{u}_2, \overset{0}{u}_3, \overset{0}{\phi}_1, \overset{0}{\phi}_2, \overset{0}{\phi}_3\}^T. \tag{14.1.2}$$

Here $u^0{}_1, u^0{}_2, u^0{}_3$ are the rigid body displacements along the directions Ox_1, Ox_2, Ox_3 respectively, and $\phi^0{}_1, \phi^0{}_2, \phi^0{}_3$ denote the rigid body rotations about the axes of the coordinate system. The superscript T denotes as usual the transpose of a vector or a matrix. By means of the principle of complementary virtual work one can find that:

$$\overset{0}{u}_N = -G_N^T \overset{0}{u}, \tag{14.1.3}$$

$$\overset{0}{u}_T = -G_T^T \overset{0}{u}. \tag{14.1.4}$$

Here $u^0{}_N$ and $u^0{}_T$ denote respectively the normal and tangential displacements of the points on the object boundary which are adjacent to the fingertips.

iii) The boundary conditions: In the normal direction to the surface of the body the compatibility conditions are written in the form :

14.1 Application XIV: Adhesive Grasping Problem in Robotics

$$\boldsymbol{u}_N + \boldsymbol{d}_N = \overset{0}{\boldsymbol{u}}_N. \qquad (14.1.5)$$

Here $\boldsymbol{u}_N$ is the vector of the normal displacements of the fingertips, and $\boldsymbol{d}_N$ is the vector of the initial openings between the fingertips and the object. The unilateral behaviour of the system can be expressed by the following unilateral contact conditions:

$$\text{if } \boldsymbol{u}_N + \boldsymbol{d}_N = \overset{0}{\boldsymbol{u}}_N, \text{ then } \boldsymbol{r}_N \geq \boldsymbol{0}, \qquad (14.1.6)$$

$$\text{if } \boldsymbol{u}_N + \boldsymbol{d}_N > \overset{0}{\boldsymbol{u}}_N, \text{ then } \boldsymbol{r}_N = \boldsymbol{0}, \qquad (14.1.7)$$

i.e. if contact occurs, then the reaction is nonnegative and if no contact occurs, then the reaction is zero.

The above relations can be written in the following linear complementarity form

$$\boldsymbol{u}_N - \overset{0}{\boldsymbol{u}}_N + \boldsymbol{d}_N = \boldsymbol{y}_N, \qquad (14.1.8)$$

$$\boldsymbol{y}_N \geq \boldsymbol{0}, \; \boldsymbol{r}_N \geq \boldsymbol{0}, \qquad (14.1.9)$$

$$\boldsymbol{y}_N^T \boldsymbol{r}_N = 0. \qquad (14.1.10)$$

Now the friction boundary conditions will be defined. By means of Coulomb's law of dry friction the tangential (frictional) forces, which exert the fingertips on the object must satisfy the following relation

$$g_i = \mu |r_{N_i}| - |r_{T_i}| \quad i = 1, \ldots, n \qquad (14.1.11)$$

$$g_i \geq 0 \qquad (14.1.12)$$

and in addition for soft finger contact the conditions

$$g_{i_T} = \mu_t |r_{N_i}| - |m_{T_i}| \quad i = 1, \ldots, n \qquad (14.1.13)$$

$$g_{i_T} \geq 0 \qquad (14.1.14)$$

where $|\cdot|$ denotes the norm in $\mathbb{R}^3$, and μ (resp. μ_t is the friction coefficient (resp. the rotational friction coeffecient) of the finger. If $|r_{T_i}| < \mu |r_{N_i}|$ (i.e. $g_i > 0$) the slip, i.e. the displacement y_{T_i} at i tangential to the boundary, must be equal to zero, and if $|r_{T_i}| = \mu |r_{N_i}|$ (i.e. $g_i = 0$), then $y_{T_i} \neq 0$, i.e. slip in the opposite direction of r_{T_i} appears. In addition, for the case of soft finger contact, if $|m_{T_i}| < \mu_t |r_{N_i}|$ (i.e. $g_{i_T} > 0$) the rotational slip $\tilde{y}_{T_i}$ must equal zero; if $|m_{T_i}| = \mu_t |r_{N_i}|$ (i.e. $g_{i_T} = 0$), then $\tilde{y}_{T_i} \neq 0$ and there appears rotational slip in the opposite direction of m_{T_i}. The friction law is further piecewise linearized by a polyhedral approximation of the friction cone from the interior (Fig. 14.1.2). Then the relations (14.1.11)÷(14.1.14) can be written as follows:

$$\boldsymbol{g} = \boldsymbol{T}_N^T \boldsymbol{r}_N + \boldsymbol{T}_T^T \boldsymbol{r}_T. \qquad (14.1.15)$$

The matrices $\boldsymbol{T}_T$ and $\boldsymbol{T}_N$ of the piecewise linearized friction law have the following form

$$\boldsymbol{T}_T = \text{diag}\,[\boldsymbol{T}_T^1, \boldsymbol{T}_T^2, \ldots, \boldsymbol{T}_T^n], \qquad (14.1.16)$$

14. Applications of the Theory of Hemivariational Inequalities in Robotics

$$T_N = \text{diag}\,[T_N^1, T_N^2, \ldots, T_N^n]. \tag{14.1.17}$$

For hard fingers it is

$$T_T^i = \begin{bmatrix} \cos a_1 & \ldots\ldots & \cos a_l \\ \sin a_1 & \ldots\ldots & \sin a_l \end{bmatrix}, \quad a_j = (j-1)\frac{2\pi}{l}$$

$$T_N^i = [\mu_1 \ldots\ldots \mu_l], \quad \mu_j = \mu\cos(\frac{\pi}{l})$$

and for soft fingers it is

$$T_T^i = \begin{bmatrix} \cos a_1 & \ldots\ldots & \cos a_l & 0 & 0 \\ \sin a_1 & \ldots\ldots & \sin a_l & 0 & 0 \\ 0 & \ldots\ldots & 0 & 1 & -1 \end{bmatrix}, \quad a_j = (j-1)\frac{2\pi}{l}$$

$$T_N^i = [\mu_1 \ldots\ldots \mu_l\ \mu_t\ \mu_t], \quad \mu_j = \mu\cos(\frac{\pi}{l}).$$

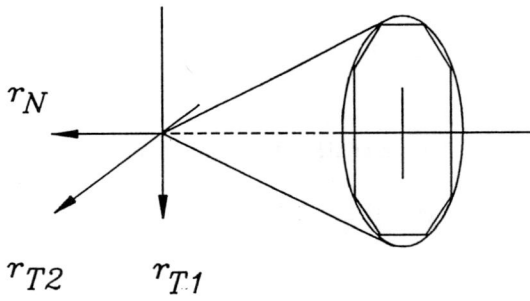

Fig. 14.1.2. Polyhedral internal approximation of the friction Cone.

Here l is the number of the surfaces of the linearised friction cone, μ is the coefficient of friction, and μ_t is the torsional friction coefficient. Now the slip can be written in the following form by considering the appropriate projections (y_T contains both the slip and the rotational slip in the case of soft finger):

$$y_T = T_T \lambda, \tag{14.1.18}$$

$$\lambda \geq 0, \tag{14.1.19}$$

where λ is a nonnegative slip parameter associated with g. Then g and λ fulfil the following orthogonality condition:

$$g^T \lambda = 0. \tag{14.1.20}$$

14.1 Application XIV: Adhesive Grasping Problem in Robotics

By means of the tangential kinematical compatibility the slip value y_T, the tangential displacements of the fingertips u_T, and u_T^0 are related by the kinematical condition

$$T_T \lambda - u_T + \overset{0}{u}_T = d_T \qquad (14.1.21)$$

where d_T denotes the initial tangential distance between the fingertip and the target point on the object boundary. In the case of soft finger the vectors in (14.1.21) contain besides the displacement terms, the rotation terms.

A linear elastic behaviour of the fingers implies that one can introduce a reaction-displacement law of the form

$$u = F r, \qquad (14.1.22)$$

where

$$u = \begin{bmatrix} u_N \\ u_T \end{bmatrix}, \quad F = \begin{bmatrix} F_{NN} & F_{NT} \\ F_{TN} & F_{TT} \end{bmatrix}, \quad r = \begin{bmatrix} r_N \\ r_T \end{bmatrix}, \qquad (14.1.23)$$

F is the symmetric flexibility matrix, F_{NN} is an $n \times n$ nonsingular matrix and F_{TT} is a $2n \times 2n$ (resp. $3n \times 3n$ for soft-finger contact) nonsingular matrix. By means of the relations (14.1.3),(14.1.4),and (14.1.22) the relations (14.1.8) and (14.1.21) become

$$y_N - F_{NN} r_N - F_{NT} r_T - G_N^T \overset{0}{u} = d_N \qquad (14.1.24)$$

$$T_T \lambda - F_{TN} r_N - F_{TT} r_T - G_T^T \overset{0}{u} = d_T. \qquad (14.1.25)$$

The equations (14.1.1), (14.1.9), (14.1.10), (14.1.15), (14.1.18), (14.1.19), (14.1.20), (14.1.24), (14.1.25) constitute a nonstandard Linear Complementarity Problem introduced in 1989 by Kostreva [Kostr] which can be written in the following matrix form

$$\begin{bmatrix} y_N \\ T_T \lambda \\ 0 \\ g \end{bmatrix} - \begin{bmatrix} F_{NN} & F_{NT} & G_N^T \\ F_{TN} & F_{TT} & G_T^T \\ G_N & G_T & 0 \\ T_N^T & T_T^T & 0 \end{bmatrix} \begin{bmatrix} r_N \\ r_T \\ \overset{0}{u} \end{bmatrix} = \begin{bmatrix} d_N \\ d_T \\ -P \\ 0 \end{bmatrix}. \qquad (14.1.26)$$

To obtain a standard L.C.P. we must eliminate the free variables r_T and u^0 from the equations (14.1.26). The equations in (14.1.26) corresponding to the free variables are the following:

$$\begin{bmatrix} T_T \lambda \\ 0 \end{bmatrix} - \begin{bmatrix} F_{TN} \\ G_N \end{bmatrix} r_N - \begin{bmatrix} F_{TT} & G_T^T \\ G_T & 0 \end{bmatrix} \begin{bmatrix} r_T \\ \overset{0}{u} \end{bmatrix} = \begin{bmatrix} d_T \\ -P \end{bmatrix}. \qquad (14.1.27)$$

Let us set

$$C = \begin{bmatrix} F_{TT} & G_T^T \\ G_T & 0 \end{bmatrix}. \qquad (14.1.28)$$

Matrix C is nonsingular if G_T has the full column rank. Then the inverse exists and is expressed as (cf. e.g. [Flet])

where

$$C^{-1} = \begin{bmatrix} B & A^T \\ A & U \end{bmatrix}, \qquad (14.1.29)$$

where

$$B = F_{TT}^{-1} - F_{TT}^{-1} G_T^T (G_T F_{TT}^{-1} G_T^T)^{-1} G_T F_{TT}^{-1} \qquad (14.1.30)$$

$$A^T = F_{TT}^{-1} G_T^T (G_T F_{TT}^{-1} G_T^T)^{-1} \qquad (14.1.31)$$

$$U = -(G_T F_{TT}^{-1} G_T^T)^{-1}. \qquad (14.1.32)$$

The above expression of C^{-1} makes sense if F_{TT}^{-1} exists. The matrix $G_T F_{TT}^{-1} G_T^T$ is positive definite if G_T^T has linearly independent columns. From (14.1.27), (14.1.29) and (14.1.32) we can express the free variables r_T and $\overset{0}{u}$ as follows:

$$r_T = -(BF_{TN} + A^T G_N)r_N + BT_T\lambda - Bd_T + A^T P \qquad (14.1.33)$$

$$\overset{0}{u} = -(AF_{TN} + UG_N)r_N + AT_T\lambda - Ad_T + UP. \qquad (14.1.34)$$

Thus (14.1.33), (14.1.34), and (14.1.26) yield the final standard L.C.P. which can be written in compact form as

$$\left.\begin{array}{c} w - Mz = b \\ w \geq 0, z \geq 0, w^T z = 0 \end{array}\right\}. \qquad (14.1.35)$$

Here

$$w = \begin{bmatrix} y_N \\ g \end{bmatrix}, z = \begin{bmatrix} r_N \\ \lambda \end{bmatrix},$$

$$b = \begin{bmatrix} d_N - F_{NT}(Bd_T - A^T P) - G_N^T(Ad_T - UP) \\ -T_T^T(Bd_T - A^T P) \end{bmatrix}$$

and $M =$

$$\begin{bmatrix} F_{NN} - F_{NT}(BF_{TN} + A^T G_N) - G_N^T(AF_{TN} + UG_N) & (F_{NT}B + G_N^T A)T_T \\ T_N^T - T_T^T(BF_{TN} + A^T G_N) & T_T^T B T_T \end{bmatrix}$$

where M is a nonsymmetric, nonnegative definite matrix. We give further a proposition concerning the existence of the solution of this L.C.P.

It is well known (cf. e.g. [Pan85]) that the L.C.P. is equivalent to a discrete variational inequality involving a nonsymmetric, nonnegative matrix M. The proof concerning the existence of the solution of this variational inequality is a special case of the proof given by G. Fichera [Fich72] for continuous semicoercive variational inequalities in Hilbert spaces. The following notation is introduced: Let $Q : \mathbb{R}^p \to \ker M$ ($\ker M = \{x \mid Mx = 0\}$) be the orthogonal projector of $\mathbb{R}^p$ onto $\ker M$. Let $P = I - Q$. We have that $\ker M \subset \ker(M + M^T)$. Let us denote by $K(M)$ the orthogonal complement of $\ker M$ with respect to $\ker(M + M^T)$, i.e. $\ker(M + M^T) = \ker M \oplus K(M)$. The following proposition holds where $V = \{z \in \mathbb{R}^p, z \geq 0\}$.

Proposition 14.1.1 The variational inequality

14.1 Application XIV: Adhesive Grasping Problem in Robotics

$$z \in V \quad (Mz + b)^T(z^* - z) \geq 0 \quad \forall z^* \in V \tag{14.1.36}$$

has a solution if the following conditions are satisfied:

i) $b^T \rho > 0$ for every $\rho \in \ker M \cap V$; \hfill (14.1.37)

ii) Let $Q_0 : \mathbb{R}^p \to K(M)$, where Q_0 is the orthogonal projector of $\mathbb{R}^p$ onto $K(M)$. For every ρ satisfying the condition $Q_0\rho \neq 0, \rho \in \ker(M + M^T) \cap V = \{(M + M^T)\rho = 0, \rho \geq 0\}$, there exists $z_\rho \in V$ such that

$$-b^T\rho + z_\rho^T M \rho < 0. \tag{14.1.38}$$

For the proof we refer to [Fich72], [Pan85].

This proposition implies the following proposition for the grasping problem with frictional unilateral contact.

Proposition 14.1.2 Suppose that for a given b (resp. for every b) (14.1.36) and (14.1.37) hold. Then we have force (resp. form) closure.

Proposition 14.1.3. If for a given b there exists some ρ which satisfy the conditions, $Q_0\rho \neq 0$, and for $\rho \in \ker(M + M^T) \cap V$, we have that

$$-b^T\rho + z^T M \rho > 0 \quad \forall z \geq 0, \tag{14.1.39}$$

then we cannot have either force or form closure.

The developed theory can be extended to the case in which the Kerr-Roth assumption is replaced by a more accurate "soft" finger relation (see e.g. [Goya89,91]). In order to achieve the greatest possible generality we use the theory of nonconvex superpotentials and we consider superpotential laws in the form

$$\{y_{T_i}, \tilde{y}_{T_i}\} \in \bar{\partial}\Phi_{T_i}(r_{T_i}, m_{T_i}; r_{N_i}). \tag{14.1.40}$$

Here $\bar{\partial}$ is the generalized gradient with respect to $\{r_{T_i}, m_{T_i}\}$ and Φ_i is an appropriate nonconvex superpotential. The law (14.1.40) contains as a special case the case of adhesive rotational contact, as well as the case of nonmonotone (zigzag) rotational and tangential friction law. If K is a nonconvex (resp. convex) set in the $\{r_{T_i}, m_{T_i}\}$ vector space and $\Phi_i = \{0$ if $\{r_{T_i}, m_{T_i}\} \in K, \infty$ otherwise $\}$, then we have the case of nonconvex (resp. convex) limit surface for the friction of the soft finger which generalizes (resp. is identical with) the theory of convex limit surface of [Goya89,91] (cf. also (2.4.15) and (2.4.17). The law (14.1.40) can be combined with the unilateral contact law (14.1.9) (14.1.10) in the normal direction or more generally with a normal compliance law of the general form

$$y_{N_i} \in \bar{\partial}\Phi_{N_i}(r_{N_i}), \tag{14.1.41}$$

where the generalized gradient is with respect to r_{N_i}. In order to treat the coupled interface conditions (14.1.40) and (14.1.41) we proceed as in Sect. 10.3. We keep r_{N_i}, $i = 1, \ldots, n$, as fixed and we solve the tangential nonmonotone problem (e.g. nonmonotone friction or adhesive contact forces) by approximating

386   14. Applications of the Theory of Hemivariational Inequalities in Robotics

the nonmonotone law by monotone ones of the friction type (as in Fig. 9.4.2). All the resulting subproblems are L.C.Ps. of the type (14.1.35), because all the subproblems are unilateral contact problems with Coulomb friction (subjected to different friction coefficients). In the next step we keep r_{T_i}, m_{T_i} as fixed and we solve the nonmonotone problem in the normal direction by solving again a number of L.C.Ps. of the type (14.1.35). This procedure is continued until convergence is achieved. In the application given further we have assumed that in the tangential direction the law of Fig. 10.3.3a holds whereas in the normal direction the law of Fig. 10.3.3b. The algorithm defined in Sect. 10.3 is again applied. Now in each substep a L.C.P. of the type (14.1.35) is solved (instead of the Q.P.P. of Sect. 10.3).

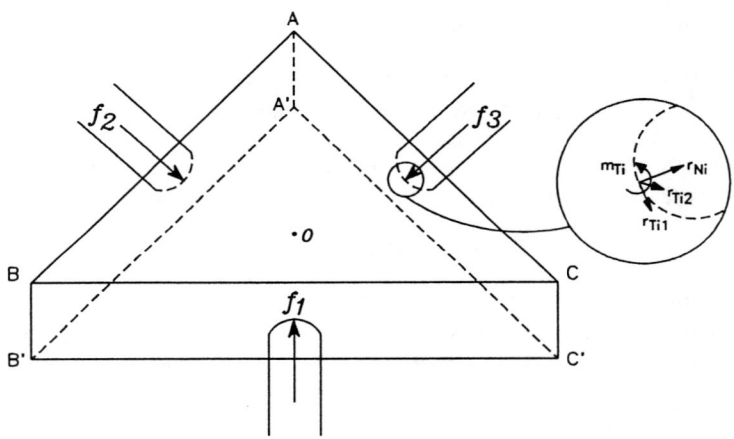

FINGERS	ADHESIVE CONTACT REACTIONS	TANGENTIAL COMPONENTS	FRICTION FORCES	RIGID BODY MOTION VALUES	
2	6.9540E+00 5.8250E+00	1	−2.7140E−01 −2.4350E−01	$u_1=$	1.0430E−05 0.8530E−05
		2	1.9270E+00 1.6350E+00	$u_2=$	8.1210E−06 6.2230E−06
		TORSION	−2.9910E+00 −2.4010E+00	$u_3=$	7.7470E−06 4.3570E−06
3	5.4140E+00 4.3230E+00	1	−1.5000E+00 −1.3500E+00	$\varphi_1=$	−1.7950E−05 −2.1000E−05
		2	1.1140E+00 1.3800E+00	$\varphi_2=$	1.8410E−05 1.6420E−05
		TORSION	−1.8410E+00 −1.6220E+00	$\varphi_3=$	−4.5920E−06 −3.4550E−06

Fig. 14.1.3. An adhesive grasping problem. Geometry and forces.

For the numerical treatments of the arised L.C.Ps. the Linear Complementarity Algorithm of Lemke is applied. For the description of this algorithm we refer to [Murty] and [Lemk]. In what follows we have compatible system of units. Let us consider the grasping of a rigid object by multifingered gripper (Fig. 14.1.3) having soft fingers. The rigid body is an orthogonal triangle with AB=BC=1 and AA' =BB'=CC'=0.2. The fingers act vertically at the centers of the lateral faces of the body and are assumed to have an adhesive contact in the normal direction and a nonmonotone frictional contact in the tangential directions. This nonmonotone frictional behaviour is the result of the action of Coulomb friction together with the action of adhesive tangential forces. Concerning finally the torque action of the fingers we adopt the simple torsional friction law (14.1.13) etc.

The soft finger contact is realized in each finger by three linear springs, one in the normal direction to the object boundary, two in the tangential direction and one angular spring with axis the normal to the object surface through the point of contact. The flexibility of the springs are, for the normal $F_{NNi} = 1.5 \times 10^{-6}$, the tangential $F_{TTi_1} = F_{TTi_2} = 6.0 \times 10^{-6}$ and for the angular one $F_{TTm_i} = 1.0 \times 10^{-5}$. We assume that the finger f_1 exerts on the object the forces $r_{N_1} = 10.0, r_{T_{11}} = r_{T_{12}} = 1.95$, a torsion torque $m_{T_1} = 4.5$; also an external force (e.g. the object weight) $P = 5.0$ is applied in the direction Ox_3. In the tangential and in the normal directions let us suppose that the action of the fingers on the object obeys the diagrams of Fig. 10.3.3a,b respectively. The tangential force diagram is normalized, i.e its ordinates are multiplied by $\mu|S_N|$, where $\mu=0.3$. The torsional friction coefficient is $\mu_t = 0.5$. In Fig. 14.1.3b we give the values of the reactions. The upper numbers in each entry correspond to simple Coulomb friction coupled with the Signorini nonpenetration condition (i.e the law 10.3.4b) in the normal direction (see e.g. [Panag93a]).

14.2 Application XV: On the Optimal Control of the Adhesive Grasping Problem in Robotics

The optimal control of the adhesive grasping problem will be reduced to a sequence of optimal control problem for the classical grasping problem, i.e. for the problem resulting if each robot hand finger is subjected to unilateral contact and Coulomb friction conditions for hard finger contact, and additionally to the torque condition (14.1.13) in the case of soft finger contact. This problem was solved in the previous section, where we showed that its solution results from the L.C.P. (14.1.35).

The optimal control theory applied to the grasping problem permits a more profound investigation of the behaviour of a gripper by the minimization of an appropriately defined performance index. The optimal control theory is closely related to the parameter identification theory, where a discrepancy between the observed and the predicted values of some variables is minimized; then some parameters of the problem are determined (the inverse problem). In the optimal control problem the performance index has a more general meaning; it may, for example express the weight, safety, or the deviation from a given state of the structure.

The static optimal control problem of the gripper is formulated by introducing a "control vector" $\chi \in \mathbb{R}^q$ and an admissible set for the controls $X_{ad} \subset \mathbb{R}^q$, which is assumed to be convex and closed. X_{ad} is defined by the "technological constraints" imposed on the control vector. The influence of the control vector on the structure is expressed linearly by a matrix $B : \mathbb{R}^q \to \mathbb{R}^q$.
To every control χ we associate a state function $z(\chi)$ through the relation

$$w - Mz = b + B\chi \quad \forall \chi \in X_{ad} \tag{14.2.1}$$

$$w \geq 0 \tag{14.2.2}$$

$$z \geq 0 \tag{14.2.3}$$

$$w^T z = 0, \tag{14.2.4}$$

or equivalently through the variational inequality

$$z \geq 0, \quad (Mz + b + B\chi)^T(z^* - z) \geq 0 \quad \forall z^* \geq 0, \chi \in X_{ad}. \tag{14.2.5}$$

We consider a "performance index" R of the form:

$$R(\chi) = \|Sz(\chi) - h\|^2 + \chi^T N \chi \tag{14.2.6}$$

where $h \in \mathbb{R}$ is the "observation vector", $S: \mathbb{R}^p \to \mathbb{R}^r$ is a matrix which relates the variable z with the observed quantity h, and N is a positive definite matrix. This form of $R(\chi)$ results from certain economy, safety, serviceability and identification criteria (cf. [Lio71]). For $N = 0$ we have a simple identification problem. The optimal control problem has now the form:
Find $J \in X_{ad}$ such that

$$R(J) = \inf\{R(\chi) | \chi \in X_{ad}\} \tag{14.2.7}$$

and is called "load optimal control problem". Since we will reduce our problem to a number of problems of the type just defined, i.e. to an optimal control problem governed by a L.C.P., we shall give now a proposition concerning the existence of solution of this last problem.

Proposition 14.2.1 Suppose that

i) $(b + B\chi)^T \rho > 0$ for every $\rho \in \ker M \cap V$ and $\chi \in X_{ad}$.

ii) Let $Q_0 : \mathbb{R}^p \to \ker(M + M^T) \ominus \ker M$. Suppose that for every ρ such that $Q_0 \rho \neq 0$ with $\rho \in \ker(M + M^T) \cap V$ there exists $z_\rho \geq 0$ such that

$$-(b + B\chi)^T \rho + z_\rho^T M \rho < 0 \quad \forall \chi \in X_{ad}. \tag{14.2.8}$$

Then the optimal control problem (14.2.1)÷(14.2.6) has at least one solution.

14.2 Application XV: Optimal Control in Robotics

Proof. Let us define a minimizing sequence χ_ν such that

$$R(\chi_\nu) \to R(J) = \inf\{R(\chi)|\chi \in X_{ad}\}. \tag{14.2.9}$$

Obviously χ_ν is bounded due to the positive definiteness of N and the relation (14.2.9). From the state relation

$$z_\nu \geq 0, \quad (M z_\nu + b + B \chi_\nu)^T(z^* - z_\nu) \geq 0 \quad \forall z^* \geq 0 \tag{14.2.10}$$

we obtain for each χ_ν at least one z_ν due to the assumptions (i) and (ii).(cf. Prop. 14.1.1). The solution z_ν will be approximated in the following way: We define $M_\alpha = M + \frac{I}{\alpha}, \alpha > 0$, and let $z_{\alpha\nu}$ be the unique solution of the inequality

$$(M_\alpha z_{\alpha\nu} + b + B \chi_\nu)^T(z^* - z_{\alpha\nu}) \geq 0 \quad \forall z^* \geq 0 \tag{14.2.11}$$

We know from the proof of the Prop. 14.1.1 (cf. [Fich72]) that $z_{\alpha\nu}$ is bounded independently of α and ν (since $||\chi_\nu|| < c$) and that as $\alpha \to \infty$, $z_{\alpha\nu} \to z_\nu$, which is a solution of (14.2.10). We note first that by the Bolzano-Weierstrass theorem and due to the boundedness of $\{\chi_\nu\}$ and $\{z_\nu\}$ we can define subsequences again denoted by $\{\chi_\nu\}$ and $\{z_\nu\}$ such that $\chi_\nu \to \chi$ and $z_\nu \to z$. Now we take in (14.2.11) the limit as $\nu \to \infty$. We have first that $z \in V$ and $\chi \in X_{ad}$ (V and X_{ad} are closed) and from the relation

$$z_\nu^T M z_\nu \leq (b + B \chi_\nu)^T(z^* - z_\nu) + z_\nu^T M z^* \tag{14.2.12}$$

we obtain that

$$z^T M z = \lim_{\nu \to \infty} z_\nu^T M z_\nu \leq (b + B \chi)^T(z^* - z) + z^T M z^* \tag{14.2.13}$$

i.e. that z and χ satisfy the initial variational inequality which is equivalent to the initial L.C.P. Now we have to verify that they are an optimal solution. Indeed

$$R(J) = \inf_{\chi \in X_{ad}} R(\chi) = \lim_{\nu \to \infty} R(\chi_\nu) = R(\chi), \tag{14.2.14}$$

which completes the proof q.e.d. If instead of the L.C.P. (14.2.1)÷(14.2.4) the state is defined by a hemivariational inequality, we refer for the existence proof to Chapt. 8. Here, however, we are interested in a numerical method for the calculation of the solution and thus we have developed an algorithm analogous to the algorithm proposed in Sect. 10.3. We have "approximated" the nonmonotone multivalued laws by monotone ones and for each "approximation" we solve the corresponding optimal control problem. Here the nonmonotone multivalued laws of Fig. 10.3.3a and Fig. 10.3.3b are assumed to hold tangentially and normally to the surface of the rigid object. Approximating them by friction type monotone diagrams, as in Fig. 9.4.2, and keeping the nonpenetration Signorini inequality in each step, leads to a sequence of L.C.Ps. of the type (14.2.1)÷(14.2.4). To explain this fact more we refer to the diagram of Fig. 10.3.5: there the subproblem concerning the unilateral contact with the rigid support and the Coulomb type friction subproblem are considered together and lead, due to the approximation of the friction cone by a pyramid, to a L.C.P., as we have shown in the previous section. Accordingly the optimal control of the adhesive grasping problem, which is an optimal control problem of a system governed by a hemivariational inequality, is reduced to the solution of a sequence of optimal control subproblems governed by L.C.Ps. Then for the numerical calculation of each subproblem we apply the direct search method [Wals]. The advantage of this procedure is that we can calculate also the rigid body displacements and rotations of the object; they result automatically from the application of Lemke's algorithm for the solution of the L.C.P. This would not be possible by applying the regularization procedure of Chapt. 8.

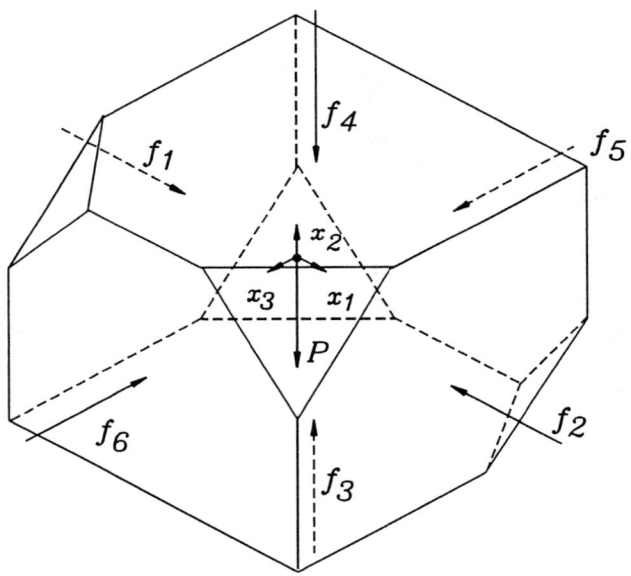

Fig. 14.2.1. On the optimal control problem for a robot gripper.

The example concerns an optimal control problem with respect to a polyhedron grasped by a multifingered gripper (Fig. 14.2.1). In this example the fingers of the gripper are assumed to be of the hard type. Here we solve the problem of the best possible positioning of the object with respect to a desired position. We consider a performance index of the form:

$$R(\chi) = \sum_i ||u_i - u_{i0}||^2 + \sum_i ||\phi_i - \phi_{i0}||^2 + \chi^T N \chi \quad i = 1,2,3 \quad (14.2.15)$$

Here χ is an additional applied force (the control), N is a positive definite matrix, and $\{u_{i0}, \phi_{i0}\}$, $i = 1,2,3$, are given. We assume that the matrix B (cf. eq. (14.2.1)) has here the following form

$$B = \begin{bmatrix} F_{NT}A^T + G_N^T U \\ -T_T^T A^T \end{bmatrix} \qquad (14.2.16)$$

in order to facilitate the calculations.

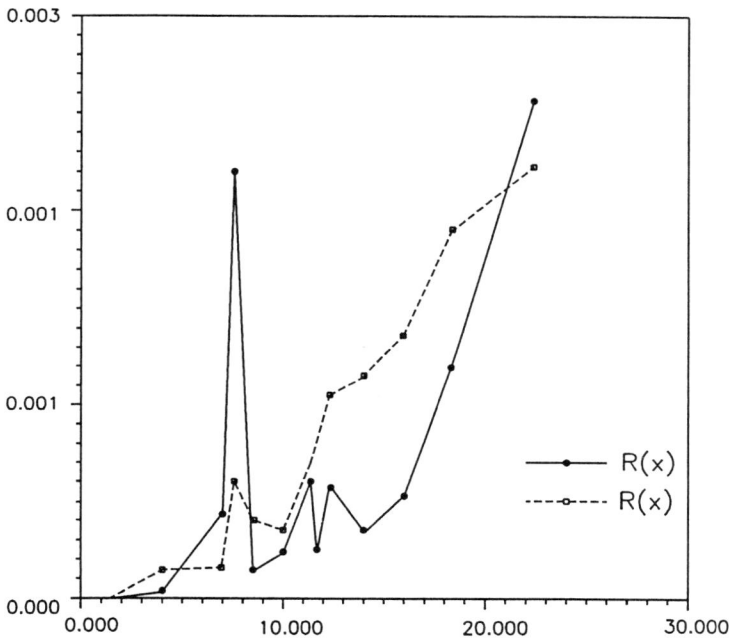

Fig. 14.2.2. The search method for the optimal control function.

In this example we assume that the control χ is an additional external force, applied by the fingertips. In Fig. 14.2.2 the curves depict the variation of the performance index as a function of the control χ. The full line curve corresponds to the case in which the fingertips are subjected simply to unilateral contact and Coulomb friction conditions and the dotted line curve corresponds to the case in which the fingertips are subjected to the interrelated nonmonotone laws of Fig. 10.3.3a,b. We assume that $\mu=0.3$ and that (in compatible units) $P=1$, and $u_{i0} = 0.01$ for $i = 1, 2, 3$ and $\phi_{i0} = 0.0$ for $i = 1, 2, 3$. Moreover we assume that N is the unit matrix and for all six fingers $F_{NN} = 1,5 \times 10^{-6}$, $F_{TT_1} = F_{TT_2} = 6,0 \times 10^{-6}$. The polyhedron was initially a cube with side length equal to 1. All forces are acting vertically to the surface planes of the cube at their initial (i.e. before the beginning of the cutting process) center. Similarly P acts in the initial center of the cube in the x_2-direction.

For the first problem, we obtain for $7.34 \leq ||\chi|| \leq 11.22$ as minimum the value $R(\chi) = 1.25 \times 10^{-4}$ which corresponds to the control $\chi = \{-3.0, 0.0, 0.0, 0.0, -6.5, 2.5\}$. For $11.22 \leq ||\chi|| \leq 12.32$ we obtain the minimum value $R(\chi) = 2.23 \times 10^{-4}$ for $\chi = \{4.2, 7.5, 0.0, 0.0, 6.3, -2.0\}^T$, and, in the case where $12.32 \leq ||\chi|| \leq 22.45$ we get as minimum the value $R(\chi) = 3.48 \times 10^{-4}$ corresponding to a control $\chi = \{5.3, 2.1, 0.0, 0.0, 17.5, -5.2\}^T$. For the numerical calculation of the solution of the optimal control problem, we have applied the systematic search method [Wals], since there are not any necessary conditions to be imposed. This method actually consists in the plotting of the performance index for several combinations of the control values. Here $||\chi||$ denotes the Euclidean norm of the vector. For the second problem we obtain for $7.34 \leq ||\chi|| \leq 11.22$ the value $R(\chi) = 1.40 \times 10^{-4}$ which corresponds to the optimal control $\chi = \{-3.45, 2.80, 0.50, 0.0, -7.20, 2.10\}$ and for $11.22 \leq \chi \leq 22.40$ the value $R(\chi) = 3.50 \times 10^{-4}$ for the optimal control $\chi = \{-4.20, 7.0, 7.50, 0.0, -8.80, -4.20\}^T$.

Note that the method cannot locate all possible solutions but only a number of them. The method presented here for the approximation of the hemivariational inequality by L.C.Ps. may be applied to all types of interface problems and for any structure. This method has the advantage of the ability to treat rigid body degrees of freedom that are not suppressed by appropriate kinematical constraints.

15. Addenda: Hemivariational Inequalities, Fractals and Neural Networks

This last chapter of the book gives some information on some recent developments in the theory of hemivariational inequalities. We deal first with fractal interfaces in the theory of hemivariatonal inequalities and secondly with the neurocomputing approach to the numerical treatment of hemivariational inequalities. The first part is important from the theoretical point of view, due to the arising noninteger dimension, and from the practical point of view as the real geometry of most physical objects is fractal. The second part is also very important because of the trend of evolution existing in the computer science towards neurocomputing. We base this chapter mainly on [Pang92a,b], [Pdp90a,b,91,92a,b,c,d], [Art], [Mis93b], [The93a] for the fractals and on [Kor], [Avd], [The91a,93b] for the neural networks.

15.1 Fractals in Mechanics. An Introduction

In all the foregoing sections of the present book we made the assumption that the boundaries and the interfaces are appropriately smooth. However, in physical problems we have boundaries and interfaces of fractal nature. Indeed the notion of fractals has been introduced in order to obtain an accurate description of the physical forms, as e.g. the geometry of cracked and crushed surfaces in deformable bodies, the geometry of diffusion fronts, the percolation patterns, the geometry of bone cracks, the geometry of metals subjected to sandblasting or to meteoritic rain etc. (cf. e.g. [Man], [Taka], [Fed], [Art], [Bar]). Here we consider according to [Man], [Wal] that a set $F \subset \mathbb{R}^n$ is a fractal set if F has fractional Hausdorff dimension dim$F$, or if dim$F$ is strictly larger than the topological dimension of F; we recall that for a plane figure the topological dimension is 2 for a solid body is 3 etc. Concerning the definition of the Hausdorff dimension see [Fal].

Here we shall investigate how the fractal nature of an interface or a boundary in a problem influences the formulation, properties and numerical treatment of hemivariational inequalities. The method we will aplly is very simple. All algorithms applied to problems having classical (i.e. nonfractal) geometries are coupled with a fixed point algorithm approximating the fractal geometry via a sequence of classical geometries.

Only a few rigorous mathematical approaches to the theory of fractals have been developed until now [Wal], [Bar],[Fal]. Out of them the method by Barnsley [Bar] seems to be the most suitable for the needs of the engineering sciences. Until now the theory of fractals was connected with mechanics only with respect to the study of the attractors of some dynamical systems as well as with respect to the detection of the fractal nature of some physical objects and the calculation of their "Hausdorff dimension". Note at this point that the classical mechanics is based on the inherent assumption of nonfractality (integral Hausdorff dimension). Thus the consideration of the fractal nature of the geometry of a mechanical problem leads to the "mechanics of fractals" which is a new type of mechanics. The author in [Pdp90a,b,91,92a,b,c,d] has examined several other connections of the theory of fractals with mechanics and especially i) the definition of mechanical quantities on fractals and ii) the fractal approximation and the solution of partial differential equations on domains with fractal boundaries. In this last area the main contributions are due to Wallin [Wal,Wal89] and to Jonsson and Wallin [Jon].

It is worth noting that every complicated geometry cannot necessarily be described through fractals ([Schol] p.241). However the method applied here consisting in the approximation of the fractal through classical curves or surfaces holds for any kinked curves or surfaces of complicated geometry, which are not fractals.

As we have mentioned before the Hausdorff dimension of a set A, denoted here dim$A$, characterizes the fractality of this set. For a rigorous mathematical definition of the Hausdorff dimension we refer to [Fal] but the reader could understand it as a notion analogous to the classical dimension. For the definition of the fractals we apply here the notion of the iterated function systems (I.F.S.):

Let $\{X, d\}$ be a complete metric space with the metric d and let us denote by $H(X)$ the space of the compact subsets of X. Then $d(A, B)$ denotes the distance between the sets $A \subset X$ and $B \subset X$ defined by

$$d(A, B) = \max_{x \in A} \min_{y \in B} d(x, y). \tag{15.1.1}$$

The space $H(X)$ endowed with the Hausdorff metric (cf. also [Sen90])

$$h(A, B) = \max\{d(A, B), d(B, A)\} \quad \forall A, B \in H(X) \tag{15.1.2}$$

is a complete metric space and it is called the space of deterministic fractals. An important notion in the theory of fractals is the notion of (deterministic) iterated function systems (I.F.S.) on X. This system is defined by n contractive mappings $w_i : X \to X$ with the contractivity factors $0 \le s_i < 1$, $i = 1, \ldots, n$, i.e.

$$d(w_i(x), w_i(y)) \le s_i d(x, y) \quad \forall x, y \in X,\ 0 \le s_i < 1. \tag{15.1.3}$$

15.1 Fractals in Mechanics. An Introduction

The following result holds [Bar]: Let $\{X; w_i, \ i = 1, \ldots, n\}$ be an I.F.S. and let us define the set-valued functions $W_i : H(X) \to H(X)$ by setting

$$W_i(B) = \{w_i(x); x \in B\} \quad \forall B \in H(X). \tag{15.1.4}$$

Moreover let

$$W(B) = W_1(B) \cup W_2(B) \cup \ldots \cup W_n(B) \quad \forall B \in H(X). \tag{15.1.5}$$

The set-valued function W is obviously also a contraction mapping on $H(X)$ with contractivity factor $s = \max\{s_1, \ldots, s_N\}$ and thus it exists a unique fixed "point" A of it according to the well-known fixed point theorem. This fixed point is a set $A \subset H(X)$ such that

$$A = W(A) = \bigcup_{i=1}^{n} W_i(A). \tag{15.1.6}$$

The set A also results from the iterations

$$A = \lim_{m \to \infty} W^{(m)}(B) \quad \forall B \in H(X), \tag{15.1.7}$$

where

$$W^{(0)}(x) = x, \ W^{(m)}(x) = W(W^{(m-1)}(X)), m = 1, 2 \ldots \tag{15.1.8}$$

are called the forward iterates of W. The set A is called the deterministic "attractor" of the I.F.S. $\{X; w_i\}$. By definition it is the deterministic fractal corresponding to the I.F.S. Two very well known fractals are the Koch curve and the Sierpienski triangle. We refer to [Bar] for the I.F.S. generating each of them and many other interesting fractals.

Moreover we consider that a fractal may be defined by means of a "fractal interpolation", i.e. interpolation of given data by a fractal "curve" or "surface". Again the fractal is defined as a fixed point of a given transformation T. Let us explain this construction with details. Suppose that in $\mathbb{R}^2$, for instance, we have a set of data $(x_i, y_i) \ i = 0, 1, \ldots, N$. We want to find a fractal interpolation function $y: [x_0, x_N] \to \mathbb{R}$, i.e. a fractal f such that $y(x_i) = y_i \ i = 0, 1, \ldots, N$. We consider the I.F.S. $\{\mathbb{R}^2; w_n, \ n = 1, \ldots, N\}$ defined by the "transformation"

$$(x, y) \to w_i(x, y) = \begin{bmatrix} a_i & 0 \\ c_i & d_i \end{bmatrix} \begin{bmatrix} x \\ y \end{bmatrix} + \begin{bmatrix} e_i \\ f_i \end{bmatrix} \quad i = 1, \ldots, N. \tag{15.1.9}$$

Let the factors d_n, called scaling factors, satisfy $0 \leq d_n < 1$; they are free parameters of the problem. Moreover we have that

$$a_i = \frac{(x_i - x_{i-1})}{(x_N - x_0)}, \quad e_i = \frac{(x_N x_{i-1} - x_0 x_i)}{(x_N - x_0)} \tag{15.1.10}$$

$$c_i = \frac{(y_i - y_{i-1})}{(x_N - x_0)} - d_i \frac{(y_N - y_0)}{(x_N - x_0)} \tag{15.1.11}$$

$$f_i = \frac{(x_N y_{i-1} - x_0 y_i)}{(x_N - x_0)} - d_i \frac{(x_N y_0 - y_N x_0)}{(x_N - x_0)} \tag{15.1.12}$$

Then the following holds [Bar]: Let F be the attractor of the I.F.S. defined by (15.1.9)÷(15.1.14). Then F is the graph of a continuous function $y \colon [x_0, x_N] \to \mathbb{R}$ interpolating the data $(x_i, y_i), i = 1, \ldots, N$. If C^0 is the set of all continuous functions $y \colon [x_0, x_N] \to \mathbb{R}$ then the sequence of functions $\tilde{y}_{m+1}(x) = (T\tilde{y}_m)(x)$, where the operator $T \colon C^0 \to C^0$ is defined by

$$T(\tilde{y}(a_i x + e_i)) = c_i x + d_i \tilde{y}(x) + f_i \quad i = 1, 2, \ldots, N \tag{15.1.13}$$

converges to the attractor F as $m \to \infty$. Furthermore if $x_0, \ldots, x_N$ are equally spaced then the $\dim F$ is given by the formula

$$\dim F = 1 + \frac{\ln(\sum_{i=1}^N |d_i|)}{\ln N}, \tag{15.1.14}$$

if the points $(x_i, y_i), i = 0, 1, \ldots, N$, do not constitute a straight line (in this case $\dim F = 1$) and if $\sum_{i=1}^N |d_i| > 1$. Note that the proper choice of the parameters d_i may make $\dim F$ very close to 1 (line-like fractal) or very close to 2 (surface-like fractal).

Fractals can also be generated by considering random I.F.S. To define them let us assign to each map w_n a constant probability p_n with $\Sigma p_n = 1$ and $p_n > 0$. Let $\Sigma(X)$ be the Borel σ-field of X and let $P(X)$ be the probability measure space on the σ-field. If $B \in \Sigma(X)$ is a Borel set we denote by χ_B the function

$$\chi_B(x) = \begin{cases} 1 & \text{if } x \in B \\ 0 & \text{if } x \notin B. \end{cases} \tag{15.1.15}$$

Let

$$K(x, B) = \sum_{n=1}^N p_n \chi_B(w_n(x)) \tag{15.1.16}$$

and let for $\mu \in P(X)$

$$M(\mu)(B) = \int_x K(x, B) d\mu(x) = \sum_{n=1}^N P_n \mu(w_n^{-1}(B)) \tag{15.1.17}$$

be the "Markov operator" on the probability measure space. Assume now that X is a compact space and let

$$A = \{f \colon X \to \mathbb{R} \,\big|\, |f(x) - f(y)| \leq d(x, y) \quad \forall x, y \in X\}. \tag{15.1.18}$$

Then the space $P(X)$ endowed with the metric [Hutch]

$$d(\mu, \mu') = \sup_{f \in A} \{\int_X f d\mu - \int_X f d\mu'\} \tag{15.1.19}$$

is a complete metric space. It can be proved [Hutch] that if the w_n's are contraction mappings, then M is a contraction mapping on $\{P(X), d(\mu, \mu')\}$ and the corresponding fixed point μ_w is called the invariant measure of the random I.F.S. The support of μ_w is the deterministic fractal of the deterministic I.F.S $\{X, w_n\}$; μ_w itself is the fractal corresponding to the random I.F.S. To construct the plotting of μ_w we can apply, for instance, a random choice method for w_n [Hutch], or we can apply to M the fixed point approximation method, i.e. $\mu_{m+1} = \mathrm{M}(\mu_m)$ (see also [Bress]).

In this chapter we will assume that the fractal F may be considered as the fixed "point" of a given nonlinear transformation. Thus we assume that it can be approximated, either using the attractor property (15.1.17) or using the corresponding fractal interpolation property (fixed point of T in (15.1.13)), by a sequence of classical, i.e. nonfractal, sets. For the purposes of mechanical problems , i.e. for the study of fractal boundaries, interfaces and stress-strain laws the deterministic fractals are sufficient.

Let us give now some information on the formulation of mechanics on fractals. This is necessary because until now the mechanical quantities were defined in the general framework of integer dimension. It arises naturally the question what happens in the case of fractal geometry. Do the same definitions hold or some changes have to be done? For instance which is the meaning of the stress tensor $\sigma = \{\sigma_{ij}\}$ for a spongy body or for the Sierpinski gasket etc.? In order to show that a new definition, at least for certain quantities, is necessary let us consider a body in $\mathbb{R}^2$ having as a boundary the Koch's curve, which is a continuous but nowhere differentiable curve having Hausdorff dimension log4/log3. Thus we cannot define the outward unit normal vector $n = \{n_i\}$ and therefore we cannot define in the same sense as in classical mechanics, e.g. the normal heat flux $q_i n_i (q = \{q_i\}$ is the flux vector) outward of the boundary, or the traction $S = \{S_i\} = \{\sigma_{ij} n_j\}$, or the normal and tangential displacements with respect to the boundary u_N and u_T. However, especially the example of the heat flux shows, that even in the case of a fractal boundary Γ the definition of the above quantity should be possible at least in an average sense; indeed every physical body having a fractal geometry radiates heat through its boundary and this heat flux can be estimated through measurements.

Let us assume that a sequence of deformable bodies $V_n \subset \mathbb{R}^3$ can be determined having integer Hausdorff dimensions such that $V_n \to V$ in the Hausdorff sense. Let Q be the physical quantity which we want to define on V. We define Q as the limit of Q_n, for $n \to \infty$, where Q_n are the same physical quantities defined on V_n. The definition of Q has a meaning if this limit exists. The last condition is important because, as we have seen before, at least in the case of Koch curve this limit may not exist and another definition is necessary. Suppose that this limit exists and that we want to define a mechanical law L between the quantities $Q_i, i = 1, \ldots, r$, on a fractal body V. Here using the limit procedure a first approach is possible. Let us first formulate on V_n the law $L(Q_{1_n}, Q_{2_n}, \ldots, V_n) = 0$ $n = 1, 2, \ldots$, where $Q_{i_n}, i = 1, 2, \ldots, r$, tend to Q_i, as $V_n \to V$ in the previously indicated sense. Then the limit, provided that it exists,

$$\lim_{n\to\infty} L(Q_{1_n},\ldots,Q_{r_n},V_n) = 0 \tag{15.1.20}$$

is by definition the mechanical law on V. However, this limit procedure is not always possible in the general case of nonlinear laws.

All the above make the following postulate necessary.

H) A mechanical law $L(\ldots)$ must be invariant with reference to the Hausdorff dimension of the body on which it applies. Note that H) holds for a law and not for any mechanical quantity, because a mechanical quantity may be defined in a different way in the case of fractal geometry. For instance, the boundary traction S_i of a linear elastic body cannot be defined as $\sigma_{ij}n_j$ in the case of the fractal boundary which is nowhere differentiable.

Let us refer now to the basic axioms of rational mechanics as stated e.g. by [Tru66] p.2 and 3. First remark is that in the concept of a body the assumption of "smooth" manifold must be suppressed. Indeed we deal with manifolds having fractal geometry. Of course there are serious mathematical questions concerning the rigorous definition of such a manifold, which here is understood only intuitively. Then the configurations of such a body will have a fractal geometry too. Moreover the deformation of a body may cause a change in its Hausdorff dimension (e.g. smoothing of a sandblasted metal surface). As stated in [Tru66] "continuum mechanics concerns contact forces"; but as we have pointed out previously in the case of fractal boundaries a new definition of the traction S_i on a fractal boundary is necessary. Analogously in the case of a spongy body a new definition of $\sigma = \{\sigma_{ij}\}$ is needed.

In order to formulate relations between body, force and motion we have to postulate the well-known principles of concervation of linear momentum and moment of momentum [Tru66]. Due to the lack of smoothness in a fractal body these two principles cannot take the well-known form of pointwise Cauchy's laws of motion. They have to be written in their integral forms over the fractal body Ω and its fractal boundary Γ. Obviously if Ω and Γ are the limits in the sense of Hausdorff of the nonfractals Ω_n and Γ_n respectively, then the local forms of the linear momentum and of the moment of momentum principles must "tend" in the limit n→∞ to the corresponding integral forms of the same principles over Ω and Γ. These principles must be satisfied independently of the material properties of the bodies and of their Hausdorff dimension. Note also that these two principles may be used to define in an average sense (due to the appearance of the integrals) the stress tensor of the body.

The results of Wallin and Jonson [Jon] make possible a more rigorous defintion of mechanical quantities on fractals. Let Ω be an open subset of $\mathbb{R}^n$ with the boundary Γ. We assume that Γ is a fractal with dimension $n-1 < d < n$. Let $W^p(\Omega)$ be the classical Sobolev space of the $L^p(\Omega)$ functions with distributional derivatives D^a up to order k in $L^p(\Omega)$, which is equipped with the classical norm $||u||_{p,k}$. A function $u \in L^1(\Omega)$ can be defined "strictly" at the point $x \in \Omega \cup \Gamma$ if the limit

15.1 Fractals in Mechanics. An Introduction

$$\tilde{u}(x) = \lim_{r \to \infty} \frac{1}{m(B(x,r) \cap \Omega)} \int_{B(x;y) \cap \Omega} u(y) d\Omega \qquad (15.1.21)$$

exists [Jon]. Here $B(x,r)$ denotes a ball in $\mathbb{R}^n$ with center at x and radius r, and $m(B(x,r) \cap \Omega)$ is the "area" of the intersection of this ball with Ω. Then we define the trace of u to Γ, i.e. the function $u|_\Gamma$, by the formula

$$u|_\Gamma(x) = \tilde{u}(x), \qquad (15.1.22)$$

at every $x \in \Gamma$, where $\tilde{u}(x)$ exists. In classical linear elasticity the displacement field u_i, $i = 1, \ldots, n$, is considered as an element of $H^1(\Omega)$, i.e. of the Sobolev space $W_1^2(\Omega)$. Then the Sobolev space $H^{1/2}(\Gamma)$ is the space for the displacements on Γ. Its dual space, the space $H^{-1/2}(\Gamma)$, is the space of the boundary tractions $S_i = \sigma_{ij} n_j$, $i,j = 1, \ldots, n$. If Γ has the fractal dimension d, then the displacement field $u_i \in H^1(\Omega)$ does not possess a trace $u_i|_\Gamma$ in $H^{1/2}(\Gamma)$ but in the Besov space $B_\beta^{2,2}(\Gamma)$ where $\beta = 1 - \frac{n-d}{2}$ [Jon]. Then the trace operator $Tr: u \to u|_\Gamma$ is, under some hypotheses which can be easily fulfilled, a bounded linear surjective operator [Wal89]

$$Tr: u \in W_k^p(\Omega) \to u|_\Gamma \in B_\beta^{p,p}(\Gamma), \qquad (15.1.23)$$

where

$$\beta = k - \frac{n-d}{p}, \qquad (15.1.24)$$

with a bounded linear right inverse. We refer to [Jon] [Wal89] for the definition of the space $B_\beta^{p,p}(\Gamma)$, $\beta > 0$, $1 \leq p$, $q \leq \infty$.

The above results imply that in a linear elastic body $\Omega \subset \mathbb{R}^3$ (resp. $\Omega \subset \mathbb{R}^2$) in which the boundary is a fractal (e.g. a fractured or fissurated boundary) the boundary displacements have the property that

$$u_i \subset B_\beta^{2,2}(\Gamma) \quad \text{with } \beta = 1 - \frac{3-d}{2} = \frac{d-1}{2} \text{ (resp. } \beta = \frac{d}{2}) \qquad (15.1.25)$$

and the boundary tractions S_i belong to the dual space $[B_\beta^{2,2}(\Gamma)]'$ if Γ has a dimension $2 < d < 3$ (resp. $1 < d < 2$).

We shall describe now the method of fractal approximation. Let us consider now a structure occupying a subset Ω of $\mathbb{R}^3$ (or $\mathbb{R}^2$) in its undeformed state and let Γ be its boundary. We assume that Ω and Γ are fractals and let $\Omega_j \to \Omega$ and $\Gamma_j \to \Gamma$ as $j \to \infty$ in the sense of the metric of Hausdorff. Moreover let us denote by g the graph of a material law which is also assumed to be a fractal and suppose that $g_j \to g$ as $j \to \infty$ in the Hausdorff metric. We denote by G (resp. by G_j) the triplet $\{\Omega, \Gamma, g\}$ (resp. $\{\Omega_j, \Gamma_j, g_j\}$). Here the following method will be applied: We consider first the classical i.e. nonfractal problems with respect to the structure G_j, $j = 1, 2, \ldots$. Let X_j be the corresponding solution, i.e. the corresponding displacement or stress field etc. Then $\lim X_j$ as $j \to \infty$ is the solution of the fractal structure.

In order to achieve the maximum of generality we shall consider a very general structural analysis problem leading to a hemivariational inequality. Obviously it includes all the other types of structural analysis problems as special cases. In the sequel we avoid on purpose the use of advanced functional analysis techniques by introducing very general assumptions. Thus we make the subject more accessible to the engineering oriented reader.

The solutions of all structures G_j, $j = 1, 2 \ldots$, and the solution of the fractal structure G will be sought on the same Hilbert space V. We may assume, for instance, that V is defined on $\tilde{\Omega}$, where $\tilde{\Omega} \supset \Omega_j$, $\tilde{\Omega} \supset \Omega$; then the operators and functions involved, e.g. in the structure G_j, must be appropriately modified on the set $\tilde{\Omega} - \Omega_j$; this would mean that some elements of the structure have zero modulus of elasticity etc. Another possibility would be to consider for each structure G_j the corresponding Hilbert space V_j and to define on Ω the Hilbert space V. This approach makes necessary the use of the function spaces on fractals according to [Jon]. We denote the norm of V by $||\cdot||$ and the duality pairing by $(\cdot, \cdot)$.

Let us consider the following problem for the triplet $\{\Omega_j, \Gamma_j, g_j\}$:
Find $u_j \in V$ such as to satisfy the hemivariational inequality

$$W^\uparrow(G_j, u_j, v - u_j) + \Phi^\uparrow(G_j, u_j, v - u_j) \geq (p(G_j), v - u_j) \quad \forall v \in V. \quad (15.1.26)$$

It is a classical (nonfractal) hemivariational inequality. Here $W^\uparrow(G_j, ., .)$ and $\Phi^\uparrow(G_j, ., .)$ denote the directional differential of Clarke-Rockafellar (cf. Sect. 1.2) and W, Φ are nonconvex superpotentials of the displacements. The superpotentials are not everywhere differentiable and may take also the values $\pm\infty$. The hemivariational inequality (15.1.26) covers all the cases of nonmonotone possibly multivalued laws and boundary conditions. The following proposition holds on the assumption that (15.1.26) has a solution $u_j \in V$. Here V is the kinematically admissible set of displacements and $p(G_j)$ is the loading vector.

Proposition 15.1.1 Suppose that as $j \to \infty$

i) $\quad p(G_j) \to p(G) \quad$ strongly in V' $\hfill (15.1.27)$

ii) $\quad W^\uparrow(G_j, u_j, -u_j) \leq -c||u_j||^2 \quad \forall u_j \in V \hfill (15.1.28)$
where $c > 0$ is a constant independent of G_j.

iii) $\quad \Phi^\uparrow(G_j, u_j, -u_j) < c \hfill (15.1.29)$
where $c > 0$ is a constant independent of G_j.

iv) As $u_j \to u \quad$ weakly in V, and for every $v \in V$

$$\limsup W^\uparrow(G_j, u_j, v - u_j) \leq W^\uparrow(G, u, v - u) \quad (15.1.30)$$

$$\limsup \Phi^\uparrow(G_j, u_j, v - u_j) \leq \Phi^\uparrow(G, u, v - u). \quad (15.1.31)$$

Then $u_j \to u$ weakly in V where u is solution of the hemivariational inequality

$$W^{\uparrow}(G,u,v-u) + \Phi^{\uparrow}(G,u,v-u) \geq (p(G), v-u) \quad \forall u \in V. \tag{15.1.32}$$

Proof. Let us put in (15.1.26) $v = 0$. From (15.1.27), (15.1.28) and (15.1.29) we have that

$$c||u_j||^2 \leq -W^{\uparrow}(G_j, u_j, -u_j) \leq (p(G_j), u_j) \tag{15.1.33}$$
$$+\Phi^{\uparrow}(G_j, u_j, -u_j) \leq c||u_j|| + c$$

Thus $||u_j|| < c$ which implies that there exists a subsequence such that $u_j \to u$ weakly in V. Now from (15.1.26) we obtain by taking the limsup of it and using (15.1.27), (15.1.30) and (15.1.31) that u fulfills (15.1.32) q.e.d.

The above proposition shows only the general method which can be applied for the numerical treatment of a structure with fractal geometry. A refinement of the assumptions of the proposition using the results of the theory of function spaces on fractal sets [Jon] and the results of the Hausdorff convergence is not yet available; the problem is still open.

15.2 Application XVI: Hemivariational Inequalities for Fractal Interfaces

Further we have studied a unilateral contact problem with nonmonotone friction, when the contact interface is of fractal nature. The fractal interface Φ is the fixed "point" of a transformation W or T resulting either from an I.F.S. or from the fractal interpolation of given data. Thus we may write that

$$\Phi = T\Phi \quad \text{and} \quad \Phi_{n+1} = T\Phi_n \quad \Phi_n \to \Phi \tag{15.2.1}$$

in the Hausdorff metric. Accordingly we shall apply for every Φ_n the algorithm of Sect. 10.3 for the splitting of the general problem into a pure unilateral contact problem with given $S_T = C_T$ and into a nonmonotone friction problem with given $S_N = C_N$. Then for each subproblem, and after the F.E. discretization, the arising Q.P.Ps. with respect to Φ_n have been formulated and solved. We repeat this procedure several times by increasing n and we claim that at the limit the solution of the fractal problem is obtained. The convergence proof is still open. Only several numerical experiments verify the correctness of the method.

As an example we consider here the structure of Fig. 15.2.1 submitted to loading in its plane. The material is linear elastic with modulus of elasticity E=2.1 $\times 10^5$t/m^2 and Poisson's ratio $\nu = 0.33$. The thickness of the plate is 0.10 m. At the interface we assume that in the normal direction the unilateral contact law of Fig. 10.3.4b holds, whereas in the tangential direction the nonmonotone friction law of Fig. 10.3.3a. This is a normalized friction law; its ordinates must be multiplied by $\mu|S_N|$, where $\mu = 0.3$ is the friction coefficient. The interface is a fractal, interpolating the data $(x_0, y_0) = (0,0)$, $(x_1, y_1) = (2.7, 4.0)$ and

$(x_2, y_2) = (8.0, 0.0)$. Then the fractal interface is the graph of the attractor of the I.F.S. $\{\mathbb{R}^2; w_1, w_2\}$, where w_1 and w_2 are the transformations

$$w_1(x,y) = \begin{bmatrix} 0.3375 & 0.0 \\ 0.5 & 0.5 \end{bmatrix} \begin{bmatrix} x \\ y \end{bmatrix} + \begin{bmatrix} 0 \\ 0 \end{bmatrix} \quad (15.2.2)$$

$$w_2(x,y) = \begin{bmatrix} 0.6625 & 0.0 \\ -0.5 & 0.5 \end{bmatrix} \begin{bmatrix} x \\ y \end{bmatrix} + \begin{bmatrix} 2.7 \\ 4.0 \end{bmatrix}. \quad (15.2.3)$$

The free parameters $d_i, i = 1, 2$ are $d_1 = d_2 = 0.5$. The structure has been discretized by triangular constant stress elements (number of elements 506) For each Q.P.P., arising in the process of approximating the nonmonotone law by monotone diagrams of the friction type, the Hildreth and d'Esopo Q.P. algorithm has been applied. The required steps for the two-level (i.e. simple unilateral contact/nonmonotone friction) algorithm for the five first approximations of the fractals $\Phi_1, \Phi_2, \ldots, \Phi_5$ are 4,4,7,10,12 respectively. It was sufficient to solve the problem for the five first approximations of the fractal interface; indeed the fourth and the fifth one give rise to almost the same stress and displacement fields.

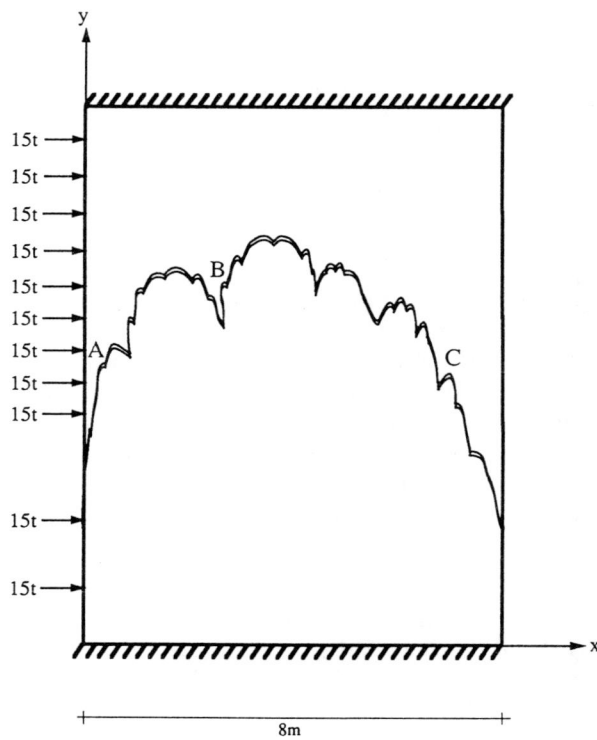

Fig. 15.2.1. Plane unilateral contact problem with nonmonotone friction for an interface of fractal type.

15.2 Application XVI: Hemivariational Inequalities for Fractal Interfaces

The whole algorithm is stable, although the complicated interfaces introduced by the approximations of the fractal interface do not offer an ideal basis for the solution of a unilateral contact problem with friction. Moreover one must be careful with the reentrant corners of the interface in connection with the F.E. discretization. Thus in order to get reliable numerical results concerning the possible stress concentrations we have increased the number of elements around the singular points. In all the approximations, the contact algorithm determines from the first steps almost completely the true contact region. Thus only a few steps are needed for the termination of the whole algorithm. The complexity of the third, fourth and fifth approximations of the fractal boundary caused an increase of the calculation time. We notice that the differences become insignificant after the third approximation of the fractal boundary. For instance between the third (resp. fourth) and the fourth (resp. fifth) approximation of Φ the differences of the maximum stress is smaller that 7% (resp. 3%) in a curve parallel to Φ and lying 10 cm inside the interface. It is also important to note that the approximation of the fractal Φ does not affect considerably the stress and displacement fields inside the body. This is compatible with the St. Venant "principle" of classical elastostatics which can be proved only for classical boundaries and for bilateral boundary conditions.

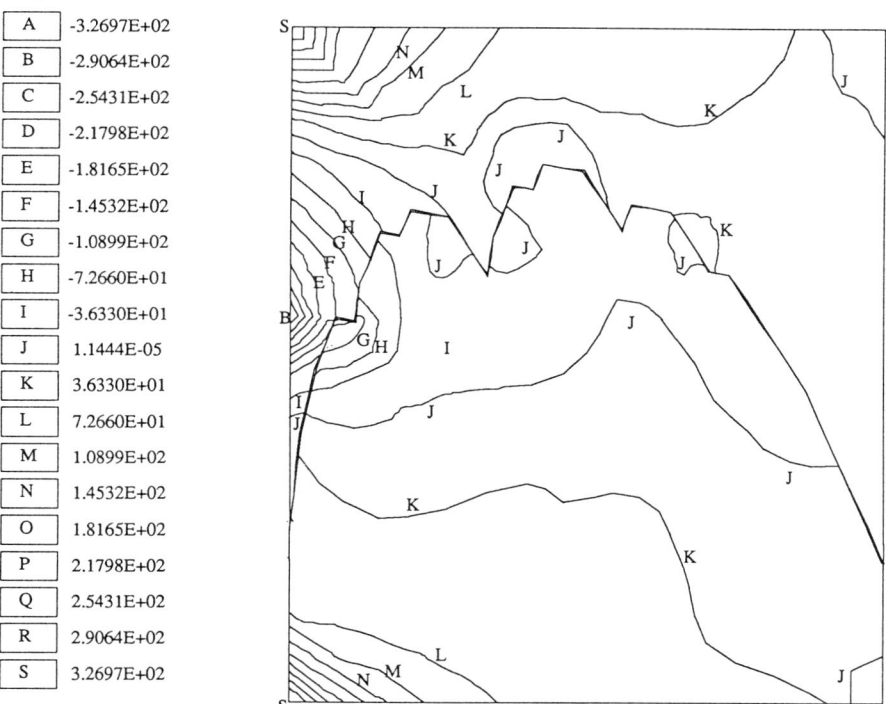

Fig. 15.2.2. Maximum stress of the Φ_5-interface (in t/m^2).

One could think also that in order to calculate the structure of Fig. 15.2.1 it would be sufficient to use a random approximation of the boundary very close to the fractal interface Φ. The numerical implementation of this case implies that the numerical results are "worse" than the results obtained by performing the approximation of the fractal boundary by $\Phi_1, \Phi_2, \ldots$ etc. In Fig. 15.2.2 and in Fig. 15.2.3 we give the variation of the maximum stress and the variations of the displacement fields respectively. Moreover in Fig. 15.2.4 (resp. Fig. 15.2.5) the maximum difference of the normal (resp. the tangential) relative displacements between the two levels of the algorithm, the normal and the tangential one for the fourth and fifth approximation of the interface are depicted with respect to k, where k is the total number of solving a purely unilateral contact and a purely nonmonotone friction problem. Note that the termination criterion was $|C_{T_i}^{k+1} - C_{T_i}^k| \leq 10^{-4}$.

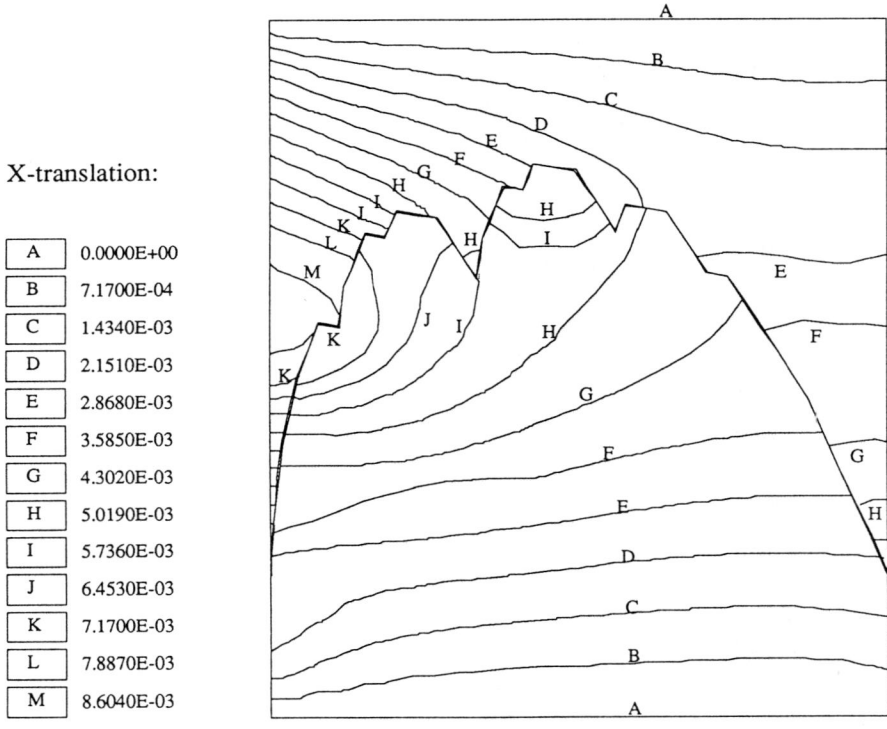

Fig. 15.2.3. X-Displacements for the Φ_5-interface (in m).

15.2 Application XVI: Hemivariational Inequalities for Fractal Interfaces

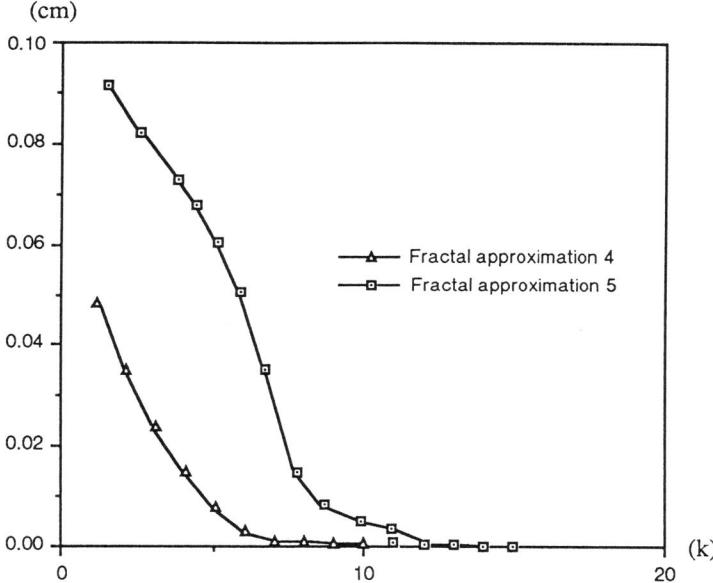

Fig. 15.2.4. Differences of the normal relative displacements between the pure contact and the pure friction subproblems with respect to k.

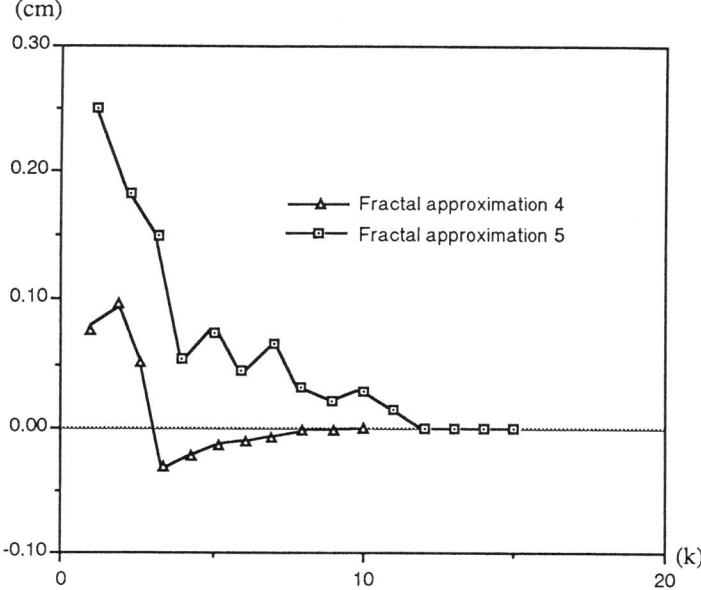

Fig. 15.2.5. Differences of the tangential relative displacements between the pure contact and the pure friction subproblems with respect to k.

15.3 The Neural Network Approach to Hemivariational Inequalities

The present section[1] is devoted to a special class of hemivariational inequalities whose the solution can be approximated solving a number of Q.P.Ps. (see also Chapt. 9, Chapt. 10 and Chapt. 11). The main result of this section is that in a neural network environment an inequality constrained Q.P.P. requires for its calculation the same computer memory as a Q.P.P. without constraints, i.e. as a linear equation system. This fact is going to constitute a major advantage of the inequality methods over other approaches as, apart from being inherently more accurate from a physical point of view, the already visible advent of neural computing hardware will, taking profit of this property, make inequality methods really efficient.

Neural network models are very efficient in computations where many assumptions must be satisfied in parallel. We take here the opportunity to comment that this is often the case in hemivariational inequalities, for instance, if the methods of Chapt. 10 and Chapt. 11 are applied. The same holds also for the quasidifferentiability. Therefore, parallel computers are very suitable for the numerical treatment of hemivariational inequality. Let us come back to neural networks. There the parallelism in contrast to the classical sequential computers, is achieved by using networks of analog neurons with nonlinear behaviour and with a high interconnectivity. The neurons are connected with links having variable weights. A neural network is defined by its node characteristics, the "learning" rules and the network topology. The "learning" rules control the improvement of the neural network performance through appropriate adaptive changes of the weights of the links. A computing machine based on the concept of neural networks has much greater fault tolerance than a classical sequential computer.Indeed neurons or links out of order do not seriously affect a great deal the whole performance of the network, as well as its learning capabilities.

The artificial neural nets "immitate" the behaviour of biological nevrous systems. The study of real biological nets can be explained by the fact that the computational power of biological nervous systems is enormous and gives answers to complicated questions in a very short time. On the other hand most digital computers do not have these abilities even for very simple questions.

The aim of the present section is to adapt to a neural network computing environment the numerical methods for hemivariational inequalities. The main idea is to use the neural network capability to solve optimization problems. Indeed this was one of the first applications, and constitutes one of the major advantages of the neural networks. The research on artificial neural networks begins in the early forty's and continues with the works of Widrow and

---

[1]This section and the next two contain some research results of the author obtained in the framework of a large scale STRIDE program on "Massively Parallel, Neurocomputing Environment with Applications in Engineering and Medicine" supported by the E.E.C and the Greek Secretary of Research and Technology.

15.3 The Neural Network Approach to Hemivariational Inequalities

Hoff [Wid], Hopfield [Hop82], Hopfield and Tank [Hop85] among others (cf. e.g. [And]). From all the research until now an important result is that the parallel analog computation in a neural network is the most natural way to solve large scale optimization problems, as those arising in fields like speech and image recognition, etc. Moreover several nonlinearities which need large computing time in a classical sequential computer may be treated very quickly in a computer based on the neural net concept. Indeed the simplest node (neuron) i of a neural network sums the weighted inputs, say $T_{ji}V_j$ (T_{ji} is the weight of the link or synapse between the i and the j neurons, V_j is the output of the j-neuron $j = 1, \ldots, n$), from all the n nodes with which it is connected and gives as an output $f_i(\sum_{\substack{j=1 \\ j \neq i}}^{n} T_{ji}V_j)$, where $f_i(.)$ is generally a nonlinear function. This output of the i-neuron is trasmitted to the other neurons and this procedure continues until a so-called "stable state" of the network is achieved, which corresponds to a local minimum of a characteristic network energy function. Here we shall "construct" a fictitious neural network apropriate for the treatment of the inequality constrained Q.P.P. resulting in the numerical treatment of hemivariational inequalities. Then the proposed neural network and the process of transmission from neuron to neuron is simulated on a digital computer in order to obtain numerical results, because neural computers are not yet available.

The fictitious analog network has parallel input and output channels and n-neurons with a large interconnectivity between them. These neurons are modelled as amplifiers having a general performance expressed through the functions f_i and the input resistor (resp. capacitor) ρ_i (resp. C_i) $i = 1, \ldots, n$. Then V_j is the output voltage of the amplifier j and u_j is the input voltage. A synapse of the neurons i, j is characterized by a conductance T_{ij} which connects the output of the neuron j with the input of the neuron i. Moreover each neuron receives an externally supplied input current I_i. The aforementioned circuit immitates fairly good the function of biological neural networks. The evolution of this circuit with the time t is given by the equations [Hop85]

$$C_i(\frac{du_i}{dt}) = \sum_{j=1}^{n} T_{ij}V_j - \frac{u_i}{R_i} + I_i \qquad (15.3.1)$$

$$V_j = F_j(u_j). \qquad (15.3.2)$$

Here $R_i^{-1} = \rho_i^{-1} + \sum_{j=1}^{N} R_{ij}^{-1}$ (parallel connection of the input resistors ρ_i and the synapses'resistors $R_{ij} = \frac{1}{T_{ij}}$). We assume further for the sake of simplicity that R_i, C_i, have the same values R, C respectively for every neuron and that the response functions f_i of all the neurons are the same and equal to f. Then for given initial values of the neuron inputs u_i at $t = 0$, the integration of (15.3.1), (15.3.2) on a digital computer gives the response of the fictitious network considered to the input values. It can be easily shown (cf. [Hop85]) that if $T_{ij} = T_{ji}$, the solution of (15.3.1), (15.3.2) converges to solutions having constant the outputs V_i of all neurons. These solutions are called stable states and stationary the quantity

$$E = -\frac{1}{2}\sum_{i,j=1}^{n} T_{ij}V_iV_j + \sum_{i=1}^{n}(\frac{1}{R_i})\int_0^{V_i} f_i^{-1}(V_i)dV_i - \sum_{i=1}^{n} I_iV_i. \quad (15.3.3)$$

E is the Liapunov function of the system. The neural network approach to an optimization problem consists in the determination of the quantities T_{ij}, I_i, R_i, and the functions f_i in (15.3.3), such that a local minimum of E, i.e. a stable state, will be identical to the solution of the minimum problem. From the solution of the differential equations (15.3.1), (15.3.2) by means of a classical differential equations solver at least one of the final stable states can be obtained on a digital computer. Roughly speaking if a neural computer would be available the calculation speed would be comparable with the electric field speed diminished by the polarization time of the electronic components of the computer. Note that the appropriate choice of the nonlinear neuron responses f_i makes possible the consideration of the inequality constraints of the problem.

Let us consider now the following two basic problems arising in the numerical treatment of hemivariational inequalities after the discretization of the problem. They are the following classical Q.P.Ps.

1. Find $x \in \mathbb{R}^n$ such that

$$\min\{\frac{1}{2}x^T A x - b^T x\} \quad \text{or} \quad Ax = b \quad (15.3.4)$$

2. Find $x \in \mathbb{R}^n$ such that

$$\min\{\frac{1}{2}x^T A x - b^T x \,|\, x \geq 0\} \quad (15.3.5)$$

Here $A = \{a_{ij}\}$ is a given symmetric matrix and $b = \{b_i\}$ is a given vector. Note that in (15.3.5) some more complicated linear constraints may result, which after ome elementary variable changes lead to the only constraint $x \geq 0$ in (15.3.5).

In order to formulate the fictitious neural network we make the following substitutions ($i, j = 1, \ldots, n$), where $A = \{a_{ij}\}$ and $b = \{b_i\}$:

$$T_{ij} = \begin{cases} -a_{ij} & \text{if } i \neq j \\ -a_{ij} + \frac{1}{R_i} & \text{if } i = j \end{cases} \quad (15.3.6)$$

$$I_i = b_i, \quad V_i = x_i, \quad V_i = f_i(u_i) = u_i \quad (15.3.7)$$

We may further assume for the sake of simplicity that $R_i = C_i = 1$. Then the circuit evolution is described by the differential equation

$$\frac{dV_i}{dt} = \sum_{j=1}^{n} T_{ij}V_j - V_i + I_i. \quad (15.3.8)$$

At the stable state we have $\frac{dV_i}{dt} = 0$ and thus (15.3.8) becomes by means of (15.3.6), (15.3.7) the matrix equation $Ax = b$.

For the minimum problem (15.3.5) the function f_i has the different form

$$V_i = f_i(u_i) = \begin{cases} u_i & \text{if } u_i > 0 \\ 0 & \text{if } u_i \leq 0 \end{cases}, \quad x_i = V_i \qquad (15.3.9)$$

Thus the solution of the circuit fulfills automatically the constraints $x_i \geq 0$, $i = 1, \ldots, n$ of the problem (15.3.5). Accordingly a solution of the differential equation (15.3.8) with given initial conditions tends to a stable state of the neural net, which is a solution of the inequality constrained Q.P.P. (15.3.5). Since the numerical result in a real neurocomputer would be obtained from the corresponding circuit evolution problem, the necessary memory needed for the calculation of the stable state is independent of the performance function f_i of the neurons. In [Kor] there is a more deep discussion of the connection between the solution of the Q.P.P (15.3.5) and the solution of the neural network circuit.

The treatment of a hemivariational inequality problem in a neural environment has the following advantages:

i) the quadratic and convex programming algorithms have substantial storage requirements for large matrices; this is not the case in a neural network treatment,

ii) the numerical solution of the circuit differential equation has better stability properties for large scale problems than the convex programming algorithms. This property together with the lack of influence of the round-off errors make the neural approach more reliable for inequality problems. Indeed numerical experimentation has proved, that even in the absence of neural computers, some inequality problems give more reliable numerical results when they are treated via the circuit differential equation and the numerical solution of an initial value problem than when they are treated via an optimization algorithm. This is especially true for hemivariational inequalities in the framework of geometric nonlinearity where we may obtain Q.P.Ps. with nonpositive definite quadratic form. Moreover the numerical experience gained in [The93b] [Kor] has shown that singular elements do not significantly influence the numerical procedure and thus the neural circuit differential equation gives around the crack tip more reliable results than the classical singular F.E.M. However this last conclusion needs further numerical experimentation.

15.4 Application XVII: D.C.B. Specimen Modelling. The Neural Network Approach

In this section we deal with the numerical solution of a hemivariational inequality arising in a delamination problem concerning a double cantilever beam (D.C.B.). The corresponding hemivariational inequality is reduced to the solution of certain Q.P.Ps. by applying the techniques of Sect. 9.4 and Sect. 10.3. The resulting Q.P.Ps. are solved by a neural network approach. The dimensions and the material properties of the model coincide with those of the 20 mm thick glass/epoxy specimen used in [Davie] (Table 1). The geometry of the upper half of the model is depicted in Fig. 15.4.1. For the discretization of the laminae,

192 constant stress quadrilateral elemenents were used. The material is linear elastic in both laminae.

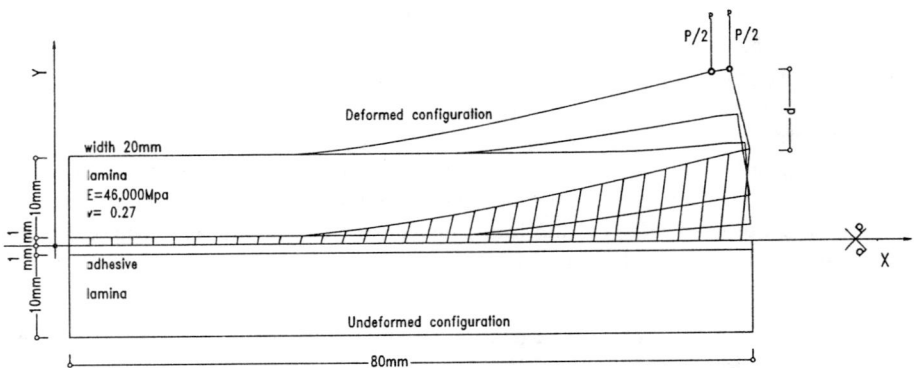

Fig. 15.4.1. Geometry of the model.

The binding layer was modeled by 32 nonmonotone joints, while the initial four nodes of the interface at the opening edge were left free. The nonmonotone laws a and b of Fig. 15.4.2 were used to describe the softening behaviour of the binding material in the normal direction to the interface. The parameters of the graphs a and b were chosen to have the same initial stiffness values (assumed to be equal to the stiffness of the material of the lamina), while they differ in the number of vertical branches and the total deformation energy of each type of joint. The area below each graph represents the energy which the corresponding joint may absorb. It is $G_a = 85.4 \, \text{J/m}^2$ and $G_b = 78.5 \, \text{J/m}^2$. We note that these numbers are of the same order of magnitude with the values of the compute a critical energy release rate given in [Davie] (Table 2). The addition of a cracking or crushing limit to the diagrams (in the form of a vertical branch) is necessary, in order to make the diagrams more realistic. In the tangential direction, the tangential forces are assumed to be zero according to the physical model of Mode I delamination [Davie].

Since fixed load conditions cause instability of the growth of delamination [Carls] the structure is submitted to displacement controlled deformation by restraining the vertical displacement of the two mounting points (the two right nodes of the upper edge). The vertical displacement is for the mounting points augmented by small increments, until the delamination opening α reaches the length of 50 mm approximately. In Fig. 15.4.3 we give the reaction-vertical displacement $(P - d)$ graph at the mounting points for the two types of binding material. The shape of both curves is similar to analogous curves of experimental origin [Carls] [Willi86]. In Fig. 15.4.4 we give the logarithmic reaction-delamination length $(P - \alpha)$ diagram.

15.4 Application XVII: D.C.B. Specimed Modelling. The Neural Network Approach   411

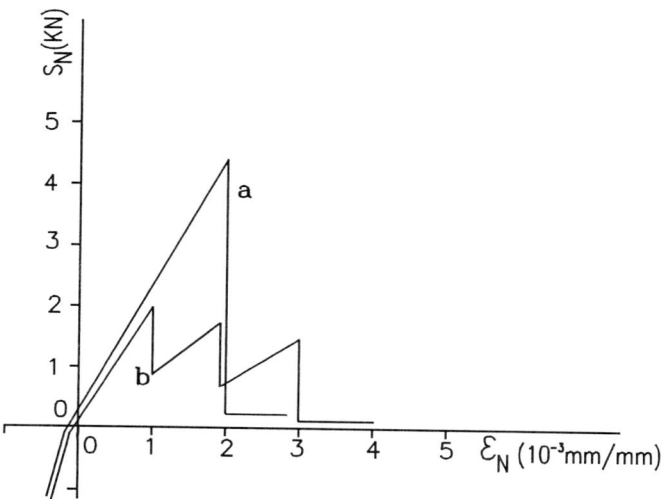

Fig. 15.4.2. The interface behaviour.

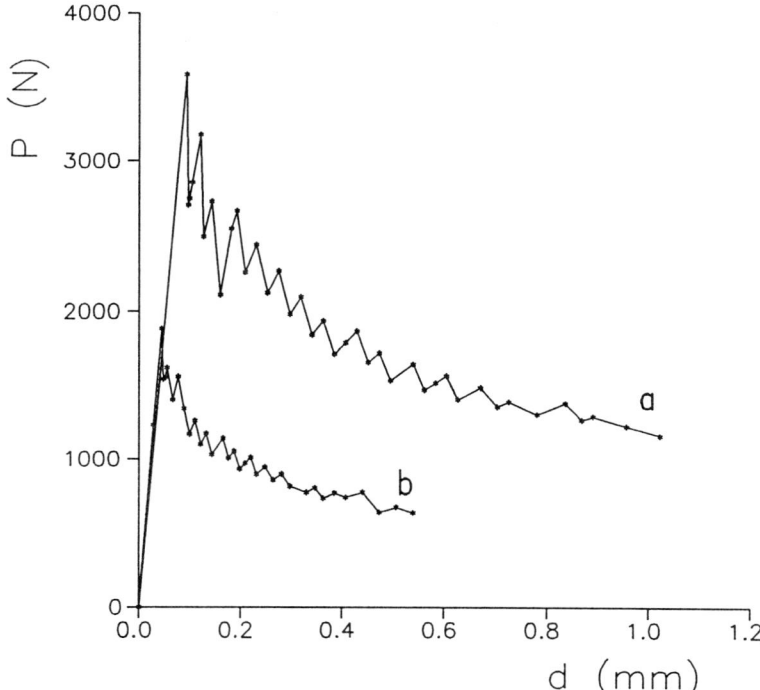

Fig. 15.4.3. Reaction P as a function of the vertical displacement d at the mounting points.

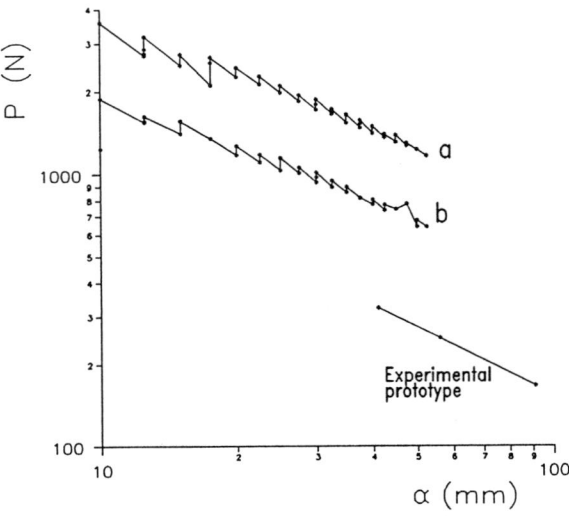

Fig. 15.4.4. Reaction P as a function of the delamination length α.

For the numerical calculations we have applied for each Q.P. subproblem the neural network approach of the previous section. The arising circuit dynamic system has been numerically solved by the Runge-Kutta fourth order method (step $\delta t = 0.5$). The neural network approach was not influenced by the round-off errors of the calculations due to its iterative character. For several random choices of the initial values, three Q.P.Ps. which we have examined have led to the same results only in $50 \div 60$ steps on a HP-720 computer. The results were the same as the results obtained by the quadratic programming algorithm of Hildreth and d'Esopo. Since the decomposition of the hemivariational inequality into Q.P. subproblems is the same, both on sequential computer and on a neural network environment, the results of the two approaches are identical.

As noticed in the previous section the neural network approach treats the problem (15.3.4), i.e a bilateral problem, and the problem (15.3.5), i.e. the simplest unilateral problem exactly by the same method. The only difference occurs in the choice of the network response function f_i. Therefore in a neural environment one does not need to distinguish between equality and inequality problems. All of them are reduced to initial value problems and their solutions are obtained by a time integration method.

15.5 Application XVIII: The Inverse Delamination Problem as a Supervised Learning Problem for a Neural Network. Extensions

The inverse or parameter identification problem in structural analysis is the problem in which the (or a) solution is prescribed and we ask for those elastic

15.5 Application XVIII: The Inverse Delamination Problem

properties and/or loading and/or geometric quantities which will give a solution identical or very close to the prescribed one. Thus the inverse delamination problem formulated as a minimum deviation problem leads to an optimal control problem governed by a hemivariational inequality. For this type of problems certain existence results are given in Chapt. 8. Further we examine the case of delamination problem and we recall that according to the numerical method of Sect. 9.3 and of Sect. 10.3 the solution of the hemivariational inequality is approximated by the solutions of appropriately defined Q.P.Ps. Here we shall present a new formulation of the parameter identification problem as a supervised learning problem of a neural network. Everything in the sequel holds for the parameter identification problems for all systems which are governed by a hemivariational inequality, or by an equivalent sequence of Q.P.Ps., or by an equivalent sequence of linear complementary problems (L.C.Ps.), as it is the case in Sect. 14.2. In the previous section we have not introduced any "learning rule" in the network, i.e. any rule for the updating of the weights of the neuronal links for a new calculation step, according to the output of the previous step(s). This is not necessary if one simply wants to solve a Q.P.P, but it becomes necessary for the treatment of the parameter identification problem.

Note that both in the case of equality and inequality contact problems the inverse problem is not easy to solve numerically using classical methods. On the contrary in a neural network environment the parameter identification problem may be treated very easily by formulating it as a "supervised learning" problem [Bea, Kha]. The learning is called "supervised" because it is guided by taking into account what we want to achieve. Any change in the elastic properties and/or the geometry of the structure gives rise to corresponding changes of the matrix A and b in problems (15.3.4) and (15.3.5). Matrix A is related to the synapses' weights T_{ij} with the eq. (15.3.6). Accordingly we introduce a learning rule of the form

$$T_{ij}^{(k+1)} = T_{ij}^{(k)} + c_i r_{ij}^{(k)} \tag{15.5.1}$$

where $T_{ij}^{(k)}$ denotes the weight of the synapse ij at the k-step of the identification procedure ($T_{ij}^{(0)} = T_{ij}$, given by (15.3.6)) $r_{ij}^{(k)}$ is the reinforcement signal [Kha] and c is the learning rate, which is a positive constant. Moreover the solution at the k-step is considered as an initial value for the calculation of the network at the $(k+1)$-step via the differential equation (15.3.8). This procedure of modifying the synapses' weights is continued until the wished result $\{x_i^\star\}$ in (15.3.4) or (15.3.5), or $\{V_i^\star\}$ in (15.3.8) is obtained. There are several types of learning rules [Bea]. We apply here the "perceptron learning algorithm", where for the adaption of the weights a learning rule analogous to the Widrow-Hoff learning rule [Bea, Kha] has been used.

The learning rule reads [Kha]

$$r_{ij}^{(k)} = (V_j^\star - V_j^{(k)}) u_i^{(k)} \tag{15.5.2}$$

where $0 \leq c \leq 1$. Here $V_j^{(k)}$ is the output of the j-neuron and $u_i^{(k)}$ the input of the i-neuron, whereas $V_j^\star$ is the wished response of the system. For more accurate calculation of c we refer to Beale and Jackson [Bea], as well as for other information concerning the perceptron algorithm. Note that the idea of this section can be extended to all other types of parameter identification and optimal control problems.

Let us solve the following problem for an interface material of the type a. Suppose that we have a series of $P - d$ measurements and we want to identify an equivalent modulus of elasticity and Poisson number variation (i.e. ϱ such that $\nu = 0.27 \pm \varrho$). The $P - d$ measurements are the following pairs: (2000 N, 0.6mm), (2100 N, 0.58 mm), (2200 N, 0.55 mm), (2500 N, 0.51 mm) and (2700 N, 0.45 mm). For each one of the Q.P.Ps. approximating (as in Sect. 9.3) the hemivariational inequality of the delamination problem we solve the corresponding parameter identification problem and we find the E and ϱ. From the solution of the Q.P.P. we proceed to the next monotone diagram (cf. the Fig. 9.3.3) and to the next identification problem. We see that we have reduced the parameter identification problem for the delamination hemivariational inequality to a sequence of parameter identification problems governed by a Q.P.P., or equivalently by a variational inequality. As we have mentioned this problem cannot be treated easily by classical methods (cf. also [Panag83]). Here for all the Q.P.Ps. we define a basic neural network, as shown in Sect. 15.3, whose properties change as E and ϱ varie according to the perceptron rules (15.5.1) and (15.5.2) and as we pass from the one monotone approximation of the nonmonotone delamination diagram to the other. The result obtained was E=48500 Mpa and $\varrho = 0.04$ and one can verify that the corresponding $P - d$ diagram passes through the given $P - d$ measurement points. We have no guarantee that the solution in unique.

The "learning" properties of neural networks may give a numerical solution to certain new problems. Let us consider a nonlinear structure whose dynamic behaviour is governed by a dynamic hemivariational inequality. Does one can determine a linear elastic structure with the same (or different) geometric characteristics whose dynamic behaviour describes fairly well the behaviour of the nonlinear structure? This means that the two structures will have at a certain time T the same stress and displacement fields. Accordingly the linear structure "must change" its elasticity characteristics at the time moments $t_0 < t_1 < \ldots < t_\rho < \ldots < T$ in order to produce finally the "same" or "very similar" stress and displacement patterns as the nonlinear structure. The problem has the same features as the control problem for "oculomotor" systems [Ritt]. Indeed, roughly speaking, one could say that the linear structure "sees" the nonlinear structure behaviour and changes appropriately its elastic properties (let us denote them by C_ρ) at the time moments t_ϱ, $\varrho = 1, 2, \ldots$. For a given loading history and a given time the neural network representing the linear structure "learns" in the training period how to obtain the values of t_ϱ and C_ϱ, $\varrho = 1, 2, \ldots$. For another loading history and another time T the same neural network is approximately trained and thus t_ϱ and C_ϱ can be determined in a shorter time. We can say that a structure "learns" from its previous linearizations, or that in a neural environment a structure acquires a kind of "memory". Note, however, that here the learning is generally "unsupervised" and therefore [Ritt] the "learning" evolution is based on a "reward function" which only shows "how well" the result obtained is compatible with the given task.

References

[Al-Fah91a] Al-Fahed, A.M.: Computation and Analysis of Robot Multifingered Gripper. Doct. Dissertation, Dept. of Civil Eng., Aristotle Univ. Thessaloniki 1991

[Al-Fah91b] Al-Fahed, A.M., Stavroulakis, G.E., Panagiotopoulos, P.D.: Hard and Soft Fingered Robot Grippers. The Linear Complementarity Approach. ZAMM **71** (1991) 257-265

[Al-Fah92a] Al-Fahed, A.M., Panagiotopoulos, P.D.: Multifingered Frictional Robot Grippers. A New Type of Numerical Implementation. Comp. and Struct. **42** (1992) 555-562

[Al-Fah92b] Al-Fahed, A.M., Stavroulakis, G.E., Panagiotopoulos, P.D.: A Linear Complementarity Approach to the Frictionless Gripper. Int. J. of Robot. Res. **11** (1992) 112-122

[Al-Fah93] Al-Fahed, A.M., Panagiotopoulos, P.D.: A Linear Complementarity Approach to the Articulated Multifingered Friction Gripper. J. of Robotic Systems (in press)

[An] Andersson, L.E.: A Global Existence Result for Quasistatic Contact Problem with Friction. LITH-MAT-R-89-00, Linköping Institute of Technology, Linköping, Sweden 1989.

[And] Anderson, J. and Rosenfeld, E.: Neurocomputing. Foundations of Research. The MIT Press, Cambridge Mass, 1988.

[And80] Andersson, T., Fredriksson, B., Allan-Persson, B.G.: The Boundary Element Method Applied to Two-Dimensional Contact Problems. In "New Developments in Boundary Element Methods", (ed. by Brebbia, C.A.) CML Publ, Southampton 1980

[And81] Andersson, T.: The Boundary Element Method Applied to Two-Dimensional Contact Problems with Friction. In "Proc. Third Intern. Seminar on Boundary Element Methods", (ed. by Brebia, C.A.). Springer-Verlag, Berlin 1981

[Ant]	Antes, H., Panagiotopoulos, P.D.: The Boundary Integral Approach to Static and Dynamic Contact Problems. Equality and Inequality Methods. Birkhäuser Verlag, Basel, Boston 1992

[Arib]	Aribert, J.M., and Abdel Aziz, U.: Calcul des poutres mixtes jusqu' à l' état ultime avec un effect de soulèvement à l' interface acier-béton. Construction métallique **4** (1985) 3-36

[Argy65]	Argyris, J.H.: Three-Dimensional Anisotropic and Inhomogeneous Elastic Media. Matrix Analysis for Small and Large Displacements. Ing. Archiv **34** (1965) 33-55

[Argy66]	Argyris, J.H.: Continua and Discontinua, Proc. 1st Conf. Matrix Meth. Struct. Mech. Wright Patterson Air Force Base, Dayton, Ohio 1965, AFFDL TR 1966 66-80

[Argy69]	Argyris, J.H., Scharpf, D.W.: Some General Considerations on the Natural Mode Technique. Aeron. J. Royal Aeron. Soc. **73** (1969) 218-226 and 361-368

[Argy]	Argyris, J.H., Mlejnek, H.P.: Die Methode der Finiten Elemente. Friedrich Vieweg & Sohn Verlag, Braunschweig Wiesbaden, 1986

[Art]	Artemiadis, N.: The Geometry of Fractals. Proc. Nat. Academy of Athens **63** (1988) 479-500 (in Greek)

[Aub77]	Aubin, J.P.: Applied Abstract Analysis. J.Wiley and Sons, New York 1977

[Aub79a]	Aubin, J.P.: Applied Functional Analysis. J. Wiley and Sons, New York 1979

[Aub79]	Aubin, J.P., Clarke, F.H.: Shadow Prices and Duality for a Class of Optimal Control Problems. SIAM J. Control Optimization **17** (1979) 567-586

[Aub84]	Aubin, J.P., Ekeland, I.: Applied Nonlinear Analysis. Wiley-Inter-Science, N.York 1984

[Aub90]	Aubin, J.P., Frankovska, H.: Set-Valued Analysis. Birkhäuser Verlag, Basel, Boston 1990

[Aub91]	Aubin, J.P.: Viability Theory. Birkhäuser Verlag, Basel 1991.

[Auch83]	Auchmuty, G.,: Duality for Non-convex Variational Principles. Journal of Differential Equations **50** (1983) pp. 80-145

[Auch89]	Auchmuty, G.,: Duality Algorithms for Nonconvex Variational Principles. Numerical Functional Analysis and Optimization, **10(3 &4)**, (1989) pp.211-264.

[Avd] Avdelas, A.V., Panagiotopoulos, P.D.: Neural Networks for Computing in Elastoplastic Analysis of structures. Meccanica 1993 (in press)

[Bane81] Banerjee, P.K., Butterfield, R.: Boundary Element Methods in Engineering Science. McGraw-Hill, London 1981

[Ban85] Baniotopoulos, C.C.: Analysis of Structures for "Complete" Constitutive Laws, Doctoral Dissertation. Scientific Annual of the Faculty of Technology of the Aristotle University, Nr. 27 of the Θ' Issue, Thessaloniki 1985

[Ban87] Baniotopoulos, C.C. and Panagiotopoulos, P.D.: A Hemivariational Approach to the Analysis of Composite Material Structures. In "Engineering Applications of New Composites" (ed. by S.A. Paipetis and G.C. Papanicolaou). Omega Publ., London 1987

[Bar83] Barbu, V.: Optimal Control of Variational Inequalities, Pitman, London 1983

[Bar88] Barnsley, M.: Fractals Everywhere. Academic Press, Boston, N. York 1988

[Bar89a] Bardzokas, D, Parton, V.Z., Theocaris, P.S.: The Plane Problem of the Theory of Elasticity for an Orthotropic Field with a Defect. Dokl. Akad. Nauk USSR **309** (1989) 1072-1077

[Bar89b] Bardzokas, D, Parton, V.Z., Theocaris, P.S.: The General Case of the Plane Problem of the Theory of Elasticity for Multiply-Connected Fields. PMM **53** (1989) 485-496

[Barb] Barbu, V.: Optimal control of variational inequalities. Pitman, London 1983

[Bat] Bathe, K.J., Mijailovich, S.: Finite Element Analysis of Frictional Contact Problems. Journal de Mécanique Théor. et Appl. **7** (special issue) (1988) 31-46

[Bath] Bathe, K.J. Finite element procedures in engineering analysis. Prentice-Hall Inc., Englewood Cliffs, N.Jersey (1987)

[Baum] Bauman, E.J.: Multilevel Optimization Techniques and Application to Trajectory Decomposition. In: Advances in Control Systems (ed. by Leondes, C.T.). Academic Press, New York (vol 6) 1968

[Baža] Bažant, Z.P., Chang, T.-P.: Nonlocal Finite Element Analysis of Strain-Softening Solids. J. Engrg Mech., ASCE, Vol. **113** (1987) pp. 89-105

[Bea] Beale, R. and Jackson, T.: Neural Computing. An Introduction. Adam Hilger, Bristol 1990

[Beck] Becker, E. and Bürger, W.: Kontinuumsmechanik. B.G. Teubner, Stuttgart 1975

[Bely] Belytschko, T., Bažant,Z.P., Hyun, Y. L., Chang, T. P.: Strain-softening Materials and Finite Element Solutions. Comput. Struct. Vol. **23**, (1986) 163-180

[Ber67] Berger, M.S.: On Von Karmánáns Equations and the Buckling of a Thin Elastic Plate. I. The Clamped Plate. Comm. Pure Appl. math. **XX** (1967) 687-719

[Ber68] Berger, M.X. and Fife, P.C.: Von Karmánáns Equations and the Buckling of a Thin Elastic Plate. II. Plate with General Edge Conditions. Comm. Pure Appl. Math. **XXI** (1968) 227-241

[Berts] Bertsekas, D.P.: Constrained Optimization and Lagrange Multiplier Methods. Academic Press, New York, 1982

[Bis86] Bisbos, C.: A Cholesky Condensation Method for Unilateral Contact Problems. S.M.Archives **11** (1986) 1-23

[Bis90] Bisbos, C.: A New Algorithm for 3-D Unilateral Frictional Contact Problems. Report for INPRO (Innovationsgesellschaft für die Produktion in der Automobilindustrie) Berlin 1990

[Bis92] Bisbos, C.: A Nash-game formulation for Frictional Unilateral Contact Problems. Proc.Contact Mechanics Int. Symp. (ed. by A. Curnier) Presses Polytechn. et Univers. Romandes, Lausanne pp. 369-390, 1992

[Bland] Blandford, G.E.; Ingraffen, A.R.; Ligget, J.A.: Two-Dimensional Stress Intensity Factor Computations Using the Boundary Element Method. Int. J. Num. Meth. Eng. **67** (1981) pp. 387-404

[Böh] Böhm, J.M.: Eine Inkrementelle Formulierung für Festkörperkontakt mit Reibung. Doct. Dissertation, RWTH Aachen 1987

[Bress] Bressloff, P.C.: Stark, J.: Neural Networks, Learning Automata and Iterated Function Systems. In: Crilly, A.J.; Earnshaw, R.A.; Jones, H. (eds.) Fractals and Chaos. Springer Verlag, Heidelberg, N.York 1991.

[Bréz72] Brézis, H.: Problèmes Unilatéraux. J. Math. pures et Appl. **51** (1972) 1-168

[Bréz73]   Brézis, H.: Opérateurs Maximaux Monotones et Semigroupes de Contractions dans les Espaces de Hilbert. North-Holland Publ. Co., Amsterdam and American Elsevier Publ. Co., New York 1973

[Brow]   Browder, F.: Variational methods for nonlinear elliptic eigenvalue problems. Bull. Amer. Math. Soc. **7** (1965) 176-183

[Budi]   Budiansky, B.: Theory of Buckling and Postbuckling Behaviour of Elastic Structures. In: Advances in Applied Mechanics (ed. by Chia-Shun Yih), Academic Press, London 1994, pp. 1-65

[Cam]   Campos, L.T., Oden, J.T., Kikuchi, N.: A Numerical Analysis of a Class of Contact Problems With Friction in Elastostatics. Comp. Meth. appl. Mech. Eng. **34** (1982) 821-845

[Carls]   Carlsson, L.A., Pipes, R.B.: Experimental Characterization of Advanced Composite Materials. (Prentice-Hall Inc., Englewood Cliff, New Jersey 1987

[Cha]   Chaboussi, J., Wilson, E.L., Isenberg J.: Finite Element for Rock Joints and Interfaces. ASCE, SM 10, **99** (1973) 833-848

[Ch]   Chang, K.C.: Variational Methods for Non-Differentiable Functionals and their Applications to Partial Differential Equations. J.Math.Anal. Appl. **80** (1981) 102-129

[Clar73]   Clarke, F.H.: Necessary Conditions for Nonsmooth Problems in Optimal Contorl and the Calculus of Variations. Ph.D.Thesis, University of Washington, Seattle 1973

[Clar75]   Clarke, F.H.: Generalized Gradients and Applications. Trans. A.M.S. **205** (1975) 247-262

[Clar81]   Clarke, F.H.: Generalized Gradients of Lipschitz Functionals. Advances in Math. **40** (1981) 52-67

[Clar83]   Clarke, F.H.: Optimization and Nonsmooth Analysis. Wiley, New York 1983

[Clec]   Clech, J.P., Lewis, J.L., Keer, L.M.: A Finite Element Technique for Determining Mode I Stress Intensity Factors:Application to no-slip Bimaterial Crack Problem. Comput. Struct., Vol **23** (1986) pp. 715-724

[Co]   Cocu, M.: Existence of Solutions of Signorini Problems with Friction. Int. J. Engng. Sci. **22** (1984) 567-575

[Com]   Comninou, J.: Interface Crack with Friction in the Contact Zone. J. Appl. Mech. **44** (1977) 780-781

[Cott] Cottreli, B.; Rice, J.R.: Slightly Curved or Kinked Cracks. Int. J. Fracture **16** (1980) pp. 155-169

[Cris82] Crisfield, M.A. Accelerated solution techniques and concrete cracking. Comp. Meth. in Appl. Mech. Eng. **33** (1982) 585-607

[Cris85] Crisfield, M.A.: New solution procedures for linear and nonlinear finite element analysis, The Mathematics of Finite Elements and Applications V, Whiteman, J.R. (ed.). Academic Press, London (1985) pp. 49-81

[Cris86] Crisfield, M.A.: Snap-through and snap-back response in concrete structures and the dangers of under-integration. Int. J. Num. Meth. in Eng. **22** (1986) 751-767

[Cris88] Crisfield, M.A., Wills, J.: Solution Strategies and Softening Materials. Comp. Meth. Appl. Mech. Eng. **66** (1988) 267-289

[Cris91] Crisfield, M.A.: Nonlinear Finite Element Analysis of Solids and Structures. J. Wiley, N. York 1991

[Cro77] Crouzeix, J.P.: Conjugacy in Quasiconvex Analysis. In: Convex Analysis and its Applications (ed. by A. Auslender). Springer-Verlag, Berlin 1977

[Cur86] Curnier, A.: A Theory of Friction, Intern. J. Solids Struct. **20** (1986) 637-647

[Cur88] Curnier, A., Alart, P.: A Generalized Newton Method for Contact Problems with Friction. J. Méc. Théor. Appl. **7** (Special issue) (1988) 67-82

[Davie] Davies, P., Benzeggagh, M.L.: Interlaminar Mode-I Fracture Testing. In: Application of Fracture Mechanics ot Composite Materials (ed. K.Friedrich). Elsevier Publishers B.V, Amsterdam 1975

[Davis] Davison, J.B., Kirby, P.A., Nethercot,D.A.: Rotational Stiffness Characteristics of Steel Beam to Column Connections. J. Constr. Steel Res. **8** (1987) 17- 54

[Dbv] Deutsche Beton-Verein: Faserbeton, (Fassung August 1992) D.B.V.-Merkblätter. Deutche Beton-Verein E.V, Wiesbaden 1992

[De] Demkowicz, L., Oden, J.T.: On some Existence and Uniqueness Results in Contact Problems with Nonlocal Friction. Nonl. Anal. **6** (1982) 1075-1093

[DelP] Del Piero, G., Maceri, F.: On the Delamination Problem of Two-Layer Plates. Proc. 2nd Meeting on Unilateral Problems in Struct. Analysis Springer Verlag, Wien, New York (1985) 1-15

[Dem83]	Demyanov, V.F., Rubinov, A.M.: On Quasidifferentiable Mappings. Math. Operationsforsch. u. Statist., (Optimization) **14** (1983) 3-21
[Dem85]	Demyanov, V.F., Vasilev, L.V.: Nondifferentiable Optimization. Optimization Software Inc., N.York 1985
[Dem86a]	Demyanov, V.F., Dixon, L.C.W. (eds): Quasidifferential Calculus. Math. Progr. Study **29**, North Holland, Amsterdam 1986
[Dem 86b]	Deyanov, V.F., Rubinov, A.M.: Quasidifferentable Calculus. Optimization Software Inc., N. York 1986
[Dem86c]	Demyanov, V.F., Polyakova, L.N., Rubinov, A.M.: Nonsmoothness and Quasidifferentiability. Math. Progr. Study **29** (1986) 1-19
[Dem89a]	Demyanov, V.F.: Codifferentiability and Codifferentials of Nonsmooth Functions. Soviet Math. Dokl. **38** (1989) 631-634
[Dem89b]	Demyanov, V.F.: Smoothness of Nonsmooth Functions. In: Nonsmooth Optimization and Related Topics. (ed. by F.H.Clarke, V.F. Demyanov and G.Giannessi), Plenum Press, N.York, London 1989
[Dru]	Drucker, D.C.: A more Fundamental Approach to Plastic Stress-Strain Relations. Proc. First U.S. Nat. Congr. on Appl. Mech. Chicago 1951
[Dub88]	Dubourg, M.C., Mouwakeh, M., Villechaise, B.: Interaction Fissure-Contact. Etude Théorique et Expérimentale. J. de Mécanique Théor. et Appl. **7** (1988) 623-643
[Dub89a]	Dubourg, M.C., Villechaise, B.: Unilateral Contact Analysis of a Crack with Friction. Eur. J. Mech. A/Solids **8** (1989) 309-319
[Dub89b]	Dubourg, M.C.: Le Contact Unilateral avec Frottement le long de Fissures de Fatigue dans les Liaisons Mécaniques. Doct. Dissertation, Institut Nat.Sc.Appl. Lyon 1989, Nr.891 SAL 0088
[Dun]	Dundurs, J., Comninou, M.: Some Consequences of the Inequality Conditions in Contact and Crack Problems. J. of Elast. **9** (1979) 131-137
[Dugi76]	Dugill J.W.: On Stable Progressively Fracturing Solids. J. of Applied Math. and Phys. **27** (1976) 423-437
[Dugi80]	Dugill J.W., Rida M.A.M.: Further Considerations of Progressively Fracturing Solids. ASCE Eng. Mech. Div. **106** (1980)
[Dunf]	Dunford, N., Schwartz, J.T.: Linear Operators Part I: General Theory. Interscience Publishers, New York 1966

[Dut] Dutta, B.K., Miti S.K., Kakodkar A.: On the Use of One Point and Two Points Singularity Elements in the Analysis of Kinked Cracks. Int. J. Num. Meth. Eng. **29** (1990) 1487-1499

[Duv71] Duvaut, G., Lions, J.L.: Un Probléme d' élasticité avec Frottement, J. de Mécanique **10** (1971) 409-420

[Duv72] Duvaut, G., Lions, J.L.: Les Inéquations en Mécanique et en Physique. Dunod, Paris 1972

[Duv80] Duvaut. G.: Equilibre d' un Solide élastique avec Contact Unilateral et Frottement de Coulomb. C.R. Acad. Sc. Paris **290** (1980) 263-265

[ECC] E.C.C.S, Composite Structures. The Construction Press, London, N. York 1981

[Eke] Ekeland, I., Temam, R.: Convex Analysis and Variational Problems. North Holland, Amsterdam and American Elsevier, New York 1976

[Ell] Ellaia, R., Hassouni, A.: Characterization of Nonsmooth Functions through their Generalized Gradients. Optimization **22** (1991) 401-416

[Euroc] Eurocode No 4: Common unified rules for composite steel and concrete structures (Draft), Commision of the European Communities, Brussels 1986

[Evan] Evans L., Gariepy, R.: Measure Theory and Fine Properties of Functions. CRC Press, Boca Raton 1992

[Fal] Falconer, K.J.: The Geometry of Fractal Sets. Cambridge Univ. Press, Cambridge 1985

[Fed] Feder, J.: Fractals. Plenum Press, N. York 1988

[Fei] Feijóo, R.A., Barbosa, H.J.C., Zouain, N.: Numerical Formulations for Contact Problems with Friction. J. de Méc. Théor. Appl. **7** (Special issue) (1988) 129-144

[Fel] Felippa, C.A., Park K.C.: Direct Time Integration Methods in Nonlinear Structural Dynamics. Comp. Meth. Appl. Mech. Eng., **17/18** (1979) 277-313

[Fels] Fels, A., Hornbogen, E., Tio, T.K., Friedrich, K., Köster, V.: Poulout Tests with Metallic Glass Ribbons in Differenta Matrices. In: G. Ondracek (ed.) Verbundwerkstoffe. Phasenverbindung und mechanische Eigenschaften, Deutsche Gesellschaft für Metallkunde Vol. **1** (1985) 185-194

[Fich63] Fichera, G.: The Signorini Elastostatics Problem with Ambiguous Boundary Conditions. Proc. Int. Conf. Application of the Theory of Functions in Continuum Mechanics, Vol. I, Tbilisi 1963

[Fich64] Fichera, G.: Problemi Elastostatici con Vincoli Unilaterali: il Problema di Signorini con Ambigue Condizioni al Contorno. Mem. Accad. Naz. Lincei, VIII **7** (1964) 91-140

[Fich72] Fichera, G.: Boundary Value Problems in Elasticity with Unilateral Constraints. In: Encyclopedia of Physics (ed. by S.Flügge) Vol. VI a/2. Springer-Verlag, Berlin 1972

[Flet] Fletcher R.: Practical Methods of Optimization. (2nd edition). J. Wiley & Sons, Chichester-N. York (1990)

[Flet82] Fletcher R.: Second order corrections for nondifferentiable optimization. In: G.A.Watson, (ed.), Numerical Analysis Dundee, Lecture Notes in Mathematics, Springer Verlag, Wien-N.York 1982

[Flo] Floegl, H., Mang, H.A.: Tension Stiffening Concept Based on Bond Slip. ASCE (ST 12) **108** (1982) 2681-2701

[For] Fortin, M., Glowinski, R.: Méthodes de Lagrangien Augmenté. Dunod, Paris 1982

[Frém83] Frémond, M.: Conditions unilatérales et non linéarité en calcul à la repture. Matem. Aplicadá e Computational Brazil **2** (1983) 237- 256

[Frém] Frémond, M.: Contact unilatéral avec adhérence: une théorie du premier gradient. Unilateral Problems in Structural Analysis - 2, (ed. by Del Pierro, G. and Maceri, F.), CISM Courses and Lectures **304**, pp. 33-45, Springer-Verlag, N.York, Wien, 1987

[Frém88] Frémond, M.: Contact with Adhesion. In Topics in Nonsmooth Mechanics (ed by J.J.Moreau, P.D.Panagiotopoulos, G. Strang), pp. 157-185 Birkhäuser Verlag, Boston, Basel, Berlin 1988

[Gas88a] Gastaldi F.: Remarks on a Noncoercive Contact Problem with Friction in Elastostatics. Publ. No. **649**, Istituto Anal. Num. C.N.R, Pavia 1988

[Gas88b] Gastaldi, F. and Martins J.A.C.: A Noncoercive Steady-Sliding Problem with Friction. Publ. No. **650**, Istituto Anal. Num. C.N.R. Pavia 1988

[Gatt] Gattesco, N.: Long-span steel and concrete beams with partial shear connection. Studi e ricerche **12** (1990) 243-266

[Ger73a] Germain, P.: Cours de Mécanique des milieux continus I. Masson, Paris 1973

[Ger73b] Germain, P.: La méthode de puissances virtuelles en mécanique des milieux continus, 1ère partie. Théorie du second gradient. J. de Mécanique **12** (1973) 235-274

[Ger73c] Germain, P.: The Method of Virtual Power in Continuum Mechanics. Part 2: Microstructure. SIAM J. Appl. Math. **25** (1973) 556-575

[Ger74] Germain, P.: The Role of Thermodynamics in Continuum Mechanics. In: Foundations of Continuum Thermodynamics (ed. by D. Domingon, M.N.R. Nina adn J.H. Whitelaw) McMillan, London 1974

[Ger83] Germain, P., Nguyen, Q.J., Suquet, P.: Continuum Thermodynamics. J.Appl. Mech. Trans. ASME **50** (1983) 1010-1020

[Gerst] Gerstle,K.H.: Effect of Connections on Frames. J.Constr.Steel Res. **10** (1988) 241-268

[Gill] Gill, P.E., Murray, W., Wright M.H.: Practical Optimization. Academic Press, New York 1981

[Gir] Girkmann, K.: Flächentragwerke. Springer-Verlag, Wien 1963

[Glow] Glowinski, R., Lions J.L., Tremolieres R.: Numerical Analysis of Variational Inequalities. Studies in Mathematics and its Applications, Vol.8, North-Holland- Elsevier, Amsterdam-New York 1981

[Gol84] Gol'dshtein, R.V., Spector, A.A.: Variational Method of Investigation of Three-Dimensional Mixed Problems of a Plane Cut in an Elastic Medium in the Presence of Slip and Adhesion of its Surfaces. PMM U.S.S.R. **47** (1984) 232-239

[Gol88] Gol'dshtein, R.V., Spector, A.A.: Variational Methods of Solution and Analysis of Spatial Contact and Mixed Problems with Friction. In: Mechanics of Deformable Solids (ed. by A.Yu.Ishlinski) Allerton Press, N.York 1988

[Goto] Goto, Y., Chen, W.F.: On the Computer Based Design Analysis for Flexibly Joined Frames. J.Constr. Steel Res. **8** (1987) 203-232

[Göp] Göpfert, A.: Mathematische Optimierung in allgemeinen Vektorräumen. B.G. Teubner, Leipzig 1973

[Goya89] Goyal, S., Ruina, A., Papadopoulos, J.: Limit Surface and Moment Function Descriptions of Planar Sliding. Proc., IEEE International Conference on Robotics and Automation, Scottsdale, AZ, 1989 pp. 794-799

[Goya91]   Goyal, S., Ruina, A., Papadopoulos, J.: Planar Sliding with dry friction: Part 1. Limit surface and moment function. Wear **17** (1991) 307-330

[Gre]   Green, A.E., Naghdi, P.M.: A General Theory of an Elastic-Plastic Continuum. Arch. Rat. Mech. Anal. **18** (1965) 251-281

[Gree]   Green, A.K., Bowyer W.H.: The testing analysis of novel top-hat stiffener fabrication methods for use in GRP ships. In Proceedings of the 1st International Conference on Composite Structures (ed. by I.H. Marshall), Applied Science, Barking, Essex 1981

[Gr]   De Groot, S.R.: Thermodynamik irreversible Prozesse. Bibliographisches Institut, Mannheim 1960

[Groot]   Grootenboer, H.J., Leijten, S.F.C.H., Blauuwendraad, J.: Numerical Models for Reinforced Concrete Structures in Plane Stress. Heron **26** (1981) 1-83

[Hal74]   Halphen, B. and Son, N.Q.: Plastic and Viscoplastic Materials with Generalized Potential. Mech. Res. Comm. **1** (1974) 43-47

[Hal75]   Halphen, B. and Son, N.Q.: Sur les Matériaux Standards Généralisés. J. de Mécanique **14** (1975) 39-63.

[Ham67]   Hamel, G.: Theoretische Mechanik. Springer-Verlag, Berlin 1967

[Hana77]   Hanafusa, H., Asad, H.: Stable prehension by a robot hand with elastic fingers. Proc. of the 7th Int. Symposium on Industrial Robots, Tokyo 1977, pp. 361-368

[Has82]   Haslinger, J., Hlaváček, I.: Approximation of the Signorini Problem with Friction by a Mixed Finite Element Method. J.Math.Anal. **86** (1982) 99-122

[Has84]   Haslinger, J., Panagiotopoulos, P.D.: The Reciprocal Variational Approach to the Signorini Problem with Friction. Approximation Results. Proc. Royal Soc. of Edinburgh **98A** (1984) 365-383

[Hasli86a]   Haslinger, J., Neittaanmaki, P., Tiihonen, T.: Shape Optimization on Contact Problems Based on Penalization of the State Inequality. Aplikace Matematiky **31** (1986) 1-88

[Hasli86b]   Haslinger, J., Neittaanmaki, P.: On Optimal Shape Design of Systems Governed by Mixed Dirichlet-Signorini Boundary Value Problems. Math. Meth. in the Appl. Sci., **8** (1986) 157-181

[Hasli89]   Haslinger, J., Panagiotopoulos, P.D.: Optimal Control of Hemivariational Inequalities. In: Control of Boundaries and Stabilization, (ed. by J.Simon), Lect. Notes in Control and Information Sciences, Vol. **1925** Springer-Verlag, N.York 1989

[Hasli93]   Haslinger, J., Panagiotopoulos P.D.: Optimal Control of Systems Governed by Hemivariational Inequalities Existence and Approximation Results. J. Nonlin. Anal., (to appear)

[Hl88]   Hlaváček, J., Haslinger, J., Nečas, J., Lovisek, J.: Solution of Variational Inequalities in Mechanics. Springer Verlag, N.York, Berlin 1988.

[Hün]   Hünlich, H., Naumann, J.: On General Boundary Value Problems and Duality in Linear Elasticity, I, II. Apl. Matematiky **23** (1978) 208-229 and **25** (1980) 11-32

[Hein70]   Heinz, C.: Das d' Alembertsche Prinzip in der Kontinuumsmechanik. Acta Mechanica **10** (1970) 110-129

[Henck]   Hencky, H.: Zur Theorie plastischer Deformationen und der hierdurch im Material hervorgerufenen Nachspannungen. ZAMM **4** (1924) 323-334

[Hill48]   Hill, R.: A Variational Principle of Maximum Plastic Work in Classical Plasticity. Quart. J.Mech. Appl. Math. **1** (1948) 18-28

[Hill50]   Hill, R.: Mathematical Theory of Plasticity. Univ. Press, Oxford 1950

[Hiri]   Hiriart-Urruty.: From Convex Optimization to Nonconvex Optimization. Necessary and Sufficient Conditions for Global Optimality. In: Nonsmooth Optimization and Related Topics, (eds. F.H. Clarke, V.F. Dem'yanov and F. Giannessi), pp. 219-239 Plenum Press, New York-London 1989

[Hop82]   Hopfield, J.J.: Neural Networks and Physical Systems with Emergent Collective Computational Abilities. Proc. of the Nat. Acad. of Sciences **79** (1982) 2554-2558

[Hop85]   Hopfield, J.J., Tank, D.W.: "Neural" Computations of Decisions in Optimization Problems. Biol. Cybern. **52** (1985) 141-152

[Howe]   Howe, R.D., Kao, I., Cutkosky M.R.: The Sliding of Robot Fingers Under Combined Torsion and Shear Loading. Proc., IEEE Conference on Robotics and Automation, Philadelphia, PA, 1988, pp. 103-107

[Hug]   Hughes, T.J.R., Taylor, R.L., Kanoknukulchai,W.: A Finite Element Method for Large Displacement Contact and Impact Problems. In: Formulations and Computational Algorithms in Finite Element Analysis (ed. by K.F.Bathe, F.T.Oden and W.Wunderich). MIT Press, Cambridge 1977

[Hult]   Hult J., Travniček L.: Carrying capacity of Fibre Bundles with Varying Strength and Stiffness. J. Méc. Théor. et Appl. **2** (1983) 643-657

[Hutch]   Hutchinson, J.F.: Fractals and Selfsimilarity. Indiana J.of Math. **30** (1981) 713-747

[Ilyu]   Ilyushin, A.A.: On the Increment of Plastic Deformation and Yield Function. Prikl. Math. Mech. **24** (1960) 663-666 (in Russian)

[Iof81]   Ioffe, A.D.: Nonsmooth Analysis: Differential Calculus of Nondifferentiable Mappings. Trans. Am. Math. Soc. **266** (1981) 1-56

[Iof82]   Ioffe A.D.: Nonsmooth Analysis and the Theory of Fans. In: Convex Analysis and Optimization. ed. by J.P.Aubin and R.B.Vinter, Res. Notes in Math. Vol. **57**, Pitman, London 1982

[Ja83]   Jarusek, J.: Contact Problems with Bounded Friction. Coercive Case. Czech. Math. J. **33** (1983) 254-278

[Ja84]   Jarusek, J.: Contact Problems with Bounded Friction. Semicoercive Case. Czech. Math. J. **34** (1984) 619-629

[Jea85]   Jean, M., Pratt, E.: A System of Rigid Bodies with Dry Friction. Int. J. Eng. Sci., 23 (1985) 497-513

[Jea87]   Jean, M., Moreau, J.J.: Dynamics in the Presence of Unilateral Contacts and Dry Friction; a Numerical Approach. In: Unilateral Problems in Structural Analysis 2 (ed. G. Del Piero and F.Maceri), CISM Courses and Lectures No **304**, pp. 151-196, Springer-Verlag, Wien 1987

[Jin]   Jin, H., Runesson, K., Samuelsson, A.: Application of the Boundary Element Method to Contact Problems in Elasticity with a Nonclassical Friction Law. In: Boundary Elements IX, Vol **2**: Stress Analysis Applications (ed. by Brebbia, C.A.; Wendland, W.L.; Kuhn, G.) Springer Verlag, Berlin 1986

[Johns]   Johnson, R.P., May, I.M.: Partial-interaction design of composite beams. The Struct. Eng. **53** (1975) 305-311

[Jon]   Jonsson, A., Wallin, H.: Function Spaces on Subsets of $\mathbb{R}^3$. Math. Rep. Vol.2, Harwood Acad. Publ., Chur, London 1984

[Jone]   Jones R.: Mechanics of Composite Materials. McGraw Hill, New York 1975

[Ju]   Ju, J.W., Taylor, R.L.: A Perturbed Lagrangian Formulation for the Finite Element Solution of Nonlinear Frictional Contact Problems. J. Méc. Théor. et Appl. **7** (Special issue) (1988) 1-14

[Ka88a] Kalker, J.J.: The Quasistatic Contact Problem with Friction for Three Dimensional Elastic Bodies. J.Méc.Theor.et Appl. **7** (Special issue) (1988) 55-66

[Ka88b] Kalker, J.J.: Contact Mechanical Algorithms. Comm. in Applied Num. Methods **4** (1988) 25-32

[Ka90] Kalker, J.J.: Three Dimensional Elastic Bodies in Rolling Contact. Kluwer Acad. Publ., Dordrecht 1990

[Kar 91] Karamanlis, I.: Buckling Problems in Composite von Karman Plates. Doct. Thesis, Aristotle University Dept. of Civil Eng. 1991

[Kar 92] Karamanlis, I., Panagiotopoulos, P.D.: The Eigenvalue Problem in Hemivariational Inequalities and its Application to Composite Plates. Journal of the Mech. Behaviour of Materials (Freund Publ. House, Tel Aviv) 1992 (to appear)

[Ker] Kerr, J., Roth, B.: Analysis of Multifingered Hands. Int. J. of Robot. Res. **4** 1986, 3-17

[Kha] Khanna Tarun: Foundations of Neural Networks. An Introduction. Addison-Wesley Pub.Co., New York 1990

[Kik] Kikuchi, N., Oden, J.T.: Contact Problems in Elasticity. A Study of Variational Inequalities and Finite Element Methods. SIAM Publ., Philadelphia 1988

[Kim 84] Kim, S.J. and Oden, J.T.: Generalized Potentials in Finite Elastoplasticity. Int. J. Engng. Sci. **22** (1984) 1235-1257

[Kim 85] Kim, S.J., Oden, J.T.: Generalized Flow Potentials in Finite Elastoplasticity - II. Examples. Int. J. Engng. Sci. **23** (1985) 515-530

[Kla84] Klarbring, A.: Contact Problems with Friction Using a Finite - Dimensional Description and the Theory of Linear Complementarity. Linköping Studies in Science and Technology, Thesis No. 20, Linköping Institute of Technology, Linköping, Sweden 1984

[Kla87] Klarbring, A.: Contact Problems with Friction by Linear Complementarity. In Unilateral Problems in Structural Analysis, 2 (ed. by Del Piero, G., Maceri, F.). CISM Courses and Lectures, No. 304 Springer Verlag, Wien, N.York 1987

[Kla88] Klarbring, A., Björkman, G.: A Mathematical Programming Approach to Contact Problem with Friction and Varying Contact Surface. Comp. and Struct. **30** (1988) 1185-1198

[Kla90a]   Klarbring, A.: Examples of Non-Uniqueness and Non-Existence of Solutions to Quasistatic Contact Problems with Friction. Ing. Arch. **60** (1990) 529-541

[Kla90b]   Klarbring, A., Mikelić, A., Shillor, M.: Duality Applied to Contact Problems with Friction. Appl. Math. Optim. **22** (1990) 211-226

[Kla90c]   Klarbring, A.: Derivation and Analysis of Rates Boundary-Value Problems of Frictional Contact. European J. of Mech. A/Solids **1** (1990) 53-85

[Kol91]   Koltsakis, E.K.: Theoretical and Numerical Study of Structures with Nonmonotone Boundary Conditions. Application to Adhesion Joints, Doct. Dissertation, Dept. of Civil Eng., Aristotle University, Thessaloniki 1991

[Kolts]   Koltsakis, E.K., Panagiotopoulos P.D.: Computational Aspects of Hemivariational Inequalities : Interfaces with Nonsmooth Softening Mechanical Behavior. Preceeding of the European Conference of New Advances in Computational Structural Mechanics 2-5 April 1991 Giens-France pp. 299-306

[Kostr]   Kostreva, M. M.: Generalization of Murty's direct algorithm to linear quadratic programming. J. of Optim. Theory and Appl. **62**, 1989, 63-76

[Kor]   Kortesis, S., Panagiotopoulos, P.D.: Neural Networks for Computing in Structural Analysis. Methods and Prospects of Applications. Int. J. Num. Eng. 1993 (in press)

[Kuma]   Kumar,V., Waldron, K.J.: Suboptimal algorithms for force distribution in multifingered grippers. IEEE Transaction on Robotics and Automation. **5** (1989) 491-498

[Kozl]   Kozlov V.V., Treshchëv D.V.: Billiards. A Genetic Introduction to the Dynamics of Systems with Impacts. Amer. Math. Soc. (Transl. of Math. Monographs Vol. **89**) Providence 1991

[Kress]   Kress, R.: Linear Integral Equations. Springer Verlag, Berlin 1989

[Kuhn]   Kuhn, J.M., Bucuner, C.D.: Effect of concrete placement on shear strength of headed studs. (ASCE) J. of Str. Div. **112** (1986) 1965-1970

[Kwa88]   Kwak, B.M., Lee, S.S.: A Complementarity Problem Formulation for Two-Dimensional Frictional Contact Problem. Comp. Struct. **28** (1988) 469-480

[Kwa91]   Kwak, B.M.: Complementarity Problem Formulation of Threedimensional Frictional Contact. ASME J. Applied Mech. **58** (1991) 134-140

[Lan]   Lanczos, C.: The Variational Principles of Mechanics. University of Toronto Press, Toronto 1966

[Land]   Landesman, E.M., Lazer, A.C.: Nonlinear Perturbations of Linear Elliptic Boundary Value Problems at Resonance. J.Math.Mech. **19** (1970) 609-623

[Lemk]   Lemke, C. E.: Some pivot schemes for the linear complementarity problem. Mathematical Programming Study. **7**, 1978, 15-35 (ed. by M. L. Balinski and R. W. Cottle)

[Léné73] Léné, F.: Sur les matériaux élastiques à energié de déformation non quadratique. Thèse de $3^{\text{ème}}$ cycle, Université Paris VI, 1973

[Léné74] Léné, F.: Sur les matériaux élastiques à energié de déformation non quadratique. J. de Mécanique **13** 9 (1974) 499-534

[Li86]   Liolios, A.A.: A Linear Complementarity Approach for the Signorini Problem with Friction. ZAMM **66** (1986) 349-352

[Li87]   Liolios, A.A.: Upper and Lower Solution Estimates in Unilateral Viscoelastodynamics. Acta Mech. **66** (1987) 275-278

[Li88]   Liolios, A.A.: Seismic Interaction Between Adjacent Structures: A Linear Complementarity Approach for the Unilateral Elastoplastic Softening Contact with Friction. In: Structural Dynamics and Earthquake Engineering, (ed. by Kounadis A.N. and Krätzig W.B.) Athens 1988

[Li89]   Liolios, A.A.: A Linear Complementarity Approach for the Non-Convex Dynamic Problem of Unilateral Contact with Friction Between Adjacent Structures. ZAMM **69** (1989) 420-422

[Li91]   Liolios, A.A.: A Numerical Estimation for the Influence of Modifications to Seismic Interaction Between Adjacent Structures. In: Earthquake Resistant Construction and Design, (ed. by Savidis S.A.), A.A. Balkema Publ., Rotterdam 1991

[Lio69]   Lions, J.L.: Quelques méthodes de résolution des problèmes aux limites non linéaires. Dunod/Gauthier-Villars, Paris 1969

[Lio71]   Lions, J.L.: Optimal control of systems governed by partial differential equations. Springer-Verlag, Berlin 1971

[Liq91]   Liqun Qi.: Quasidifferentials and maximal normal operators. Math. Progr. **49** (1991) 263-271

[Lui]   Lui, E.M., Chen, W.F.: Steel Frame Analysis with Flexible Joints. J. Constr. Steel Res. **8** (1987) 161-202

[Ma68]   Maier, G.: A Quadratic Programming Approach for Certain Classes of Nonlinear Structural Problems. Meccanica **3** (1968) 121-130

[Ma71]   Maier, G.: Incremental plastic analysis in the presence of large displacements and physical instabilizing. Int. J. Solids Structures. **7** (1971) 345-372

[Ma78]   Maier, G.: Future Directions in Engineering Plasticity. In: Engineering Plasticity by Mathematical Programming (ed. by Cohn, M.Z., Maier, G. and Grierson, D.E.) Pergamon Press, New York 1978

[Man]   Mandelbrot, B.: The Fractal Geometry of Nature. W.H. Freeman and Co., New York 1972

[Mand]   Mandel, J.: Plasticité classique et viscoplasticité. CISM Course No. **97**. Springer-Verlag, Wien 1972

[Mar86]   Martins, J.A.C.: Dynamic Frictional Contact Problems Involving Metallic Bodies. Ph.D.Dissertation, University of Texas at Austin 1986

[Mar87]   Martins, J.A.C., Oden, J.T.: Existence and Uniqueness Results for Dynamic Contact Problems with Nonlinear Normal and Friction Interface Laws. Nonlinear Anal. **11** (1987) 407-428

[Marke]   Markenscoff, X., Ni, L., Papadimitriou, Ch.: The Geometry of Grasping. Int. J. of Robot. Res. **9** 1990 61-74

[Mason]   Mason, M. T., Salisbury, J.K.: Robot Hands and the Mechanics of Manipulation. MIT Cambridge Press, 1985

[Maug80] Maugin, G.A.: The Method of Virtual Power in Continuum Mechanics. Application to Coupled Fields. Acta Meccanica **35** (1980) 1-70

[Michal] Michalowski, R., Mroz, Z.: Associated and Non-associated Sliding Rules in Contact Friction Problems. Arch. of Mech. (Arch. Mech. Stosowanej) **30** (1978) 259-276

[Mign76] Mignot, F.: Contrôle dans ses inéquations variationnelles elliptiques. J.Funct. Anal. **22** (1976) pp. 130-185

[Mign84] Mignot, F., Puel, J.P.: Optimal control in some variational inequalities. SIAM J. Control and Optim. **22** (1984) 466-476

[Mis92a] Mistakidis, E., Panagiotopoulos, P.D.: On the Numerical Treatment of Nonmonotone (zigzag) Friction and Adhesive Contact Problems with Debonding. Approximation by Monotone Subproblems, 1992, Int.J.Comp. and Struct. (to appear)

[Mis92b] Mistakidis, E.: Theoretical and Numerical Study of Structures with Nonmonotone Boundary and Constitutive laws. Algorithms and Applications, Doct. Dissertation, Dept. of Civil Eng. Aristotle Univ. Thessaloniki (1992)

[Mis93] Mistakidis, E., Panagiotopoulos, P.D.: On the Approximation of Nonmonotone Multivalued Problems by Monotone Subproblems. (Submitted for publication)

[Mis93a] Mistakidis, E., Thomopoulos, K., Avdelas, A., Panagiotopoulos, P.D.: Shear Connectors in Composite Beams: A New Accurate Algorithm. (Submitted for publication)

[Mis93b] Mistakidis, E., Panagiotopoulos, P.D., Panagouli O.K.: Fractal Surfaces and Interfaces in structures. Methods and Algorithms. Chaos, Solitons and Fractals **2** (1992) 551-574

[Mit83] Mitsopoulou, E.: Unilateral Contact, Dynamic Analysis of Beams by a Time-Stepping Quadratic Programming Procedure. Meccanica **18** (1983) 254-265

[Mit87] Mitsopoulou, E.N., Doudoumis, I.N.: A Contribution to the Analysis of Unilateral Contact Problems with Friction. Solid Mech. Arch. **12** (1987) 165-186

[Mit91] Mitsopoulou, E., Panagiotopoulos, P.D., Zervas P.A.: Dynamic Boundary Integral "Equation" Method for Unilateral Contact Problems. Eng. Anal. with Bound. Elements **8** (1991) 192-199

[Mit93] Mitsopoulou, E., Panagiotopoulos, P.D., Zervas P.A.: A Boundary Integral Equation Approach to Dynamic Inequality Problems and Applications. Eng. Anal. with Bound. Elements (to appear)

[Mof] Moffat,U.R, Dowling, P.J.: The longitudinal bending behaviour of composite box girder bridges having incomplete interaction. The Structural Engineer **56B** (1978) 53-60

[Mon] Monteiro Marques, M.D.P.: Inclusões Differenciais e Choques inelasticos. Doct. Dissertation, Faculty of Sciences, Univ. of Lisbon, 1988

[Mor67] Moreau, J.J.: Fonctionnelles Convexes. Séminaire sur les équations aux Dérivées Partielles. Collège de France, Paris 1967

[Mor68]   Moreau, J.J.: La Notion du Surpotentiel et les Liaisons Unilatérales on Elastostatique. C.R.Acad.Sci. Paris **167A** (1968) 954-957

[Mor70]   Moreau, J.J.: Sur les lois de frottement, de plasticité et de viscosité. C.R. Acad. Sc. Paris **271A** (1970) 608-611

[Mor86]   Moreau, J.J.: Une Formulation du Contact Frottement Sec; Application au Calcul Numérique. C.R. Acad. Sci. Paris, Sér. II, **302** (1986) 799-801

[Mor88a]   Moreau, J.J., Panagiotopoulos, P.D.(eds): Nonsmooth Mechanics and Applications, CISM Vol. **302**, Springer Verlag, Wien 1988

[Mor88b]   Moreau, J.J., Panagiotopoulos, P.D., Strang, G.(eds): Topics in Nonsmooth Mechanics, Birkhäuser Verlag, Basel, Boston 1988

[Mor88c]   Moreau, J.J.: Unilateral Contact and Dry Friction in Finite Freedom Dynamics. In: Nonsmooth Mechanics and Applications (ed. by J.J.Moreau and P.D.Panagiotopoulos),CISM Vol. **302**, Springer Verlag, Wien, N.York 1988

[Mos]   Moser, K.: Faserkunststoffverbund, VDI Verlag, Düsseldorf 1992

[Mot86]   Motreanu, D.: Existence for minimization with Nonconvex Constraints, J.Math. Anal. Appl. **117** (1986) 128-137

[Mot93]   Motreanu, D., Panagiotopoulos, P.D.: Hysteresis: The Eigenvalue Problem for Hemivariational Inequalities. Models of Hysteresis (ed. by A. Visintin) Pitman Research Notes in Mathematics, Longman, Harlow (1993)

[Moys]   Moyson, E., van Gemert, D.: Experimentalle Prüfung der Laminattheorié für Faserverstárhte Verbundwerkstoffe. In: G. Ondracek (ed.) Verbundwerkstoffe. Phasenverbindung und mechanische Eigenschaften, Deutsche Gesellschaft für Metallkunderbe **1** (1985) 99-115

[Mroz73]   Mroz, Z.: Mathematical Models of Inelastic Material Behaviour. Univ. of Waterloo Press, Waterloo 1973

[Mroz92]   Mroz, Z., Stupkiewicz, S.: Constitutive Modelling of Slip and Wear in Elastic, Frictional Contact. In: Proc. Contact Mech. Int. Symp. (ed. by A. Curnier) Presses Polyt. et Univers. Romandes, Lausanne, pp. 133-156 (1992)

[Murty]   Murty, K.G.: Linear Complementarity. Linear and Nonlinear Programming. Heldermann Verlag, Berlin 1988

[Nan89a]   Naniewicz, Z.: On Some Nonconvex Variational Problems Related to Hemivariational Inequalities. Nonlin. Anal. **13** (1989) 87-100

[Nan89b] Naniewicz, Z., Wozniak, C.Z.: On the Quasi-Stationary Models of Debonding Processes in Layered Composites. Ing. Archiv **60** (1989) 31-40

[Nan88] Naniewicz, Z.: On Some Nonmonotone Subdifferential Boundary Conditions in Elastostatics. Ing. Archiv **58** (1988) 403-412

[Nether] Nethercot, D.A.,Chen, W.F.: Effects of Connections on Columns. J.Constr.Steel Res. **10** (1988) 201- 240

[Nec80] Nečas, J., Jarusek, J., Haslinger, J.: On the Solution of the Variational Inequality to the Signorini Problem with Small Friction. Bulletino U.M.I. **17B** (1980) 796-811

[Nguy] Nguyen, V-D.: Constructing stable grasps. Int. J. of Robot. Res. **8** (1989) 26-36

[Nits67] Nitsiotas G.: Zur Bildung und Berechnung statisch bestimmter Tragwerke mit Verbindungen aus Seilen. Bauingenieur **42** (1967) 412-414

[Nits71] Nitsiotas G.: Die Berechnung statisch unbestimmter Tragwerke mit einseitigen Bindungen. Ing.Archiv **41** (1971) 46-60

[Nits78] Nitsiotas, G.: Elastostatics. Linear Theory Vol. I, II, Thessaloniki 1978 (in Greek)

[Ode81] Oden, J.T., Pires, E.: Contact Problems in Elastostatics with Non-Local Friction Laws. TICOM Report 81-12, University of Texas at Austin, 1981

[Ode83] Oden, J.T., Pires, E.B.: Nonlocal and Nonlinear Friction Laws and Variational Principles for Contact Problems in Elasticity. ASME J. Appl. Mech. **50** (1983) 67-76

[Ode85] Oden, J.T., Martins, J.A.C.: Models and Computational Methods for Dynamic Friction Phenomena. Comp. Meth. Appl. Mech. Eng. **52** (1985) 527-634

[Oehl] Oehlers, D.J., Johnson, R.D.: The strength of stud shear connections in composite beams. The Structural Engineer **65B** (1987) 44-48

[Olsz] Olszak, W., Mroz, Z., Perzyna, P.: Recent Trands in the Developments of the Theory of Plasticity. Pergamon Press, Oxford and PWN-Polish Sci. Publ. Warszawa 1963

[Ord] Ordracek G. (ed): Verbundwerkstoffe. Phasenverbindung und mechanische Eigenschaften, Deutsche Gesellschaft für Mefallkunde, Band 1, 1985.

[Pagan]   Pagano, N.J.: Stress Fields in Composite Laminates. Int. J.Solids Struct. Vol. **14** (1978) 385-400

[Paip]   Paipetis, S.A., Papanicolaou, G.C. (eds): Engineering Applications of New Composites. Omega Scientific. Oxford 1988

[Palm]   Palmer, A.C., Maier, G., Drucker, D.C.: Normality Relations and Convexity of Yield Surfaces for Unstable Materials of Structural Elements. Trans. ASME **24** (1967) 464-470

[Pan75]   Panagiotopoulos, P.D.: A Nonlinear Programming Approach to the Unilateral Contact – and Friction – Boundary Value Problem in the Theory of Elasticity. Ing. Archiv. **44** (1975) 421-432

[Pan76]   Panagiotopoulos, P.D.: A Variational Inequality Approach to the Inelastic Stress-Unilateral Analysis of Cable Structures. Comp. and Struct. **6** (1976) 133-139

[Pan76a]   Panagiotopoulos, P.D.: Convex Analysis and Unilateral Static Problems. Ing. Archiv **45** (1976) pp. 55-68

[Pan78]   Panagiotopoulos, P.D: Variational Inequalities and Multilevel Optimization Techniques. Comp. and Struct. **8** (1978) 649-650

[Pan80]   Panagiotopoulos, P.D., Talaslidis, D.: A Linear Analysis Approach to the Solution of Certain Classes of Variational Inequality Problems in Structural Analysis. Int. J. Solids and Struct. **16** (1980) 991-1006

[Pan81]   Panagiotopoulos, P.D.: Non-Convex Superpotentials in the Sense of F.H. Clarke and Applications. Mech. Res. Comm. **8** (1981) 335-340

[Pan82]   Panagiotopoulos, P.D.: Non-Convex Energy Functionals. Application to Non-convex Elastoplasticity. Mech. Res. Comm. **9** (1982) 23-29

[Pan83a]   Panagiotopoulos, P.D.: Nonconvex Energy Functions. Hemivariational Inequalities and Substationarity Principles. Acta Mechanica **42** (1983) 160-183

[Pan83b]   Panagiotopoulos, P.D.: A Boundary Integral Inclusion Approach to Unilateral B.V.Ps in Elastostatics. Mech. Res. Comm. **10** (1983) 91-96

[Pan84]   Panagiotopoulos, P.D., Baniotopoulos, C.C.: A Hemivariational Inequality and Substationarity Approach to the Interface Problem. Theory and Prospects of Applications, Engineering Analysis **1** (1984) 20-31

[Pan85]   Panagiotopoulos, P.D.: Inequality Problems in Mechanics and Applications. Convex and Nonconvex Energy Functions. Birkhäuser Verlag, Basel, Boston 1985. Russian Translation MIR Publ. Moscow 1989

[Pan87a]   Panagiotopoulos, P.D., Koltsakis, E.K.: Interlayer Slip and Delamination Effect: A Hemivariational Inequality Approach. Canadian Society for Mech. Engineering 11 (1987) 43-52.

[Pan87b]   Panagiotopoulos, P.D.: Ioffe's Fans and Unilateral Problems: A New Conjecture. In: Unilateral Problems in Structural Analysis 2, (ed. by G. del Piero, F.Maceri), CISM Courses and Lectures 304. Springer Verlag, Wien, N.York 1987

[Pan87c]   Panagiotopoulos, P.D., Koltsakis, E.K.: Hemivariational Inequalities for Linear and Nonlinear Elastic Materials. Meccanica 22 (1987) 65- 75

[Pan88]   Panagiotopoulos, P.D.: Nonconvex Superpotentials and Hemivariational Inequalities. Quasidifferentiability in Mechanics. In: Nonsmooth Mechanics and Applications (ed. by J.J. Moreau, P.D. Panagiotopoulos), CISM Courses and Lectures Nr. 302, Springer Verlag, Wien, N.York 1988

[Pan88a]   Panagiotopoulos, P.D., Stavroulakis, G.: A Variational-hemivariational Inequality Approach to the Laminated Plate Theory under Subdifferential Boundary Conditions. Quart. of Appl. Math. XLVI (1988) 409-430

[Pan89a]   Panagiotopoulos, P.D.: Semicoercive Hemivariational Inequalities. On the Delamination of Composite Plates. Quart. of Appl. Math., **XLVII** (1989) 611-629

[Pan89b]   Panagiotopoulos, P.D., Haslinger, J.: Optimal Control of Systems Governed by Hemivariational Inequalities. In: Mathematical Models for Phase Change Problems, (ed. by J.F.Rodriques) Birkhäuser Verlag, Basel, Boston 1989

[Pan 90]   Panagiotopoulos, P.D., Stavroulakis, G.: The Delamination Effect in Laminated von Karman Plates under Unilateral Boundary Conditions. A Variational-Hemivariational Inequality Approach. J. of Elasticity 23 (1990) 69-96

[Pan91]   Panagiotopoulos, P.D.: Coercive and Semicoercive Hemivariational Inequalities. Nonlin. Anal. 16 (1991) 209-231

[Pan92a]   Panagiotopoulos, P.D., Haslinger, J.: On the Dual Reciprocal Variational Approach to the Signorini-Fichera Problem. Convex and Nonconvex Generalizations. ZAMM 72 (1992) 497-506

[Pan92b]   Panagiotopoulos, P.D., Stavroulakis, G.: New Types of Variational Principles Based on the Notion of Quasidifferentiability. Acta Mechanica **94** (1992) 171-194

[Pan92c]   Panagiotopoulos, P.D.: Adhesive Joints and Interfaces of Linear Elastic Bodies in Loading and Unloading. Semicoercive Hemivariational Inequalities. J. of Elasticity **28** (1992) 29-54

[Pan93a]   Panagiotopoulos, P.D., Mistakidis, E., Koltsakis, E.: Debonding and Sliding in Adhesively Bonded Cracks - A Numerical Algorithm (to appear)

[Pana80]   Panagiotopoulos, P.D.: Time-Space Unilateral Variations and Variational Inequalities in Relativistic Mechanics. ZAMM **60** (1980) 264-265

[Pana81]   Panagiotopoulos, P.D.: Dynamic and Incremental Variational Inequality Principles, Differential Inclusion and their Applications to Co-Existent Phases Problems. Acta Mechanica **40** (1981) 85-107

[Pana81a]   Panagiotopoulos, P.D.: Non-convex Superpotentials in the Sense of F.H. Clarke and Applications. Mech. Res. Comm. **8** (1981) 335-340

[Pana83]   Panagiotopoulos, P.D.: Nonconvex Energy Functions. Hemivariational Inequalities and Substationarity Principles. Acta Mechanica **42** (1983) 160-183

[Pana83a]   Panagiotopoulos, P.D.: Une génèralization non-convexe de la notion du surpotentiel et ses applications. C.R. Acad. Sc. Paris **296II** (1983) 1105-1108

[Pana85a]   Panagiotopoulos, P.D.: Nonconvex Problems of Semipermeable Media and Related Topics. ZAMM **65** (1985) 29-36

[Pana85b]   Panagiotopoulos, P.D.: Hemivariational Inequalites and Substationrity in the Static Theory of von Kármán Plates. ZAMM **65** (1985) 219-229

[Pana87]   Panagiotopoulos, P.D.: Multivalued Boundary Integral Equations for Inequality Problems. The Convex Case. Acta Mechanica **70** (1987) 145-167

[Pana88a]   Panagiotopoulos, P.D.: Hemivariational Inequalities and their Applications. In: Topics in Nonsmooth Mechanics (ed. by J.J.Moreau, P.D.Panagiotopoulos and G. Strang) Birkhäuser Verlag, Boston 1980

[Pana88b]   Panagiotopoulos, P.D.: Nonconvex Superpotentials and Hemivariational Inequalities. Quasidifferentiablility in Mechanics. In: Nonsmooth Mechanics and Applications (ed. by J.J. Moreau and P.D.

Panagiotopoulos) CISM Lect. Notes, Vol. **302**, Springer Verlag, Wien, N.York 1988

[Pana88c] Panagiotopoulos, P.D.: Variational Hemivariational Inequalities in Nolinear Elasticity. The Coercive Case. Aplikace Matematiky (now Applications of Mathematics) **33** (1988) 249-268

[Pana89] Panagiotopoulos, P.D.: Boundary Integral Equations for Inequality Problems. The Nonconvex Case. Acta Mechanica **72** (1989) 152-168

[Pana91] Panagiotopoulos, P.D.: The B.I.E.M. for Inequality Problems. Math. Comput. Modelling **15** (1991) 257-267

[Panag77] Panagiotopoulos, P.D.: Optimal control in the Unilateral Thin Plate Theory, Archives of Mechanics **29** (1977) 25-39

[Panag83] Panagiotopoulos, P.D.: Optimal Control and Parameter Identification of Structures with Convex and Nonconvex Strain Energy Density. Applications to Elastoplasticity and to Contact Problems. Solid Mech. Archives **8** (1983) 363-411

[Panag84] Panagiotopoulos, P.D.: Optimal Control of Structures with convex and nonconvex energy densities and variational and hemivariational Inequalities. Engng. Struct. **6** (1984) 12-18

[Panag88] Panagiotopoulos, P.D., Liolios A.A.: On the Dynamic of Inelastic Shocks. A New Approach. J. Proc of the Greek-German Seminar in Structural Dynamics and Earthquake Engineering, (ed. by A.N. Kounadis and W.B. Krätzig) Publ. of the Hellenic Soc. Theor. Appl. Mech., Athens 1988 pp.12-18

[Panag89] Panagiotopoulos, P.D., Haslinger, J.: Optimal Control of Systems governed by Hemivariational Inequalities. In: Math. Models for Phase Change Problems, (ed. by J.F.Rodrigues), Vol. **88**, Birkhäuser Verlag, Basel Boston ISNM, 1989

[Panag90] Panagiotopoulos, P.D.: Optimal Control of Systems Governed by Hemivariational Inequalities. Necessary Conditions. In: Int. Series of Num. Math. Vol. **95**, (ed. by K.H.Hoffmann and J.Sprekels), Birkhäuser Verlag, Basel, Boston 1990

[Panag91] Panagiotopoulos, P.D.: Optimal Control of Systems Governed by Variational-Hemivariational Inequalities. Proc. 4th U.P.S.A., Unilateral Problems in Structural Analysis, Capri, June 1989, (ed. by F.Maceri and G.Del Piero) Birkhäuser Verlag, Basel, Boston 1991

[Panag92] Panagiotopoulos, P.D., Haslinger, J.: Optimal Control and Identification of Structures Involving Multivalued Nonmonotonicities. Existence and Approximation Results. European J.Mech. A/ Solids **11** (1992) 425-445

[Panag93]   Panagiotopoulos, P.D., Panagouli, O.K., Mistakidis, E.S.: Fractal Geometry and Fractal Material Behaviour in Solids and Structures. Arch. Appl. Mech. (former Ing. Archiv) **63** (1993) 1-24

[Panag93a]   Panagiotopoulos, P.D., Al-Fahed, A.M.: On the Robot Hard Grasping and Related Problems. Optimal Control and Identification. Int. J. of Robot. Res. (in press)

[Panagi]   Panagiotopoulos, P.D., Koltsakis, E.K.: The Nonmonotone Skin Effects in Plane Elasticity Obeying to Linear Elastic and subdifferential Material Laws, ZAMM **70** (1990) 13-21

[Pang92a]   Panagouli, O.K., Panagiotopoulos, P.D., Mistakidis, E.S.: On the Numerical Solution of Structures with Fractal Geometry: The F.E. Approach. Meccanica **27** (1992) 263-274

[Pang92b]   Panagouli, O.K.: Fractal Geometry, in Structural Analysis. Doct. Dissertation, Dept. of Civil Eng., Aristotle University, Thessaloniki 1992

[Pard]   Pardalos, P.M., Rosen, J.B.: Constrained Global Optimization: Algorithms and Applications, Lecture Notes is Computer Science Vol. **268**, Springer Verlag, N. York 1987

[Pdp90a]   Panagiotopoulos, P.D.: The Mechanics of Fractals. Proc. Acad. of Athens **65** (1990) 185-212 (in Greek)

[Pdp90b]   Panagiotopoulos, P.D.: On the Fractal Nature of Mechanical Theories. ZAMM **70** (1990) 258-260

[Pdp91]   Panagiotopoulos, P.D.: Fractal Approximation in the Theory of Elasticity. ZAMM **71** (1991) T658-T659

[Pdp92a]   Panagiotopoulos, P.D.: Fractals and Fractal Approximation in Structural Mechanics. Meccanica **27** (1992) 25-33

[Pdp92b]   Panagiotopoulos, P.D.: Fractal Geometry in Solids and Structures. Int. J. Solids and Structures, **29** (1992) 2159-2175

[Pdp92c]   Panagiotopoulos, P.D., Mistakidis, E.S., Panagouli, O.K.: Fractal Interfaces with Unilateral Contact and Friction Conditions. Comp. Meth. in Appl. Mech. Eng. **99** (1992) 395-412

[Pdp92c]   Panagiotopoulos, P.D., Panagouli, O.K.: Fractal Interfaces in Structures: Methods of Calculations. Comp. and Struct. **45** (1992) 369-380

[Poly86]   Polyakova, L.N.: On minimizing the sum of a convex function and a concave function. Math. Progr. Study **29** (1986) 69-73

[Pow] Powell, M.J.D.: Convergence properties of algorithms for nonlinear optimization. SIAM Review, 4281 **44** 1(1986) pp. 487-500

[Prag57] Prager, W.: On Ideal-locking Materials. Trans. Soc. Rheol. 1(1957) 169-175

[Prag58] Prager, W.: Elastic Solids of Limited Compressibility. Proc. 9th Int. Congress Appl. Mech. Brussels 1958, Vol. **5**

[Prag] Prager, W.: Problems of Network Flow. Z.A.M.P. **16** (1965) 185-193

[Prze] Przemieniecki, J.S., Theory of Matrix Structural Analysis. McGraw-Hill Book Co., New York 1968

[Rabi] Rabinowitz, P.H.: Minimax Methods in Critical Point Theory with Applications to Differential Equations. CBM Reg. Conf. Ser. in Math. no.**65**, Amer. Math. Soc., Providence 1986

[Rauch] Rauch, J.: Discontinuous Semilinear Differential Equations and Multiple Valued Maps. Proc. Amer. Math. Soc. **64** (1977) 277-282

[Reck] Reckling, K.A.: Plastizitätstheorie und ihre Anwendung auf Festigkeitsprobleme. Springer-Verlag, Berlin 1967

[Rekt] Rektorys, K.: The Method of Discretization in Time and Partial Differential equation. Reidel Publ. Co., Dordrecht 1982

[Riks] Riks, E.: The application of Newton's method to the problem of elastic stability. J.Appl.Mech. **39** (1972) 1060-1066

[Ritt] Ritter, H., Martinez, T., Schulten, K.: Neural Computations and self-Organizing Maps. Addison-Wesley Publ. Co. N.York 1992

[Rock60] Rockafellar, R.T.: Extension of Fenchel's Duality Theorem for Convex Functions. Duke Math. J. **33** (1960) 81-90

[Rock68] Rockafellar, R.T.: Integrals which are Convex Functionals. Pacific J. Math. **24** (1968) 525-539

[Rock70] Rockafellar, R.T.: Convex Analysis. Princeton Univ. Press, Princeton 1970

[Rock79] Rockafellar, R.T.: La théorie des Sous-Gradients et ses Applications à l'optimization. Fonctions Convexes et Non-convexes, Les Presses de l' Université de Montréal, Montréal 1979

[Rock80] Rockafellar, R.T.: Generalized Directional Derivatives and Subgradients of Non-convex Functions. Can.J.Math. **XXXII** (1980) 257-280

[Roman] Roman I., Harlet H., Marom G.: Stress intensity factor measurements in composite sandwich structures. In: I.H. Marshal, ed., Proc. 1st Conf. on Composite Structures (Applied Scince Publishers, London, 1981 pp. 633-645

[Rub86] Rubinov, A.M., Yagubov, A.A.: The space of star-shaped sets and its applications in nonsmooth optimization. Math. Progr. Study **29** (1986) 176-202

[Sal] Salencon, J., Tristán-Lopez, A.: Analyse de la stabilité des talus en sols cohérents anisotropes. C.R. Acad. Sc. Paris, **290B** (1980) 493-496

[Schel] Schellekens, J.C.J., De Borst, R.: Application of Linear and Nonlinear Fracture Mechanicm Options to Free Edge Delamination in Laminated Composites. Heron **36** (1991) 37-48

[Schol] Scholz, C.H., Mandelbrot, B. (editors): Fractals in Geophysics. Birkhäuser Verlag, Boston Basel 1989

[Schw] Schwartz, M.M.: Composite Materials Handbook. McGraw-Hill, New-York 1984

[Sen90] Sendov, Bl.: Hausdorff Approximations. Kluwer Acad. Publ., Dordrecht 1990

[Shi] Shi Shuzhong: Optimal control of Strongly Monotone Variational Inequalities. SIAM J. Control and Optim. **26** (1988) 274-290

[Spe82a] Spector, A.A.: Variational Methods of Analysis for Certain Classes of Spatial Problems of Contact Between Elastic Bodies with Friction. Dokl. Acad. Nauk. SSR **265** (1982) 111-117

[Spe82b] Spektor, A.A.: Variational Methods of Investigation of Certain Classes of Three-Dimensional Problems of Contact Between Elastic Bodies in the Presence of Friction. Dokl. Akad. Nauk SSR **265** (1982) 592-596

[Spe85] Spektor, A.A.: Variational Method of Solving Three-Dimensional Contact Problems of the Nonstationary Interaction Between Elastic Solids with Friction. Dokl. Akad. Nauk. SSSR **285** (1985) 865-870

[Spe87] Spektor, A.A.: Variational Methods in Three-Dimensional Problems of Non-stationary Interaction of Elastic Bodies with Friction. PMM U.S.S.R., **51** (1987) 56-62.

[Stav91] Stavroulakis, G.E.: Analysis of Structures with Interfaces. Formulation and Study of Variational- Hemivariational Inequality Problems. Doct. Dissertation, Dept. of Civil Eng., Aristotle University, Thessaloniki 1991

[Stav91a] Stavroulakis G.E., Panagiotopoulos P.D., Al-Fahed, A.M.: On the Rigid Body Displacements and Rotations in Unilateral Contact Problems and Applications. Comp. and Struct. **40** (1991) 599-614

[Stav93a] Stavroulakis, G.E.: Convex Decomposition for Nonconvex Energy Problems in Elastostatics and Applications. European J. of Mech. A /Solids **12** (1993) 1-20

[Stav93b] Stavroulakis, G.E., Panagiotopoulos, P.D.: Convex Multivalued Decomposition Algorithms for Nonmonotone problems. Intern.J. for Num. Meth. in Eng. (to appear)

[Stein] Stein, E., Wagner, W., Wriggers, P.: Grundlagen nichtlinearer Berechnugsverfahren in der Strukturmechanik, Nichtlineare Berechnungen im Konstruktiven Ingenieurbau, Stein, E. (ed.), pp. 1-48, Springer Verlag, Berlin, N. York 1989

[Strod] Strodiot, J.J., Nguyen, V.H.: On the Numerical Treatement of the Inclusion $0 \in \partial f(x)$. In: Topics in Nonsmooth Mechanics. (ed. by Moreau,J.J. Panagiotopoulos, R.D. Strand G.), Birkhäuser Verlag, Basel, Boston 1988

[Ta] Tato, Y.: Signorini's Problem with Friction in Linear Elasticity. Japan J. of Appl. Math. **4** (1987) 237-268

[Taka] Takayasu, H.: Fractals in the Physical Sciences. Manchester Univ. Press, Manchester 1990

[Tabak] Tabak, D., Kuo, B. C.: Optimal Control by Mathematical Programming. Prentice-Hall, New Jersey, 1971

[Tel] Telega J.J.: Topics on Unilateral Contact Problems of Elasticity and Inelasticity. In: Nonsmooth Mechanics and Applications (ed. by Moreau, J.J. Panagiotopoulos, P.D.), CISM Courses and Lectures No. **302**, Springer Verlag, Wien 1988

[The86] Theocaris, P.S., Makrakis, G.N.: The Kinked Crack Solved by Mellin Transform. J. Elasticity **16** (1986) 393-411

[The87] Theocaris, P.S., Makrakis, G.N.: Crack Kinking in Anti-plane Shear Solved by the Mellin Transform. Int. J. Fracture **34** (1987) 251-262

[The91] Theocaris, P.S., Panagiotopoulos, P.D.: On Debonding Effects in Adhesively Bonded Cracks - A Boundary Integral Approach. Arch. of Appl. Mech. (former Ing. Archiv) **61** (1991) 578-587

[The91a] Theocaris, P.S., Panagiotopoulos, P.D.: Neural Networks and Artificial Intelligence in Fracture Mechanics (in Greek), Proc. Nat. Acad. Athens **66** (1991) 373-400

[The92]	Theocaris, P.S., Panagiotopoulos, P.D.: On the Consideration of Unilateral Contact and Friction in Cracks. The Boundary Integral Method. Int. J. Num. Meth. Eng. **35** (1992) 1697-1708
[The93]	Theocaris, P.S., Panagiotopoulos, P.D., Bisbos C.: Unilateral Contact, Friction and Related Interaction in Cracks. The Direct Boundary Integral Method. Int. J. Solids and Structures 1993 (in press)
[The93a]	Theocaris, P.S., Panagiotopoulos, P.D.: Cracks of Fractal Geometry with Unilateral Contact and Friction Interface Conditions, Int. J. on Fracture 1993 (in press)
[The93b]	Theocaris, P.S., Panagiotopoulos, P.D.: Neural Networks for Computing in Fracture Mechanics. Methods and Prospects of Applications, Comp. Meth. Applied Mech. 1993 (in press)
[Thom]	Thompson, J.M.T., Hunt, G.W.: A General Theory of Elastic Stability. J.Wiley and Sons, London 1973
[Tol]	Toland, J.F.: A Duality Principle for Non-convex Optimization and the Calculus of Variations. Arch. Rat. Mech. Anal. **71** (1979) 41-61
[Tont]	Tonti, E.: A Systematic Approach to the Search for Variational Principles. In: Variational Methods in Engineering (ed. by C.A.Brebbia and H.Tottenham) Vol. I, Southampton Univ. Press, Southampton 1973
[Tru65]	Truesdell, C., Noll, W.: The Non-linear Field Theories of Mechanics. In: Encyclopedia of Physics. Vol. III/3 (ed. by S. Flügge) Springer-Verlag, Berlin 1965
[Tru66]	Truesdell, C.: Six Lectures on Modern Natural Philosophy. Springer-Verlag, Berlin 1966
[Tuy86]	Tuy, H.: A General Deterministic Approach to Global Optimization via D.C. Programming. In: FERMAT Days 85: Mathematics for Optimization, (ed. H.-B. Hiriart-Urruty), pp. 273-303, North-Holland, Amsterdam 1986
[Tuy87]	Tuy, H.: Convex Programs and Additional Reverse Convex Constraint. J. of Optim. Th. and Appl. **52** (1987) 463-486
[Tzaf91]	Tzaferopoulos, M.A.: Numerical Analysis of Structures with Monotone and Nonmonotone, Nonsmooth Material Laws and Boundary Conditions: Algorithms and Applications, Doct. Dissertation, Aristotle University, Dept. of Civil Eng., Thessaloniki 1991
[Tzaf91a]	Tzaferopoulos, M.Ap., Panagiotopoulos, P.D.: Analysis of steel frames with nonmonotone flexible joints. In: Proc. 1st Nat.Conf. on Steel Structures (ed. by A.N.Kounadis), Athens 1991 pp. 130-140

[Tzaf93]   Tzaferopoulos M.A., Panagiotopoulos P.D.: Delamination of Composites as a Substationarity Problem: Numerical Approximation and Algorithms. Comp. Meth. Appl. Mech. (to appear)

[Wal]   Wallin, H.: Interpolating and Orthogonal Polinomials on Fractals. Constr. Approx. **5** (1989) 137-150

[Wal89]   Wallin, H.: The Trace to the Boundary of Sobolev Spaces on a Snowflake. Rep. Dept. Math. Univ. of Umea, Sweden 1989

[Wals]   Walsh, G. R.: Methods of Optimization. John Wiley and Sons, New York, 1975

[Walt]   Walter, W.: Differential-und Integral- Ungleichungen. Springer Verlag, Berlin 1964

[Warg75]   Warga, J.: Necessary Conditions without Differentiability Assumptions in Optimal Control. J. Dif. Equations **18** (1975) 41-62

[Warg76a]   Warga, J.: Necessary Conditions without Differentiability Assumptions in Unilateral Control Problems. J.Dif. Equations **21** (1976) 25-38

[Warg76b]   Warga, J.: Derivate Containers, Inverse Functions and controllability. In: Calculus of Variations and Control Theory (ed. by D.L. Russell) Acad. Press New York 1976

[Wid]   Widrow, B., Hoff, M.: Adaptive Switching Circuits. 1960 IRE WESCON Convention Record, New York IRE, 96-104

[Willi82]   Williams, J.G., Rhodes, M.D.: Effect of Resin on Impact Damage Tolerance of Graphite/Epoxy Laminates. In "Proceedings of the 6th International Conference on Composite Materials, Testing and Design", (ed. by I.M. Daniel). ASTM Special Technical Publication **787**, ASTM Philadelphia (1982) pp.450-480

[Willi86]   Williams, J.F., Stouffer, D.C., Ilič, S., Jones, R.: An Analysis of Delamination Behaviour. Comp. Struct. **5** (1986) 203-216

[Womer]   Womersley, R.S., Fletcher, R.: An algorithm for composite nonsmooth optimization problems. J.Optim. The. Appl. **48**, (1986) 493-523

[Wri]   Wriggers, P., Simo, J.C., Taylor, R.L.: Penalty and Augmented Lagrangian Formulations for Contact Problems. In: Proceedings of the International Conference on Mumerical Methods in Engineering - Theory and Applications. NUMETA '85, (ed. by J. Middleton and G.N. Pande), Vol. I, 1985, pp. 97-106

[Yvon]	Yvon, J.P.: Etude de quelques problèmes de contrôle pour des systèmes distribués. Thèse de Doctorat d' Etat, Université Paris VI, 1973
[Zer]	Zervas, P.A.: Seismic Behaviour of Frame Structures with Unilateral Contact Conditions. Doct. Dissertation, Dept. of Civil Eng., Aristotle Univ., Thessaloniki 1992
[Zh88]	Zhong, W.X., Sun, S.M.: A Finite Element Method for Elasto-Plastic Structures and Contact Problems by Parametric Quadratic Programming. Int. J. for Num. Meth. in Eng. **26** (1988) 2723-2738
[Zh89]	Zhong, W.X., Sun, S.M.: A Parametric Quadratic Programming Approach to Elastic Contact Probems with Friction. Comp. and Struct. **32** (1989) 37-43
[Zhuk]	Zhukov, A.: Plastic Deformation of Isotropic Metals in Combined Loading. Izv. Akad. Nauk. SSSR. OTN **12** (1956) 72-87 (in Russian)
[Zieg62]	Ziegler, H.: A Possible Generalization of Onsager's Theory. Advances in Solid Mechanics. Acad. Press, New York 1962
[Zieg63]	Ziegler, H.: Some Extremum Principles in Irreversible Thermodynamics with Application to Continuum Mechanics. Progress in Solid Mechanics IV, North Holland Publ. Co., Amsterdam 1963
[Zieg81]	Ziegler, H.: Discussion of some Objections to Thermomechanical Orthogonality. Ing. Archiv. **50** (1981) 149-164
[Zie80]	Zienkiewicz, O., Wood, W. L., Taylor, R. L.: An Alternative Single-Step Algorithm for Dynamic Problems. Earthq. Eng. and Struct. Dyn. **8** (1980) 31-40

Subject Index

adhesion 249
adhesive contact 105,116,271
adhesive grasping problem 377
adhesive interface 53
adhesive joint 331
adhesive material 53,124,143,
adhesively 176,177,248
affine hull 3,6
ambiguous degrees of freedom 362
approximation 151,155,230
attractor 394,396,402
Besov space 399
Betti 138,139,362
bifurcation 123
bilateral contact 175,369
bilateral problem 149,412
bimodulus 70,71
bone crack 393
boundary conditions 24,33,53
boundary integral equation 143,147,151
brittle 53,79,111
Brouwer 158,164
bundle 241,242,318
Clarke 122
combinatorial character 253,91
composite beams 306
composite 78,80,254
conjugate 19,20,346,
convex analysis 26,41,94
convex hull 3,15,20
convex problems 351
convex programming 352,409
convex superpotentials 44,65
Coulomb friction 261,271,372
crack 111,361,373
criterion 55,67,75,129

damping coefficient 363
debonding 109,143,248
decomposition method 345
decreasing branch 239,245,318
delamination 105,106,111
detachment 48,259,294,
differential 13,14,161
diffusion front 393
direct method 135,138
discretization algorithm 365
dislocation 83,136,138
dissipation 82,83
distribution 289,296
Drucker 129,68
duality 18
Dunford-Pettis 158,227
dynamic problems 121
Egoroff 160,197,204
eigenvalue problems 121,190
energy 167
epigraph 4,5,20
equilibrium path tracing method 251
extension 4,58,63
F-superpotential 126
fan 126,127,243
fiber 70,74,247
fixed point 158,162,167
fractal interface 402,403
fractal interpolation 395,401
fractal set 393,401
fractals 393
fracture 376
friction boundary conditions 381
friction 269,277,374
generalized gradient 10,122,125
gradient 9

granular 53,74,83
graph 191
Green's operator 139,146
Gâteaux 6,14,122
Hamilton 204
Hausdorff 98,394,397
hemivariational 16,17,103
hidden 83
Hildreth and d' Esopo 367
Hill 87
Hooke's elasticity tensor 68
hysteresis 121
identification problems 223
impact 361,362,365
indicator 15,40,42
indirect B.I.E.M. 144,372
indirect method 135
inequalities 101,103,114
inequality problem 99,260,349
interfacial debonding 351
iterated function system 394
Kirchhoff shearing force 50
Kuhn-Tucker 240
L.C.P. 377,383,384
Lagrange multiplier 17,327
Lagrangian 33,350,372
laminated plates 109,176
layers of forces 138
learning rule 413
Liapunov function 408
Lipschitz function 63,72,76
lipschitzian ,17,42,63
local maximum 16,43,140
local minimum 15,17,241
locking 55,66,78
m-step linear difference operators 363
matrix 34,75,78
maximal monotone 23,24,58
meteoritic rain 393
monotonicity 121,140,172
multifunction 11,23,90
multiplier 17,321,327
multivalued 9,,53,90
neural network 406,407,413
neurocomputing 333,406

neuron 406,407,413
nonconvex optimization algorithm 251
nonconvex programming 252
nonconvex superpotentials 51,72,81
nonconvex superpotential 52,90,91
nonmonotone friction 269,284,371
nonmonotone stress-strain law 300
nonmonotone 52,53,101
normal algorithm 272
normal cone 6,9,42
objectivity 81
Onsager 84,85
optimal control 226,387
path following method 297
pattern 186,375,393
penalty method 372
perceptron 414
percolation pattern 393
perturbed Lagrangian 372
principle of complementary virtual work 119
principle of virtual work 38,40,127
projection 70,71,330
propagation 334
proper 13,14,281
quadratic programming problem 353,367
quasidifferentiability 28,95
quasidifferentials 24
quasivariational inequality 378
regularity 63,201,204
regularization 172,243,390
rigid 71
robotics 377,387
saddle point 140,16
sandblasting 393
sawtooth 300,78
Scanlon 53
semicoercive 155,162,384
semirigid connections 337
shear connectors 306
shocks 362,363
Signorini 47,61,377
stability 240
stable state 407,408,409
stamp problem 351,353
star-shaped set 75,98

subdifferential 3,6,96
subgradient 5
substationarity 15,43, 116
substationary point search 317
superpotential 39,41,80
supervised learning 412
surface 50,68,129
tangential algorithm 272,278
thermoelasticity 212
trial and error method 240,372
unilateral boundary conditions 105
unilateral contact 46,47,265
unilateral problem 104,412
unloading 65,69,122
V-superpotential 126
variational equality 100,104,134
variational inequalities 40,49,104
variational inequality problem 260,349
variational-hemivariational inequalities 126
virtual power 36,37,200
virtual work 38,92,127
viscoplastic 72
Winkler 46,47,265
work 38,68,100
yield condition/criterion 51,60,98
Zorn 23